AUSTRALIAN SEAFOOD HANDBOOK

an identification guide to domestic species

editors

G. K. YEARSLEY, P. R. LAST AND R. D. WARD

contributing authors

J. ANDREW, R. K. DALEY, N. G. ELLIOTT, P. R. LAST, B. D. MOONEY,
P. D. NICHOLS, N. V. RUELLO, P. VIRTUE, R. D. WARD AND G. K. YEARSLEY

GPO Box 1538
Hobart
Tas. 7001
Tel: (03) 6232 5222
Fax: (03) 6232 5000
Email: seafood@marine.csiro.au

National Library of Australia Cataloguing-in-Publication entry

Australian seafood handbook: an identification guide to domestic species.

Bibliography.
Includes index.
ISBN 0 643 06194 0.
ISBN 0 643 06195 9 (waterproof).
ISBN 0 643 06200 9 (leather bound).

1. Seafood—Australia—Handbooks, manuals, *etc.*
2. Seafood—Australia—Identification.
3. Seafood—Australia—Pictorial works.
I. Yearsley, G. K. II. Last, P. R. III. Ward, R. D. (Robert D.). IV. CSIRO. Division of Marine Research.

641.3920994

Funded by the Fisheries Research and Development Corporation and CSIRO Marine Research

Design by Gordon Yearsley

Layout by Gordon Yearsley, Ross Daley and Antonia Hodgman

Cover design and layout by Antonia Hodgman

Cover photograph of West Australian dhufish, *Glaucosoma hebraicum,* by Thor Carter

Paper stock: New Era Satin 115 gsm
Synthetic stock: Yuppo plastic 100 gsm (waterproof)

Printed by Courtney Colour Graphics, Melbourne

. . . adapted from *Fish on a Blue and White Plate*
W. B. Gould 1845

In the Night while Calm we fish'd with Hook and Line, and caught good Store of Fish, viz. Snappers, Breams, Old-Wives and Dog-fish . . . We caught also a Monk-fish of which I brought Home a Picture.

Captain William Dampier, *Voyage to New Holland*, 1703

Foreword

When it comes to the riches of the sea, Australians are lucky! We live on the world's biggest island, and we love the seas around us, our golden beaches and noble rivers. We enjoy their bountiful harvest—seafood that's second to none in quality and variety.

There is no food so mouth-watering, so promising, as the sizzling fillets of a fish you caught only an hour ago. And because so many of the Australians who catch, handle, transport and prepare seafood are using the best techniques available, what's served onto our plates in restaurants can bring us the same pleasurable experience.

We need this book, because it fills an important gap in making sure that these pleasurable experiences are the best they can be.

In my 50-odd years as a fisherman and restaurateur, my biggest disappointment has been to see how many people miss out on tasting the full range of our wonderful seafood. Most Australians aren't familiar with fish species or how to handle and cook them. The fishing industry needs to do more to educate and help our consumers—and even people within the industry itself. One of the things we've needed for years has been a reliable guide on how to identify seafood.

This book fits the bill very well. Preparing it was a gigantic task: the CSIRO team had to collect detailed, accurate information on the main seafood species eaten locally, and to take colour photographs of each one. Their dedication really shows through.

I'm very proud that three features of this book make it a 'world first': it describes fillets, displays protein fingerprints (products of error-free genetic testing to identify species), and gives information on nutritional value.

Once this book becomes well-known and well-used in the commercial and recreational sectors of the industry, many more people will be confident about seafood varieties. Consumer demand for seafood will increase—and will open the way for greater enjoyment of our healthy product.

I congratulate the people—CSIRO staff and many others—who worked so hard to prepare this book, and the Fisheries Research and Development Corporation for its leadership and funding of the project that brought this excellent book to reality.

Peter Doyle
'Doyles on the Beach', Sydney

Contents

Introduction

G. K. Yearsley, P. R. Last and R. D. Ward

Australian seafood

Australia boasts some of the world's finest and most diverse seafood. From the Patagonian toothfish of the deep Southern Ocean to prawns caught near coastal mangroves in far north Queensland, the variety is enormous. Over 600 species of finfish and shellfish, both marine and freshwater, are caught and sold in Australia for local and overseas consumption.

Some of these—such as red emperor, whitings and wrasses—are well known to fishers and have been popular foodfishes for many decades. Others are relatively new to Australian palates but have gained rapid consumer acceptance. For example, the main fishing grounds for orange roughy and oreos were only discovered in the late 1980s, but these species are now among Australia's most popular foodfishes. Still others—Atlantic salmon, oysters and marron to name a few—are produced by a rapidly expanding aquaculture industry.

There are two main reasons for the enormous variety of Australian seafood. Firstly, Australia is responsible for some 16 million square kilometres of ocean and other aquatic areas, housing one of the most diverse biotas on earth. These areas encompass a plethora of tropical, temperate and Antarctic habitats, including freshwater lakes, rivers, estuaries, beaches, coastal bays, reefs, the continental shelf, and oceanic waters nearly six kilometres deep. The oceanic area, called the Exclusive Economic Zone (EEZ), extends 200 nautical miles off the coast. Together with the seas surrounding Australia's various island territories, these huge aquatic domains are home to hundreds of seafood species, many of which are harvested or farmed for human consumption.

Secondly, Australia's seafood diversity is well matched by its cultural diversity. The many cultural groups have preferences for certain seafoods. Consequently, some species that were previously non-targeted or were discarded are now utilised. For example, jellyfishes and sea urchins may not be high on the shopping list of those of northern European descent but are highly regarded by many Asians. The availability of new products and the influence of the different cultures has created opportunities for imaginative and experimental preparation and cooking by all Australians.

Australians are beginning to understand the benefits of eating seafood. Apart from its varied appearance, taste and texture, it is generally widely available, quick and easy to prepare, and very healthy. Seafood is stacked with key long-chain omega-3 fatty acids—which help prevent coronary heart disease and rheumatoid arthritis—and is generally low in cholesterol. It is also high in vitamins, protein and minerals although its nutritional value varies greatly between species and groups.

Australian seafoods, which are marketed under about 300 names, can be divided into four main groups: cartilaginous fishes (19 marketing names), bony fishes (223), crustaceans (30) and molluscs (25). There are also marketing names for three lesser invertebrate groups. Given that many of these marketing names cover a group of species (rather than a single species), the total number of domestic seafood species marketed in Australia is about 600.

Of the biomass of species produced (about 220 000 tonnes in 1997–98), about two-thirds is made up of cartilaginous and bony fishes. However, these groups account for only about a quarter of the total value of Australian fisheries ($1.86 billion in 1997–98). Most of the remainder comes from rocklobsters (20 per cent), prawns (20 per cent) and pearl oysters (10 per cent). About three-quarters of Australia's current fisheries earnings are from exports.

In addition to domestic products, Australia imports vast quantities of seafood from Asia, New Zealand, North America, South Africa, South America, Europe and elsewhere. It is estimated that over 60 per cent of seafood eaten in Australia is imported.

Seafood names

Long before Europeans settled Australian shores, the indigenous people had established names for several common seafood species. A few of these names—such as barramundi—are still used today. However, Europeans and other ethnic immigrants later applied names from their homelands to local species (e.g. gemfish, *Rexea solandri*, was called 'hake' and the name 'cod' was applied to a variety of unrelated fish groups) or coined new names (e.g. 'teraglin' for the jewfish, *Atractoscion aequidens*). Predictably, many species ended up with more than one name in different parts of the country (some species have a dozen or more regional names), while some quite unrelated species were given the same common name (e.g. both the dory, *Zeus faber*, and the butterfish, *Scatophagus multifasciatus*, have been called 'John dory'). Fishers, often from small, coastal regions, frequently use locally entrenched common names that date back to the 1800s. As a result, a plethora of names is used for Australian seafoods even today. This is confusing for both consumers and those in the fishing industry.

Marketing names vary from state to state, and in some cases from shop to shop in the same city or town. To complicate matters, fish are usually sold as fillets (which are more difficult to identify than whole fish), and imported species are sold alongside domestic seafoods with little or no indication of product origin. Consumers are concerned that the seafood they purchase may not be correctly or adequately labelled and this can lower their confidence in the Australian Seafood Industry.

To help overcome these problems, Australian seafood marketing names were standardised nationally through a joint Industry and Government review group, the Seafood Marketing Names Review Committee, which reports to the Federal and State governments through the Standing Committee on Fisheries and Aquaculture. Over the past 15 years, this group has reviewed all available names of Australia's seafood species. Apart from imported species (to be treated in a companion publication), this handbook supersedes a 1995 interim reference, *Marketing Names for Fish and Seafood in Australia*. Details of the role of the Review Committee, and the process by which a marketing name can be approved or changed, can be obtained from the organisations listed in Appendix A.

The problem of identifying fillets has been overcome by forensic techniques. Regulatory and policing authorities can use these to detect substitution or misrepresentation, increasing consumer confidence in names used by vendors. These techniques depend on genetic variation between species. Only a small piece of fillet or invertebrate muscle—one centimetre cube or less—is needed to identify a species. One of the most reliable techniques—protein fingerprinting—was chosen for this book because it is simple to use and can be used outside the laboratory.

Aim of this handbook

Millions of Australians—consumers, commercial fishers, processors, retailers, chefs, scientists, managers and recreational fishers—are in some way connected with the fishing industry and require at least a basic understanding of the different types of seafoods. However, given such a rich diversity of edible species, identification can be daunting even for experts. The main aim of this handbook is to provide an affordable, easy-to-use identification guide to all major

Australian domestic seafood species, including fish fillets, and to link each species to its approved marketing name. The Fisheries Research and Development Corporation funded the research on which this handbook is based to help eliminate the confusion over fish names in Australia. Handling, preparing and marketing, and a guide to imported species and their identification, will be treated in two later publications.

The first edition of this handbook will be expanded and refined in the future. Although much of its content is original, published literature and anecdotal information from specialists were often relied on for size, depth range and distributional data. Glaring contradictions and omissions were noted in the literature and better data are needed for many species. Consequently, comments and information from industry representatives, and readers who have a good knowledge of their local seafood species, are welcome. Please use the contact details given on the reverse title page. Future editions may also become available on CD-ROM or on-line on the internet.

How to use this handbook

G. K. Yearsley, P. R. Last and R. D. Ward

2

Introduction

This handbook is designed to help readers identify Australia's main domestic seafood species, whether whole, in fillet form, or shelled. Externally visible identifying features are provided for most species. In addition, a simple genetic technique called 'protein fingerprinting' is used to further distinguish each species. This information, along with data on the nutritional oil compositions of many species, is detailed in separate chapters. We have tried to avoid using technical terms but do so occasionally to avoid the repetitive use of long explanations. An illustrated glossary is provided to acquaint the reader with these terms. Related literature and references cited in the text are listed in three sections: (A) seafood, (B) protein fingerprinting, and (C) oil composition.

The use of names

Use of this handbook requires an understanding of common, marketing and scientific names. While a particular species may have a number of common names, it has only one approved marketing name in Australia and only one valid scientific name internationally. For example, the common name of one of Australia's premier foodfishes, the snapper, varies depending on its size and the part of the country where it is caught: cockney, pink snapper, red bream, snapper and squire are some of its common names. However, it has only one approved marketing name in Australia (snapper), and only one valid scientific name internationally (*Pagrus auratus*).

Common and marketing names

Literally thousands of common names have been used for Australia's 600 or so seafood species. To minimise confusion, commercial species should be sold only under their approved marketing name, demarcated as blue headings at the top of each page in the seafood entries (Chapters 4–8).

There are two main types of marketing names—unique and group names. 'Unique' marketing names cover only one species, whereas 'group' marketing names cover two or more species (sometimes entire groups of related species known as families). Some group marketing names, which are also known as 'core' marketing names, are used as a default marketing name for a number of similar species, some of which also have a unique marketing name. The flatheads (family Platycephalidae) provide examples of each marketing name type. 'Dusky flathead' is a unique marketing name that applies to only one species of flathead. 'Tiger flathead', however, is a group marketing name that applies to two almost identical species. 'Flathead' is an even broader group marketing name that can be used for all members of the flathead family. As such, 'flathead' is also a core marketing name, a default group for all flathead species, including those with unique marketing names. Hence, flatheads, including the dusky and tiger flatheads, can also be marketed simply as 'flathead'. However, vendors can usually gain a better price by using the correct unique name when one exists. In this handbook, group marketing names are identified by the letter 'G' at the bottom, inner margin of each page where applicable.

Invalid marketing names for a species are listed herein as 'previous names'. Generally, this list gives old or erroneous marketing names rather than including all other common names used for the species. However, other frequently used common names are included for some of the individual species covered by group marketing names. Some valid names have been selected for historical reasons. For example, Australian salmons (*Arripis trutta* and *A. truttaceus*) are unrelated to the true salmons (Salmonidae). However, the group name is well known to the Australian public and is unlikely to cause confusion.

Occasionally, different species are erroneously marketed under the same name. The previous name of one species (or group of species) may be the approved marketing name of another. For example, 'striped butterfish' (*Scatophagus multifasciatus*) has been sold under the approved marketing name of 'John dory' (*Zeus faber*). Extremely confusing scenarios have eventuated for 'seaperch', 'seabream' and 'trevally', with several unrelated species sold under these names.

In some cases, a group marketing name is the same as an individual species' common name. For example, 'yabby', which is the marketing name of several types of yabby (*Cherax* species), is also the most frequently used common name of one of these (*C. destructor*).

Scientific names and classification

Classification is a means of cataloguing organisms in a hierarchical manner resembling a family tree. At the top of the tree are major groups called Kingdoms. At the bottom of the tree, each species is given a unique scientific name.

Each Kingdom is divided into smaller groups called phyla (singular phylum). Similarly, each phylum is divided into classes, classes into orders, orders into families, families into genera and genera into species. Each rank or level in this hierarchy is called a taxon (plural taxa). For example, here are the taxonomic groupings into which the southern bluefin tuna is classified:

Kingdom:	Animalia (animals)
Phylum:	Chordata (chordates)
Class:	Osteichthyes (bony fishes)
Order:	Perciformes (perch-like fishes)
Family:	Scombridae (mackerels and tunas)
Genus:	*Thunnus*
Species:	*maccoyii*

The term 'group' is used exclusively in this book to refer to taxonomic levels above a species. Sometimes, those studying classification (taxonomists) subdivide such groups. For example, an order may be subdivided into a number of infraorders, or one family may be subdivided into a number of tribes or subfamilies. Sometimes groups are aggregated. For example, a number of families may be combined to produce a superfamily. Certain rules and conventions apply to the endings of higher-level group taxa. For example, order names end in '-iformes', family names in '-idae' and subfamily names in '-inae'.

The combination of the genus (plural genera) and species (plural species) parts constitute the species name. While a species has only one species name, a genus may contain any number of closely related species. To continue with the example of southern bluefin tuna (*Thunnus maccoyii*), there are five other members of the genus *Thunnus* in Australian waters.

Species names should always be italicised with only the first letter of the genus in upper case. The genus name precedes the species name but may be abbreviated to its initial letter if no ambiguity results. For example, once southern bluefin tuna has been introduced as *Thunnus maccoyii* it may be referred to as *T. maccoyii*. If a species has not been scientifically described, it is referred to by its genus followed by the abbreviation 'sp.' Groups of species are referred to by 'spp'.

The author of a species (the person who described it) and the date it was described follows the scientific name. For species in this handbook, such details are listed in Appendix B. Parentheses around the author and date indicate that the species is no longer classified in its original genus. A later worker (or the same worker in a later publication) transferred the species to another genus.

Order of presentation of groups and species

Biological identification guides can be ordered in several different ways. To someone familiar with the names, an alphabetical arrangement by common or scientific name is easy to use. However, comparisons are difficult because similar species (members of the same family) are not necessarily placed near each other. To enable comparisons to be made, guides often order species scientifically (systematically), following an internationally recognised order based on our knowledge of their evolutionary lineage. Obviously, while this is useful to biologists, it can be extremely confusing for the lay-person.

This handbook uses a combination of alphabetic and systematic methods, making it user-friendly to the non-specialist while still allowing comparisons of closely related species. Broad groups (cartilaginous fishes, bony fishes, crustaceans, molluscs and other invertebrates) are separated by chapter. Within chapters, families and family groups (two or more closely related families) are ordered alphabetically by common name. Within families, individual species or species groups are again ordered alphabetically this time by their marketing name.

Information for a particular species or family can be found in a number of ways:

— your species can be found using a visual inspection of each finfish or shellfish image;

— if you know the general group to which the species belongs but are unsure of its correct name, you can quickly scan the contents pages (family and family-group common names are listed)—the abbreviated family name is given with its component groups in brackets;

— if you know the correct family or family-group common name, scanning the contents will be unnecessary—the target group can be found alphabetically in the relevant chapter and each chapter is colour-coded to assist navigation through the guide;

— once in the correct group, your species can be found from an alphabetic search of marketing names within the section;

— you can search the index of common and marketing names (marketing names in bold text), or the index of scientific names (family and other higher-level taxa names, only included when specifically mentioned in the text, are in capitals)—in both indexes, numbers in bold refer to the page where that species or group is featured, numbers prefixed with 'P' refer to the relevant protein fingerprint figure, and those prefixed with 'O' refer to oil composition;

— you can also check Appendix B which lists all families systematically and provides cross-referenced page numbers. Group marketing names are highlighted in bold text and ordered alphabetically by scientific name. Component species are indented and also ordered alphabetically. Only those species mentioned in the text are included.

The protein fingerprinting and oils chapters are arranged in the same way as the main identification chapters—alphabetically by family (or family group) common name. Page numbers for information on the protein fingerprints and oil compositions are listed on the feature page for each species where applicable (usually in the fillet box for fishes and at the end of the 'Remarks' section for invertebrates).

Seafood treatments

Chapters 4–8 form the basis of this handbook. Every approved Australian seafood marketing name—except 'tilefish' (Malacanthidae)—is included. Other species, currently caught at low levels or covered by a group marketing name, may warrant inclusion in future editions. Abbreviated, coloured headings at the foot of the page are used to flag sections of similar species, genera or families. The groups treated in each section are listed sequentially on the contents pages. The commonly used Australian name is sometimes different from its international name (e.g. 'mackerels' is used rather than 'tunas'). Similarly, family group names are selected based on local use and to minimise confusion. For example, members of the family Lutjanidae are known as 'seaperches' and 'snappers'. However, as 'snapper' can be confused with one of Australia's premium temperate fishes (*Pagrus auratus*), 'seaperch' is used as both a collective group marketing name and the section name.

Unique marketing names are treated on a single page. Group marketing names cover one or two pages; the first (or only) treats the group as a whole, showing and naming a representative species. Where the fillet picture is of a different species, its species name is provided in the fillet box. If two pages are used, the second contains specific information (common name, distribution, distinguishing features and size) on three individual species. The species pictured on the first page is usually repeated on the second. In groups such as 'ray', however, where there is much morphological variation, the species pictured on the first page is replaced on the second page with another member of the group.

Various other aspects of each species and group of species are discussed under the headings below. The following abbreviations are used for distributional information: NSW—New South Wales; Vic.—Victoria; Qld—Queensland; WA—Western Australia; SA—South Australia; Tas.—Tasmania; and NT—Northern Territory.

Identifying features

The main features (characters) by which a species or group of species can be most easily identified are listed and highlighted on each figure as corresponding numbers to show each feature's location. The number may be located near, rather than on, a particular feature to make it visible. Internal features, and others that cannot be seen on the photograph, are identified by a number within a circle. Features that separate closely related species are listed first. Usually, each number refers to the same character on each species within a family or group of families but this is not possible with larger groups. Character traits are not necessarily provided for every species in the group.

Comparisons

Comparisons are made with closely related species—those in the same genus and family and sometimes those in closely related families. In special cases, species with a similar marketing name are also compared (e.g. 'red emperor' is compared with 'emperor') or when group-names could be confusing.

Fillet

Fillet identification is far more difficult than whole fish identification, as many useful features such as fin shape and mouth position are missing. However, other features can be used. These are provided along with, for most species, a photograph of the outer, skinned fillet.

Very small species, such as 'whitebait', are marketed whole rather than as fillets so a photograph is not included. Similarly, large fishes, such as tunas, are marketed as cutlets or steaks rather than fillets. In such cases, a cutlet or steak photograph is usually provided. The identifying features used here are discussed under the heading 'Characteristics of fillets' below. More comprehensive information helping users identify fillets will be published in a separate technical report. The fillet section, for obvious reasons, is omitted from the invertebrate chapters (6–8).

Apart from informal comments in the 'Remarks' section, edible qualities are not discussed. All the seafood consumed in Australia is highly esteemed by at least a few palates and some species are generally highly regarded. However, preferences for seafood vary greatly within the community. Future editions of the handbook should contain taste comparisons and information on changes that occur during cooking.

Size

Weights are expressed for whole animals in kilograms (kg) or grams (g). Multiply the kilogram value by 2.2 to convert to pounds. Lengths for sharks, bony fishes and some invertebrates are expressed as total length, unless stated otherwise. Total widths are provided for rays and some invertebrates such as abalone and crabs. All length and width measurements are expressed in centimetres (cm).

Maximum sizes and weights recorded in the literature are largely unreliable. Weights are sometimes based on non-scientifically based 'guesstimates'. Pounds have sometimes been confused with metric units accentuating the maximum weight of species. Lengths, which are based on either standard, fork or total lengths, are rarely specified resulting in quoted sizes often being smaller than the true size of the species. Often only length or only weight data has been recorded from outsized specimens. Consequently, reported maxima for the two types of measurement rarely correspond, leading to gross over- or understatements of either size measure. Also, maximum sizes of seafood species rarely reflect the normal size marketed.

We have evaluated the size information, making use of the literature, anecdotes and museum specimens, to provide a provisional review of maximum and common (average) sizes for each species. For many species, these data remain inadequate; the authors would appreciate receiving any new information to refine these statistics.

Habitat

This section includes information on the species' environment. Its general habitat—whether it lives mainly in freshwater, estuaries, near the coast or offshore—is further qualified by its preferred habitat (whether it lives on the continental shelf, continental slope, or is pelagic or bottom-dwelling, *etc.*). The depth at which the species is found is provided when available.

Distribution

The Australian distribution of each species (or group of species) is shown on a map. Dark blue shading is used for marine distribution while light blue shading shows freshwater and estuarine distribution. For marine species, shading touching the coastline signifies a continental shelf distribution; wider shading suggests a distribution over both the continental shelf and slope. Shading separated from the coastline is used for species that occur on the continental slope only. The letter 'I' indicates an international distribution (rather than occurring only in Australia) and the product may be imported. Species without an 'I' are endemic.

Fishery

This section summarises how and where commercial and recreational fisheries for the species occur. The value of the fishery may be mentioned. Quotas, fishing restrictions, and size and bag limits are not included as they often differ between states and change frequently; they should be checked with Commonwealth and State authorities.

Remarks

This informal section may include such items as a discussion of old or anomalous marketing and scientific names, migrating behaviour, worldwide distribution, potential danger to humans, flesh taste and texture, and anecdotal information.

Characteristics of fillets

The edible qualities of different types of seafood can be assessed a number of ways. Features of the flesh, such as general appearance, moisture level, flavour, texture, colour and shelf life, have all been used as measures of quality. Similarly, these features, along with many others, can be used to identify the fillets of fish groups and, in some cases, species.

Fishes as a group come in all shapes and sizes, which remains apparent in the form of their fillets. Their overall shape and muscle structure, extent of fat deposits, connective tissue colour, and the presence or absence of remnant processing features such as scale pockets, lateral line, swim bladder, belly flap, sensory pores, and pin bones can be used to assist species identification.

This handbook includes pictures and simple descriptions of fillets for most species, and a glossary section defining and illustrating the key features (Figs 3.8–3.12). The use of fillets as an identification tool will be treated comprehensively in more technical supplementary publications. However, in this handbook, we have attempted to demonstrate some of the variability that exists within fish fillets, giving examples where necessary.

Fillet shape

Possibly the most obvious feature of a fillet is its basic appearance. Short, deep-bodied fishes, such as dories, usually produce short, strongly tapering fillets (Fig. 3.9A). Long, slender fishes, such as eels, usually produce long fillets with little taper (Fig. 3.9G). There is a variety of intermediate forms between these extremes (Figs 3.9B–F). These forms are generally consistent within species and may enable discrimination between close relatives. Larger fishes are often marketed as cutlets (Fig. 3.8C) and rays as flaps (Fig. 3.8D).

Other aspects of shape, such as thickness, are important in characterising fillets. The fillets of round-bodied fishes, such as tunas, are usually thick with a convex outer (external) surface. These contrast strongly with pronounced laterally compressed (e.g. dories) or depressed (e.g. flounders) fishes, both of which produce almost flat fillets.

Filleters may vary slightly in the way they cut, but the lower angle of the fillet is nonetheless surprisingly consistent within a species. This angle may vary from almost a right angle (oreo dories) to obtuse angles exceeding 135 degrees (flatheads). However, removal of the belly flap can dramatically change the appearance of the fillet. Some species are marketed with or without the belly flap, depending on the flap's thickness and the ease of separating rib bones.

Similarly, the presence or absence of remnants of the swim bladder, its position and type can be useful in discriminating not only groups but also members of groups. For example, the swallowtail (*Centroberyx lineatus*) has a thick swim bladder that penetrates as a conical tube into the tail (well beyond the origin of the anal fin). Other redfishes (*Centroberyx* species) have a more delicate swim bladder that lies within the main body cavity. While the swim bladder remnants are often trimmed from fillets, the conical remnant is usually present in swallowtail fillets.

Another structure, a remnant of the caudal peduncle, may be present, although its absence can be due to the method of filleting.

Outer features

A key identifying feature on the outer (external) surface of a fillet is the skin. This may be removed, depending on its thickness, revealing other characters such as muscle band structure and features of the integument underlying the skin. The evidence of scale pockets, and their size and ability or inability to remain intact after scaling, are important characters. Scales and pockets may be deeply embedded (e.g. lings, *Genypterus* species), indistinct (e.g. leather-jackets, Monacanthidae) or totally absent (e.g. catfishes, Ariidae). When absent, the skin is sometimes covered in minute sensory pores. Similarly, the position and clarity of the lateral line can assist in the identification of some groups, such as the trevallies (Carangidae), whose lateral-line scales may be enlarged to form scutes. In the tunas (tribes Sardini and Thunnini), fleshy keels are present on the caudal peduncle which remain evident on carcasses. Sharks are covered in small spiny scales known as denticles.

Skinned fillets usually have traces of integument remaining that reflect the colour of the skin and its connective tissue (Fig. 3.9D). The most common tones are translucent, white, pink, and silver, but the integument colour of the upper and lower halves of the fish may be different. Similarly the sliminess of the outer part can vary from almost dry to very slimy.

The arrangement of W-shaped, vertical muscle bands (myomeres, Fig. 3.8A), and the number and form of exposed, outer pin bones in fillets are useful in distinguishing species. Each myomere is separated from the next by a partition of the membranous skeleton known as a 'myoseptum' (Fig. 3.8A). Upper and lower muscle masses of the body are separated by a 'horizontal septum' (Fig. 3.8A), referred to in the tissue descriptions as HS, appearing as a midline (corresponding to the midpoint of the 'W'). When present, the outer pin bones are either exposed or embedded slightly in the anterior part of this septum (Fig. 3.11A).

The muscles above the horizontal septum form the 'epaxial mass', those below, the 'hypaxial mass' (Fig. 3.8A). Myosepta appear either 'V' or 'W' shaped in each mass. The relative heights of these masses, the lengths of the outer arms of the 'V's and 'W's in each mass, the position of the horizontal septum, and the number of epaxial myomeres, are very useful in identifying species. Similarly, the position of a line drawn through the posterior angle of the 'V's in each mass in relation to the horizontal septum varies greatly in fishes. In many perch-like fishes, the epaxial line (referred to in the tissue descriptions as EL), which runs parallel to the dorsal profile of the fillet, is convex in shape (Fig. 3.9A). Alternatively, it may converge with the HS towards the tail end (Fig. 3.9B,F) or can be almost parallel (Fig. 3.9G). The orientation of the hypaxial line (HL) can vary similarly.

In most fishes, the myosepta are distinct, but in some species (Fig. 3.9E) they can be difficult to see, the myomeres either being strongly connected or the myosepta covered with fat.

The horizontal septum may also be concealed by another type of muscle, the red or dark muscle band. The colour of this band has been used as an index of fillet freshness. Reddish muscle is characteristic of very fresh fish whereas a brownish band with gaping flesh (Fig. 3.12C) usually indicates the opposite. The width and strength of this band also vary greatly even between related species. In prolonged swimmers, such as tunas and trevallies, it is continuous, broad and very thick (Fig. 3.9D), whereas in non-migratory fishes, such as the wrasses, it is barely detectable (Fig. 3.9B). Thin red muscle bands often appear as 1–3 discontinuous bands that may taper towards the caudal region (Fig. 3.8A). In garfishes the red muscle band is uniform in width and continuous (Fig. 3.11D).

Inner features

The inner surface of a fillet, which is the internal part closest to the backbone, is equally useful for identifying fish species. As the muscle fibres are cut when filleting, the flesh colour is clearer than on the outer surface. Fresh fishes vary considerably in this character (Fig. 3.10A–J), but it is reasonably consistent within a species. However, old or poorly handled fillets tend to become yellowish or brownish, and this must be considered when assessing colour. Some groups are distinctive—for example, the reddish flesh of most tunas, the orange flesh of salmons (Salmonidae), the green flesh of batfish (*Platax* species), and the bluish-white flesh of the grass whiting (*Haletta* species). Others, such as the garfishes (Hemiramphidae) and some whitings (Sillaginidae), have almost translucent flesh with greyish 'veins'. Colour reproduction in this handbook is as accurate as possible but some slight variations can occur.

Similarly, the membrane of the gut cavity wall (the peritoneum) and the myosepta can vary from transparent to silver, white or black (Fig. 3.8A,B). Other useful characters include fat deposits (Fig. 3.8A), the flakiness of the muscle, and the remnants of radial muscles (associated with the dorsal and anal fins and arranged vertically). In some fishes, such as the dories, these usually remain even on trimmed fillets. Other fishes such as trevallies may have pockets or pits in the flesh formed by outgrowths of the skeleton (Fig. 3.12B). Parasites are sometimes hard to see but usually occur only in specific fish hosts.

Analogous to the outer pin bones are the various rib and intermuscular bones of the inner surface (Fig. 3.12B). These inner pin bones, particularly the ribs, are often removed by filleting; when present, they are useful for distinguishing groups and species. In some fishes (e.g. garfishes) the non-rib pins extend well along the fillet and cannot easily be removed from the fillet. Their size and number are important characters.

Protein fingerprinting

Chapter 9 presents protein fingerprint information on 380 species. The principles of the technique are outlined, together with a recommended strategy for checking identification using these methods. Technical details are briefly presented. The main body of the chapter shows a stylised protein fingerprint of each species, together with common variants; the accompanying text highlights the salient identifying features of the members of each group, especially with respect to species separation.

Oil composition

Chapter 10 contains summary oil data on selected Australian seafood species—oil (fat) content of the edible portion of the species, content of major polyunsaturated fatty acids and the average levels of saturated, monounsaturated and polyunsaturated fatty acids. Full oil composition details are provided in other publications. As in other chapters of the handbook, we have attempted to demonstrate the variability that exists between species and to demonstrate this using histograms of the three fatty acids that are important in our diet.

Appendix A: Marketing names contacts

Appendix A lists the contact details of organisations that can provide information on the Seafood Marketing Names Review Committee and the process the Committee follows to amend or add a marketing name.

Appendix B: Table of species

The table in Appendix B includes all the marketing names and species names mentioned in the text. One column heading requires explanation (the others are discussed in various sections above):

CAAB

Each species (including those covered by a group marketing name) and each group marketing name has been assigned a numeric code based on a system known as 'Codes for Australian Aquatic Biota' (CAAB). These codes are used widely to link data on Australian fisheries and on marine biota generally. The coding system was recently upgraded from six to eight digits (Yearsley *et al.* 1997). Invertebrate groups are still in the transition phase and are therefore prefixed with 00. Fishes are prefixed with 37.

Glossary

G. K. Yearsley and P. R. Last

3

AA: arachidonic acid, 20:4(n-6), a nutritionally valuable omega-6 fatty acid present in Australian seafood

abdomen: the part of the body containing the digestive and reproductive organs—in crustaceans it consists of several segments (Fig. 3.5)

adipose fin: a small fleshy fin, without fin rays, usually situated behind the dorsal or anal fins in some fishes (Fig. 3.3)

adrostral ridge: in prawns, a ridge that runs alongside the rostrum, sometimes nearly reaching the posterior margin of the carapace

allozyme: alternative enzyme forms encoded by different alleles of a gene

anal fin: the unpaired fin on the underside of the body behind the anus in fishes (Fig. 3.3)

antenna (pl. antennae): the more lateral of the segmented, paired appendages on the head of crustaceans, usually long and whip-like (Fig. 3.5) but sometimes short, broad and flattened

antennule: the more medial (central) of the segmented, paired appendages on the head of crustaceans (Fig. 3.5), usually shorter than the antennae

anterior: the front or head end (Fig. 3.3)

anus: external opening of the digestive system (Fig. 3.3)

aperture: the entrance or opening to the internal cavity of a single-shelled mollusc (Fig. 3.6A)

arm: the sucker-bearing appendages surrounding the mouth of cephalopods (Fig. 3.7)

bar: a broad, more or less vertical, line of a different colour to the main (or adjacent) body colour

barbel: a slender, fleshy, tentacle-like sensory structure on the head of some fishes (Fig. 3.3)

basal: the region of a projection (often a fin) nearest the body

belly flap: post-filleting remnant of the muscle of the abdominal region or belly of a fish (Fig. 3.8B)

benthic: living on the sea floor

benthopelagic: free-swimming near the sea floor

bill: a usually long and thin anterior extension of one or both of a fish's jaws

blotch: an irregularly shaped area that differs in colour to the surrounding colour

body length: in crustaceans, the distance from the anterior margin of the carapace (near the eyes) to the end of the extended tail fan

breast: in fishes, the ventral surface of the body below the pectoral fins (Fig. 3.3)

bycatch: the component of the catch excluding the targeted species

canine tooth: an enlarged, conical tooth (for holding prey)

carapace: the hard, external shell of crustaceans that covers the 'body' (i.e. the head and thorax but not the abdomen) (Fig. 3.5)

carapace length: in crustaceans, the distance from the anterior margin (near the eyes) to the posterior margin of the carapace

caudal fin: a fish's tail fin (Fig. 3.3)

caudal peduncle: the posterior part of a fish's body from the posterior end of the anal-fin base to the base of the caudal fin (Fig. 3.3)

cervical groove: in prawns, a groove that runs obliquely across the carapace from the ventral anterior third, posteriorly and towards the dorsal midline (Fig. 3.5)

cheek: in fishes, the fleshy area in front of the preoperculum

cloaca: a common opening for digestive, urinary and reproductive tracts in many fishes (Fig. 3.3)

compressed: flattened laterally, from side to side

concave: arched, curved inwards (opposite of convex)

conical teeth: teeth shaped like a cone

continental shelf: the shelf-like part of the ocean floor beside continents and extending from the coast to a depth of about 200 m; the area at about 200 m is called the shelf break

continental slope: the sloping, often steep, part of the ocean floor bordering the continental shelf and extending to a depth of about 2000 m; divided into the upper slope (200–700 m), mid-slope (700–1500 m) and lower slope (below 1500 m)

convex: arched, curved outward (opposite of concave)

corselet: a band of specialised scales encircling the pectoral region of the body

ctenoid scale: a scale with a spiny hind margin

cusp: a projection on a tooth

cuttlebone: calcified, surfboard-shaped internal shell of cuttlefishes

cycloid scale: a smooth-edged scale without spines along its hind margin

deciduous: easily shed or rubbed off, usually referring to scales

demersal: living on or near the bottom of the ocean

depressed: flattened from top to bottom

DHA: docosahexaenoic acid, 22:6(n-3), a nutritionally valuable omega-3 fatty acid present as the main polyunsaturated fatty acid in Australian seafood; considered a primary building block of the body

disc: the combined head, trunk and enlarged pectoral fins of some sharks and rays with depressed bodies (Fig. 3.2)

dorsal: the upper surface of the body (or head) (Fig. 3.3)

dorsal fin: the unpaired fin or fins along the upper surface of the back or tail of fishes; in bony fishes often divided into an anterior spinous portion and a posterior soft-rayed portion, which may be separate fins (Fig. 3.3)

EEZ: Exclusive Economic Zone, the oceanic area controlled by and surrounding Australia; it extends 200 nautical miles from the coast and includes ocean around island territories

EPA: eicosapentaenoic acid, 20:5(n-3), generally the second most abundant nutritionally valuable omega-3 fatty acid present in Australian seafood

epaxial line: in a fish fillet, an imaginary line drawn through the posterior angle of the 'V' formed by the myomeres in the upper (epaxial) muscle mass (Fig. 3.8A)

filament: a thread-like process or appendage

fin: in cephalopods, membranous extensions of the body that assist in locomotion, steering and stabilisation (Fig. 3.7)

fin element: each spine or ray in a fish's fin

fin ray: see *ray*

fin spine: see *spine*

finlet: a small fin-like structure behind the anal and/or dorsal fins of some fishes (Fig. 3.3)

flagellum (pl. flagella): segmented, usually whip-like terminal section of an antenna or antennule in crustaceans

foot: the large muscular mass on which single-shelled molluscs move (Fig. 3.6A)

fork length: the length of a fish from the tip of the snout to the fork of the caudal fin

forked: a common shape of a fish's caudal fin, one with a deeply concave or excavated hind margin

free rear tip: posterior tip of a fin closest to the fin's posterior point of attachment (Fig. 3.1)

fusiform: spindle-shaped, tapering to both ends

genus (pl. genera): a natural grouping of closely related species

gill arch: a cartilaginous or bony arch supporting the gills and gill rakers of fishes (Fig. 3.4); the gill arch closest to the interior surface of the operculum is the 'first gill arch'

gill membrane: the skin on each side of the head that encloses the gill chamber during respiration

gill rakers: the peg-like structures along the front edge of the gill arches in fishes (Fig. 3.4)

gill slit: a long, narrow gill opening in sharks and rays (Fig. 3.1)

groove: well-defined furrow

hepatic ridge: in prawns, a ridge that runs posteriorly near the anterior ventral margin of the carapace, sometimes beginning close to the anterior margin (Fig. 3.5)

hinge: where two valves of a bivalve mollusc are joined, on the dorsal margin (Fig. 3.6B)

horizontal septum: line of connective tissue running centrally along the length of a fish fillet and separating the muscle into the upper (epaxial) and lower (hypaxial) masses (Fig. 3.8A)

hypaxial line: in a fish fillet, an imaginary line drawn through the posterior angle of the 'V' formed by the myomeres in the lower (hypaxial) muscle mass (Fig. 3.8A)

incisor tooth: a cutting tooth with a flat chisel-shaped tip

integument: tissue between the skin and underlying muscle mass—remnants may be present on skinned fish fillets (Fig. 3.11A)

ischial spine: in prawns, a spine projecting from the third segment (counting from the body) of a leg

isthmus: the fleshy area on the undersurface of the head between the gill openings

keel: a fleshy or bony ridge (Fig. 3.3)

lateral: refers to the sides

lateral line: a line or row of pored scales or sensory organs along the side of a fish (Fig. 3.3)

LC-PUFA: long-chain (C20) polyunsaturated fatty acids, including DHA and EPA

lip: in abalones, the outer edge of the foot

lobe: a usually rounded outgrowth

lunate: shaped like a crescent moon, usually referring to a fish's caudal fin

mandible: the lower jaw (Fig. 3.3)

mantle: the muscular integument surrounding the internal organs of molluscs; in cephalopods it is greatly strengthened (Fig. 3.7)

mantle length: in squids and cuttlefishes, the distance from the dorsal anterior margin of the mantle to its most posterior point

maxilla: a paired bone forming the lateral margin of the upper jaw (Fig. 3.3)

melanophores: black pigment cells in skin or internal membranes (Fig. 3.12B)

membrane: the thin layer of tissue covering part of an animal or connecting the fin elements in fishes

molar tooth: a blunt and rounded grinding tooth

morphology: the physical form and structure of an animal

MUFA: monounsaturated fatty acids, containing one carbon-carbon unsaturated centre, generally with *cis* configuration—e.g. oleic acid [18:1(n-9)c], which is abundant in olive oil

multicuspid: having two or more cusps (on a tooth)

myomeres: segmentally arranged blocks of muscle in the bodies of fishes (Fig. 3.8A)

myoseptum: line of connective tissue separating the bundles of muscle or myomeres (Fig. 3.8A)

nape: in fishes, the region of the head above and behind the eyes, before the dorsal fin (Fig. 3.3)

nostril: external opening of the nasal organs, usually pore-like in fishes (Fig. 3.3)

omega-3: series of polyunsaturated fatty acids with two or more *cis*-unsaturated centres, separated from each other by one methylene group and having the first unsaturated centre three carbons from the end methyl

omega-6: series of polyunsaturated fatty acids with two or more *cis*-unsaturated centres, separated from each other by one methylene group and having the first unsaturated centre six carbons from the end methyl

opercular spine: a spine on the posterior margin of the operculum in fishes (Fig. 3.3)

operculum: in fishes, the bony flap covering the gills (Fig. 3.3); in single-shelled molluscs, a hard, plate-like structure that partly or completely closes the aperture when retracted into the shell (Fig. 3.6A)

origin: of a fish's fin, the most anterior point of a fin base

pectoral fins: paired fins just behind or just below the gill opening of fishes (Fig. 3.3)

pelagic: free-swimming in the open ocean

pelvic fins: paired fins on the underside of the body between a fish's mouth and anus (Fig. 3.3)

pen: cartilaginous, feather-shaped internal shell of squids

peritoneum: membranous lining of the internal belly flap or body cavity housing the internal organs (Fig. 3.8B)

pin bones: rows of thin, usually exposed bones along or near the horizontal septum of the outer (Fig. 3.11A) and inner (Fig. 3.12B) surfaces of a fish fillet

plankton: small animals and plants that float or drift with ocean currents

posterior: the hind or tail end (Fig. 3.3)

precaudal pit: in sharks, a transverse or longitudinal notch on the dorsal or ventral surface just in front of the caudal fin (Fig. 3.1)

preopercular spine: a spine on the posterior margin of the preoperculum in fishes (Fig. 3.3)

preoperculum: the main bone forming the anterior part of the operculum in fishes (Fig. 3.3)

protein fingerprint: the pattern of bands seen after a tissue sample is subjected to gel electrophoresis and the gel stained with a protein-specific stain

protrusible: a condition of the jaws in which the mouth projects forward, downward or upward as a tube when the mouth is opened

PUFA: polyunsaturated fatty acids, containing more than one carbon-carbon unsaturated centre

ray: a flexible, often branched and segmented structure that supports a fish's fin (Fig. 3.3)

red muscle: relatively dark muscle evident as a flattened wedge extending outwards over the horizontal septum on the outer surface of a fish (Fig. 3.8A)

rostral spine: a spine on the rostrum of a crustacean (Fig. 3.5), sometimes referred to as a rostral 'tooth'

rostrum: a forward-projecting, thick extension on the carapace of a crustacean (Fig. 3.5)

rounded: usually referring to a fish's fin shape, one with an evenly convex margin

run: electrophoresis procedure whereby proteins are separated in a gel medium on the basis of their size and electrical charge; in a fast run the proteins move further and become more widely separated from one another than in a slow run

SAT: saturated fatty acids without carbon-carbon unsaturation—e.g. myristic (14:0) and palmitic (16:0) acids; generally more common in animal than plant fats

scale: a small membranous or horny modification of the skin of many fishes

scale pockets: remnant structures on a fish's skin after scaling—provide evidence of the size and shape of scales

scute: a bony plate or enlarged, ridged scale (Fig. 3.3)

seamount: a mountain or hill on the seabed

shelf: see *continental shelf*

shelf break: see *continental shelf*

shoulder: the upper side of a fish just behind the head and usually above the pectoral fin (Fig. 3.3)

skin fold: an area where skin is bent over upon itself, forming a fleshy ridge

skin nodules: wart-like structures beneath the skin of some fishes (Fig. 3.11C)

slope: see *continental slope*

snout: the part of the upper head in front of the eyes of fishes (Fig. 3.3)

soft ray: see *ray*

species: a group of actually or potentially interbreeding animal or plant populations that are reproductively isolated from other such groups

spicule: minute, plate-like, calcareous structure that provides support

spine: a sharp projecting point; in fishes, a firm, undivided and unsegmented fin support (Fig. 3.3)

spinous: spine-like or bearing or composed of spines

spiracle: a respiratory opening behind the eye in sharks and rays (Figs 3.1, 3.2)

spire: the pointed or raised part of a single-shelled mollusc, opposite the aperture (Fig. 3.6A)

spot: a regularly shaped or rounded marking (usually small) that differs in colour from the adjacent area

standard length: in fishes, the distance from the snout tip to the last caudal vertebra

stinging spine: the large, serrated, sword-like bony structure on the tail of some rays (Fig. 3.2), sometimes abbreviated to 'sting'

stripe: a more or less horizontal line or marking of a different colour to the main (or adjacent) colour

suborbital: area below the eye

sucker: suction-cup structure on the arms and tentacles of cephalopods

swim bladder: a sac (usually gas-filled) in the body cavity beneath the backbone of many fishes—developed as part of the alimentary canal for buoyancy compensation (Fig. 3.8B)

tail fan: the terminal section of most crustaceans' bodies, consisting of a central telson and two flattened appendages either side (Fig. 3.5)

telson: the central unit of a crustacean's tail fan

tentacle: in some cephalopods, one of two long, sucker-bearing arms originating near the mouth

thorn: a sharp tooth-like structure on the skin of a skate or ray (Fig. 3.2)

total length: of fishes, the greatest distance from the tip of the snout to the end of the tail

truncate: with a straight margin, terminating abruptly; usually refers to fin shape

trunk: that part of a fish (other than the fins) between the head and the tail

tube foot: external muscular extension of the water vascular system of echinoderms, usually occurring in rows

valve: each shell of a mollusc (Fig. 3.6)

ventral: refers to the lower surface or underside of the body (or head) (Fig. 3.3)

ventral fins: see *pelvic fins*

vomer: in fishes, a bone forming the front part of the roof of the mouth in the nasal region, frequently toothed

Figure 3.1—Structural features of a generalised shark.

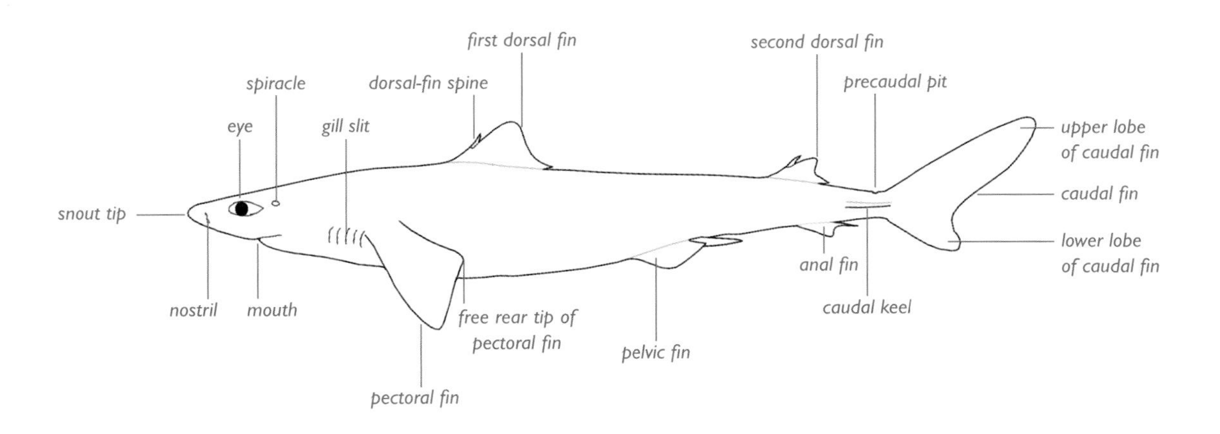

Figure 3.2—Structural features of a generalised ray.

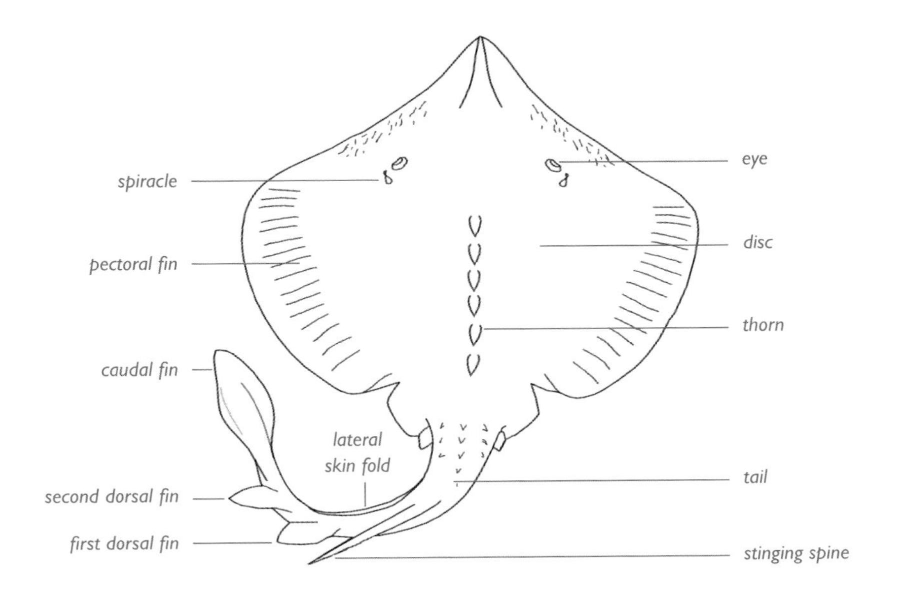

Figure 3.3—Structural features of a generalised bony fish.

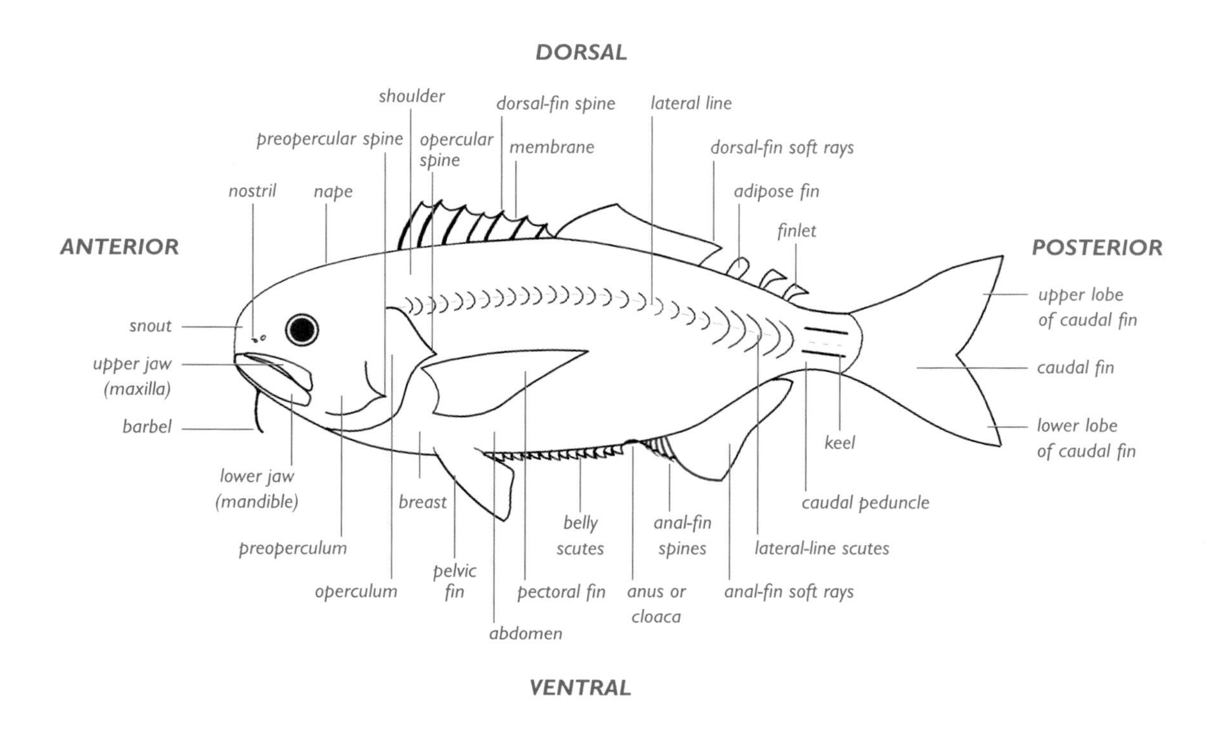

Figure 3.4—Structural features of a generalised bony fish outer gill-arch.

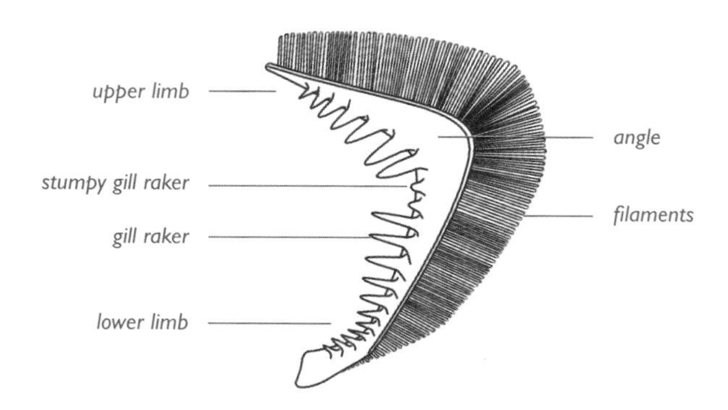

Figure 3.5—Structural features of a generalised crustacean.

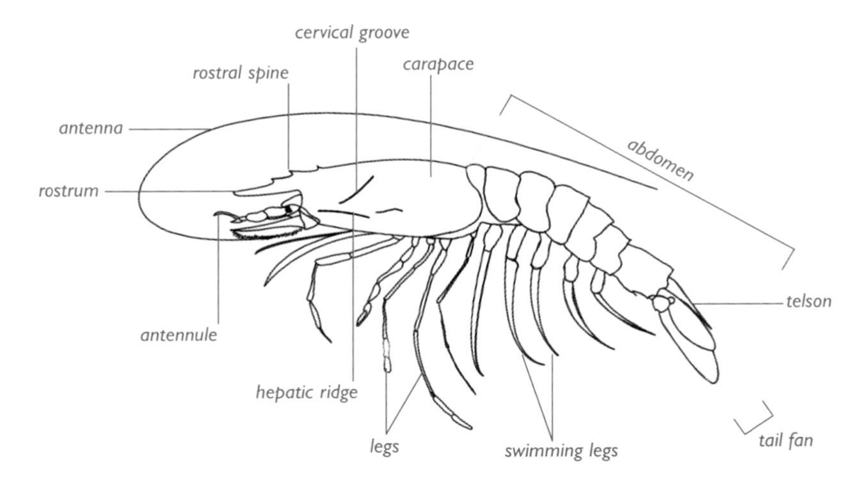

Figure 3.6—Structural features of a generalised (A) single-shelled mollusc; and (B) bivalve mollusc.

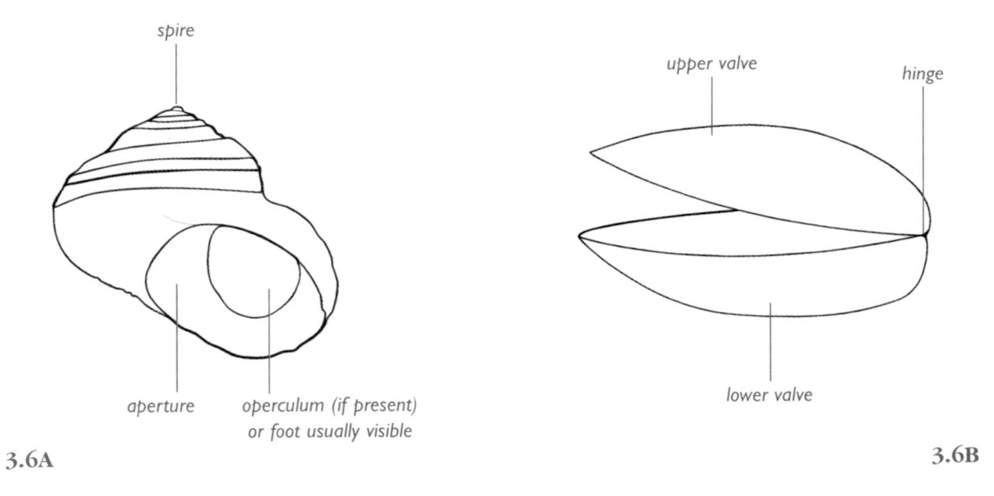

Figure 3.7—Structural features of a generalised cephalopod.

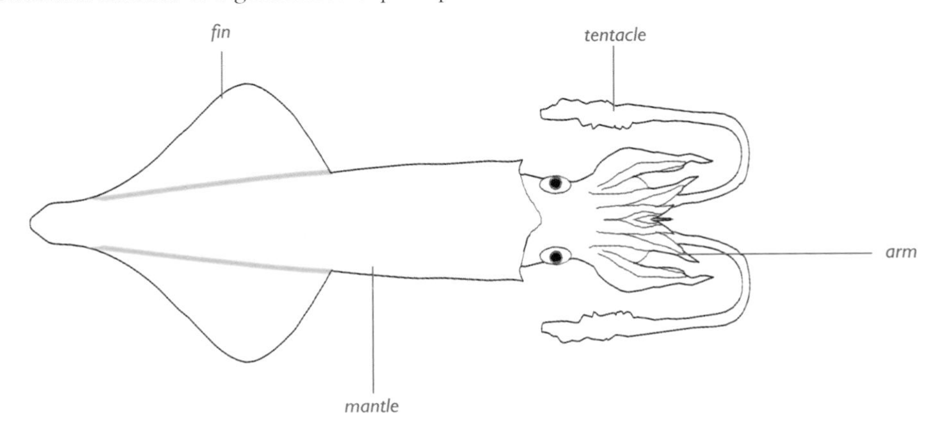

Figure 3.8—Fillet definitions: (A) outer fillet—cardinal fish (*Epigonus telescopus*); (B) inner fillet—cardinal fish (*E. telescopus*); (C) cutlet—yellowfin tuna (*Thunnus albacares*); (D) ray flap—skate (*Raja* sp.).

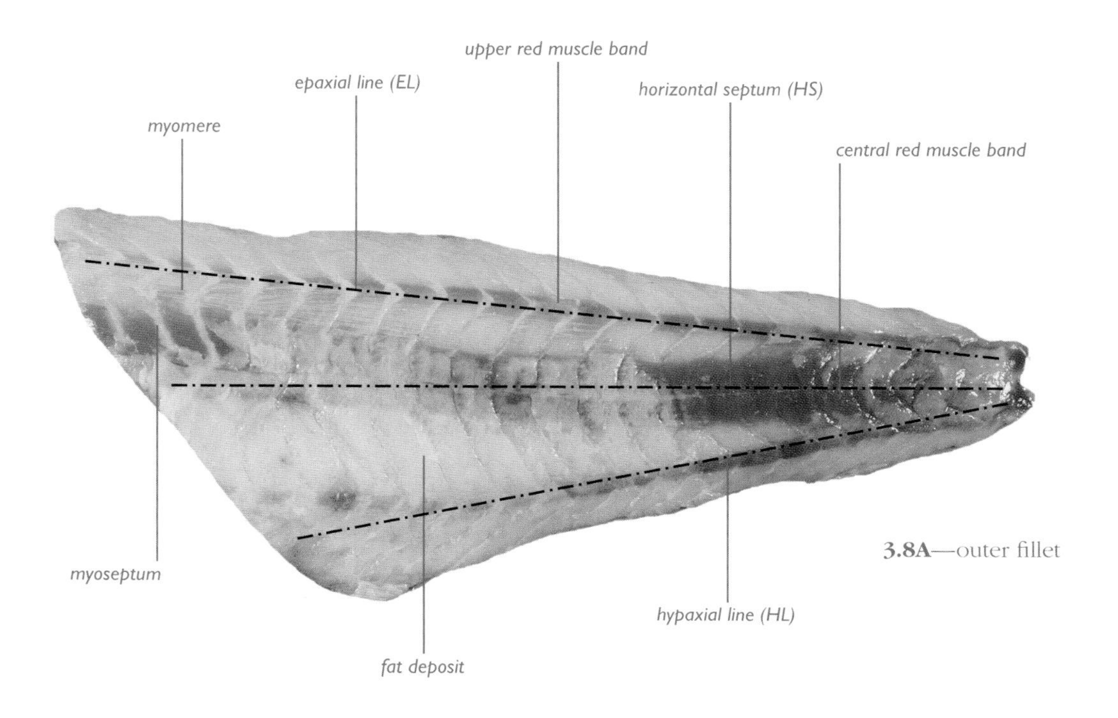

3.8A—outer fillet

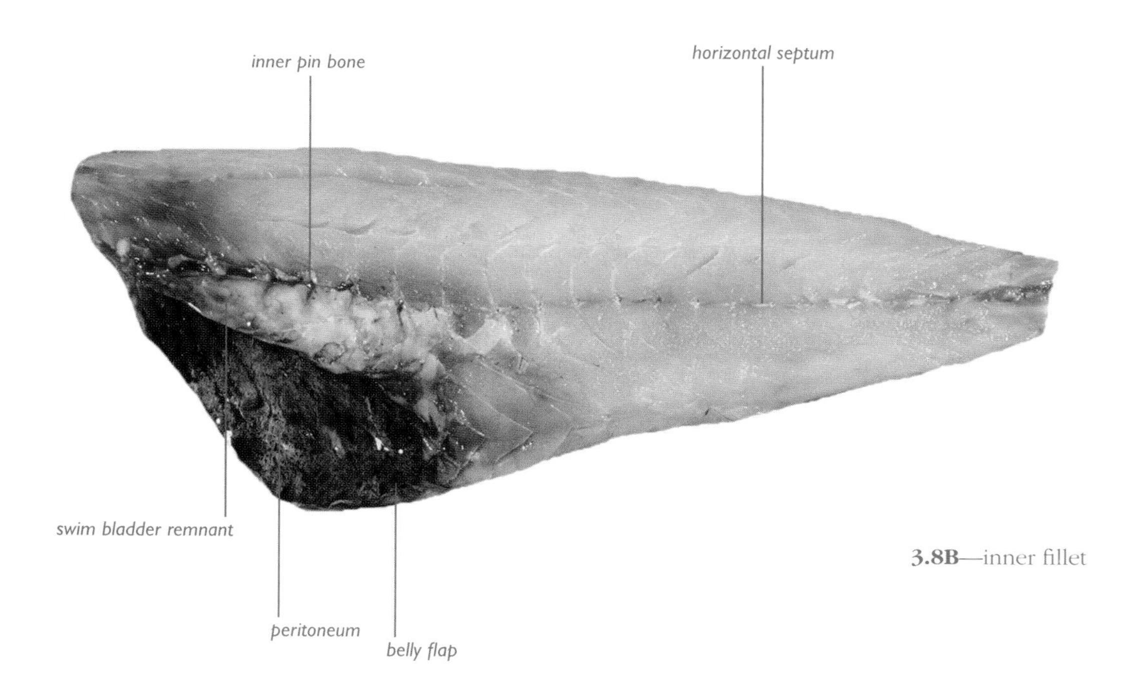

3.8B—inner fillet

Figure 3.8—Fillet definitions, continued.

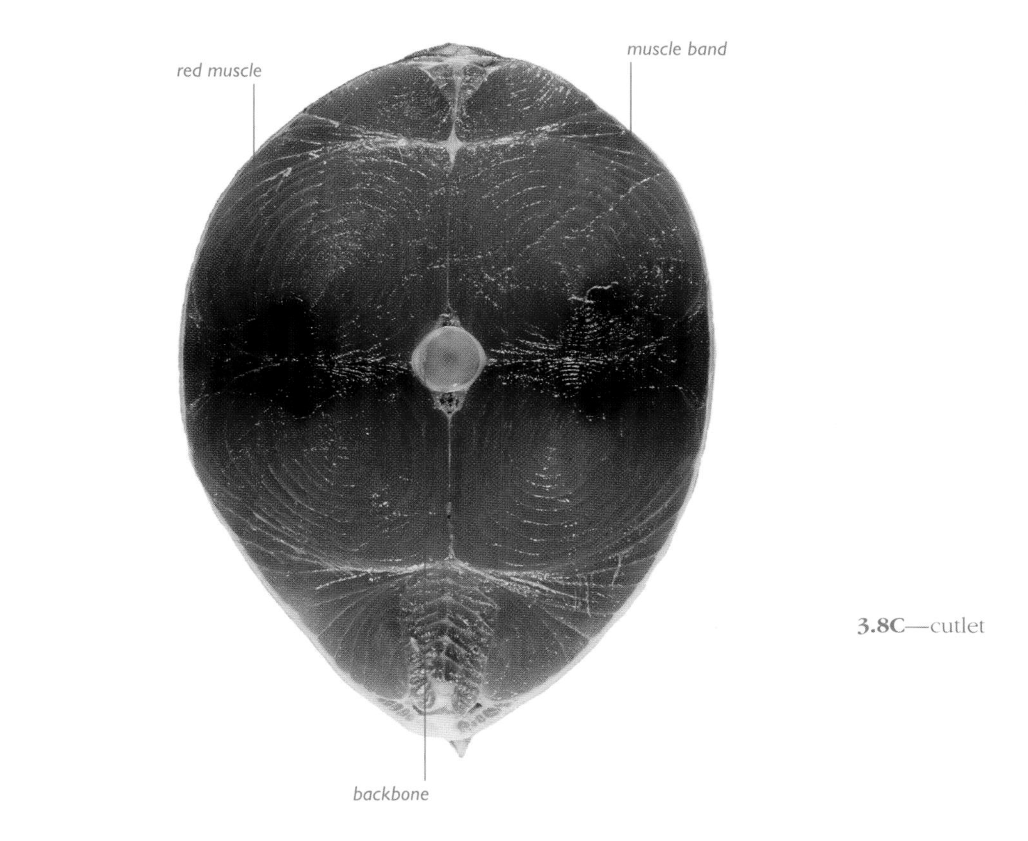

3.8C—cutlet

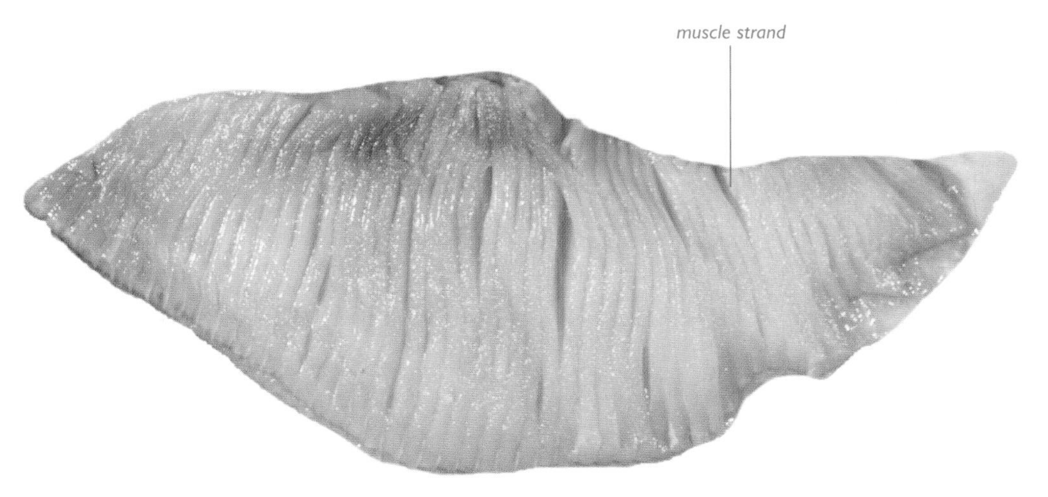

3.8D—ray flap

Figure 3.9—Fillet shapes: (A) deep bodied, short—John dory (*Zeus faber*); (B) deep, elongate, evenly tapering—baldchin groper (*Choerodon rubescens*); (C) elongate, deep, abruptly tapering—redspot emperor (*Lethrinus lentjan*); (D) bottle-shaped—blue-spotted trevally (*Caranx bucculentus*); (E) moderately deep, elongate, gentle taper—mullet (*Mugil argentea*); (F) slender, pronounced taper—tiger flathead (*Neoplatycephalus richardsoni*); (G) long, slender, barely tapering—longfin eel (*Anguilla reinhardtii*).

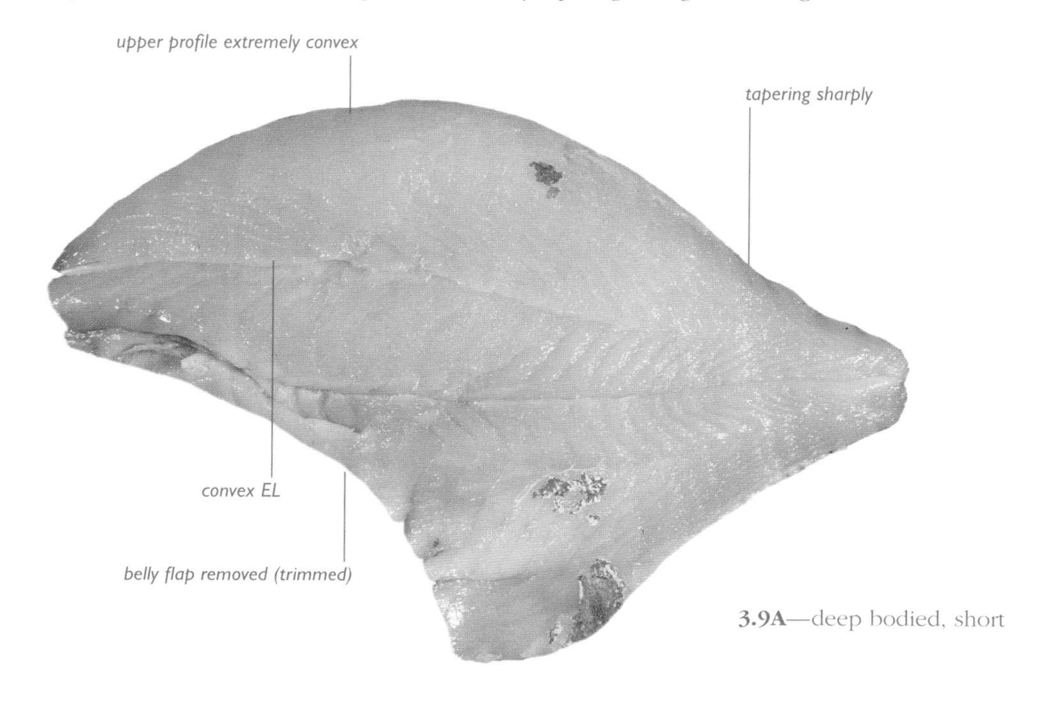

3.9A—deep bodied, short

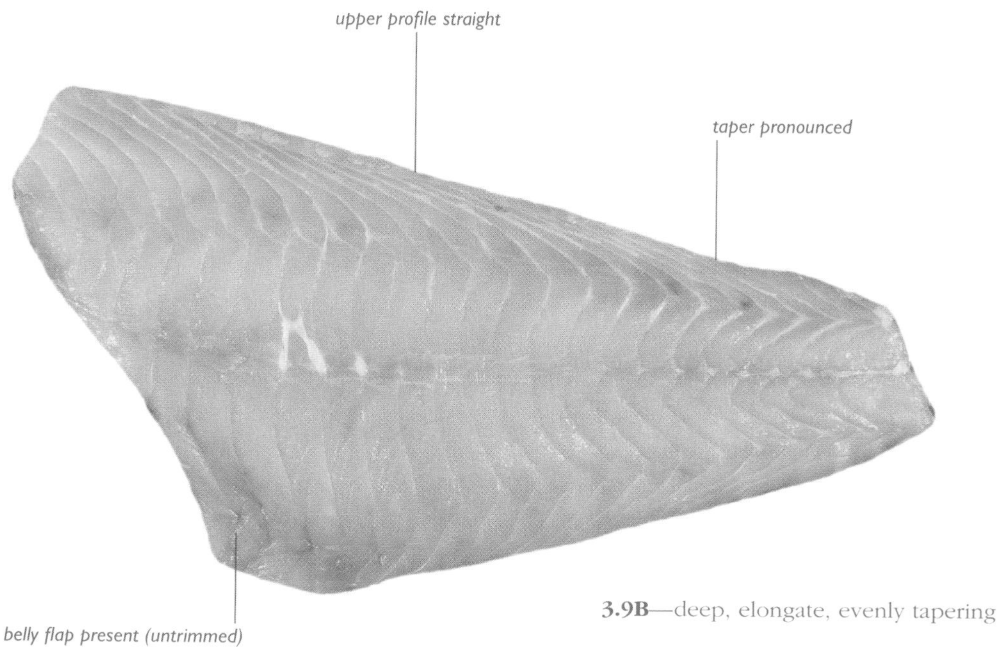

3.9B—deep, elongate, evenly tapering

Figure 3.9—Fillet shapes, cont.

3.9C—elongate, deep, abruptly tapering

3.9D—bottle-shaped

Figure 3.9—Fillet shapes, cont.

3.9E—moderately deep, elongate, gentle taper

3.9F—slender, pronounced taper

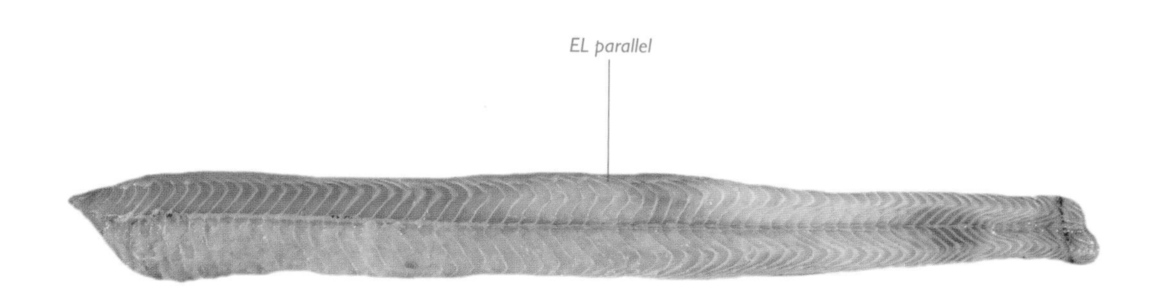

3.9G—long, slender, barely tapering

Figure 3.10—Fillet colours: (A) milky white—rudderfish (*Centrolophus niger*); (B) pearly white—orange roughy (*Hoplostethus atlanticus*); (C) off-white yellowish—Ray's bream (*Brama brama*); (D) off-white greyish—garfish (*Hyporhamphus quoyi*); (E) bluish-white—grass whiting (*Haletta semifasciata*); (F) pale pinkish—honeycomb rockcod (*Epinephelus quoyanus*); (G) reddish-brown—yellowfin tuna (*Thunnus albacares*); (H) brown—Australian salmon (*Arripis truttaceus*); (I) orange—Atlantic salmon (*Salmo salar*); (J) green—batfish (*Platax batavianus*).

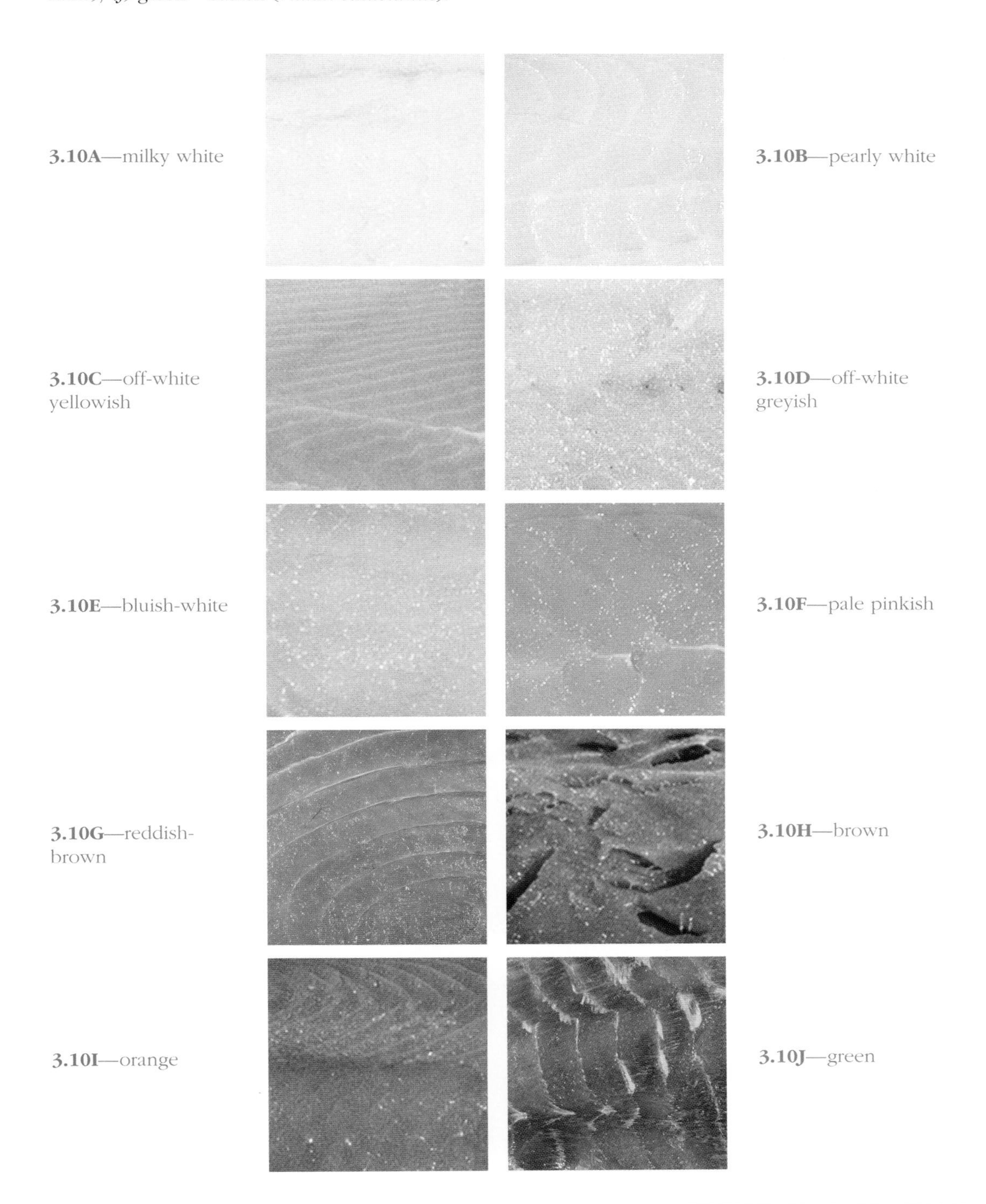

Figure 3.11—Other features, outer fillet: (A) morwong (*Nemadactylus macropterus*); (B) Ray's bream (*Brama brama*); (C) rudderfish (*Centrolophus niger*); (D) garfish (*Hyporhamphus quoyi*); (E) rock flathead (*Platycephalus laevigatus*); (F) greeneye dogfish (*Squalus* sp.).

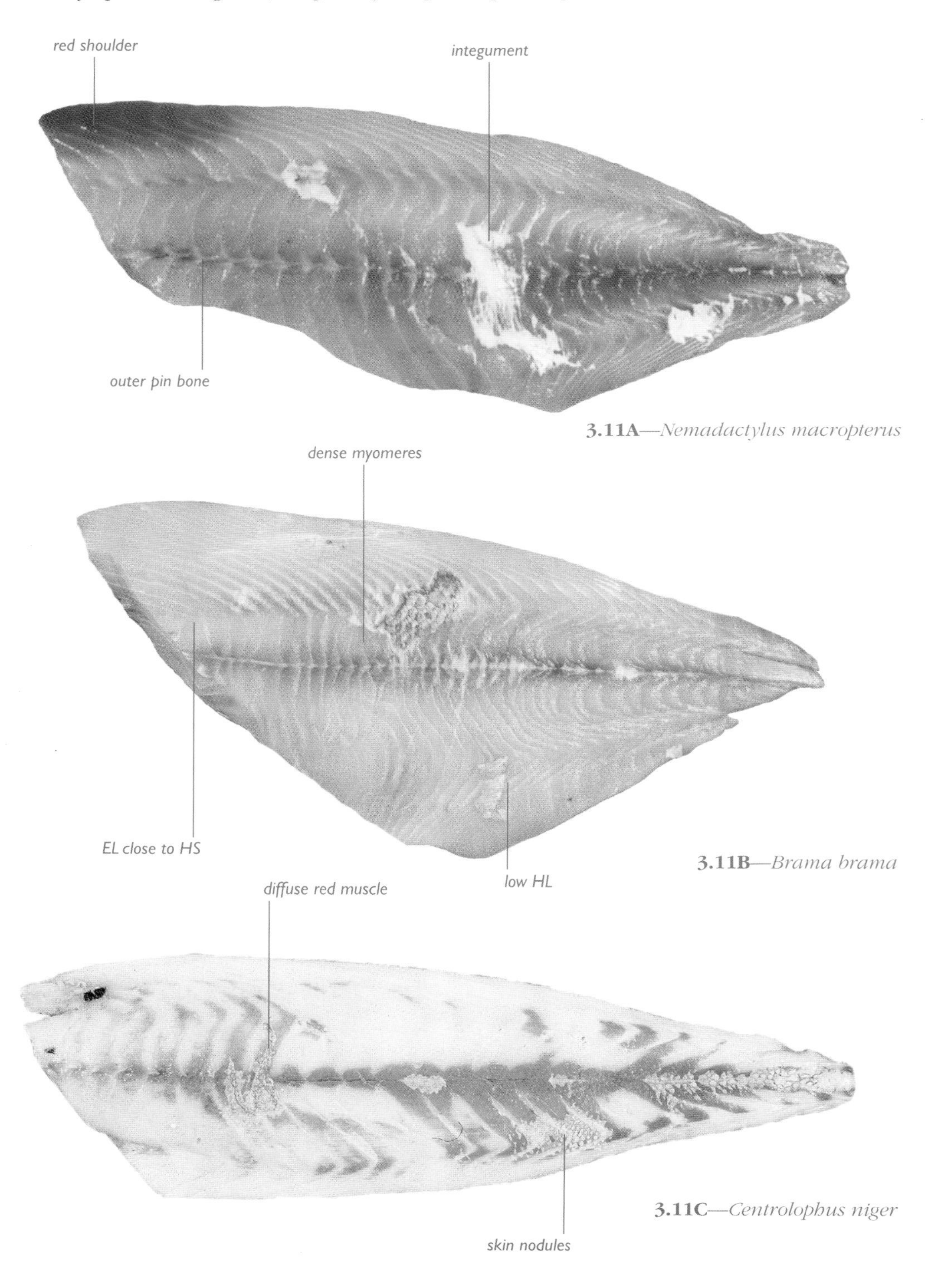

3.11A—*Nemadactylus macropterus*

3.11B—*Brama brama*

3.11C—*Centrolophus niger*

Figure 3.11—Other features, outer fillet, cont.

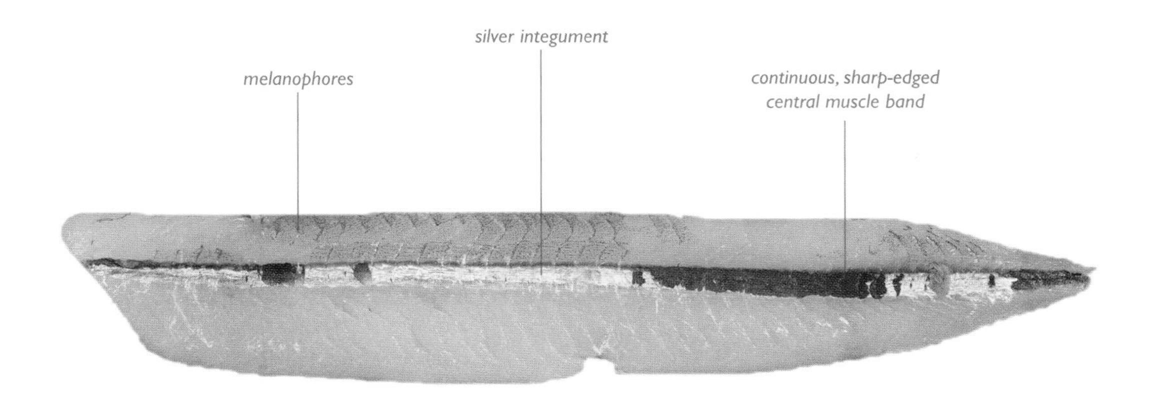

3.11D—*Hyporhamphus quoyi*

3.11E—*Platycephalus laevigatus*

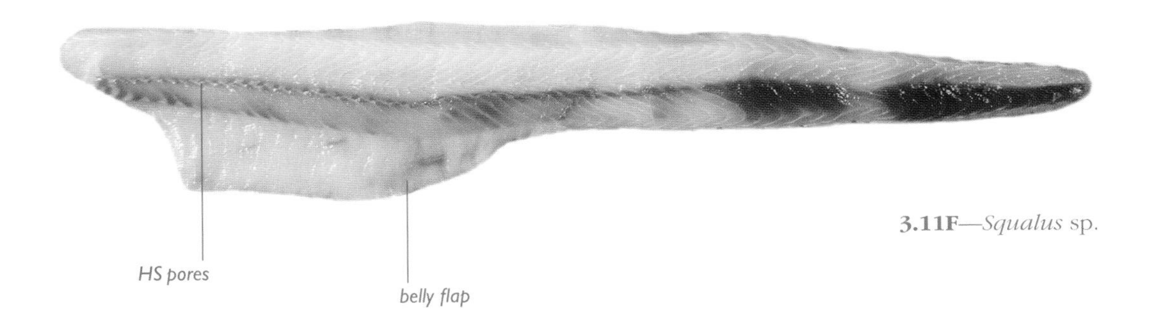

3.11F—*Squalus* sp.

Figure 3.12—Other features, inner fillet: (A) black jewfish (*Protonibea diacanthus*); (B) silver trevally (*Pseudocaranx dentex*); (C) Australian salmon (*Arripis truttaceus*).

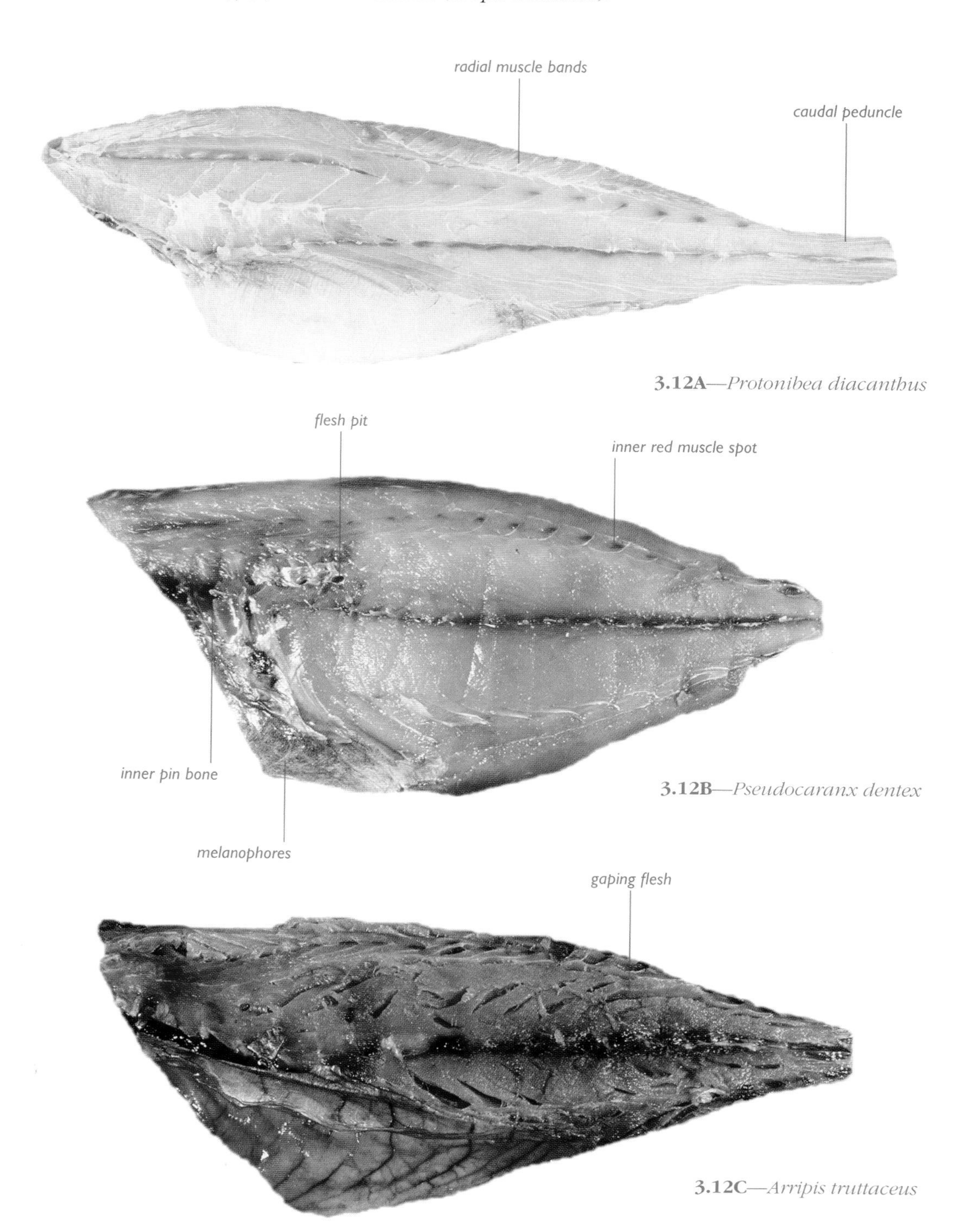

3.12A—*Protonibea diacanthus*

3.12B—*Pseudocaranx dentex*

3.12C—*Arripis truttaceus*

Cartilaginous fishes

P. R. Last and G. K. Yearsley

Angel shark

Squatina species

Previous names: monkfish, ornate angel shark

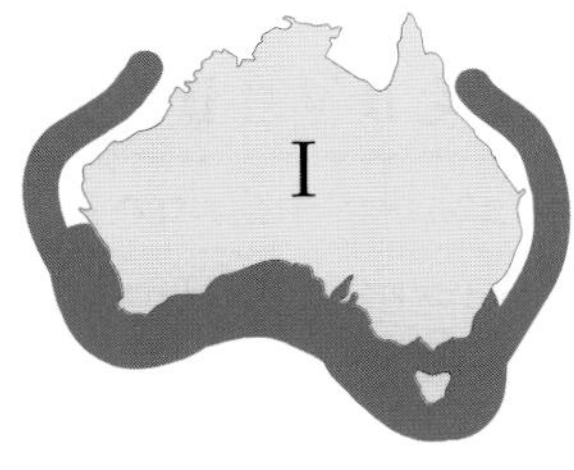

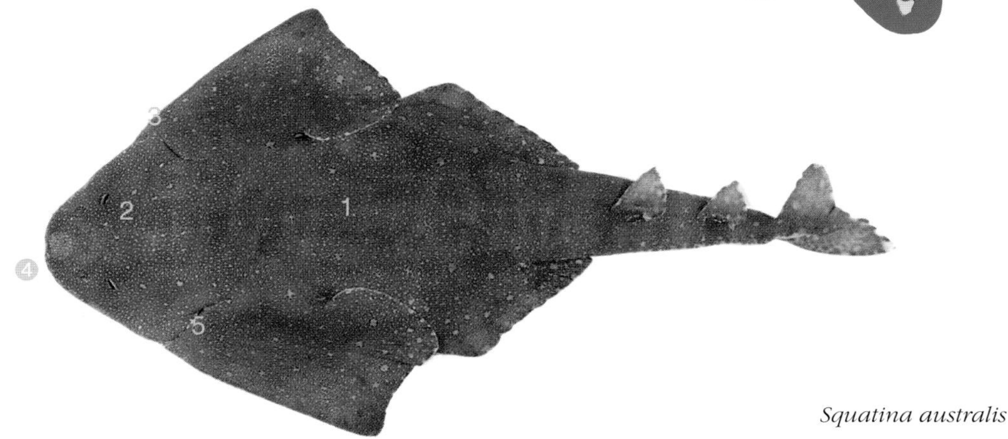

Squatina australis

Identifying features: ① body depressed (flattened); ② eyes on top of head; ③ gill slits on side of head; ④ barbels around mouth; ⑤ pectoral fins not joined to head.

Comparisons: Angel sharks (Squatinidae), with their flattened bodies, wide pectoral fins, and eyes and spiracles on top of the head, look more like rays than sharks. However, unlike rays, their gills are on the side of the head (rather than on the undersurface) and the pectoral fins are not joined to the head to form a disc (the gill slits are anterior of the pectoral-fin origin). No other sharks have this body shape. Four angel shark species occur locally and the three taken off southern and southeastern Australia are landed most regularly.

Fillet: *S. australis* long, slender, tapering slightly, yellowish-white. Outside with discontinuous red muscle band pronounced posteriorly, convex EL, weakly converging HL; HS through upper half of fillet with belly flap on; integument silvery translucent. Inside flesh medium; belly flap deep, rectangular when present; peritoneum silvery white to translucent. Always skinned.

Size: To about 150 cm and possibly nearly 20 kg (commonly to 100 cm and about 8.5 kg).

Habitat: Marine; bottom-dwelling, capable of burying into sand and mud. Found inshore to the upper continental slope in depths to 400 m.

Fishery: Caught mainly as bycatch using demersal trawl and Danish seine nets. Also occasionally taken using gillnets in the southern shark fishery.

Remarks: Diet consists of small bony fishes and shellfishes. The flap-like pectoral fins are removed and sold as steaks.

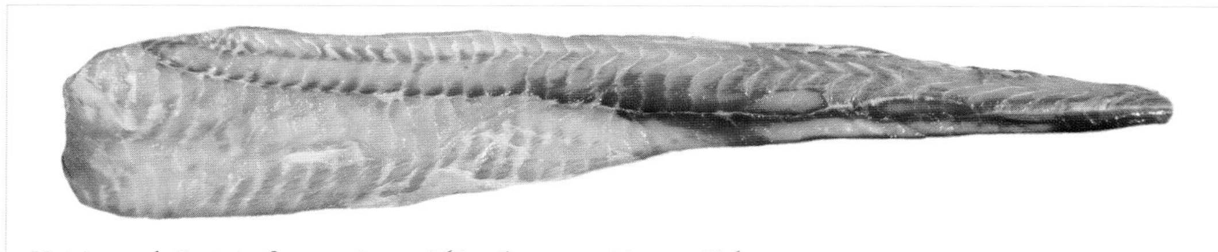

Untrimmed. Protein fingerprint p. 360; oil composition p. 396.

Endeavour dogfish

Centrophorus harrissoni, C. moluccensis & C. uyato

Previous names: Harrison's dogfish, southern dogfish, tough-skin dog

Centrophorus moluccensis

Identifying features: ① body greyish; ② snout rather long; ③ rear tip of pectoral fin usually long and pointed; ④ teeth in upper jaw differing in shape to those of lower jaw; ⑤ no anal fin; ⑥ hard spines at front of both dorsal fins.

Comparisons: Dogfishes (Squalidae) can be easily distinguished from other commercial sharks by their body shape, lack of an anal fin, and the presence of strong spines in front of each dorsal fin. Endeavour dogfishes (*Centrophorus* species) closely resemble another small group of edible dogfishes, the greeneye dogfishes (*Squalus* species, p. 36), but have the inner edges of the pectoral fin extended into a pointed flap and the teeth in each jaw are dissimilar.

Fillet: *C. moluccensis* long, slender, barely tapering, yellowish-white. Outside with discontinuous red muscle band pronounced posteriorly, parallel EL, parallel HL; HS just above midline of fillet; integument white. Inside flesh fine; belly flap absent; pores along midline; peritoneum silvery white to translucent. Always skinned.

Size: To at least 110 cm and at least 6 kg (commonly to 80 cm and about 2.5 kg).

Habitat: Marine; demersal on the continental shelf and upper slope in depths of 50–1400 m but more typically 200–650 m.

Fishery: Small targeted dropline fishery and bycatch of deepwater trawling mostly off southern Australia. A sharp decline in recent trawl catches suggests susceptibility to overfishing.

Remarks: Marketed mainly in Sydney and Melbourne as fresh headed and gutted carcasses.

Untrimmed. Protein fingerprint p. 360.

Greeneye dogfish

Squalus species

Previous names: doggie, greeneye, grey spiny dogfish, grey spurdog, spiny dogfish

Squalus sp.

Identifying features: ① body greyish, central margin of caudal fin with dark patch; ② snout rather long; ③ dorsal-fin origin forward of or over rear tip of pectoral fin; ④ rear tip of pectoral fin angular (not long and pointed); ⑤ teeth in both jaws similar in shape; ⑥ no anal fin; ⑦ hard spines at front of both dorsal fins.

Comparisons: Commonly caught with the Endeavour dogfishes (*Centrophorus* species, p. 35), but distinguishable as the inner edges of the pectoral fins are angular (rather than being extended into a narrow, pointed flap) and the teeth in each jaw are almost identical in size and shape (rather than being distinctly different).

Fillet: *Squalus* sp. long, slender, tapering slightly, yellowish-white. Outside with continuous moderate red muscle band, parallel EL, parallel HL; HS near midline of fillet; integument translucent above, white below. Inside flesh fine; belly flap absent; peritoneum silvery white to translucent. Always skinned.

Size: To about 100 cm and about 5 kg (commonly 60–90 cm and 1–4 kg).

Habitat: Marine; demersal, sometimes inshore on the inner continental shelf but mostly deeper on the upper slope in depths of 200–600 m.

Fishery: Currently taken as bycatch of demersal trawl fisheries mainly off temperate Australia; also caught with droplines off South Australia. Small fishes, which comprise much of the catch, are generally discarded.

Remarks: Marketed mainly in Sydney as fresh headed and gutted carcasses. All members of the genus *Squalus* can be sold as 'greeneye dogfish' but the two main commercial species, spikey dogfish (*S. megalops*, p. 38) and white-spotted dogfish (*S. acanthias*, p. 39), have separate marketing names.

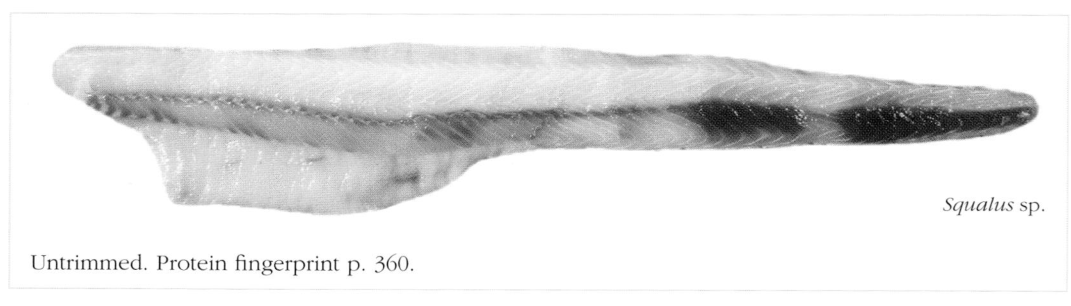

Squalus sp.

Untrimmed. Protein fingerprint p. 360.

Roughskin dogfish

Centroscymnus & *Deania* species

Previous names: brier shark, golden dogfish, Owston's dogfish, Plunket's dogfish, velvet dogfish

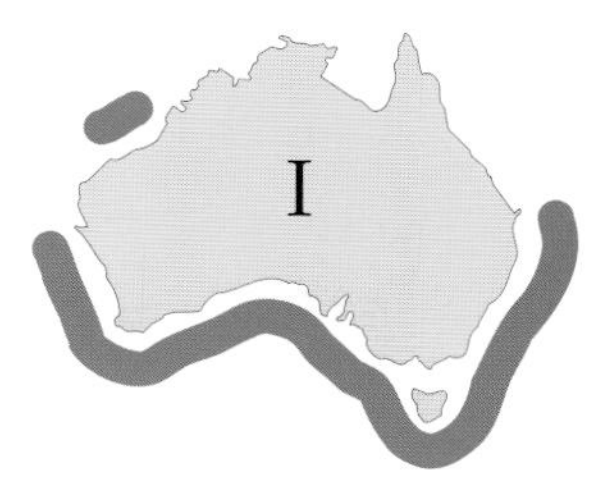

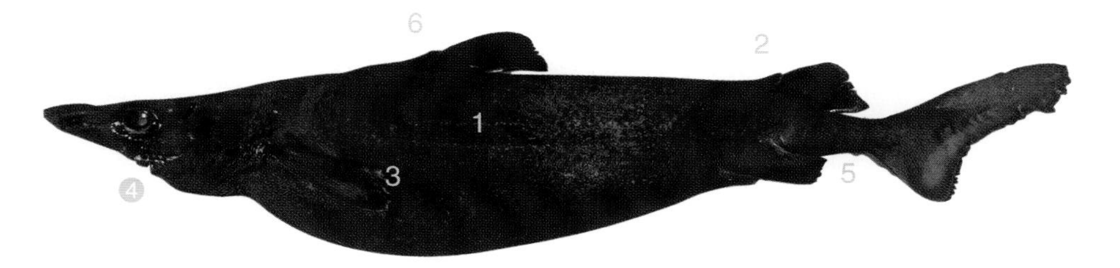

Centroscymnus crepidater

Identifying features: 1 body dark brown or blackish; 2 second dorsal-fin spine directly above anal fin; 3 rear tip of pectoral fin rounded; 4 teeth in upper jaw differing in shape to those of lower jaw; 5 no anal fin; 6 hard spines at front of both dorsal fins (sometimes short).

Comparisons: Distinctive, medium-sized, dark, soft-bodied dogfishes with six species from two genera marketed locally. The brier sharks (*Deania* species) have a low, long-based dorsal fin and a characteristic, long, flattened snout. The black or brownish velvet dogfishes (*Centroscymnus* species) are represented by four primary species. Compositional structure of current landings has not been investigated.

Fillet: *C. crepidater* long, slender, barely tapering, pearly white. Outside with discontinuous feeble red muscle band, parallel EL, parallel HL; HS just above midline of fillet; integument translucent. Inside flesh fine; belly flap very deep when present; peritoneum silvery white to translucent. Always skinned.

Size: To 115 cm and 9 kg (commonly to 100 cm and 8 kg).

Habitat: Marine; demersal on the mid-continental slope in depths of 500–1500 m.

Fishery: Bycatch of deepwater dropline and demersal trawl operations, including the fishery for orange roughy (*Hoplostethus atlanticus*, p. 229) off southern Australia.

Remarks: Once discarded, now marketed in increasing quantities in the Sydney Fish Market as fresh headed and gutted carcasses. The livers are used to produce valuable 'squalene' oil.

Deania calcea

Untrimmed. Protein fingerprint p. 360.

Spikey dogfish

Squalus megalops

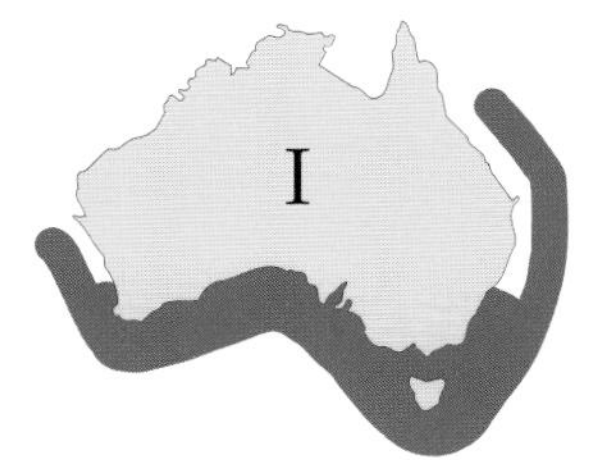

Previous names: dogshark, piked dogfish, piked spurdog, shortnose spurdog, Tasmanian dogfish

Identifying features: ① body greyish-brown, margin of caudal fin whitish; ② snout short; ③ dorsal-fin origin forward of or over rear tip of pectoral fin; ④ rear tip of pectoral fin angular (not long and pointed); ⑤ teeth in both jaws similar in shape; ⑥ no anal fin; ⑦ hard spines at front of both dorsal fins.

Comparisons: Plain-coloured, resembling several less commercially important species of the genus *Squalus* marketed under the collective name 'greeneye dogfish' (p. 36). Unlike most of these species, it has a short snout and the hind margin of the caudal fin is distinctly white edged (rather than black in the fork). The other main commercial *Squalus* species, the white-spotted dogfish (*S. acanthias*, p. 39), is covered in white spots (rather than plain).

Fillet: Long, slender, tapering slightly, yellowish-white. Outside with continuous pronounced red muscle band, parallel EL, parallel HL; HS through upper half of fillet with belly flap; integument white. Inside flesh fine; belly flap usually present; peritoneum silvery white to translucent. Always skinned.

Size: To 62 cm and about 1.2 kg (commonly to 55 cm and 0.9 kg).

Habitat: Marine; demersal on soft-bottoms of the continental shelf and upper slope to depths of 500 m.

Fishery: Bycatch of inshore gillnetting and trawling mainly off southern Australia.

Remarks: Marketed in small quantities in major seafood markets as fresh headed and gutted carcasses.

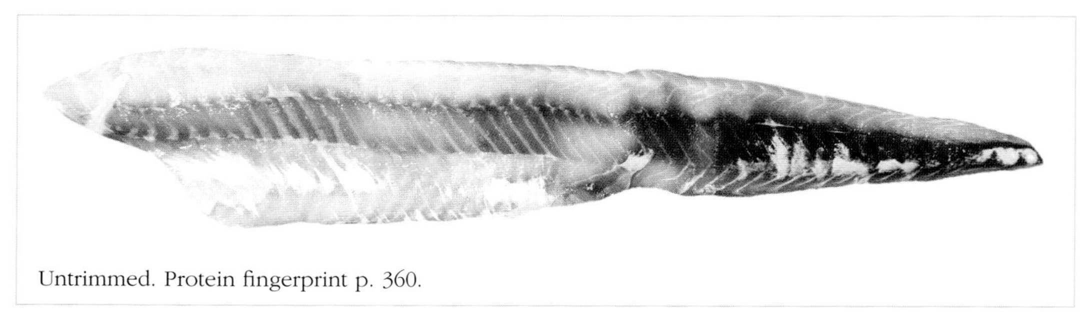

Untrimmed. Protein fingerprint p. 360.

White-spotted dogfish

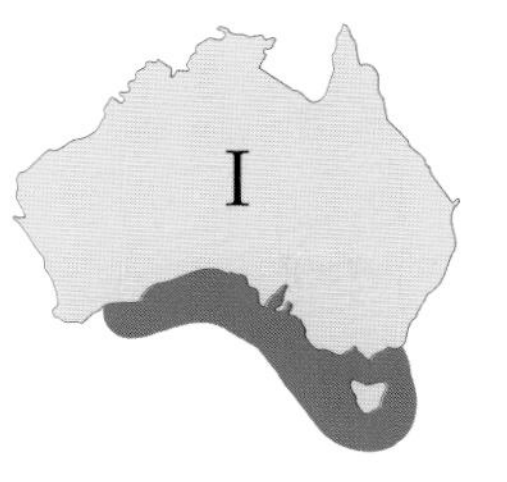

Squalus acanthias

Previous names: piked dogfish, spiny dogfish, spotted spiny dogfish, white-spotted spurdog

Identifying features: 1 body bluish-grey with scattered white spots; 2 snout rather long; 3 dorsal-fin origin distinctly behind rear tip of pectoral fin; 4 rear tip of pectoral fin angular (not long and pointed); 5 teeth in both jaws similar in shape; 6 no anal fin; 7 hard spines at front of both dorsal fins.

Comparisons: A bluish body covered in prominent white spots makes it easily distinguishable from other *Squalus* species. Unlike the Endeavour dogfishes (*Centrophorus* species, p. 35), the inner pectoral-fin margin is not extended lengthways into a flap.

Fillet: Long, slender, tapering slightly, yellowish-white. Outside with continuous pronounced red muscle band, parallel EL, parallel HL; HS through upper half of fillet with belly flap; integument white. Inside flesh fine; belly flap usually present; peritoneum silvery white to translucent. Always skinned.

Size: To 100 cm and about 5 kg (commonly 50–85 cm and 0.6–1.8 kg).

Habitat: Marine; demersal inshore over soft-bottoms off beaches, venturing into large southern Australian bays and estuaries to breed.

Fishery: One of the most abundant sharks, but only forms a small bycatch component of local inshore gillnets and Danish seiners.

Remarks: The flesh, which is considered to be coarser than the main species of the southern shark fishery, is marketed in small quantities as fresh headed and gutted carcasses. Consumed fresh internationally but also processed (smoked, salted, marinated or made into fish cakes). Also used in oil, leather and petfood industries.

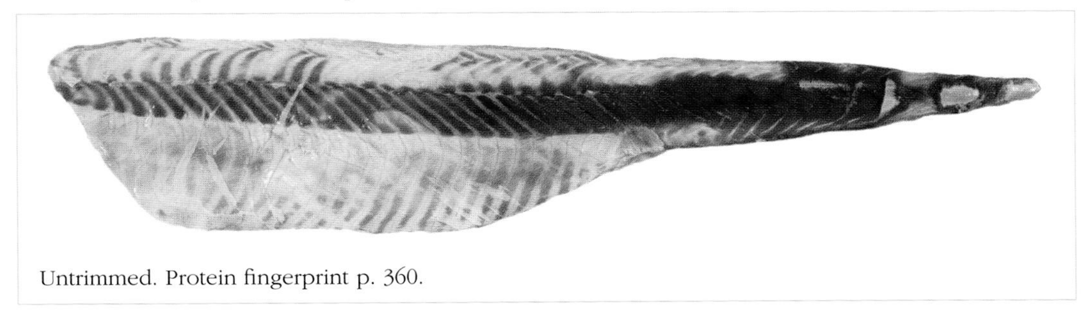

Untrimmed. Protein fingerprint p. 360.

Elephant fish

Callorhinchus milii

Previous names: elephant shark, reperepe, white fillets, whitefish

Identifying features: 1 hoe-shaped structure at snout tip; 2 second dorsal fin with a short base; 3 caudal fin long and pointed; 4 strong spine at front of first dorsal fin; 5 silvery, scaleless skin; 6 paddle-like pectoral fins.

Comparisons: Unusual silvery fish with a prominent, hoe-shaped snout and large, paddle-like pectoral fins. Also differs from the related ghostsharks (Chimaeridae, p. 41) in having a much shorter-based second dorsal fin and a shark-like caudal fin (rather than an eel-like tail ending in a rubbery filament).

Fillet: Moderately elongate, tapering slightly, white. Outside lacking red muscle, weakly converging EL, converging HL; HS through middle of fillet; integument silvery white. Inside flesh medium; belly flap thin, shape irregular; peritoneum translucent with black patches. Sometimes skinned.

Size: To at least 110 cm and 9 kg (commonly 80–100 cm and 2.5–7.2 kg).

Habitat: Marine; bottom-dwelling on sand and mud of the continental shelf in depths to 200 m. Migrates into inshore bays in late summer to breed.

Fishery: Caught mainly by demersal trawlers and Danish seiners, and in large-mesh gillnets off Tasmania and Victoria. Occasionally taken inshore by line fishers in South Australia.

Remarks: Formerly sold in southern markets as 'white fillets', its firm flesh is brilliant white in colour when fresh.

Untrimmed. Protein fingerprint p. 360.

Ghostshark

Chimaeridae

Previous names: spookfish, white fillets, whitefish

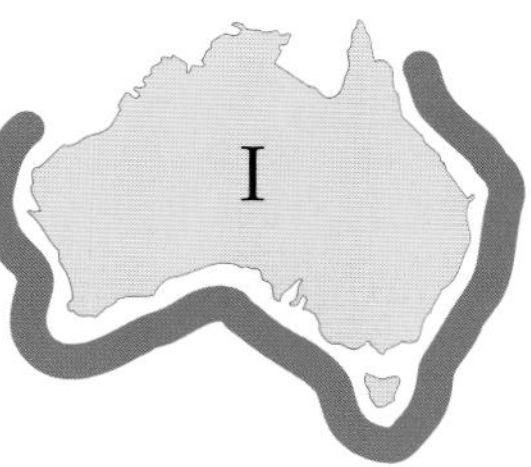

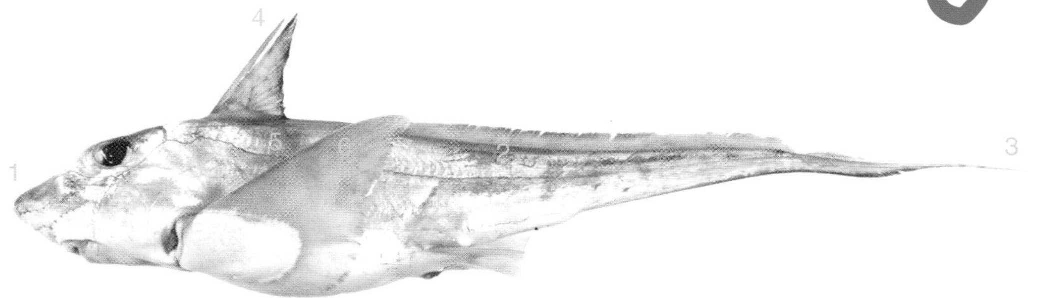

Hydrolagus lemures

Identifying features: ① snout tip rounded or pointed; ② second dorsal-fin with a long, low base; ③ caudal fin small, mostly with a fleshy filament at its tip; ④ strong spine at front of first dorsal fin; ⑤ scaleless skin; ⑥ paddle-like pectoral fins.

Comparisons: Distinctive silvery to black fishes that are unlike any other shark in appearance. They lack a hoe-shaped snout (it is rounded or pointed) typifying the related elephant fish (*Callorhinchus milii*, p. 40) and have a much longer second dorsal-fin base, and an eel-like tail ending in a rubbery filament (rather than a shark-like caudal fin).

Fillet: *Hydrolagus ogilbyi* long, slender, tapering slightly, white. Outside lacking red muscle, weakly converging EL, weakly converging HL; HS through middle of fillet; integument white. Inside flesh fine; belly flap as short, triangular lobe when present; no peritoneum. Sometimes skinned.

Size: To at least 135 cm and 10.5 kg (commonly 70–100 cm and 0.6–3.0 kg).

Habitat: Marine; wide ranging bottom-dwellers on soft substrates of the outer continental shelf and slope between 120–1500 m depth.

Fishery: Caught mainly as bycatch by deepwater demersal trawlers off southern Australia.

Remarks: The ghostsharks make up a third major group of fishes, along with 'bony fishes' and 'sharks and rays'. Ten species are caught in varying quantities and their flesh quality is likely to be similar. Like elephant fish, formerly sold in southern markets as 'white fillets'.

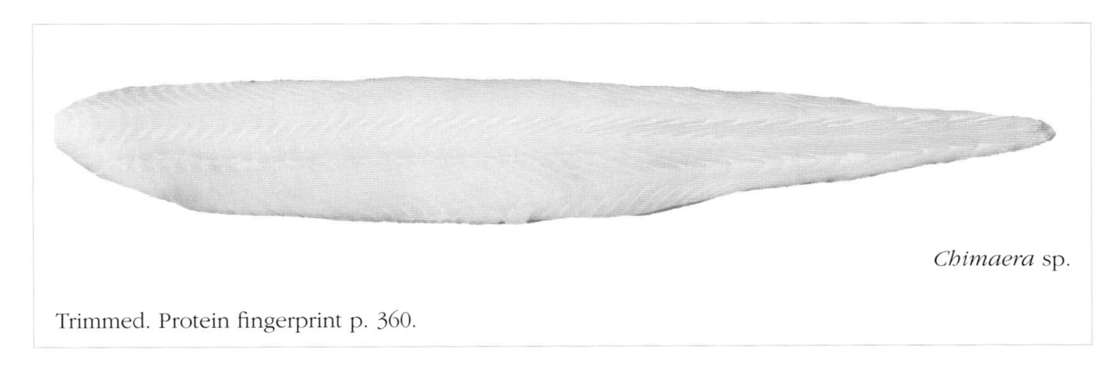

Chimaera sp.

Trimmed. Protein fingerprint p. 360.

Gummy shark

Mustelus species

Previous names: Australian smooth hound, sweet William

Mustelus antarcticus

Identifying features: 1 no barbel in front of nostril; 2 second dorsal fin only slightly smaller than first; 3 small lobe on outer edge at top of caudal fin; 4 teeth in both jaws flattened; 5 plain-coloured or with tiny, white spots; 6 first dorsal-fin origin more or less above pectoral-fin rear tip; 7 no furrow (precaudal pit) at base of upper caudal-fin lobe.

Comparisons: Hound sharks (Triakidae) resemble their larger relatives the whaler sharks (Carcharhinidae) in body shape but lack a distinctive furrow at the base of the upper lobe of the caudal fin. The teeth of the gummy sharks, which are flattened to form crushing plates, make them readily distinguishable from other hound sharks.

Fillet: *M. antarcticus* long, slender, tapering slightly, yellowish-white. Outside with moderate red muscle band, parallel EL, weakly converging HL; HS through upper half of fillet; integument white. Inside flesh medium; belly flap distinct, not lobe-like; pores along midline; peritoneum white. Always skinned.

Size: To 177 cm and 25 kg (commonly 100–120 cm and about 5 kg).

Habitat: Marine; demersal on the continental shelf and upper slope from shallow estuaries to depths of about 400 m.

Fishery: Historical fishery with more than 1000 tonnes landed annually off southern Australia using mainly bottom-set longlines and gillnets.

Remarks: Three species occur in Australian seas but only the southern gummy shark (*M. antarcticus*) is caught in quantity. Marketed as fresh headed and gutted carcasses and consumed as 'flake'.

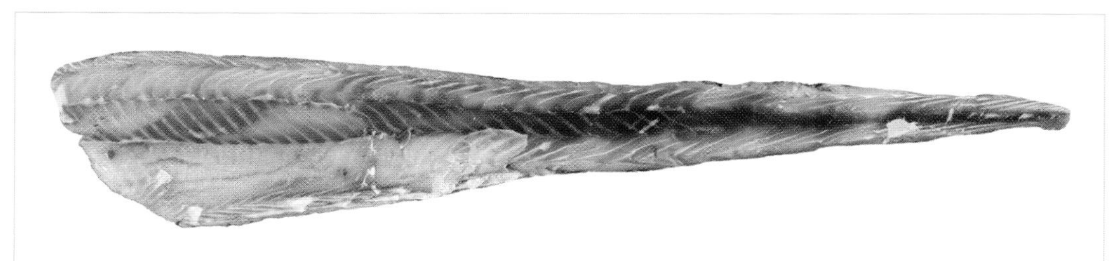

Untrimmed. Protein fingerprint p. 360; oil composition p. 396.

School shark

Galeorhinus galeus

Previous names: snapper shark, tope

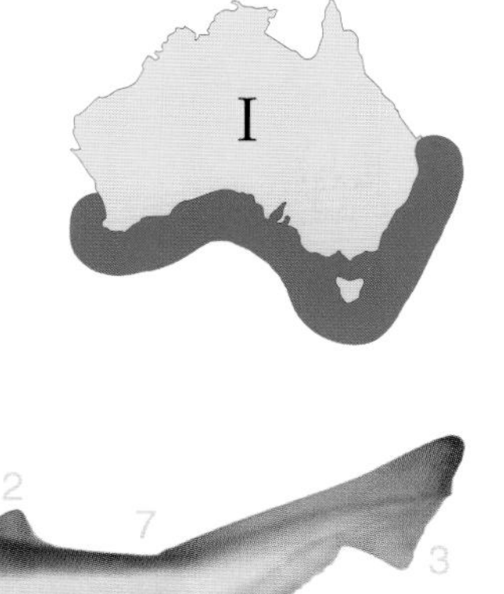

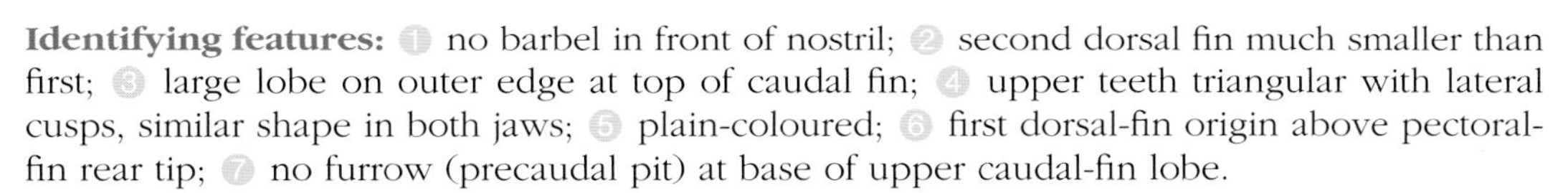

Identifying features: 1 no barbel in front of nostril; 2 second dorsal fin much smaller than first; 3 large lobe on outer edge at top of caudal fin; 4 upper teeth triangular with lateral cusps, similar shape in both jaws; 5 plain-coloured; 6 first dorsal-fin origin above pectoral-fin rear tip; 7 no furrow (precaudal pit) at base of upper caudal-fin lobe.

Comparisons: Similar in general appearance to the whaler sharks (Carcharhinidae) but lacks a precaudal pit and has a more distinctive lobe at the top of the caudal fin.

Fillet: Long, slender, tapering slightly, yellowish-white. Outside with incomplete red muscle band, parallel EL, parallel HL; HS through upper half of fillet; integument white. Inside flesh medium; belly flap distinct, not lobe-like; peritoneum silvery white. Always skinned.

Size: To 180 cm and 33 kg (commonly 100–130 cm and 6–12 kg).

Habitat: Marine; demersal and midwater over the continental shelf and upper slope from inshore to at least 600 m and probably deeper. Young sharks use bays and estuaries as nursery areas.

Fishery: Important target species of the southern shark fishery with annual catches frequently exceeding 2000 tonnes annually. Predominantly taken in depths of less than 200 m using bottom-set longlines and gillnets. Also taken as bycatch in demersal trawl and Danish seine nets.

Remarks: Wide ranging shark with some specimens tagged and released off New Zealand recaptured in Australia. Marketed as fresh headed and gutted carcasses and sold by fish-and-chip shops as 'flake'.

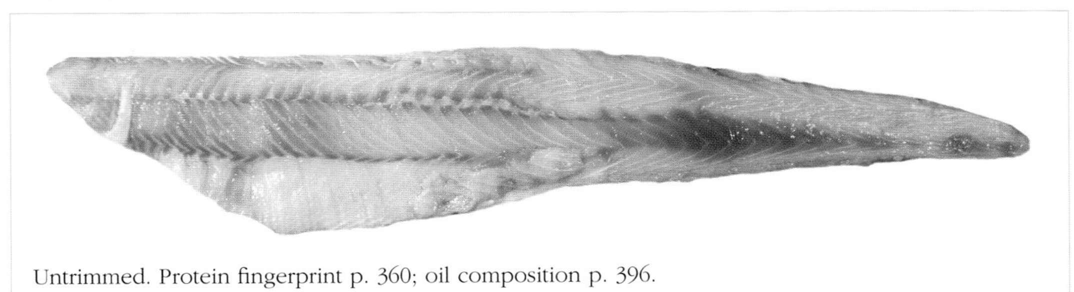

Untrimmed. Protein fingerprint p. 360; oil composition p. 396.

Whiskery shark

Furgaleus macki

Previous names: reef shark, sundowner

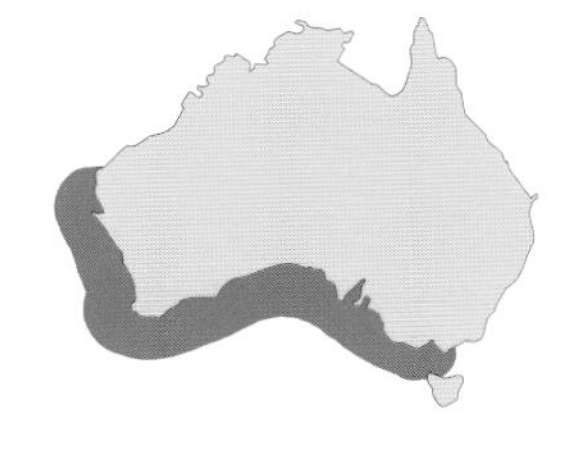

Identifying features: ① barbel in front of each nostril; ② dorsal fins of similar size; ③ large lobe on outer edge at top of caudal fin; ④ plain-coloured or with dark blotches; ⑤ upper teeth blade-like with lateral cusps, lower teeth pointed and without cusps; ⑥ first dorsal-fin origin just behind pectoral-fin rear tip; ⑦ no furrow (precaudal pit) at base of upper caudal-fin lobe.

Comparisons: Distinctive hound shark (Triakidae) in having a prominent fleshy barbel on each nostril. Young fish have characteristic dark blotches over the body which are faint or often indistinguishable in adults.

Fillet: Long, slender, tapering slightly, dark pink. Outside with moderate red muscle band, parallel EL, parallel HL; HS through upper half of fillet; integument greyish-white. Inside flesh fine; belly flap distinct, not lobe-like; peritoneum white to translucent. Always skinned.

Size: To 160 cm and likely to exceed 15 kg (commonly to 130 cm and 11 kg).

Habitat: Marine; bottom-dwelling on reefy or weedy substrates from the inner continental shelf to depths of about 220 m.

Fishery: Limited entry fishery—using bottom-set longlines, handlines, droplines, and gillnets—centred mainly off southwestern Australia where adults are caught with the gummy shark (*Mustelus antarcticus*, p. 42) and the dusky shark (*Carcharhinus obscurus*, p. 54).

Remarks: Bottom-feeder, with a distinct preference for octopuses as food. Marketed as fresh headed and gutted carcasses.

Untrimmed. Protein fingerprint p. 360.

Guitarfish (page 1 of 2)

Rhinobatidae & Rhynchobatidae

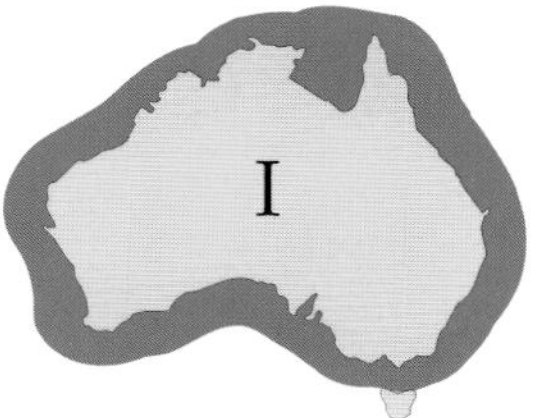

Previous names: banjo shark, fiddler ray, shark ray, shovelnose ray, white-spotted shovelnose ray

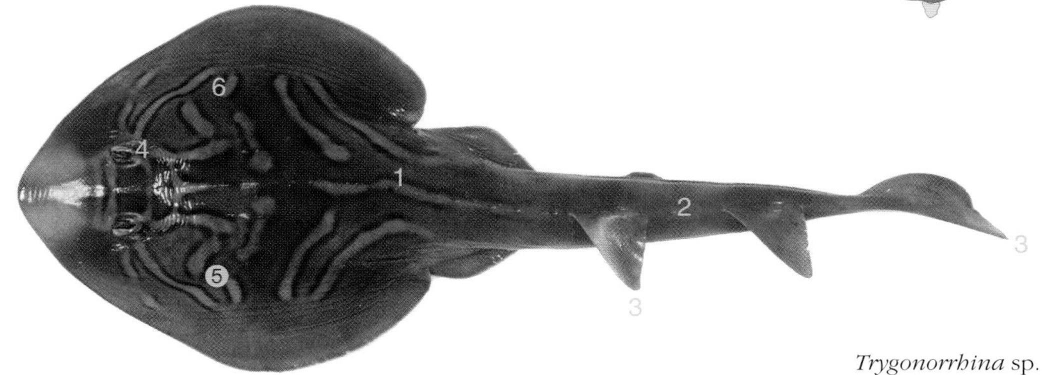

Trygonorrhina sp.

Identifying features: ① body firm; ② tail long, broad and flattened; ③ dorsal and caudal fins well developed; ④ eyes on top of head; ⑤ gill slits underneath head; ⑥ pectoral fins joined to head to form a disc.

Comparisons: Rays are related to sharks but differ in having their head and pectoral fins joined to form a flattened 'disc' with their eyes on the dorsal surface and the mouth and gills on the undersurface. The guitarfishes include two main families that have a distinctive round or shovel-shaped anterior disc and a broad shark-like tail with tall dorsal fins and well-developed caudal fin. They differ from the skates (Rajidae, p. 49), which have larger discs and much thinner tails, and several other groups marketed as 'ray' (pp 47–48) whose members have venomous stinging spines on the tail.

Fillet: *Trygonorrhina* sp. long, broad, tapering slightly, pale pink. Outside with very pronounced, deep, continuous red muscle band, weakly converging EL, weakly converging HL; integument white. Inside flesh fine; belly flap distinct, not lobe-like; peritoneum white to translucent. Always skinned.

Size: To about 300 cm and more than 220 kg (commonly to 150 cm and exceeding 50 kg).

Habitat: Marine; bottom-dwelling, capable of burying into sand and mud. Found mainly inshore but occasionally to depths of 200 m.

Fishery: Caught mainly as bycatch by demersal trawlers but also by seine and gillnets.

Remarks: The flesh is highly regarded in Asia but is under-utilised locally. The fins are often used as a major ingredient of shark-fin soup. Consequently, they have been subjected to the wasteful practice of 'finning' in which the ray's body is discarded.

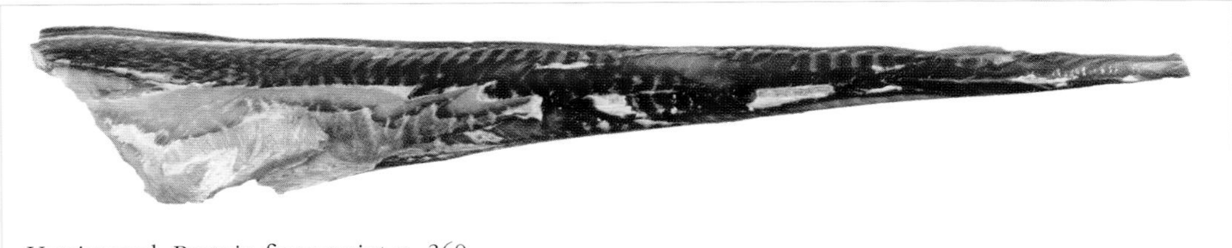

Untrimmed. Protein fingerprint p. 360.

Guitarfish (page 2 of 2)

Rhinobatidae & Rhynchobatidae

Trygonorrhina sp.

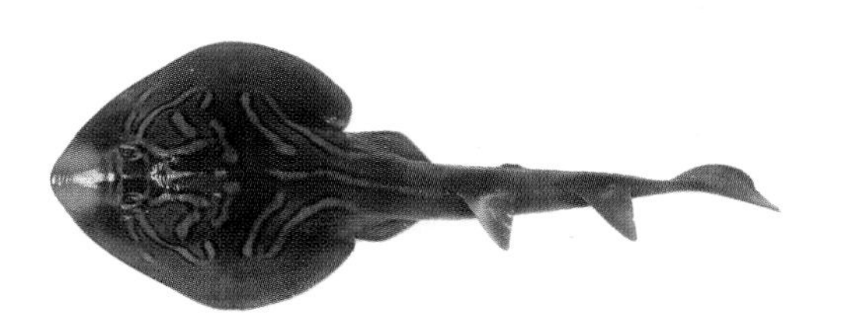

Remarks: Commonly called 'fiddler ray', this species is distributed from Moreton Bay (Qld) south to at least Twofold Bay (NSW) in depths to 100 m. Best distinguished by an oval-shaped disc with a pattern of dark-edged bands. Reported to 120 cm and about 15 kg. A closely related species occurs from eastern Bass Strait (including northern Tasmania) to Lancelin (WA).

Rhynchobatus australiae

Remarks: Commonly called 'white-spotted guitarfish', this species is distributed on the continental shelf around northern Australia between Coffs Harbour (NSW) and Fremantle (WA) to depths of about 100 m. Best distinguished in having a relatively small, wedge-shaped disc, a black spot at the base of each pectoral fin, the first dorsal fin situated over the pelvic fins, and a pronounced lower caudal-fin lobe. To at least 300 cm and more than 220 kg.

Aptychotrema rostrata

Remarks: Commonly called 'shovelnose ray', this species occurs off eastern Australia between Moreton Bay (Qld) and Jervis Bay (NSW) from inshore to at least 50 m depth. Best distinguished in having a relatively small, wedge-shaped disc, multiple dusky spots on the upper surface, the first dorsal fin situated well behind the pelvic fins, and an indistinct lower caudal-fin lobe. Reaches at least 120 cm and up to at least 10 kg. A closely related species is found between Bass Strait (Vic.) and Port Hedland (WA).

Ray (page 1 of 2)

Dasyatididae, Gymnuridae, Myliobatididae & Urolophidae

Previous names: butterfly ray, eagle ray, stingaree, stingray

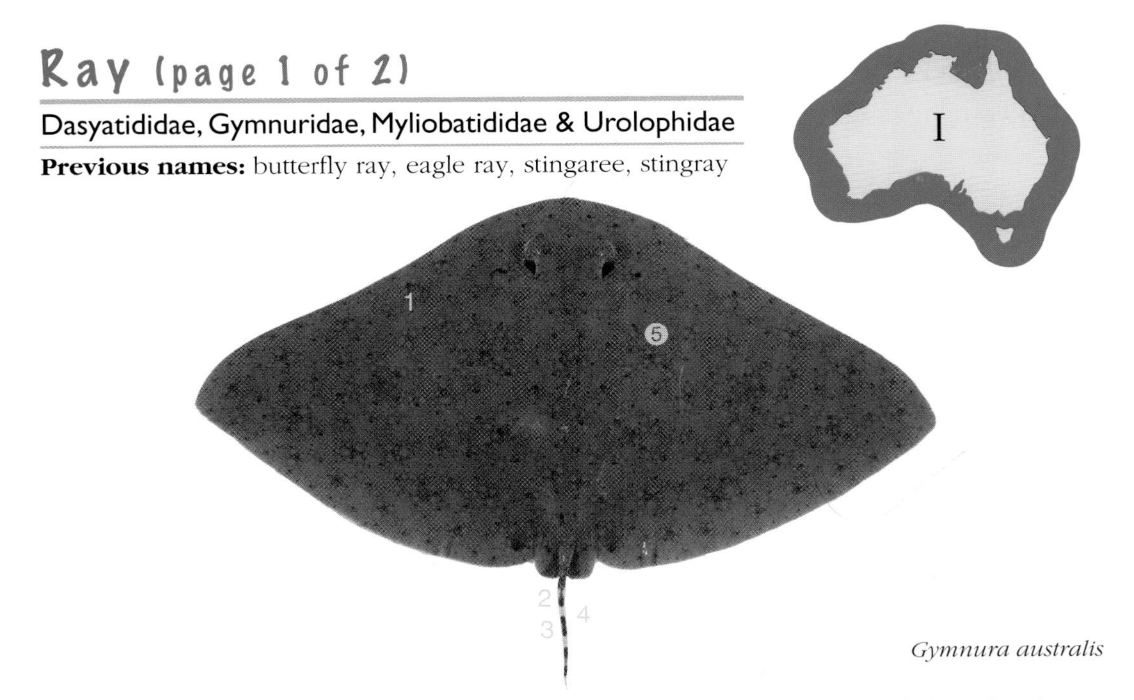

Gymnura australis

Identifying features: ① pectoral fins joined to head to form a large flattened disc; ② tail very slender to whip-like; ③ stinging spine usually present on tail; ④ dorsal fins often absent (sometimes 1 small fin before sting); ⑤ gill slits underneath head.

Comparisons: Includes the stingrays and their relatives, most of which have a venomous stinging spine on the tail. Premium food groups of rays, skates (Rajidae) and guitarfishes (Rhynchobatidae and Rhinobatidae) have separate marketing names. Skates (p. 49) have a large disc but their thin, short tail lacks a sting and usually has 2 small dorsal fins. They also have rows of sharp spines along the upper surface and a low caudal fin. Guitarfishes (pp 45–46) also lack a sting and have relatively smaller discs and stronger tails.

Fillet: *Myliobatis australis* flap-like, outer edge pointed; fine muscle strands extending cross-wise from inner edge; median cartilage along inner edge; yellowish-white; no peritoneum; integument silver to translucent. Skin medium, greenish-grey dorsally, almost white below.

Myliobatis australis

'Flap'. Protein fingerprint p. 360.

Size: Includes some of the largest fishes, with a disc width exceeding 600 cm (over 1 tonne) and a total length exceeding 500 cm. Typically marketed at less than 100 cm disc width and 50 kg.

Habitat: Marine; mainly inshore on the continental shelf in both tropical and temperate regions to depths of 200 m. Mostly bottom-dwelling but a few are exclusively pelagic.

Fishery: Caught as bycatch, mostly by demersal trawling, seining, gillnetting and trap nets.

Remarks: The pectoral fins are sold as 'ray flaps'. The flesh of small individuals is generally more tender than that of large adults.

Ray (page 2 of 2)

Dasyatididae, Gymnuridae, Myliobatididae & Urolophidae

Myliobatis australis

Remarks: Commonly called 'eagle ray', this species is distributed around southern Australia between Moreton Bay (Qld) and Jurien Bay (WA), including Tasmania, mainly inshore but to depths of 85 m. Best distinguished in having pointed, wing-like pectoral fins, a short bulbous head and long, whip-like tail with a short stinging spine near its base. Reaches at least 120 cm and more than 100 kg.

Himantura toshi

Remarks: Commonly called 'black-spotted whipray', this species is distributed off New Guinea and around northern Australia between Clarence River (NSW) and Port Hedland (WA) in depths of 10–140 m. Best distinguished in having an angular disc lightly or moderately covered in black spots, and a long, whip-like tail with a sting but without a caudal fin and skin folds along its upper and lower surface. To at least 180 cm and more than 10 kg.

Urolophus paucimaculatus

Remarks: Commonly called 'sparsely-spotted stingaree', this widely distributed species occurs around southern Australia between Crowdy Head (NSW) and Lancelin (WA) from inshore to at least 150 m depth. Best distinguished in having a rounded disc usually covered with a few white spots, a greyish U-shaped bar between the eyes, and a short tail with a barbed stinging spine followed by a lobe-like caudal fin. Reaches at least 44 cm and about 2 kg.

Skate

Rajidae

Previous names: skate flap, skate wing

I

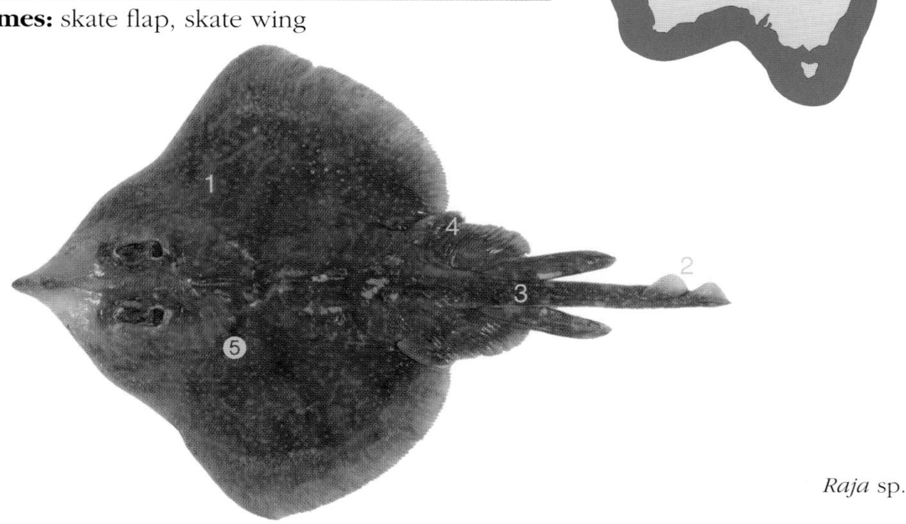

Raja sp.

Identifying features: ① pectoral fins joined to head to form a very large, round or angular disc; ② tail narrow, usually with 2 small dorsal fins near tip; ③ upper tail usually with rows of sharp thorns but lacking a stinging spine; ④ pelvic fins divided into 2 lobes; ⑤ gill slits underneath head.

Comparisons: Skates, unlike related fishes sold as 'ray' (pp 47–48), have a short, slender tail with rows of sharp thorns along the upper surface and 2 small dorsal fins (otherwise absent). They also lack the stinging spine of the rays. Guitarfishes (Rhynchobatidae and Rhinobatidae, pp 45–46) also lack a stinging spine but have a relatively smaller disc and stronger tail.

Fillet: *Raja* sp. flap-like, outer edge rounded; gaping muscle strands extending cross-wise from inner edge; median cartilage along inner edge; yellowish-white; peritoneum white with dark flecks. Skin thick, dark brownish to grey dorsally, white with dark blotches below.

Size: To 240 cm and about 100 kg (commonly landed at 50–80 cm and 0.8–1.8 kg).

Habitat: Marine; demersal on soft bottoms in both temperate and tropical latitudes. Found inshore on the continental shelf and down the deep slope to depths of at least 1500 m.

Fishery: Caught mainly as bycatch of demersal trawlers in temperate southern waters. Mostly discarded but landings are increasing steadily to reflect a more adventurous market.

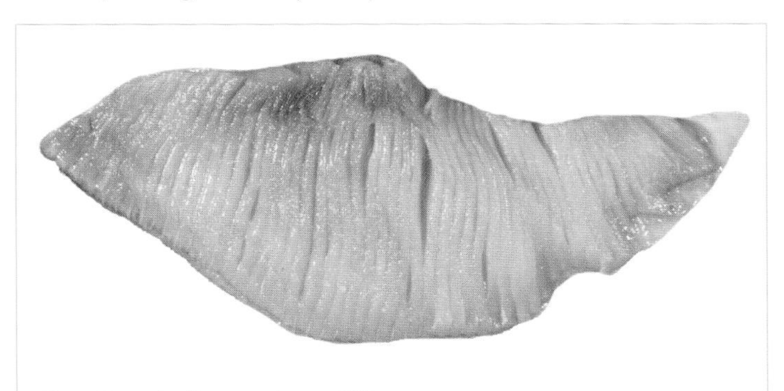

Flap. Protein fingerprint p. 360.

Remarks: Skates are utilised more frequently overseas where the flesh is considered of excellent quality and often commands high prices. However, the sustainability of skate fisheries is questionable as experience elsewhere indicates that they are susceptible to overfishing.

Sawshark

Pristiophorus species

Previous names: common sawshark, doggie, sawdog, southern sawshark

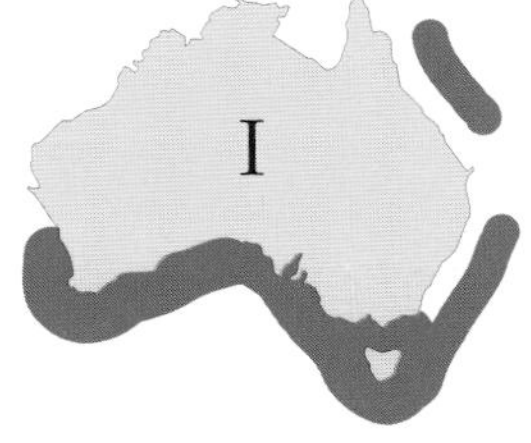

Pristiophorus cirratus

Identifying features: ① saw-like structure on snout; ② barbels present; ③ gills located on side of head.

Comparisons: Sawsharks (Pristiophoridae) resemble members of a non-commercial ray group, the sawfishes (Pristidae), in having a blade-like snout armed along its edges with sharp, tooth-like cusps. However, unlike sawfishes, the gills are on the side of the head (rather than on the undersurface), the pectoral fins are not joined to the head to form a disc (the gill slits are anterior of the pectoral-fin origin) and they have a pair of long sensory barbels on the snout. No other sharks have a saw-shaped snout.

Fillet: *P. cirratus* long, slender, tapering slightly, brownish-white. Outside with moderate red muscle band, parallel EL, parallel HL; HS through upper half of fillet anteriorly; integument yellowish-white. Inside flesh fine; belly flap distinct, not lobe-like; pores along midline; peritoneum silvery white to translucent. Always skinned.

Size: To at least 134 cm and 4.4 kg (commonly to 100 cm and about 1.7 kg).

Habitat: Marine; bottom-dwelling, burying or resting on sand and mud. Found from inshore in large estuaries to the upper continental slope in depths to 630 m.

Fishery: Caught mainly as a bycatch of demersal trawl nets, Danish seines, and gillnets off southeastern Australia and in the Great Australian Bight.

Remarks: Four species occur locally, of which the common (*P. cirratus*) and southern (*P. nudipinnis*) sawsharks form the basis of the Australian fishery. Diet consists mainly of small demersal fishes and shellfishes, which they detect in the sediments using their barbels. Marketed as fresh headed and gutted carcasses.

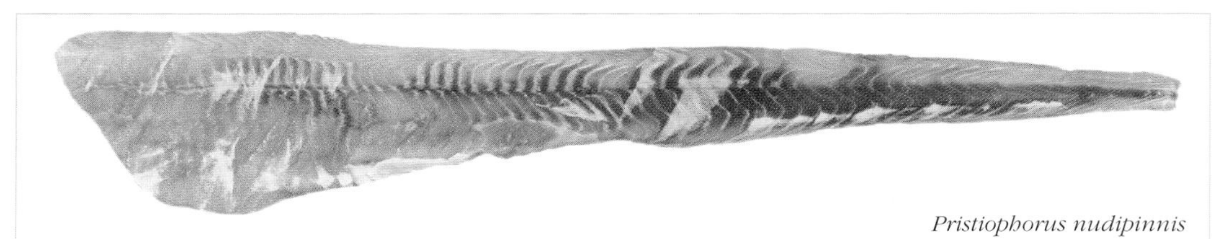

Pristiophorus nudipinnis

Untrimmed. Protein fingerprint p. 361.

Broadnose shark

Notorynchus cepedianus

Previous names: ground shark, sevengill shark, spottie, Tasmanian tiger shark

Identifying features: ① 7 gill slits; ② broad, blunt snout; ③ single dorsal fin; ④ black and white speckling on skin.

Comparisons: Only commercial shark with a single dorsal fin and more than 5 gill slits on each side of head. Other six-gilled members of the family in the region are rarely marketed. Sometimes referred to as 'tiger shark'. However, its only resemblance to the true tiger shark (*Galeocerdo cuvier*) found in warmer waters to the north is in head shape and shape of the lower jaw teeth.

Fillet: Long, slender, gradually tapering, pale-white to milky. Outside with broad red muscle band, parallel EL, weakly converging HL; HS just above midline of fillet; integument white. Inside flesh fine; belly flap as separate lobe when present; pores along midline; peritoneum white. Always skinned.

Size: To 300 cm and possibly 100 kg (commonly to 250 cm and about 70 kg).

Habitat: Marine, demersal; common in bays and lower reaches of estuaries and extending offshore to the mid-continental shelf to depths of 130 m.

Fishery: Caught off southern Australia, mainly using longlines and shark nets offshore and, in smaller quantities, gillnets inshore.

Remarks: Found in all oceans; a small fishery exists off California (USA). Apart from its flesh, which is of good quality, it is also caught for its skin and liver. Locally, it appears to be most common off Tasmania and Victoria. One of the most common large sharks in large bays and estuaries of these States, such as the Derwent River and Port Phillip Bay.

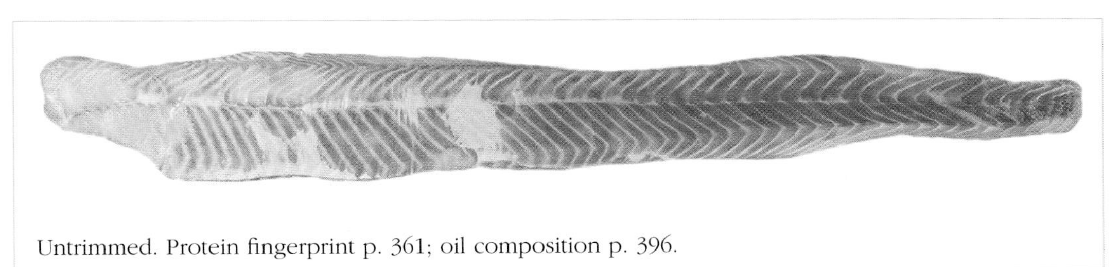

Untrimmed. Protein fingerprint p. 361; oil composition p. 396.

Blacktip shark

Carcharhinus, Loxodon & Rhizoprionodon species

Previous names: tropical shark, whaler shark

Carcharhinus dussumieri

Identifying features: 1 body may be greyish but never bluish; 2 first dorsal-fin origin usually over or at rear tip of pectoral-fin; 3 pectoral fin short or medium-length; 4 furrow (precaudal pit) at base of upper and lower caudal-fin lobes; 5 second dorsal fin much smaller than first; 6 lobe on outer edge at top of caudal fin very small.

Comparisons: Whaler shark species sold as 'blacktip shark' have a classic, streamline body shape with relatively small second dorsal and anal fins, and long upper caudal-fin lobe. They are often difficult to identify, even for specialists. Apart from the bronze whaler sharks (*C. brachyurus* and *C. obscurus*, p. 54), two tropical species with black fin tips (*C. tilstoni* and *C. sorrah*) make up the bulk of the fishery. Unlike the related hound sharks (Triakidae), they have crosswise furrows (precaudal pits) at the origin of both upper and lower caudal-fin lobes.

Fillet: *C. dussumieri* moderately elongate, tapering slightly, off-white brownish. Outside with pronounced red muscle band, parallel EL, almost parallel HL; HS through upper half of fillet; integument white. Inside flesh medium; belly flap distinct, not lobe-like; peritoneum silvery white translucent. Always skinned.

Size: To about 360 cm and 350 kg (commonly to 150–200 cm and about 25–50 kg).

Habitat: Marine; mainly pelagic over the continental shelf, frequently off reefs and inshore in bays and estuaries, but several species venture into the open ocean.

Fishery: Caught using pelagic gillnets, drift longlines and handlines, and as gillnet and trawl bycatch, in tropical Australia.

Remarks: More than 20 species of whaler sharks (Carcharhindae) occur in Australian seas. All are edible and most are marketed. Three temperate species, the blue whaler shark (*Prionace glauca*, p. 53) and the two bronze whaler sharks, are the most important commercially and have separate marketing names. Usually marketed as frozen trunks.

Untrimmed. Protein fingerprint p. 361; oil composition p. 396.

Blue whaler shark

Prionace glauca

Previous names: blue whaler, great blue

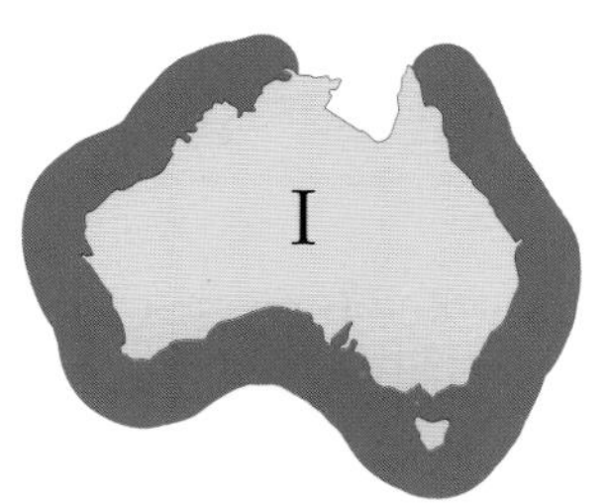

Identifying features: ① body indigo blue; ② upper teeth narrow and bent sideways; ③ first dorsal-fin origin well behind pectoral-fin rear tip; ④ pectoral fin very long; ⑤ furrow (precaudal pit) at base of upper and lower caudal-fin lobes; ⑥ second dorsal fin much smaller than first; ⑦ lobe on outer edge at top of caudal fin very small.

Comparisons: Distinctive among whaler sharks in having a slender body with a long pointed snout (otherwise more heavily built with short blunt snout), brilliant bluish body colour (versus grey-blue, grey or brownish), relatively longer pectoral fins, and a more posteriorly located first dorsal fin.

Fillet: Long, slender, gradually tapering, pale pink. Outside fatty, with moderate discontinuous red muscle band, parallel EL, weakly converging HL; HS just above midline of fillet; integument transparent. Inside flesh medium; belly flap distinct, not lobe-like; pores along midline; peritoneum transparent. Always skinned.

Size: To about 380 cm and at least 220 kg (commonly to 220 cm and about 40 kg).

Habitat: Marine; mainly pelagic in the upper open ocean. Ventures inshore in bays and large estuaries of southern Australia on a seasonal basis.

Fishery: Caught mainly well offshore in temperate seas as bycatch of tuna longline fishery.

Remarks: Available in small quantities in main city fish markets. Resource probably underexploited but long-term viability unknown. The flesh is considered to be softer than some other shark species.

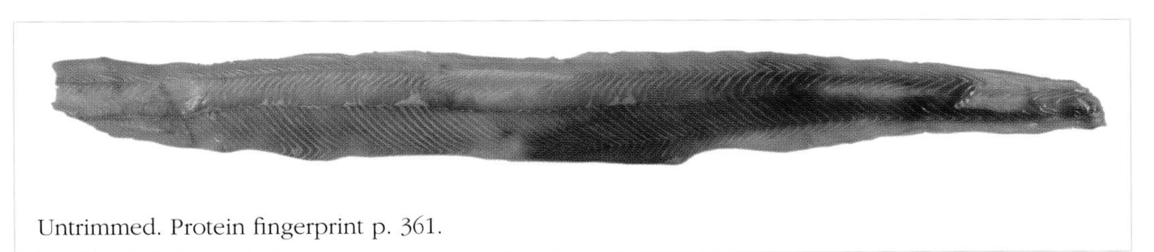

Untrimmed. Protein fingerprint p. 361.

Bronze whaler shark

Carcharhinus brachyurus & *C. obscurus*

Previous names: black whaler, cocktail shark, copper shark, dusky shark

Carcharhinus obscurus

Identifying features: ① plain bronze or greyish in colour; ② upper teeth narrow and bent sideways towards their tip or broadly triangular with serrated edges; ③ first dorsal-fin origin almost over pectoral-fin rear tip; ④ furrow (precaudal pit) at base of upper and lower caudal-fin lobes; ⑤ second dorsal fin much smaller than first; ⑥ lobe on outer edge at top of caudal fin very small.

Comparisons: Consists of two similar species, the bronze whaler (*C. brachyurus*) and dusky (*C. obscurus*) sharks. The upper-jaw teeth of the bronze whaler are more slender and strongly curved and the upper surface of the body tends to be more bronze-coloured than grey. These species are very difficult to distinguish from large tropical whalers caught in northern waters.

Fillet: *C. obscurus* moderately elongate, pronounced taper, pale pink. Outside with weak red muscle band posteriorly, parallel EL, converging HL; HS through upper half of fillet; integument pale grey above, white below. Inside flesh fine; belly flap distinct, weak lobe present; peritoneum translucent white. Always skinned.

Size: To about 360 cm and at least 350 kg (commonly to 250 cm and about 180 kg).

Habitat: Marine; mainly pelagic over the inner continental shelf but known from the slope to depths of about 400 m.

Fishery: Caught mainly by gillnets and set lines as a component of the temperate shark fishery, and as bycatch of demersal trawls.

Remarks: Often marketed as fillets or sectioned carcasses making identification difficult without protein tests. Higher priced than tropical whaler sharks.

Untrimmed. Protein fingerprint p. 361; oil composition p. 396.

Bony fishes

P. R. Last, G. K. Yearsley and N.V. Ruello

Australian herring

Arripis georgianus

Previous names: ruff, ruffie, tommy rough, tommy ruff

Identifying features: 1 prominent black tips on caudal fin; 2 scales rough to touch; 3 pectoral fin typically greyish without black blotch at its base; 4 upper surface with golden spots and/or vertical bars; 5 single continuous dorsal fin with a notch after last fin spine; 6 anal fin with 3 spines, much shorter-based than dorsal fin; 7 dorsal fin with 9–10 spines, 13–14 soft rays.

Comparisons: Should not be confused with the true herrings (Clupeidae) in which the dorsal fin is much shorter. Resembles the Australian salmons (p. 57) in body shape but has rough scales (rather than smooth) and fewer dorsal-fin soft rays (13–14 rather than 15–19).

Fillet: Moderately deep, rather elongate, tapering slightly, convex above, yellowish-white. Outside with continuous, intermediate central red muscle band; convex EL, converging HL; HS along middle of fillet; integument silvery white. Inside flesh medium; belly flap sometimes present; peritoneum white with melanophores. Sometimes skinned, skin thin, scale pockets large and defined.

Size: To about 45 cm and at least 1 kg (adults usually 17–25 cm and 0.1–0.2 kg).

Habitat: Coastal marine; pelagic, often in large schools. Prefers clear water along the open coast but also occurs in estuaries and bays.

Fishery: Caught in greatest quantity off southern Western Australia with smaller quantities taken off South Australia as bycatch of the garfish and whiting net-fisheries. Several catching methods are used, including haul nets, gillnets and seine nets. Also a popular recreational species.

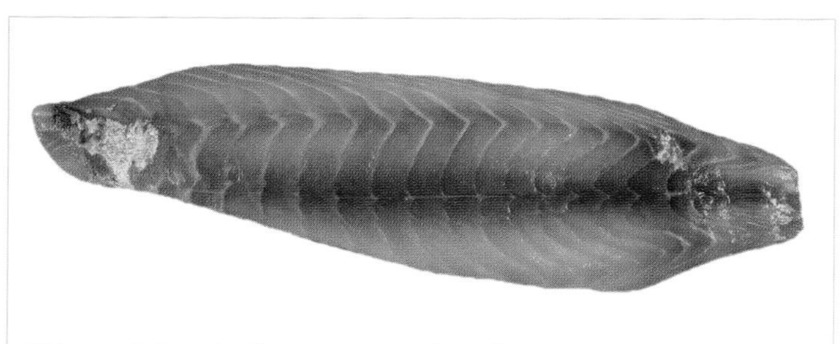

Trimmed. Protein fingerprint p. 361; oil composition p. 397.

Remarks: Marketed whole or as fillets (fresh, frozen, or smoked). Most popular with consumers in South Australia.

Australian salmon

Arripis trutta & *A. truttaceus*

Previous names: bay trout, black back, buck salmon, cocky salmon, colonial salmon, kahawai, salmon, salmon trout

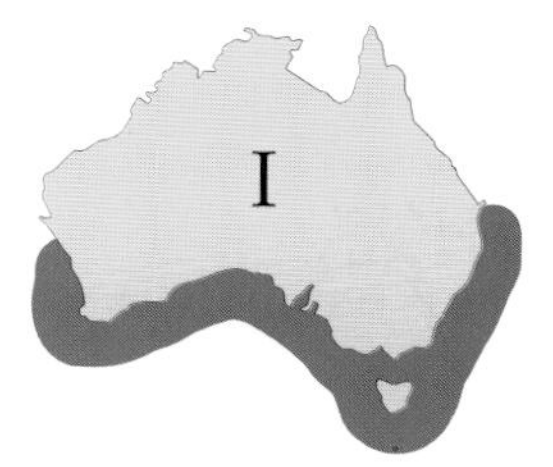

Arripis trutta

Identifying features: 1 no black tips on caudal-fin lobes; 2 scales smooth to touch; 3 pectoral fin distinctly yellowish with black blotch at its base; 4 upper surface with golden spots and/or vertical bars; 5 single continuous dorsal fin with a notch after last fin spine; 6 anal fin with 3 spines, much shorter-based than dorsal fin; 7 dorsal fin with 9 spines, 15–19 soft rays.

Comparisons: Differ from true salmons (Salmonidae), which have a much shorter dorsal fin. Most similar to the Australian herring (*Arripis georgianus*, p. 56) but have smoother scales and more dorsal-fin soft rays (15–19 rather than 13–14). The tailor (*Pomatomus saltatrix*, p. 272) is similar in shape but has 2 separate dorsal fins (rather than 1 continuous fin) and lacks spots.

Fillet: *A. truttaceus* moderately deep, rather elongate, tapering slightly, weakly convex above, pale pinkish to brownish. Outside with broad, continuous, pronounced central red muscle band; convex EL, converging HL; HS along middle of fillet; integument silvery above, whiter below. Inside flesh coarse; belly flap sometimes present; peritoneum translucent. Sometimes skinned, skin thin, scale pockets large and well defined.

Size: To about 96 cm and 10.5 kg (commonly 40–75 cm and 0.9–5 kg).

Habitat: Coastal marine; pelagic, often occurring in huge shoals.

Fishery: Caught mainly using beach or purse seines along the southern coast, often with the aid of spotter planes. Haul nets are used to capture small fish inshore. Sought-after recreationally; among the most highly regarded sport fishes in South Australia.

Arripis truttaceus

Untrimmed.
Protein fingerprint p. 361; oil composition p. 397.

Remarks: Two temperate species—eastern (*A. trutta*) and western (*A. truttaceus*). The eastern is smaller and has more rakers on the first gill arch (33–40 versus 25–31). Small fish are better to eat than the adults, which tend to be dryer and are usually canned or used as rocklobster bait.

Batfish

Ephippidae

Previous name: moonfish

Platax batavianus

Identifying features: ① head taller than long, with very short snout; ② anal fin with 3 spines; ③ dorsal fin with 5–9 spines, 18–40 soft rays; ④ body deep and compressed; ⑤ continuous dorsal fin, spinous portion often distinct; ⑥ mouth small with band of small multicuspid teeth in each jaw; ⑦ body often with dark bands (faint or absent in some adults).

Comparisons: Body shape superficially resembles the brightly coloured butterfly fishes (Chaetodontidae) and angel fishes (Pomacanthidae).

Fillet: *Platax batavianus* deep, short, upper profile very convex, bottle-shaped, greenish. Outside with feeble, diffuse central red muscle band; convex EL, weakly converging HL; HS just above middle of fillet; integument silvery white. Inside flesh medium; belly flap sometimes present; peritoneum silvery white with melanophores. Usually skinned, skin thick, scale pockets small and defined.

Size: To about 75 cm and 6 kg (adults usually 35–60 cm and 1.4–4.5 kg).

Untrimmed. Protein fingerprint p. 361.

Habitat: Marine; inshore in coastal bays and estuaries, but also extend offshore to outer reefs.

Fishery: Small quantities reach northern markets as bycatch of other fisheries.

Remarks: The general shape of each species changes with size, and marketed species vary widely in appearance. Although considered fair to good eating in some areas, batfishes are not generally favoured as food because of their feeding habits; most eat vegetation which gives their coarse flesh a weedy smell and taste.

Butterfish

Scatophagus species

Previous names: dory, john dory, johnny dory, old maid, scat

Scatophagus multifasciatus

Identifying features: 1 head not distinctly taller than long, snout of medium length; 2 anal fin with 4 spines; 3 dorsal fin with 11–12 spines, 16–18 soft rays; 4 body deep and compressed; 5 continuous dorsal fin, spinous portion distinct; 6 mouth small with band of small multicuspid teeth in each jaw; 7 body with dark bands or spots.

Comparisons: Distinctive fishes with a deep, almost oval body, small head and mouth, and characteristic colour pattern. They have more dorsal-fin and anal-fin spines and more elongate heads than the batfishes (Ephippidae, p. 58).

Fillet: *S. multifasciatus* deep, short, upper profile very convex, bottle-shaped, pale pink to reddish-brown. Outside with feeble, continuous central red muscle band; convex EL, weakly converging HL; HS just above middle of fillet; integument silvery, often patchy. Inside flesh medium; belly flap rarely present; peritoneum silvery white to translucent. Usually skinned, skin thick, scale pockets small and indistinct.

Size: To about 40 cm and 1.6 kg (adults usually 25–35 cm and 0.5–0.9 kg).

Habitat: Marine; coastal bays and estuaries in shallow water.

Fishery: Caught as bycatch in seine nets and traps in the shallows of northern Australia.

Trimmed. Protein fingerprint p. 361; oil composition p. 397.

Remarks: Sometimes incorrectly called 'John dory' in Queensland. At least two species, which differ in the extent of dark banding and spotting on their body. Make excellent aquarium fishes but should be handled carefully as the fin spines contain venom. The flesh is good when fresh but becomes soft and buttery if poorly handled.

Red bullseye

Priacanthus species

Previous names: bigeye, bigeye redfish, bullseye

I

Priacanthus hamrur

Identifying features: ① very large eyes; ② skin thick with tiny, embedded, spiny scales; ③ 1 continuous dorsal fin with 10 spines, 11–15 soft rays; ④ anal fin with 3 spines, 10–16 soft rays; ⑤ mouth large with lower jaw protruding; ⑥ body usually reddish.

Comparisons: The reddish colour, distinctive body form, sandpapery skin and large eyes distinguish these fishes from all others. Nine species occur in Australian seas, of which three are most commonly marketed as 'red bullseye': the common red (*P. macracanthus*), the spotted-fin (*P. tayenus*) and the crescent-tail (*P. hamrur*) bullseyes.

Fillet: *P. hamrur* deep, short, tapering prominently, very convex above, pale pinkish. Outside with intermediate, continuous, central red muscle band; weakly converging EL, weakly converging HL; HS along middle of fillet; integument pinkish to white. Inside flesh fine; belly flap sometimes present; peritoneum black to translucent. Usually skinned, skin medium, scale pockets small and defined.

Size: To at least 45 cm and 1.7 kg (adults usually 25–35 cm and 0.3–0.7 kg).

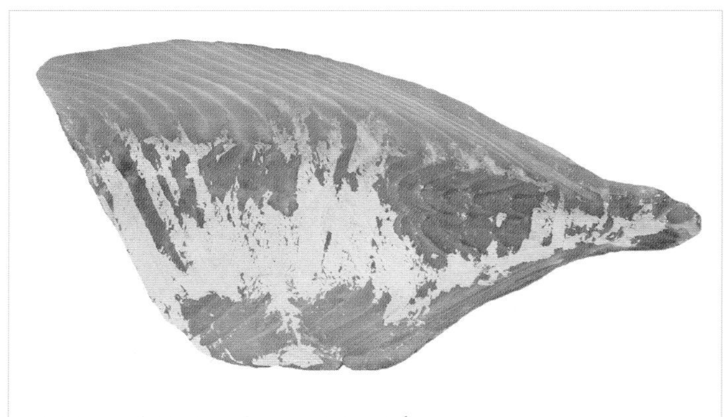

Untrimmed. Protein fingerprint p. 361.

Habitat: Marine; demersal, coastal and on coral reefs. Also offshore to continental shelf margin (about 200 m).

Fishery: Bycatch species of trawling and trapping.

Remarks: Frequently shelter in caves and under ledges during the day, becoming more active and feeding at night. Not highly regarded as foodfishes in Australia as the flesh is coarse and lacks a distinctive taste.

Marlin

Istiophoridae

Previous names: sailfish, spearfish

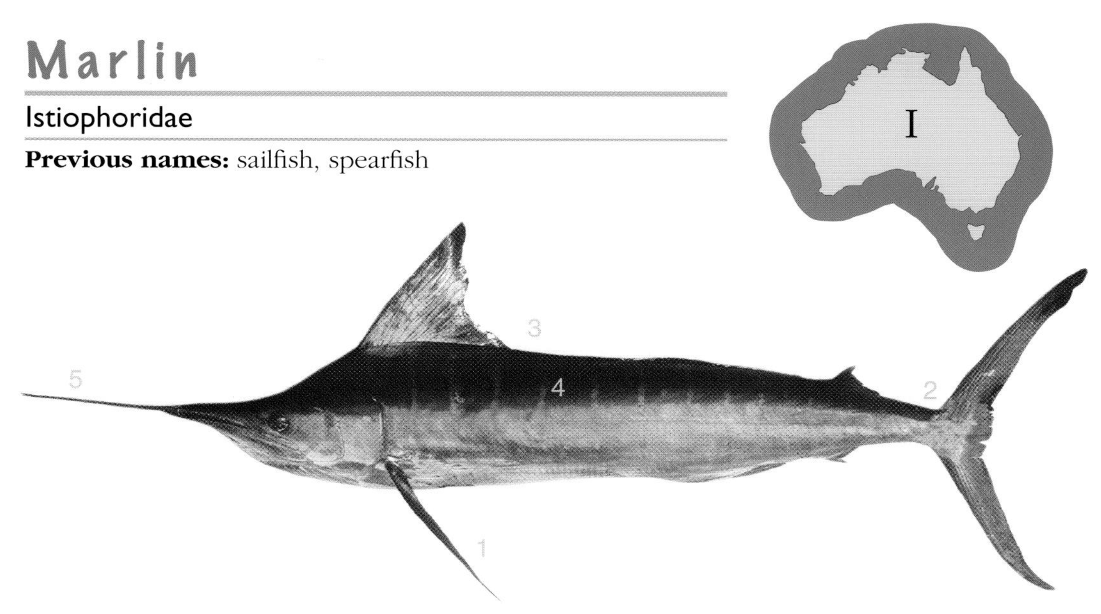

Tetrapturus audax

Identifying features: 1 rigid pelvic fin; 2 2 keels on each side of caudal peduncle; 3 dorsal fins close together; 4 small bony scales on body; 5 long bill, rounded in cross-section.

Comparisons: Easily identified by their strong elongate bodies with the snout extended forward as a spear-like bill. Five species are taken locally, the largest being the blue (*Makaira mazara*) and black (*M. indica*) marlins. The pectoral fins of the black marlin are rigid (unlike those of the other species) and cannot be folded back against its sides. Striped marlin (*Tetrapturus audax*), the most frequently landed of the marlins, has a much taller anterior dorsal-fin lobe than its larger relatives.

Fillet: Too large for filleting. Usually sold as steaks or cutlets. Flesh reddish or orange, medium coarseness.

Size: To at least 450 cm and 906 kg (typically marketed at 10–120 kg processed weight depending on the species).

Habitat: Marine, open ocean nomads; seasonally approach warm temperate and tropical coasts.

Fishery: Important gamefishes, caught mainly as tuna longline bycatch. The bulk of the commercial catch is taken by Japanese vessels. Angling culture, in particular off Queensland, has shifted from taking trophy fish to tag-and-release. No longer landed commercially in some markets due to state regulations. However, small quantities are still sold in major centres.

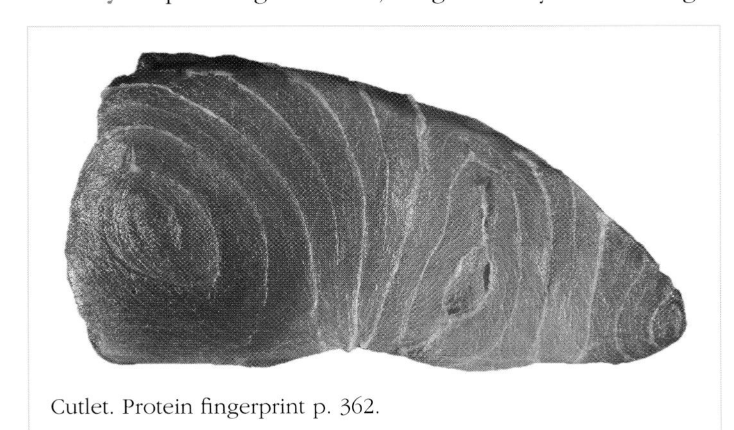

Cutlet. Protein fingerprint p. 362.

Remarks: Likely to disappear as local seafood due to their perceived importance to sport fisheries and tourism.

Swordfish

Xiphias gladius

Previous name: broadbill swordfish

Identifying features: 1 no pelvic fins; 2 1 keel on each side of caudal peduncle; 3 2 dorsal fins, widely separated in adults but continuous in juveniles; 4 no scales on body of adults; 5 long bill, horizontally flattened in cross-section.

Comparisons: Adults resemble marlin (Istiophoridae, p. 61) in general appearance but have a quite different bill shape and fin arrangement. Like marlin, young swordfish have a continuous dorsal fin and spiny scales on the skin but differ at all growth stages by lacking pelvic fins. Along with marlin, swordfish appear in the market finned, headed and gutted. In this form they are more difficult to distinguish but the skin of swordfish lacks the bony scales typical of marlin.

Fillet: Too large for filleting. Usually sold as steaks or cutlets. Flesh pale to pink, usually paler than marlins and tunas. The texture of the flesh is comparatively fine.

Size: To about 450 cm and 540 kg (typically marketed at about 100–160 kg).

Habitat: Marine, nomadic in the open ocean; thought to migrate to the surface at night and mainly descend to depths of 600 m or so during the day.

Fishery: Caught by domestic tuna longliners mainly off the east coast, but catches are increasing off Western Australia. Taken more broadly in the Exclusive Economic Zone by Japanese vessels. Annual catch about 1000 tonnes. Can be targeted using lightsticks attached to squid bait at the surface.

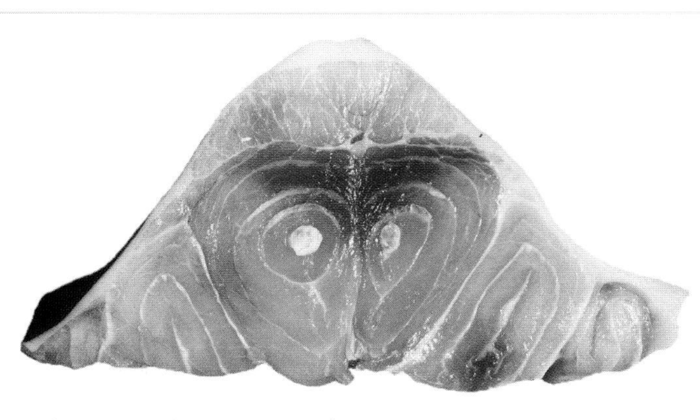

Cutlet. Protein fingerprint p. 362; oil composition p. 397.

Remarks: Swordfish go through many shape changes as they grow. Popular foodfish in major urban centres, although quality is variable and mercury levels are high in some large individuals.

Bigspine boarfish

Pentaceros decacanthus

Previous names: diamond fish, yellow boarfish

Identifying features: ① mouth small, teeth present on inside roof; ② greyish-yellow, no black bands on sides; ③ snout short; ④ dorsal-fin and anal-fin spines very robust; ⑤ dorsal fin with 11 spines, 12–14 rays; ⑥ central body scales relatively large; ⑦ head rough and bony.

Comparisons: Distinguished from other commercial domestic species by plain colour and distinctive body shape. Another deepwater relative, the pelagic armourhead (*Pseudopentaceros richardsoni*), which is targeted in other parts of the Indo–Pacific, has more spines (14–15) and fewer rays (8–9) in the dorsal fin. Pelagic armourhead are rarely caught in the Australian EEZ but appear to be abundant on nearby seamounts.

Fillet: Moderately deep, rather short, very convex above, yellowish-white. Outside with feeble, discontinuous central red muscle band; parallel EL, converging HL; HS below middle of fillet; integument translucent above, white below. Inside flesh fine; belly flap removed, belly cavity extremely long, extending about two-thirds of fillet length; peritoneum black. Usually skinned, skin very thick, scale pockets small and well defined.

Size: To at least 35 cm and 1.0 kg (typically landed at 25–35 cm and 0.5–1.0 kg).

Habitat: Marine; demersal on the outer continental shelf and upper slope in 80–550 m depth.

Fishery: Patchy throughout domestic range but common off Western Australia.

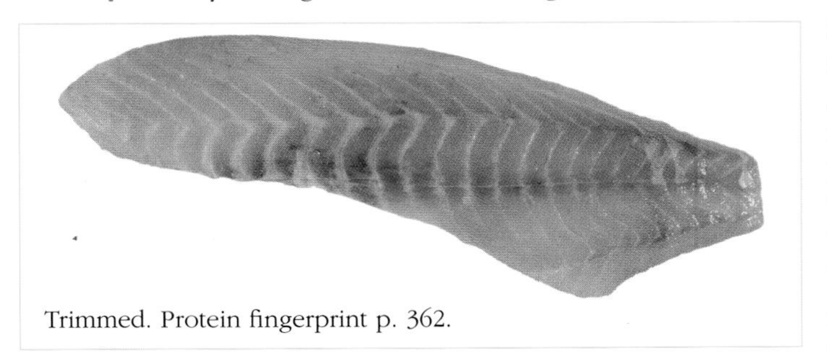

Trimmed. Protein fingerprint p. 362.

Remarks: The name 'boarfish' is less well known in Western Australia where poor recovery rates, unfamiliarity with the product, and a perception that the name 'boarfish' is unattractive have led to marketing difficulties. Elsewhere (e.g. in Melbourne), boarfishes are held in great esteem.

Blackspot boarfish

Zanclistius elevatus

Previous names: boarfish, longfin boarfish, onespot boarfish, short boarfish

Identifying features: 1 mouth small, no teeth present on inside roof; 2 dark bands below dorsal fin and through eye; 3 snout moderately long; 4 black spot on dorsal fin; 5 head profile rising steeply with prominent bumps over eye and on shoulder; 6 dorsal fin tall and sail-like; 7 dorsal fin with 5–8 spines, 25–29 soft rays; 8 head rough and bony.

Comparisons: Distinctive boarfish with a short and deep body, a triangular sail-like dorsal fin, and a large black spot on the hind third of the dorsal fin. The similar southwestern Australian short boarfish (*Parazanclistius hutchinsi*) has a more rounded sail-like dorsal fin.

Fillet: Very deep, short, very convex above, off-white to yellowish. Outside with feeble, discontinuous central red muscle band; convex EL, converging HL; HS along middle of fillet; integument silvery white. Inside flesh fine; belly flap removed; peritoneum white with melanophores. Usually skinned, skin medium, scale pockets small and well defined.

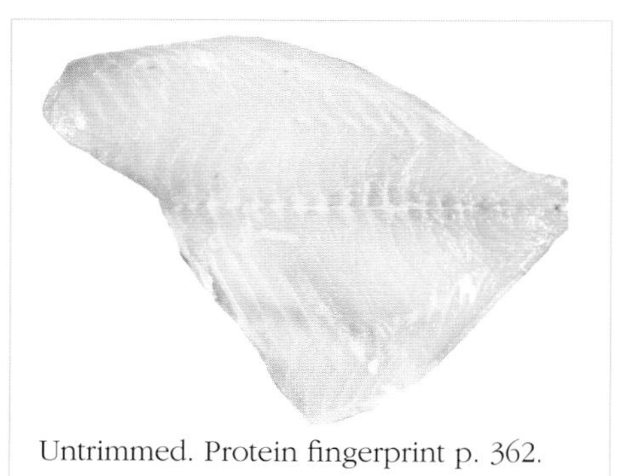

Untrimmed. Protein fingerprint p. 362.

Size: Reaches about 40 cm and 1.0 kg (commonly 20–30 cm and 0.3–0.7 kg).

Habitat: Marine; mainly demersal on the continental shelf deeper than 30 m but recorded from the upper continental slope to 500 m depth.

Fishery: Taken in small quantities by trawlers off southeastern Australia.

Remarks: Associated with hard bottoms in shallow water where it lives in crevices and caves. Trawl catches likely to be opportunistic. The flesh is highly regarded.

Boarfish

Pentaceropsis recurvirostris

Previous names: duckfish, longnose, longnose boarfish, long-snout boarfish

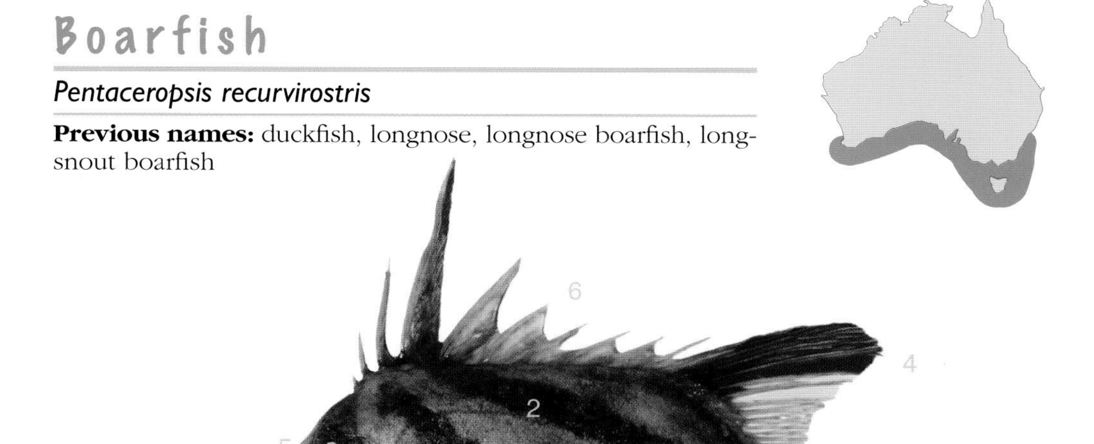

Identifying features: ① mouth small, no teeth present on roof inside; ② dark bands below dorsal fin and through eye; ③ snout very long and slender; ④ soft dorsal fin greatly enlarged, without black spot; ⑤ head profile strongly arched and rising steeply; ⑥ dorsal fin with 10–11 spines, 14–15 soft rays; ⑦ central body scales relatively small; ⑧ head rough and bony.

Comparisons: Distinctive temperate fish with an exceptionally long tubular snout, tall fins, strong spines and prominently banded colour pattern. This pattern is superficially similar to that of the giant boarfishes (*Paristiopterus* species, p. 67) but the latter have a much thicker snout, larger mouth and less elevated soft dorsal fin.

Fillet: Moderately deep, rather elongate, slightly convex above, off-white to yellowish. Outside with feeble, continuous central red muscle band; parallel EL, converging HL; HS along middle of fillet; integument silvery white. Inside flesh fine; belly flap usually removed; peritoneum white with melanophores. Usually skinned, skin thick, scale pockets small and defined.

Size: To at least 70 cm and 4.5 kg (typically 40–55 cm and 1.0–2.0 kg).

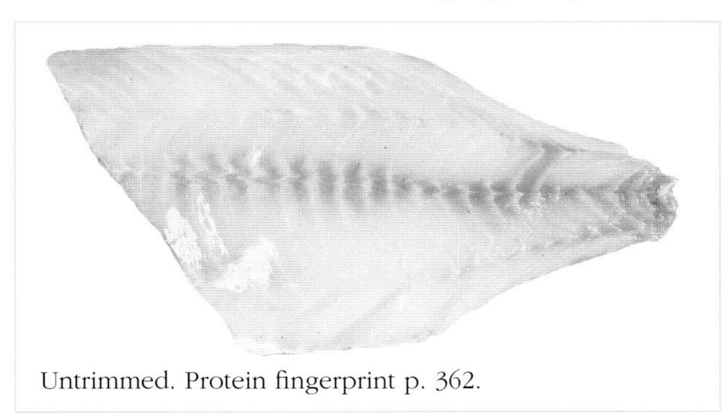

Untrimmed. Protein fingerprint p. 362.

Habitat: Marine, demersal; mainly coastal and inner continental shelf but taken to depths of at least 260 m.

Fishery: Trawled over the continental shelf in the Great Australian Bight. Caught inshore using gillnets.

Remarks: Eats small invertebrates using its long snout to pry among crevices in search of crabs, worms and brittle stars. The flesh is highly regarded in southeastern states.

Conway

Oplegnathus woodwardi

Previous names: hoofjaw, horseshoe-jaw, knifejaw

Identifying features: ① teeth in each jaw bonded together into a parrot-like 'beak'; ② prominent black bands below dorsal fin, on caudal peduncle, and through eye; ③ snout short, not extended forward; ④ dorsal fin long-based, relatively low; ⑤ upper head profile rising moderately; ⑥ dorsal fin with 11 spines, 11 soft rays; ⑦ central body scales relatively small.

Comparisons: Distinctive fish that resembles some of the boarfishes in colouration. Known more widely both in Australia and overseas as 'knifejaw' due to its beak-like teeth. Parrotfishes (Scaridae) have similar teeth but are more brightly coloured and have large scales.

Fillet: Moderately deep, rather elongate, very convex above, off-white to brownish. Outside with broad intermediate, diffuse central red muscle band; convex EL, converging HL; HS along middle of fillet; integument white. Inside flesh medium; belly flap usually removed; peritoneum silvery white to translucent. Usually skinned, skin thick, scale pockets small.

Size: To at least 48 cm and 2.2 kg (typically 30–40 cm and 0.5–1.5 kg).

Habitat: Marine; demersal on the continental shelf and upper slope in depths of 50–400 m.

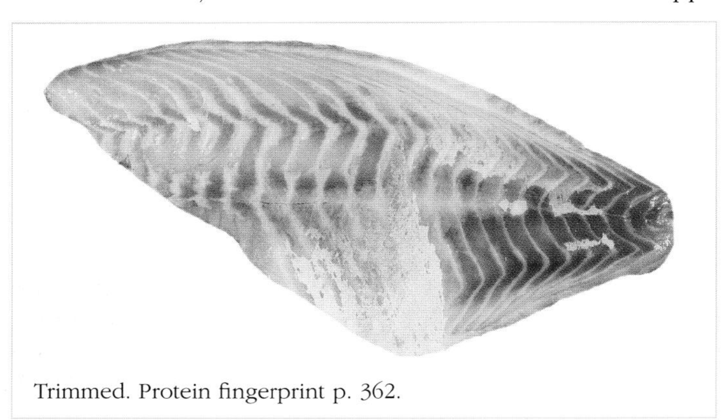

Trimmed. Protein fingerprint p. 362.

Fishery: Taken in greatest quantities by trawlers, and as shark-net bycatch, in the Great Australian Bight.

Remarks: Black bands most pronounced on juveniles and may become faint on adults. When mature, the second last band extends on to the dorsal and anal fins as a black blotch. Uses beak-like teeth to crush shells and sea urchins. The flesh is highly regarded.

Giant boarfish

Paristiopterus gallipavo & *P. labiosus*

Previous names: boarfish, duckfish, penfish, sowfish, yellow-spotted boarfish

I

Paristiopterus labiosus

Identifying features: ① mouth rather small, no teeth on roof inside; ② dark bands mostly present below dorsal fin and through eye; ③ snout long, beak-like; ④ soft dorsal fin distinctly shorter than longest spines, without black spot; ⑤ upper head profile rising; ⑥ dorsal fin with 7–8 spines, 16–18 soft rays; ⑦ central body scales relatively small; ⑧ head rough and bony.

Comparisons: Two species, which are similar in body shape with a long beak-like snout and relatively long tapering body, are marketed as 'giant boarfish' (this common name is usually used for one of them—*Paristiopterus labiosus*). The dominant western species, the brown-spotted boarfish (*P. gallipavo*), has 3 anal-fin spines (rather than 2 in *P. labiosus*) and its dorsal-fin spines are long and slender (rather than shorter and thicker).

Fillet: *P. labiosus* moderately deep, rather elongate, convex above, pale pinkish. Outside with broad intermediate, diffuse central red muscle band; convex EL, converging HL; HS along middle of fillet; integument white. Inside flesh fine; belly flap usually removed; peritoneum white with melanophores. Usually skinned, skin thick, scale pockets small and defined.

Size: To about 90 cm and 12 kg (typically 45–60 cm and 1.0–2.5 kg).

Habitat: Marine; demersal on the continental shelf and slope in depths of 10–260 m.

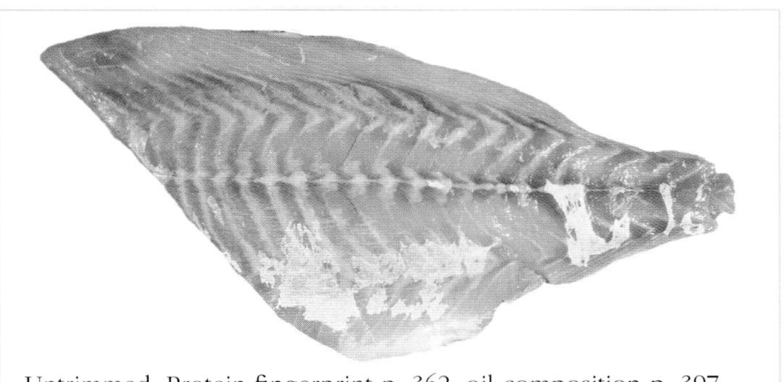

Untrimmed. Protein fingerprint p. 362; oil composition p. 397.

Fishery: Caught mainly in southern trawl fisheries.

Remarks: Adult male *P. labiosus* are purplish brown with a rash of yellowish spots, which has led to confusion between the two species. Contrary to much of the published literature, the spots on *P. gallipavo* are dark brown rather than yellow. Their flesh is very tasty.

Redbait

Emmelichthys species

Previous name: pearlfish

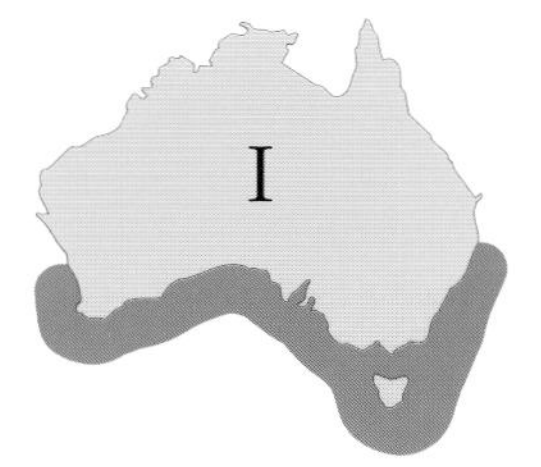

Emmelichthys nitidus

Identifying features: ① short spines between main lobes of dorsal fin; ② body very slender; ③ sides, belly and fins pinkish; ④ mouth protrusible, capable of considerable forward extension; ⑤ hind bone of upper jaw scaly and greatly enlarged; ⑥ caudal fin deeply forked and capable of being folded in scissor-like fashion.

Comparisons: Most similar to rubyfishes (*Plagiogeneion* species, p. 69) but have a more elongate body and the last spines of the dorsal fin are very short giving the appearance of the spinous and soft rayed parts of the fin being separate. Often caught with jack mackerels (*Trachurus* species, p. 267) but redbaits lack scutes along the caudal peduncle. The greatly enlarged posterior part of the upper jaw (maxilla), which is not overlapped by the bony area beneath the eye when the mouth is closed, is characteristic of bonnetmouths.

Fillet: *E. nitidus* slender, rather elongate, tapering gently, weakly convex above, brownish. Outside with continuous, very pronounced central red muscle band, very deep at peduncle; converging EL, converging HL; HS along middle of fillet; integument greyish above, white below. Inside flesh fine; belly flap sometimes present; peritoneum white translucent with melanophores. Sometimes skinned, skin thick, scale pockets small and well defined.

Size: To about 36 cm and 0.4 kg (adults usually 20–34 cm and 0.1–0.3 kg).

Habitat: Marine; pelagic over the continental shelf but most common in shallower water, depths of 20–100 m.

Fishery: Taken mainly in the bycatch of purse seiners targeting jack mackerels off south-eastern Australia. Occasionally large catches are taken from aggregations and the annual landings may exceed 1000 tonnes.

Untrimmed. Protein fingerprint p. 362.

Remarks: Presently used for fish meal but have potential for greater use as bait for the tuna fishery.

Rubyfish

Plagiogeneion species

Previous name: cosmopolitan rubyfish

Plagiogeneion macrolepis

Identifying features: ① dorsal fin continuous; ② body not slender; ③ sides, belly and fins reddish-yellow; ④ mouth protrusible, capable of considerable forward extension; ⑤ hind bone of upper jaw scaly and greatly enlarged; ⑥ caudal fin deeply forked and capable of being folded in scissor-like fashion.

Comparisons: Rubyfishes are related to redbaits (*Emmelichthys* species, p. 68) but have a deeper body and the dorsal fin is continuous, although there is a low notch where the membranes of the spines and rays meet. The greatly enlarged posterior part of the upper jaw (maxilla), which is not overlapped by the bony area beneath the eye when the mouth is closed, distinguishes bonnetmouths from other similar fishes.

Fillet: *P. macrolepis* moderately deep, rather elongate, tapering rapidly, weakly convex above, pale pinkish. Outside with continuous, intermediate central red muscle band, very deep at peduncle; converging EL, converging HL; HS just below middle of fillet; integument greyish above, white below. Inside flesh fine; belly flap sometimes present; peritoneum white translucent with melanophores. Sometimes skinned, skin thick, scale pockets moderate, well defined.

Size: To 65 cm and 3.0 kg (commonly 30–35 cm and 0.4–0.6 kg).

Habitat: Marine; demersal to semi-pelagic over the outer continental shelf and upper slope in depths of 50–550 m.

Untrimmed. Protein fingerprint p. 362.

Fishery: Small commercial quantities taken in the Great Australian Bight by demersal trawlers in 130–220 m. Also taken as bycatch of the blue-eye trevalla dropline fishery off the south-east.

Remarks: Two species occur in Australian seas; the western rubyfish (*P. macrolepis*) is most frequently marketed.

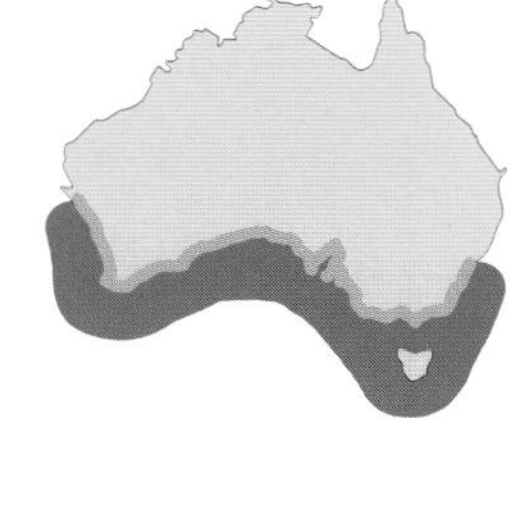

Black bream

Acanthopagrus butcheri

Previous names: bream, golden bream, silver bream, southern bream

Identifying features: ① upper body silvery to olive brown, often with blackish hue; ② anal and pelvic fins greyish-brown; ③ small black spot at base of pectoral fin; ④ 4–5 vertical rows of scales between front of dorsal fin and lateral line; ⑤ dorsal fin with 10–13 spines, 10–13 soft rays; ⑥ second anal-fin spine distinctly larger than other 2; ⑦ peg-like canines at front and rounded or flattened molar teeth along back of jaws.

Comparisons: Often confused with the yellowfin bream (*A. australis*, p. 75) which is usually paler and has yellowish-white anal and pelvic fins. The similar pikey bream (*A. berda*, p. 72) is also dark bodied but lacks a dark spot at the top of the pectoral-fin base.

Fillet: Moderately deep, rather elongate, upper profile convex, tapering rapidly, greyish with dark veins. Outside with pronounced, continuous central red muscle band; parallel EL, converging HL; HS along middle of fillet; integument silvery white. Inside flesh medium; belly flap rarely present; peritoneum silvery translucent. Usually skinned, skin medium, scale pockets large and well defined.

Size: To at least 55 cm and 3.6 kg (adults usually 30–45 cm and 0.4–1.5 kg).

Habitat: Estuarine; usually in brackish or freshwater but in higher salinities in Western Australia.

Fishery: Caught mainly in Victorian waters, particularly from the Gippsland Lakes where the fishery has existed for more than a century. Also commonly taken off southern Western Australia. Main methods include haul seining and gillnetting.

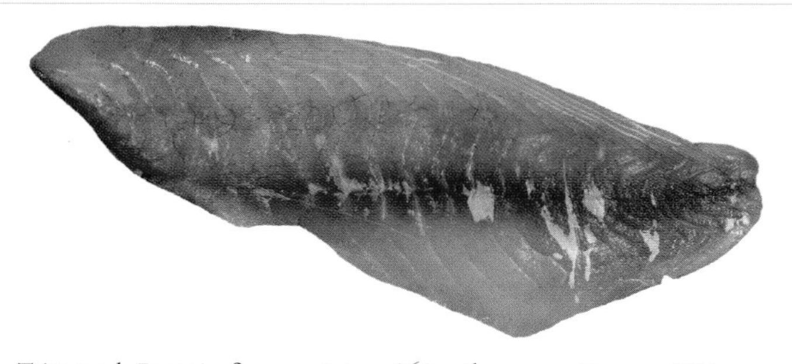

Trimmed. Protein fingerprint p. 362; oil composition p. 397.

Remarks: Marketing range overlaps with the yellowfin bream leading to occasional misnaming. One of southern Australia's main angling species. Fine table-fish.

Frypan bream

Argyrops spinifer

Previous names: frypan snapper, longfin snapper, longspine snapper

Identifying features: ① upper body reddish; ② leading dorsal-fin spines elongated; ③ body very deep, forehead raised; ④ about 8 vertical scale rows between front of dorsal fin and lateral line; ⑤ dorsal fin with 11–12 spines, 10–11 soft rays; ⑥ second and third anal-fin spines enlarged; ⑦ small canines at front and rounded or flattened molar teeth along back of jaws.

Comparisons: Distinctive member of the bream family with deep, almost oval reddish body, and long filamentous spines in the dorsal fin.

Fillet: Deep, short, upper profile very convex, tapering rapidly, yellowish-white, no dark veins. Outside with intermediate, continuous central red muscle band; convex EL, converging HL; HS just below middle of fillet; integument silvery white. Inside flesh fine; belly flap rarely present; peritoneum silvery white. Usually skinned, skin medium, scale pockets moderate and defined.

Size: To 58 cm and about 4.0 kg (adults usually 30–45 cm and 0.6–1.5 kg).

Habitat: Marine; mainly on the inner continental shelf to at least 150 m depth.

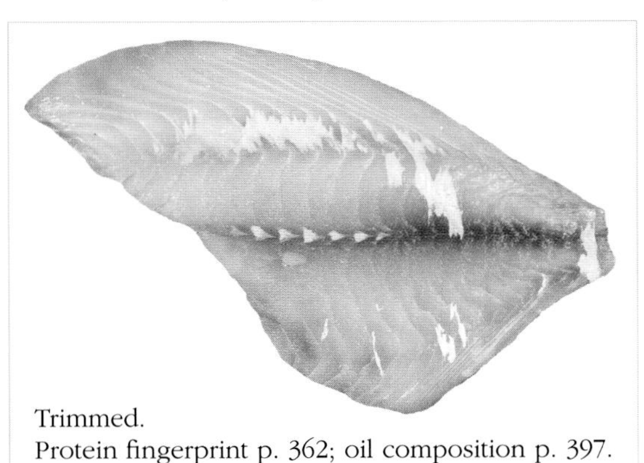

Trimmed.
Protein fingerprint p. 362; oil composition p. 397.

Fishery: Mainly trawled, lined and trapped off northwestern Australia. Also taken by line in sections of the Great Barrier Reef.

Remarks: Has been marketed as 'frypan snapper' presumably due to its similarity to the southern snapper (*Pagrus auratus*, p. 73). 'Seaperches' (Lutjanidae) are also called 'snappers' in parts of the world. To avoid further confusion, members of this group (Sparidae) should be called 'breams'. A special case has been made for snapper as the name has been in use locally for more than a century. Highly regarded foodfish.

Pikey bream

Acanthopagrus berda

Previous names: black bream, bream

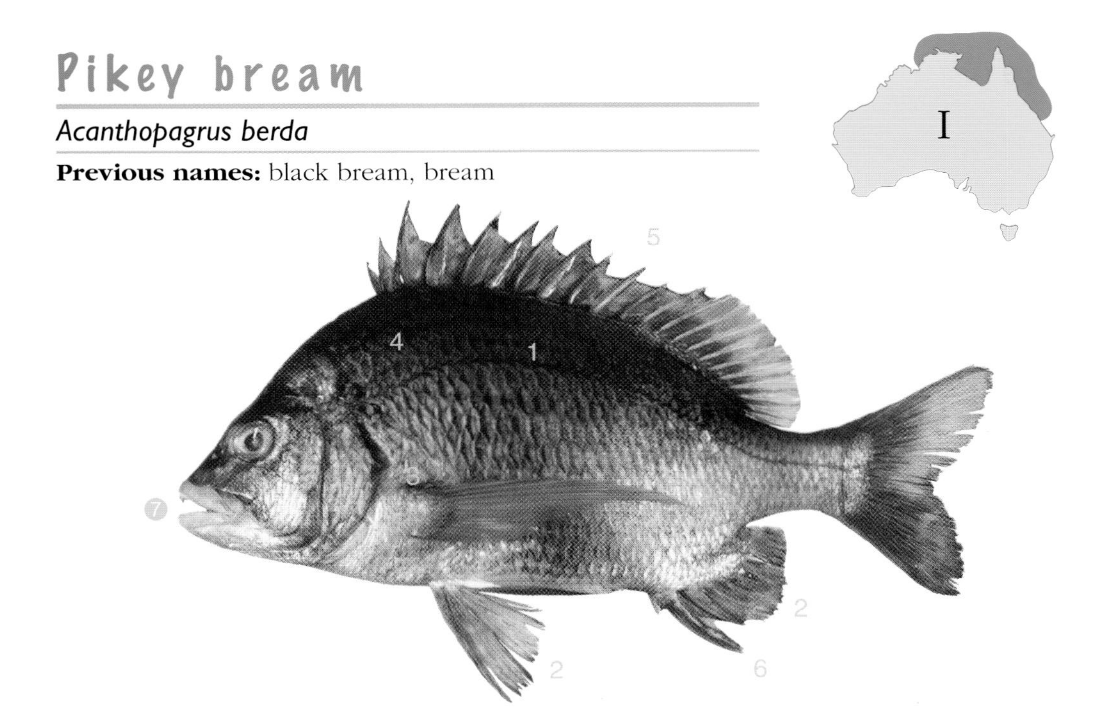

Identifying features: ① upper body greyish to olive brown, often with blackish hue; ② anal and pelvic fins greyish, often dark; ③ no black spot at base of pectoral fin; ④ 4 vertical rows of scales between front of dorsal fin and lateral line; ⑤ dorsal fin with 11–12 spines, 10–13 soft rays; ⑥ second anal-fin spine distinctly larger than other 2; ⑦ peg-like canines at front and rounded or flattened molar teeth along back of jaws.

Comparisons: Apart from the absence of a black spot above the pectoral-fin base and a more pointed head with eyes located relatively closer to mouth, it most closely resembles the yellowfin bream (*A. australis*, p. 75) and the black bream (*A. butcheri*, p. 70).

Fillet: Deep, rather elongate, upper profile very convex, tapering rapidly, pale pinkish with some dark veins. Outside with intermediate, continuous central red muscle band; convex EL, converging HL; HS along middle of fillet; integument white. Inside flesh fine; belly flap rarely present; peritoneum silvery white to translucent. Usually skinned, skin thick, scale pockets large and irregular.

Size: To at least 66 cm and about 4.0 kg (typically landed at 30–40 cm and about 0.6–1.0 kg).

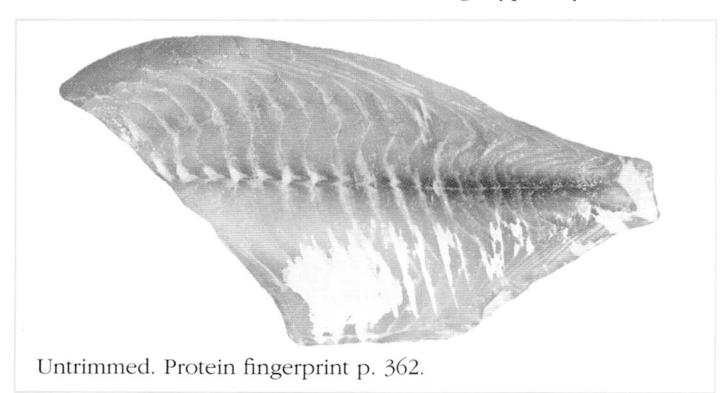

Untrimmed. Protein fingerprint p. 362.

Habitat: Marine, coastal and estuarine; common around river mouths.

Fishery: Small catch component of coastal net-fisheries. Frequently caught by anglers.

Remarks: Distribution overlaps with yellowfin bream in Queensland. Colour depends largely on habitat, with fish from muddy upper reaches of rivers dark. Good table-fish.

Snapper

Pagrus auratus

Previous names: cockney, pink snapper, pinkies, red bream, squire

Identifying features: ① upper body varies from pale pink to reddish, mostly with bluish spots; ② no dorsal-fin spines greatly elongated; ③ forehead with prominent humps in adults; ④ 8–12 vertical rows of scales between front of dorsal fin and lateral line; ⑤ dorsal fin with 12 spines, 9–10 soft rays; ⑥ second and third anal-fin spines enlarged; ⑦ enlarged canines at front and smaller rounded or flattened molar teeth along back of jaws.

Comparisons: Well known, attractive fish that is distinguished from other commercial breams by its pinkish colouration with blue spots, and body shape. The frypan bream (*Argyrops spinifer*, p. 71) is somewhat similar but has a more rounded body and taller dorsal-fin spines.

Fillet: Moderately deep, rather elongate, upper profile convex anteriorly, tapering gradually to deep peduncle, pale pinkish without dark veins. Outside with intermediate, continuous central red muscle band; convex EL, converging HL; HS just below middle of fillet; integument translucent above, silvery white below. Inside flesh coarse; belly flap rarely present; peritoneum silvery white to translucent. Usually skinned, skin medium, scale pockets large and defined.

Size: To at least 130 cm and 19.5 kg (mostly landed at 38–90 cm and about 0.8–8 kg).

Habitat: Marine; coastal and continental shelf to about 200 m. Juveniles mostly estaurine.

Fishery: Fished throughout its range using a variety of methods including traps, trawling, longlines, handlines, and gillnets. Aquaculture trials are continuing. Very popular recreational species.

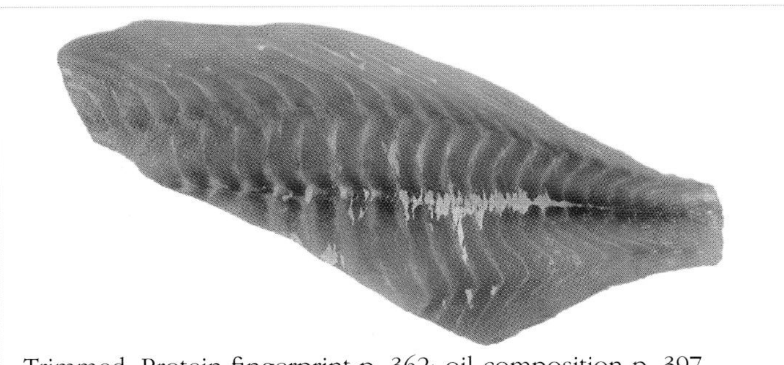

Trimmed. Protein fingerprint p. 362; oil composition p. 397.

Remarks: The marketing name 'snapper' has been retained on a historical basis. This fish should not be confused with the true snappers or seaperches (Lutjanidae). One of Australia's most highly regarded foodfishes.

Tarwhine

Rhabdosargus sarba

Previous names: bream, silver bream

Identifying features: ① silvery with yellow lines matching scale rows; ② anal and pelvic fins varying from pale yellow to almost orange; ③ no black spot at base of pectoral fin; ④ 7–8 vertical rows of scales between front of dorsal fin and lateral line; ⑤ dorsal fin with 11 spines, 13–15 soft rays; ⑥ second and third anal-fin spines similarly enlarged; ⑦ chisel-shaped canines at front and rounded or flattened molar teeth along back of jaws (last one greatly enlarged).

Comparisons: Distinctive bream with a rounded head and thin wavy yellow lines along silvery sides. The greatly enlarged tooth at the back of each jaw is diagnostic locally. Often caught with the yellowfin bream (*Acanthopagrus australis*, p. 75) but in addition to those features noted above it has 2 enlarged anal-fin spines (rather than only the middle one enlarged).

Fillet: Deep, rather elongate, bottle-shaped, pale pinkish with some dark veins. Outside with intermediate, continuous central red muscle band; convex EL, converging HL; HS along middle of fillet; integument translucent above, white below. Inside flesh medium; belly flap sometimes present; peritoneum black. Usually skinned, skin medium, scale pockets moderate, defined.

Size: To about 50 cm and 2 kg (typically much smaller, 20–30 cm and about 0.2–0.5 kg).

Untrimmed.
Protein fingerprint p. 362; oil composition p. 397.

Habitat: Coastal marine; also estuaries and offshore reefs to depths of about 35 m.

Fishery: Caught mainly by handlines, haul nets, traps and gillnets, and as bycatch of inshore trawling.

Remarks: Smaller fish occur in estuaries with larger fish captured in nearshore waters, off beaches and around rocky reefs. Delicate flesh that must be treated well to preserve its flavour.

Yellowfin bream

Acanthopagrus australis

Previous names: black bream, bream, seabream, silver bream, surf bream

Identifying features: (1) upper body silvery to olive brown, often with blackish hue; (2) anal and pelvic fins yellowish-white; (3) small black spot at base of pectoral fin; (4) 4–5 vertical rows of scales between front of dorsal fin and lateral line; (5) dorsal fin with 10–12 spines, 10–13 soft rays; (6) second anal-fin spine distinctly larger than other 2; (7) peg-like canines at front and rounded or flattened molar teeth along back of jaws.

Comparisons: Estuarine forms are darker than those from the sea and closely resemble the black bream (*A. butcheri*, p. 70). However the anal and pelvic fins of the yellowfin bream are yellowish-white rather than greyish-brown. The pikey bream (*A. berda*, p. 72) is dark bodied but lacks a dark spot at the top of the pectoral-fin base.

Fillet: Deep, rather elongate, upper profile convex, tapering rapidly, pale pinkish with some dark veins. Outside with intermediate, continuous central red muscle band; convex EL, converging HL; HS along middle of fillet; integument white. Inside flesh medium; belly flap rarely present; peritoneum silvery white to translucent. Usually skinned, skin medium, scale pockets large and well defined.

Size: To at least 55 cm and 4.0 kg (typically landed at 25–35 cm and about 0.2–1.2 kg).

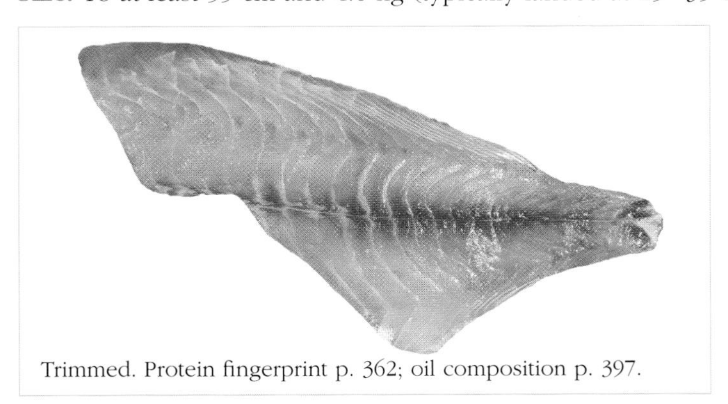

Trimmed. Protein fingerprint p. 362; oil composition p. 397.

Habitat: Coastal marine; also estuaries and inshore reefs to depths of about 35 m.

Fishery: Caught mainly in seine nets, gillnets and traps in estuaries and off beaches along the central east coast south to Victoria. Very significant recreational species.

Remarks: Extremely popular with anglers. The flesh is highly regarded.

Cardinal fish

Epigonus telescopus

Previous name: sea bass

Identifying features: ① very large eyes; ② very long caudal peduncle; ③ brown body with loose scales; ④ 2 anal-fin spines; ⑤ 2 widely spaced dorsal fins with short bases; ⑥ most of mouth toothless or with short conical teeth.

Comparisons: Distinctive robust, elongate fish with small dorsal fins, large eyes and brownish colouration. Its delicate mouth is capable of protruding forward and lacks sharp teeth. No other species with this combination of features is marketed locally.

Fillet: Moderately deep, somewhat elongate, tapering gently, almost straight above, pale pinkish. Outside with pronounced, continuous, central red muscle band; weakly converging EL, converging HL; HS along middle of fillet; integument whitish. Inside flesh medium; belly flap sometimes present; peritoneum black to translucent. Usually skinned, skin thick, scale pockets large and defined.

Size: To at least 65 cm and 3.0 kg (adults usually 40–60 cm and 0.8–2.5 kg).

Habitat: Marine; close to bottom on the continental slope and nearby seamounts in depths of 300–800 m.

Fishery: Caught mainly by trawling and droplining over rough bottom, mainly near seamounts.

Remarks: More than 120 species of cardinal fishes live in Australian seas and most of these are tropical in shallow water. Most are small, and although some are sold for food in Asian markets, they are more valuable locally as aquarium fishes. Small, deepwater cardinal fishes (*Epigonus* species), often rich in number, are among the discarded bycatch of southern trawl fisheries. However, the largest of these, *E. telescopus*, has received increasing acceptance in southern markets because of its fine eating qualities.

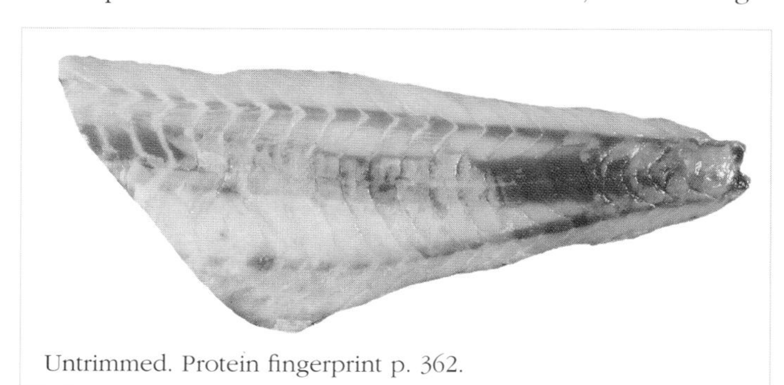

Untrimmed. Protein fingerprint p. 362.

European carp

Cyprinus carpio

Previous names: Asian carp, carp, common carp, German carp

Identifying features: ① mouth small with 2 barbels at each corner; ② dorsal fin continuous with a long base, its last spine with a serrated edge; ③ scales large (33–40 along lateral line); ④ pelvic fin well behind pectoral-fin base; ⑤ no adipose fin; ⑥ head bony, without scales.

Comparisons: Resembles the goldfish (*Carassius auratus*, p. 78) but has barbels at the corner of the mouth and the edge of the last rigid spine in the dorsal fin is not saw-like.

Fillet: Moderately elongate, barely tapering, pale pinkish. Outside with intermediate, continuous red muscle band; parallel EL, weakly converging HL; HS along middle of fillet; integument patchy, silvery white. Inside flesh medium; belly flap sometimes evident, not lobe-like; peritoneum silvery white translucent. Usually skinned, skin medium, scale pockets large and well defined.

Size: Locally to 85 cm and 15 kg (commonly about 35–60 cm and 0.6–4 kg). Elsewhere reported to 120 cm and 60 kg.

Habitat: Freshwater; soft muddy bottoms in lakes, ponds and slow-flowing rivers.

Fishery: Taken in drainages of southeastern mainland states using electrofishers, drum nets, gillnets, and beach seines.

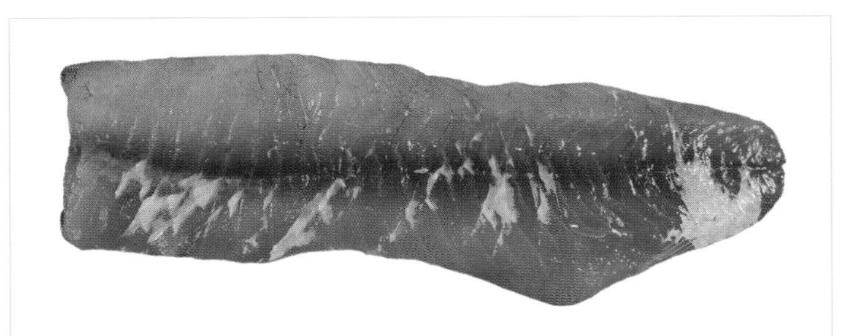

Trimmed. Protein fingerprint p. 362; oil composition p. 397.

Remarks: Native of Asia, subsequently introduced to Europe and now found on all continents except Antarctica. Introduced to Australia in the late 1800s. Now possibly the world's most widely distributed freshwater fish. Allegedly destructive to native waterways. Also used for petfood and rocklobster bait.

Goldfish

Carassius auratus

Previous names: crucian carp, golden carp, native carp

Identifying features: 1 mouth small, without barbels at each corner; 2 dorsal fin continuous with a long base, its last spine with a serrated edge; 3 scales large (26–34 along lateral line); 4 pelvic fin well behind pectoral-fin base; 5 no adipose fin; 6 head bony, without scales.

Comparisons: Similar to the European carp (*Cyprinus carpio*, p. 77) but lacks barbels at the corner of the mouth and the last rigid spine of the dorsal fin has a fine saw-like edge.

Fillet: Very deep, short, tapering sharply, slightly convex above, red to pale pink. Outside with intermediate, continuous red muscle band, more pronounced posteriorly; parallel to convex EL, strongly converging HL; HS just above middle of fillet; integument patchy, white. Inside flesh fine; belly flap usually evident, very thick; peritoneum silvery to translucent. Usually skinned, skin thin, scale pockets large and well defined.

Size: Locally to 40 cm and at least 1 kg (commonly about 20–30 cm and 0.3–0.6 kg). Elsewhere, reported to reach 5 kg.

Habitat: Freshwater; introduced from Japan and China where it occurs in still habitats such as ponds and riverine backwaters. Extremely tolerant of both warm and cold water and now widespread locally.

Untrimmed. Protein fingerprint p. 362; oil composition p. 397.

Fishery: Small scale, based mainly on wild catches from dams and ponds.

Remarks: Colour can be plain golden or olive-green, or brightly mottled red, white and black. Bred in fish farms as an aquarium or pond fish. Flesh has little taste but is popular in Asian cuisine.

Tench

Tinca tinca

Previous names: none

Identifying features: ① mouth small with a thin barbel at each corner; ② dorsal fin continuous with a short base; ③ scales very small (about 100 along lateral line); ④ pelvic fin well behind pectoral-fin base; ⑤ no adipose fin; ⑥ head bony, without scales.

Comparisons: Distinctive member of the carp family with a short dorsal-fin base and small barbels on the edge of its mouth.

Fillet: Moderately elongate, tapering slightly, yellowish-white. Outside with pronounced and continuous red muscle band; convex EL, weakly converging HL; HS along middle of fillet; integument white. Inside flesh fine; belly flap usually evident, not lobe-like; peritoneum silvery white to translucent. Usually skinned, skin thick, scale pockets small and well defined.

Size: Locally to 70 cm and 9 kg (commonly about 30 cm and 0.6 kg). Elsewhere, reported to 120 cm and 60 kg.

Habitat: Freshwater; introduced from Europe in the 1800s and widely distributed in waterways of southeastern Australia, including Tasmania. Associated with weedy habitats or in deep holes in still, freshwater lakes and ponds and slow-flowing rivers. Can live in brackish water but rarely found in main channels of rivers.

Untrimmed. Protein fingerprint p. 362.

Fishery: Very minor commercial species based on small incidental catches from freshwater.

Remarks: Fair sport fish of little regard locally. Often schools in shaded areas during the day, becoming more isolated and active at night. Very hardy. Firm, white flesh has little taste.

Catfish

Arius species

Previous names: blue catfish, forktail catfish, salmon catfish

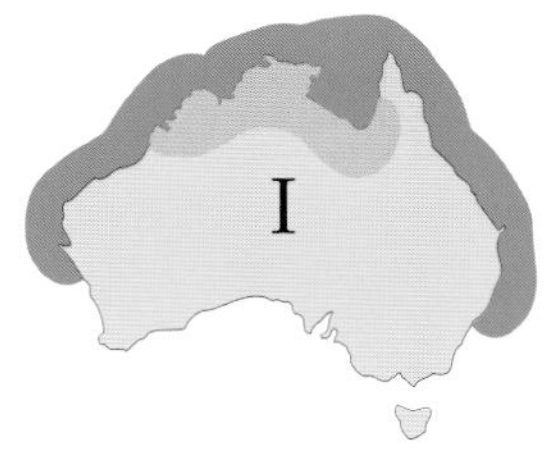

Arius thalassinus

Identifying features: ① tail forked; ② head flattened with 2–3 pairs of mouth barbels; ③ no scales; ④ large spine at the front of dorsal and pectoral fins; ⑤ adipose fin present.

Comparisons: At least four species from this large tropical family reach the marketplace. The species are very similar to each other and often need to be identified by a specialist. They differ from other native catfishes in having a forked tail rather than an eel-like tail.

Fillet: *A. thalassinus* relatively deep, moderately elongate, tapering slightly, pale pinkish. Outside with pronounced, continuous, central red muscle band; convex EL, strongly convex HL (almost reaching HS anteriorly); HS along middle of fillet; integument patchy, silvery. Inside flesh fine; belly flap sometimes evident, not lobe-like; peritoneum silvery white to translucent. Usually skinned, skin medium, no scale pockets.

Size: To 130 cm and 28 kg (commonly 40–60 cm and 0.6–3 kg).

Habitat: Primarily freshwater, estuarine and coastal marine, with some species extending on to the outer continental shelf.

Fishery: Caught near the coast and in rivers by traps, handlines and gillnets. Incidental catch by prawn trawling.

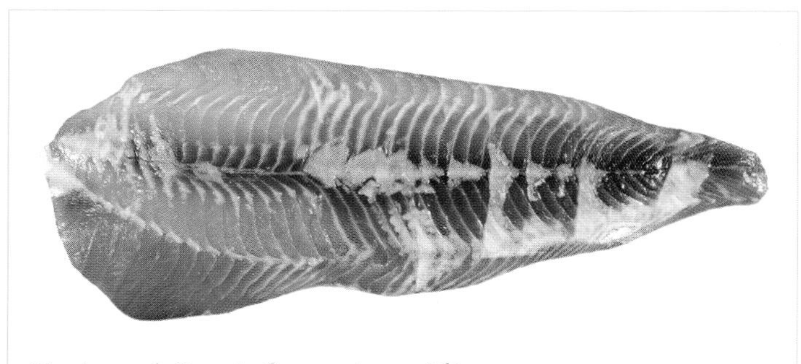

Untrimmed. Protein fingerprint p. 362.

Remarks: Important food-fishes in Asia but, despite their fine food qualities, presently under-utilised in Australia. Eggs and larvae are incubated in the mouths of the male parent. Their fecundity is low making them susceptible to overfishing. However, several of these species have good aquaculture potential.

Cobbler

Cnidoglanis macrocephalus

Previous names: catfish, estuary catfish, South Australian catfish

Identifying features: ① body slender; ② first dorsal fin with 3–5 (mostly 4) rays; ③ gill membranes continuously joined to isthmus along ventral midline of head; ④ nasal barbels sometimes extending past eye; ⑤ tail eel-like; ⑥ head flattened, with 4 pairs of barbels around mouth and pronounced fringe on chin; ⑦ concealed spine at the front of dorsal and pectoral fins; ⑧ no adipose fin.

Comparisons: Similar to the freshwater catfish (*Tandanus tandanus*, p. 82) but with a more slender body, fewer soft rays in the dorsal fin, and longer nasal barbels. Differs from the fork-tail catfishes in its eel-like body.

Fillet: Relatively long, slender, tapering clearly, yellowish-white. Outside with feeble, continuous, central red muscle band; parallel EL, strongly convex HL (almost reaching HS anteriorly); HS along middle of fillet; integument patchy, white. Inside flesh medium; belly flap rarely evident, not lobe-like; peritoneum silvery white to translucent. Usually skinned, skin very thick, no scale pockets.

Size: To 76 cm and 2.5 kg (commonly 40–60 cm and 0.4–1.0 kg).

Habitat: Marine bays, inlets and lower to mid-reaches of estuaries. Often remains concealed in holes or beneath rocky ledges during daylight.

Fishery: Targeted at night off southwestern Australia mainly in estuaries using gillnets and haul nets, and using traps in the Swan River estuary. Smaller bycatch off South Australia.

Remarks: Excellent table-fish mostly consumed in Perth. Fast growing, reaching maturity at about 40 cm in 3 years. Taken by recreational fishers using gillnets and angling. Should be handled carefully as the spines can inflict a painful sting.

Untrimmed. Protein fingerprint p. 362; oil composition p. 398.

Freshwater catfish

Tandanus tandanus

Previous names: catfish, dewfish, eeltail catfish, freshwater jewfish, tandan

Identifying features: ① body rather deep; ② first dorsal fin with 6 rays; ③ gill membranes not joined to isthmus along entire ventral midline of head; ④ nasal barbels capable of extending to eye; ⑤ tail eel-like; ⑥ head flattened, with 3 pairs of barbels around mouth and no fringe on chin; ⑦ concealed spine at the front of dorsal and pectoral fins; ⑧ no adipose fin.

Comparisons: Resembles the cobbler (*Cnidoglanis macrocephalus*, p. 81) but has a deeper body, more soft rays in the dorsal fin, and shorter nasal barbels. Differs from the forktail catfishes in its eel-like body.

Fillet: Relatively deep, moderately elongate, tapering slightly, pale pinkish. Outside with intermediate, continuous, central red muscle band; convex EL, strongly convex HL (almost reaching HS anteriorly); HS along middle of fillet; integument patchy, translucent white. Inside flesh fine; belly flap rarely evident, not lobe-like; peritoneum silvery white to translucent. Usually skinned, skin very thick, no scale pockets.

Size: To 90 cm and 6.8 kg (commonly to 50 cm and 1.8 kg).

Habitat: Freshwater; bottom-dweller of freshwater lakes and slow-moving rivers.

Fishery: Once a valuable commercial species, wild catches are now significantly smaller.

Remarks: Excellent aquarium fish but is vulnerable in temperatures lower than 4°C. Reportedly prone to fungal infection at low temperatures. Builds impressive nests, appearing as a depression up to 2 m across. Will breed in ponds and has expanding aquaculture potential. The white, firm, tasty flesh makes good eating, although quality may depend on prey eaten. The fin spines are venomous.

Untrimmed. Protein fingerprint p. 362.

Silver cobbler

Arius midgleyi

Previous names: Lake Argyle catfish, Lake Argyle silver cobbler, Ord River catfish

Identifying features: (1) snout tip blunt, almost straight, when viewed from above; (2) barbel of upper jaw short, just or barely reaching opercular margin; (3) no rakers on the innermost edge of the first and second gill arches; (4) lateral line never branching at tail base; (5) tail forked; (6) head flattened with 2–3 pairs of mouth barbels; (7) large spine at the front of dorsal and pectoral fins.

Comparisons: Of the freshwater forktail catfishes, most similar to the salmon catfish (*A. leptaspis*), which has a more rounded head and a longer barbel at the corner of the upper jaw (extending well past operculum). Other freshwater *Arius* species, including the blue catfish (*A. graeffei*), have rakers on the innermost edge of all gill arches. Master's catfish, *A. mastersi*, has a butterfly-shaped bone before the dorsal-fin spine.

Fillet: Relatively deep anteriorly, moderately elongate, tapering strongly, pale pinkish. Outside with pronounced, continuous, central red muscle band; weakly converging EL, strongly convex HL (almost reaching HS anteriorly); HS along middle of fillet; integument white. Inside flesh medium; belly flap sometimes evident, not lobe-like; peritoneum silvery white to translucent. Usually skinned, skin very thick, no scale pockets.

Size: To 130 cm and 28 kg (commonly about 60–90 cm and 3–7 kg).

Habitat: Freshwater, riverine, rarely venturing into brackish water.

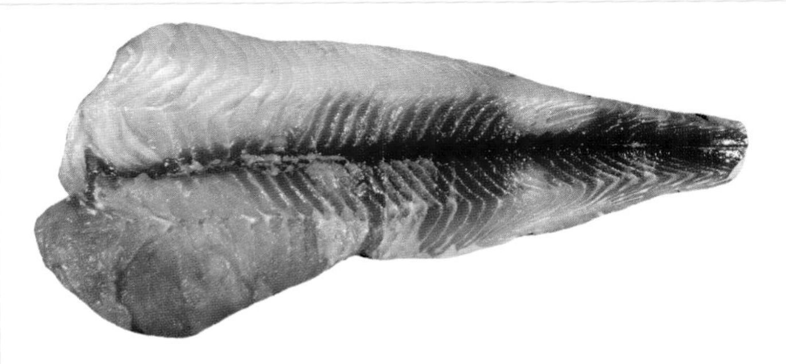

Untrimmed. Protein fingerprint p. 362.

Fishery: Target species of an important gillnet fishery in Lake Argyle (WA) since 1978 but caught in smaller quantities across northern Australia.

Remarks: The flesh is of excellent quality and fetches a high price in Darwin and Perth. Fast growing and likely to have aquaculture potential. Can also be marketed under the general name 'catfish'.

Blue grenadier

Macruronus novaezelandiae

Previous names: blue hake, hoki, New Zealand whiptail, whiptail

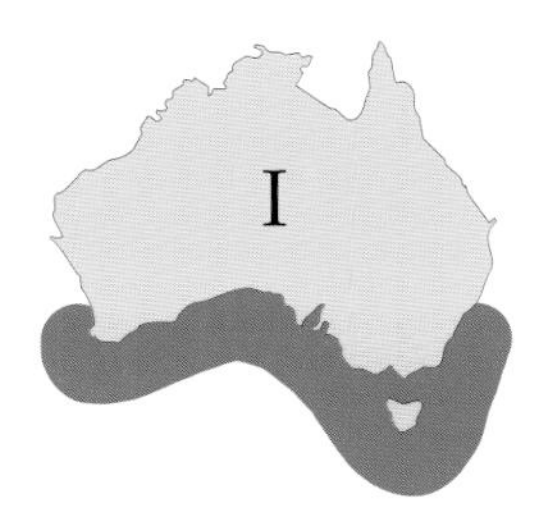

Identifying features: ① no separate caudal fin; ② hind dorsal-fin rays much longer than anal-fin rays; ③ body silvery blue; ④ pelvic-fin base beneath or slightly behind pectoral-fin base; ⑤ scales easily removed; ⑥ first dorsal fin separated slightly from second dorsal fin.

Comparisons: Important fish, differing from its commercial cod-like relatives, which include the southern hake (*Merluccius australis*, p. 86), in having an almost eel-like tail. Another, less common eel-tail hake, the silver grenadier (*Lyconus* sp.), has long fang-like canines at the tip of the upper jaw that overhang the lower jaw when the mouth is closed (rather than canines that are short and concealed).

Fillet: Long, rather slender, tapering slightly, pale pinkish. Outside with pale, pronounced, discontinuous, central red muscle band; parallel EL, parallel HL; HS along middle of fillet; integument silvery. Inside flesh coarse; belly flap sometimes evident, not lobe-like; peritoneum black. Often skinned, skin thin, scale pockets small and barely detectable.

Size: To 115 cm and about 6 kg (commonly 60–100 cm and averages 1–3.5 kg).

Habitat: Marine; demersal on the upper continental slope mainly in depths of 200–700 m. Juveniles venture inshore, occasionally into large estuaries.

Fishery: Fished widely in the South East Trawl Fishery with largest catches made along the continental slope off western Bass Strait and Tasmania.

Remarks: A type of 'grenadier hake' (family Macruronidae). These fishes are confined to the Southern Hemisphere and should not be confused with the widely distributed true hakes (family Merlucciidae), of which the southern hake is marketed locally. Mainly sold as fillets or cutlets but ideal for surimi. The soft flesh has a short shelf life and requires careful handling.

Trimmed. Protein fingerprint p. 363; oil composition p. 398.

Ribaldo

Mora moro & *Lepidion* species

Previous names: deepsea cod, ghost cod, giant cod, googly-eyed cod

Mora moro

Identifying features: ① skin pale greyish to whitish; ② body robust and soft; ③ middle rays of anal fin much shorter than those adjacent (may appear as 2 fins); ④ eyes large; ⑤ ventral fins thread-like and placed well forward of pectoral fins; ⑥ very short barbel at tip of lower jaw.

Comparisons: Belong to a group known as morid cods with at least three species marketed under this name. All are soft bodied with whitish skin, a bulky head and trunk, and 2 distinct lobes in the anal fin. The main species is true ribaldo (*Mora moro*). The less common giant cods (*Lepidion* species), which can grow to huge proportions, are characterised by having an extremely long anterior ray in the dorsal fin.

Fillet: *M. moro* moderate length, rather deep, tapering slightly, white. Outside with pale, feeble, discontinuous, central red muscle band; convex EL, weakly converging HL; HS along middle of fillet; integument white. Inside flesh medium; belly flap rarely evident, not lobe-like; peritoneum black. Often skinned, skin medium, scale pockets small and irregular.

Size: To at least 183 cm and 45 kg (commonly 40–70 cm and 1.5–5 kg).

Habitat: Marine; demersal on the continental slope mainly in depths of 300–900 m but taken to at least 1200 m.

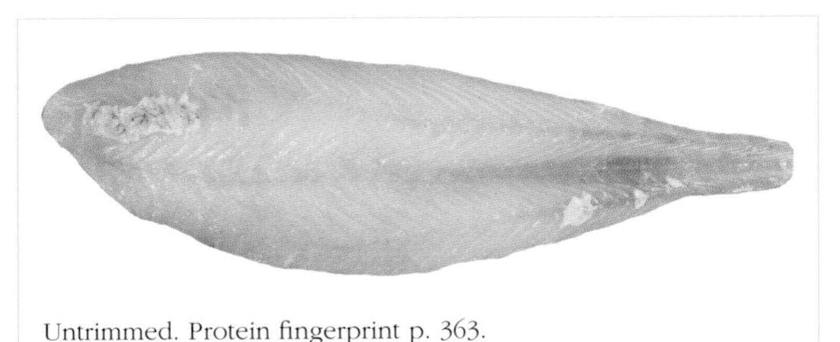

Untrimmed. Protein fingerprint p. 363.

Fishery: Mainly trawled as bycatch but also taken in small quantities by dropline. Often discarded. Resource size unknown, but likely to be small.

Remarks: Members of the family Moridae. Ribaldo flesh is very soft and should be eaten as fresh as possible.

Southern hake

Merluccius australis

Previous names: blue whiting, hake

Identifying features: 1 separate caudal fin; 2 second dorsal fin and anal fin deeply indented or notched; 3 body pale greyish to white; 4 pelvic-fin base beneath or slightly before pectoral-fin base; 5 scales easily removed; 6 first dorsal fin separated slightly from second dorsal fin.

Comparisons: Deep notches in the dorsal and anal fins, as well as its characteristic body form, enable it to be distinguished from all other cod-like fishes marketed locally. Has also been confused with hapuku (*Polyprion* species, p. 220) but is less heavily built, has a much longer anal fin, and lacks thick dorsal-fin spines. The southern blue whiting (*Micromesistius australis*), which is imported from New Zealand as fillets, is similar in shape but has 3 dorsal fins (rather than 2).

Fillet: *M. capensis* moderate length, uniform depth, barely tapering, upper profile straight, yellowish-white. Outside with intermediate, continuous red muscle band; parallel EL, parallel HL; HS along middle of fillet; integument silvery grey above, whitish below. Inside flesh medium; belly flap rarely present; peritoneum white to translucent. Usually skinned, skin medium, scale pockets small and indistinct.

Size: To 126 cm and at least 14 kg (commonly 70–100 cm and 2–9 kg).

Habitat: Marine; demersal on the continental slope mainly in 400–1000 m but usually shallower than 800 m.

Fishery: Caught occasionally as bycatch of trawlers targeting blue grenadier (*Macruronus novaezelandiae*, p. 84).

Remarks: The only true hake taken in Australian waters. It is not common, being confined to the southern limits of the Exclusive Economic Zone. This range constriction may be related to a lack of suitable habitat at the right latitude as it is caught in quantity well south of New Zealand. The flesh has good eating qualities. The fillet image and description are of a closely related South African import, *Merluccius capensis*.

Merluccius capensis

Trimmed. Protein fingerprint p. 363.

Southern rock cod

Lotella & Pseudophycis species

Previous names: bearded rock cod, bearded southern rock cod, beardy, red cod

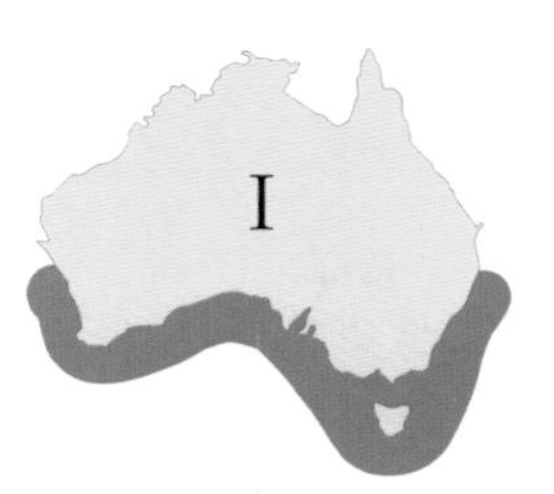

Pseudophycis bachus

Identifying features: ① skin reddish, silvery pink or brownish; ② body elongate and soft; ③ middle rays of anal fin not appreciably shorter than those adjacent; ④ eyes moderate or small; ⑤ ventral fins thread-like and placed well forward of the pectoral fins; ⑥ barbel at tip of lower jaw.

Comparisons: Typical morid cods, with a long soft body, large head and mouth, a chin barbel, long dorsal and anal fins, and a small caudal fin. They differ from deeper water relatives (such as ribaldos, p. 85) in being reddish, brownish or pinkish (rather than pale grey or white) and there is no deep notch in the anal fin.

Fillet: *P. barbata* moderate length, rather deep anteriorly, tapering rapidly, yellowish-white. Outside with pale, intermediate, continuous, central red muscle band; convex EL, weakly converging HL; HS along middle of fillet; integument silvery white. Inside flesh coarse; belly flap sometimes evident, almost triangular; peritoneum greyish to black. Usually skinned, skin thin, scale pockets small and irregular.

Size: To at least 70 cm and 5.9 kg (commonly 40–50 cm and 0.8–1.5 kg).

Habitat: Marine; coastal and inner continental shelf to depths of 160 m. Most abundant in less than 60 m depth.

Pseudophycis barbata

Untrimmed. Protein fingerprint p. 363.

Fishery: Caught in small quantities by Danish seines, gillnets and by lining inshore.

Remarks: The red cod (*Pseudophycis bachus*) is the main species marketed. It was an important commercial fish off southern Australia in the 1800s but is less so now. The flesh is very soft with little taste.

John dory

Zeus faber

Previous name: kuparu

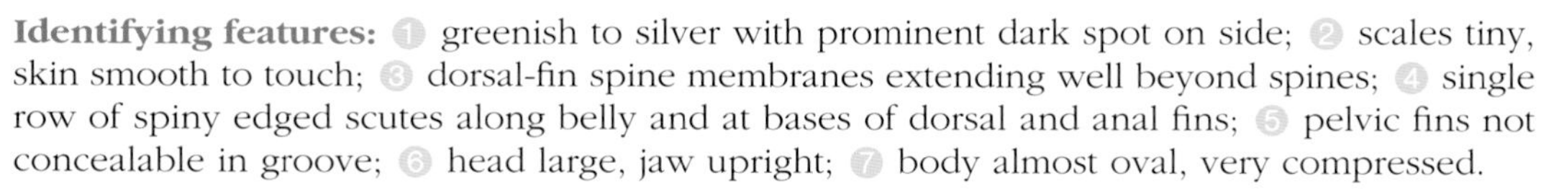

Identifying features: ① greenish to silver with prominent dark spot on side; ② scales tiny, skin smooth to touch; ③ dorsal-fin spine membranes extending well beyond spines; ④ single row of spiny edged scutes along belly and at bases of dorsal and anal fins; ⑤ pelvic fins not concealable in groove; ⑥ head large, jaw upright; ⑦ body almost oval, very compressed.

Comparisons: Dories (Zeidae) are distinct from all other fishes, with the possible exception of their deepwater relatives the oreos (Oreosomatidae) from which they differ in having smoother, paler skin and a smaller eye. The John dory, with its dark fingerprint spot and long filamentous dorsal fin, is an unmistakable member of the group.

Fillet: Very deep, short, tapering sharply, extremely convex above, yellowish-white. Outside without central red muscle band; parallel EL, weakly converging HL; HS well below middle of fillet, EL closer to HS than dorsal margin; integument silvery grey. Inside flesh fine; belly flap usually absent; peritoneum translucent. Rarely skinned, skin medium, scale pockets small and barely detectable.

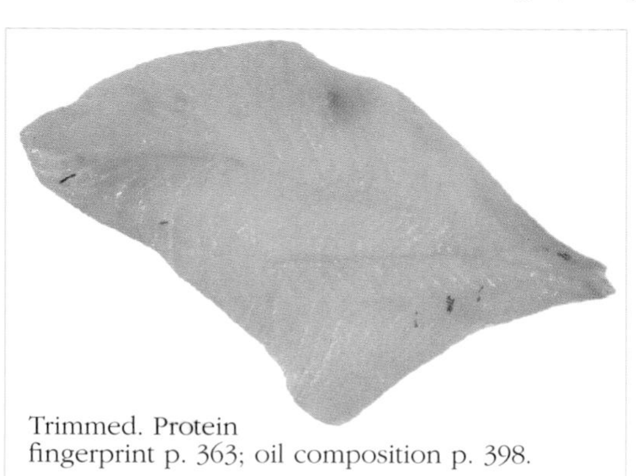

Trimmed. Protein fingerprint p. 363; oil composition p. 398.

Size: Reaches 75 cm and 3.5 kg (commonly 30–45 cm and 0.5–1.5 kg).

Habitat: Marine; demersal on the continental shelf from estuaries to depths of 200 m.

Fishery: Caught by trawlers and Danish seiners mainly in the south-east. Also taken in haul nets in bays and by recreational anglers.

Remarks: One of Australia's most highly priced and regarded foodfishes.

King dory

Cyttus traversi

Previous names: horsehead, lookdown dory, McCulloch's dory

Identifying features: 1 silver with no large dark spot on side; 2 scales small, skin rough to touch; 3 dorsal-fin spines short (shorter than longest soft rays); 4 slightly enlarged scales on belly resemble a zip, no bony spines at bases of dorsal and anal fins; 5 pelvic fins concealable in a groove; 6 head large, jaw directed vertically; 7 body almost round, very compressed.

Comparisons: Often caught with the mirror dory (*Zenopsis nebulosus*, p. 90) but has a different dorsal-fin configuration and is scaled. Closer in many features to the silver dory (*C. australis*, p. 91) but is silver (rather than pinkish), deeper and heavier bodied, and has a much shorter first dorsal fin.

Fillet: Very deep, short, tapering sharply, extremely convex above, yellowish-white. Outside with feeble, broken central red muscle band (as flecks); parallel EL, weakly converging HL; HS well below middle of fillet, EL much closer to HS than dorsal margin; integument silvery. Inside flesh fine; belly flap usually absent; peritoneum silvery white to translucent. Rarely skinned, skin thin, scale pockets indistinct.

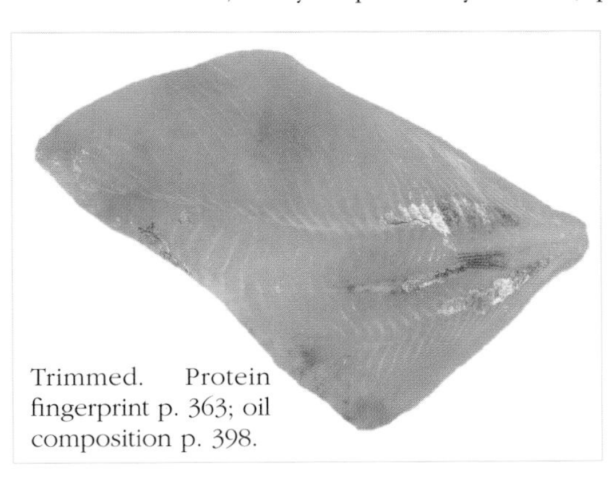

Trimmed. Protein fingerprint p. 363; oil composition p. 398.

Size: To 65 cm and 6 kg (commonly 40–55 cm and about 1.5–3.2 kg).

Habitat: Marine; demersal on the upper continental slope in depths of 200–1000 m but most common in 500–600 m.

Fishery: Taken mainly as bycatch of trawlers targeting blue grenadier (*Macruronus novaezelandiae*, p. 84) but also targeted.

Remarks: Commonly called 'lookdown dory', probably due to the unusually high placement of its eye in regard to its mouth.

Mirror dory

Zenopsis nebulosus

Previous name: silver dory

I

Identifying features: 1 silver with faint dark spot on side; 2 no scales (except on lateral line); 3 dorsal-fin spine membranes extending only slightly beyond spines; 4 single row of spiny edged scutes along belly and with greatly enlarged plates at bases of dorsal and anal fins; 5 pelvic fins not concealable in a groove; 6 head large, jaw almost upright; 7 body almost oval, very compressed.

Comparisons: Difficult to confuse but most closely resembles John dory (*Zeus faber*, p. 88). However, it differs in having a scaleless skin, a fainter spot on the side, larger bony plates at the bases of the vertical fins, and relatively shorter membranes in the first dorsal fin.

Fillet: Very deep, short, tapering sharply, extremely convex above, pale pinkish. Outside with feeble, discontinuous central red muscle band; parallel EL, converging HL; HS well below middle of fillet, EL much closer to HS than dorsal margin; integument silvery grey. Inside flesh fine; belly flap usually absent; peritoneum silvery white to translucent. Rarely skinned, skin thin, scale pockets absent.

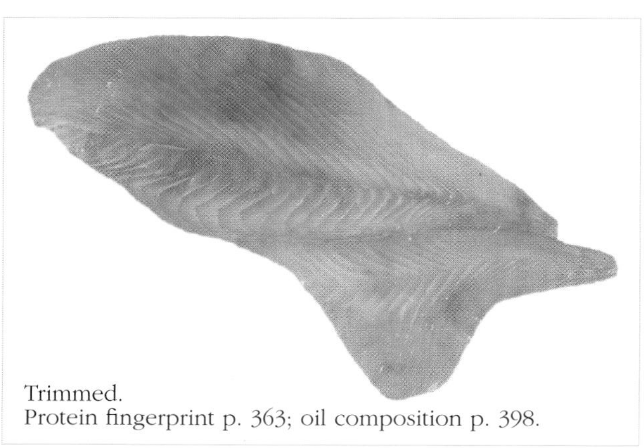

Trimmed.
Protein fingerprint p. 363; oil composition p. 398.

Size: To 70 cm and almost 3 kg (commonly 40–50 cm, 0.7–1.2 kg).

Habitat: Marine; demersal on the outer shelf and the upper slope in 50–800 m depth but most common in 200–500 m.

Fishery: Trawled in moderate quantities off the south-east, with smaller catches in the Great Australian Bight and off western Australia.

Remarks: As with the John dory, catches of this fine foodfish are presently regulated.

Silver dory

Cyttus australis

Previous names: red dory, sun dory

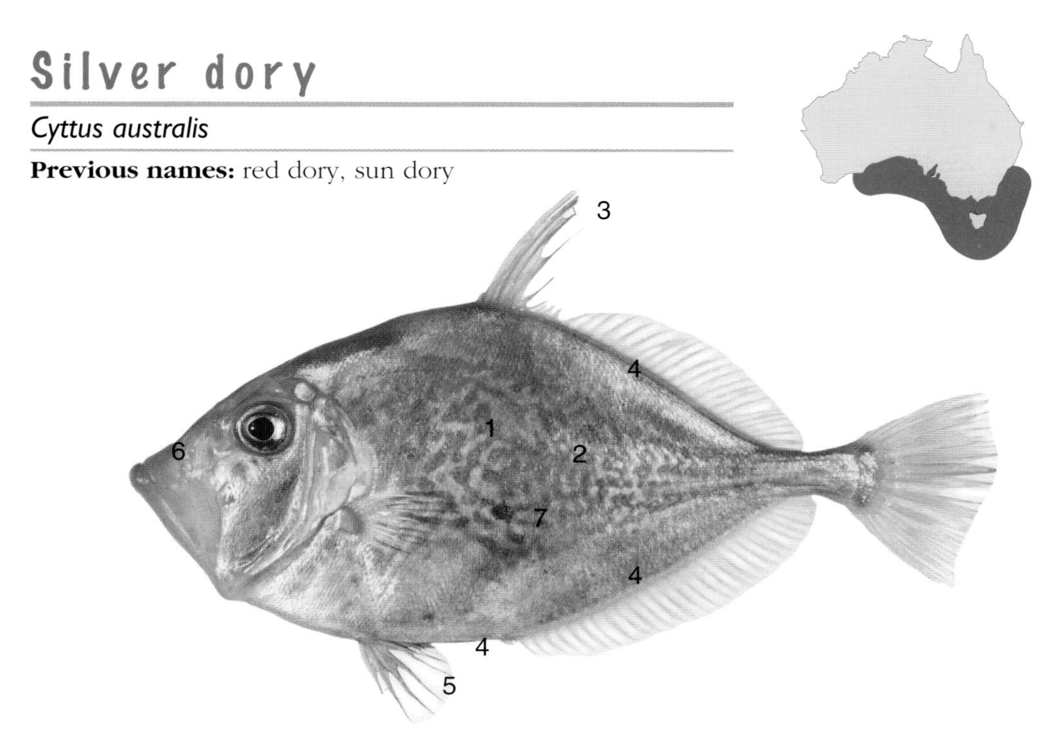

Identifying features: 1 sides silvery pink without large dark spot; 2 scales small, skin rough to touch; 3 dorsal-fin spines long (much longer than longest soft rays); 4 no enlarged scales on belly or bony spines at bases of dorsal and anal fins; 5 pelvic fins concealable in a groove; 6 head large, jaw directed vertically; 7 body deep, very compressed.

Comparisons: Caught with the John dory (*Zeus faber*, p. 88) but the two are very different in appearance and unlikely to be confused. More similar to the king dory (*C. traversi*, p. 89) and a smaller non-commercial relative, the New Zealand dory (*C. novaezealandiae*), but unlike these species it has a more elongate body and a relatively much taller first dorsal fin.

Fillet: Very deep, short, tapering sharply, extremely convex above, yellowish-white. Outside without central red muscle band; parallel EL, weakly converging HL; HS well below middle of fillet, EL much closer to HS than dorsal margin; integument silvery grey. Inside flesh fine; belly flap usually absent; peritoneum silvery white to translucent. Rarely skinned, skin thin, scale pockets small and defined.

Size: To 50 cm and 1.5 kg (commonly 30–43 cm and about 0.4–0.9 kg).

Habitat: Marine; demersal on the continental shelf and upper slope in 10–350 m depth but most common in 50–200 m.

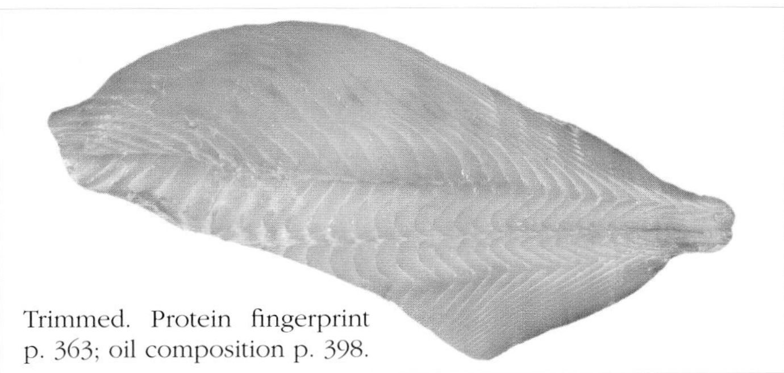
Trimmed. Protein fingerprint p. 363; oil composition p. 398.

Fishery: Taken mainly as bycatch of inshore trawl and Danish seine fisheries in the south-east.

Remarks: Fillets sometimes confused in markets with the more highly regarded John dory but its flesh, while tasty, is cheaper and has a shorter shelf life.

Luderick

Girella tricuspidata

Previous names: black bream, blackfish, darkie, nigger

Identifying features: ① greyish to greyish-brown with very thin darker bands; ② mouth small with incisor teeth at front forming cutting edge; ③ about 48–51 scales in the lateral line; ④ single, un-notched dorsal fin with 14–16 spines, 11–12 soft rays; ⑤ pelvic fin well behind pectoral fin; ⑥ small hind bone of upper jaw (maxilla) concealed by bony area below eye when mouth closed.

Comparisons: Distinguished by a concealed maxillary bone of the upper jaw and front incisors forming a distinct cutting edge. Resembles breams (Sparidae) in shape but has a smaller mouth and chisel-like teeth in the front of the jaw rather than broad canines and molars. The zebra fish (*G. zebra*) has much broader banding and much smaller scales (72–80 in lateral line).

Fillet: Deep, rather elongate, almost bottle-shaped, white with brownish tinge. Outside with pronounced, continuous central red muscle band; parallel EL, weakly converging HL; HS along middle of fillet; integument translucent above, silvery white below. Inside flesh coarse; belly flap rarely present; peritoneum black. Usually skinned, skin thick, scale pockets moderate and defined.

Size: To 71 cm and 4 kg (commonly 25–45 cm and about 0.4–1.2 kg).

Habitat: Marine; coastal from estuaries and bays to exposed rocky coasts.

Fishery: Caught throughout its range using a variety of methods such as beach seines, gillnets, haul nets and pound nets. Sometimes targeted, but mostly as bycatch of the bream fishery. Popular angling species.

Untrimmed. Protein fingerprint p. 363; oil composition p. 398.

Remarks: Belongs to a group of fishes with a generalised perch-like appearance known as 'drummers' (Kyphosidae). Should be processed soon after capture to prevent spoilage, but eating qualities are good when fresh.

Sweep

Scorpididae

Previous names: banded sweep, mado sweep, sea sweep, silver sweep

Scorpis lineolatus

Identifying features: 1 greyish to bluish, plain or with broad darker bands; 2 mouth small with teeth in broad bands, outer teeth enlarged with some curved inward; 3 lateral line with 90 or more scales; 4 single dorsal fin with 9–11 spines, 16–28 soft rays; 5 pelvic fin behind pectoral fin; 6 small hind bone of upper jaw (maxilla) not concealed by bony area below eye when mouth closed.

Comparisons: Sweeps (Scorpididae) are closely related to drummers (Kyphosidae). Some resemble a large group of small aquarium fishes known as butterfly fishes (Chaetodontidae) in form and colour pattern. Two main species are marketed: sea sweep (*Scorpis aequipinnis*) and silver sweep (*S. lineolatus*). Sea sweep have a faint but distinctive banded pattern and a yellow chin, whereas silver sweep are plain-coloured. The related mado (*Atypichthys strigatus*) has distinctive dark stripes.

Fillet: *S. lineolatus* deep, rather elongate, almost bottle-shaped, milky white. Outside with pronounced, continuous central red muscle band; convex EL, converging HL; HS along middle of fillet; integument white. Inside flesh fine; belly flap rarely present; peritoneum black. Usually skinned, skin thin, scale pockets small and irregular.

Size: To about 60 cm and 3.5 kg (commonly 20–35 cm and about 0.2–1.0 kg).

Untrimmed. Protein fingerprint p. 363; oil composition p. 398.

Habitat: Marine; coastal, occur in the vicinity of reefs, often in rough water.

Fishery: Sea sweep, the largest of the group, is caught off the southern coast using nets and lines. Most of the silver sweep marketed are caught off New South Wales using traps and purse seines. Popular angling species.

Remarks: The firm flesh provides very good eating.

Conger eel

Conger verreauxi & *C. wilsoni*

Previous names: southern conger eel, northern conger eel

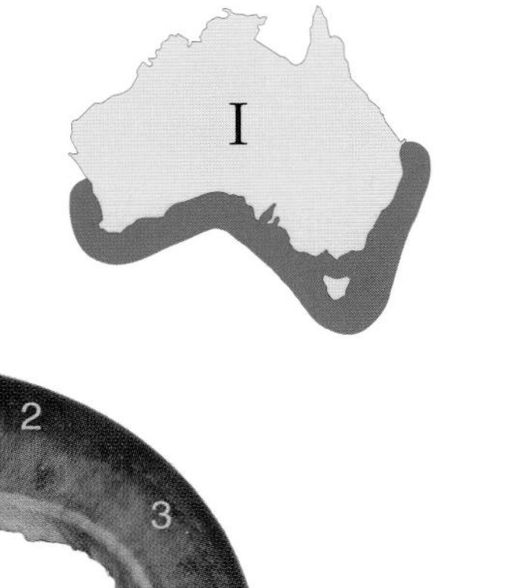

Conger verreauxi

Identifying features: (1) dorsal-fin origin over tip of pectoral fin; (2) back uniform grey, brownish or bluish-black; (3) no scales; (4) lips fleshy; (5) body long and tubular; (6) gill openings partly in front of pectoral fins.

Comparisons: Differ from freshwater eels in lacking scales and having a very long dorsal fin. Two species reach local markets. A more robust, southern species (*C. verreauxi*) is most common off Tasmania and Victoria, whereas the more slender, northern conger (*C. wilsoni*) is found in warmer seas off New South Wales and Western Australia. They differ slightly in fin-ray counts and positions. Smaller relatives occur in deepwater offshore.

Fillet: *C. verreauxi* long, slender, barely tapering, yellowish-white. Outside with intermediate, discontinuous red muscle band, parallel EL, converging HL; HS well above midline of fillet anteriorly; integument white. Inside flesh medium; belly flap indistinct; peritoneum evident, silvery with melanophores. Always skinned.

Size: To 200 cm and 20 kg (typically to 150 cm and 6.0 kg).

Habitat: Marine; mainly rocky habitats inshore where they live in caves and crevices during the day moving out to feed at night. Common reef dwellers, mostly solitary.

Fishery: Secondary species caught in lobster pots, traps and by lining, mostly in the south. Little recreational use but often seen by divers and sometimes speared.

Remarks: Have a fearsome reputation but will not bite unless provoked. Extremely strong and difficult to remove from pots. Also found off New Zealand in similar habitats. Flesh is firm, bordering on rubbery. Marketing research is required.

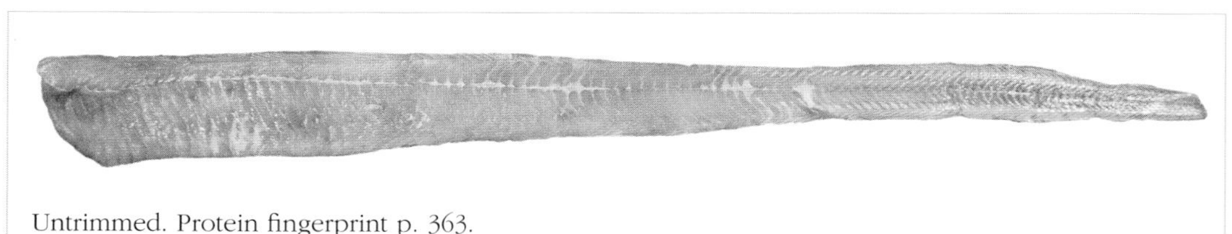

Untrimmed. Protein fingerprint p. 363.

Longfin eel

Anguilla reinhardtii

Previous names: freshwater eel, river eel, spotted eel

Identifying features: ① dorsal-fin origin well forward of anus; ② back yellowish-brown with darker spots and blotches; ③ scales tiny and embedded; ④ lips fleshy; ⑤ body long and tubular; ⑥ gill openings partly in front of pectoral fins.

Comparisons: Similar to the shortfin eel (*A. australis*, p. 96) in general form but distinguished by its longer dorsal fin. Marketed fish (taken from freshwater) have a distinctive mottled colour pattern that is lost for a uniform silvery colouration during their sea migration to spawn. The marine conger eels (*Conger* species, p. 94) have a much longer dorsal fin.

Fillet: Long, slender, barely tapering, pale pinkish. Outside with feeble, continuous intermediate red muscle band, parallel EL, parallel HL; HS through middle of fillet; integument translucent. Inside flesh fine; belly flap indistinct; peritoneum evident, silvery white to translucent. Always skinned.

Size: To 160 cm and 16.3 kg (typically to 100 cm and 3.0 kg).

Habitat: Freshwater; prefers rivers to still habitats. Adults migrate in the open sea to spawning grounds.

Fishery: Caught mainly using traps in New South Wales and Queensland as fyke nets are largely prohibited.

Remarks: Largest of the freshwater eels. A voracious feeder, it has been observed feeding on a variety of small aquatic fauna including ducks. Adults will move over farm walls during the spawning run. Thought to spawn in the northern Coral Sea, near New Caledonia. Young eels take almost a year to drift back to the rivers of southeastern Australia. The flesh is considered to be more oily than that of the shortfin eel.

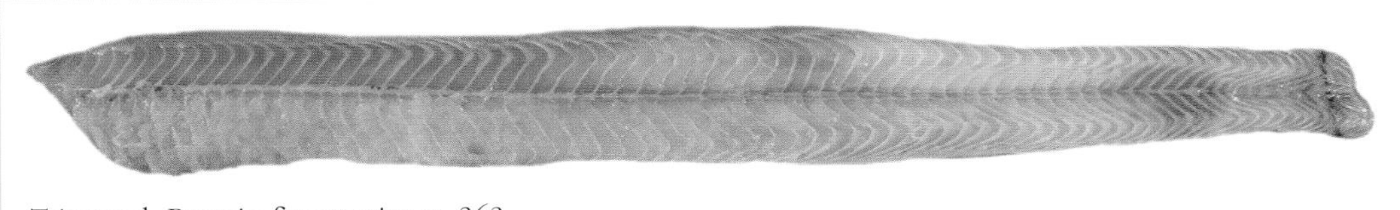

Trimmed. Protein fingerprint p. 363.

Shortfin eel

Anguilla australis

Previous names: freshwater eel, river eel, silver eel

I

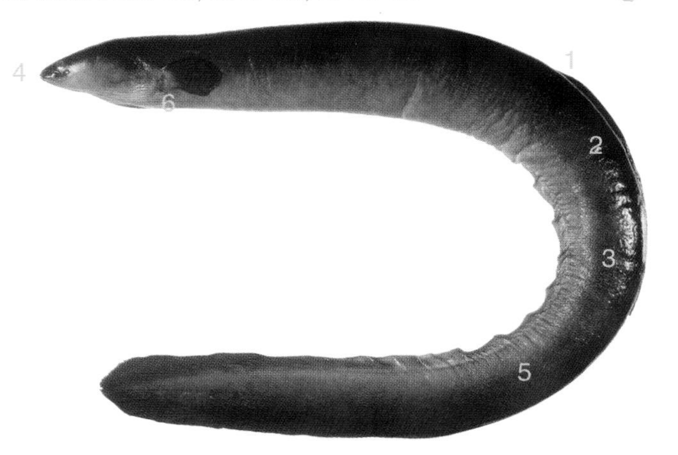

Identifying features: ① dorsal-fin origin almost over anus; ② back uniform olive green in colour; ③ scales tiny, embedded; ④ lips fleshy; ⑤ body long and tubular; ⑥ gill openings partly in front of pectoral fins.

Comparisons: Eels are distinct among fishes and can usually be identified from cutlets as well as whole fish by their skin texture and colour. Two freshwater eels reach the marketplace and can be readily distinguished by colour and the origin of the dorsal fin. The conger eels have similarly bulky bodies but lack scales and their dorsal fin starts much further forward. Several tropical and deepwater eels also occur in Australian seas but these differ in the features above.

Fillet: Long, slender, barely tapering, yellowish-white. Outside with feeble, discontinuous red muscle band, weakly converging EL, weakly converging HL; HS just above middle of fillet; integument black above, silvery grey below. Inside flesh medium; belly flap indistinct; peritoneum evident, silvery white to translucent with melanophores. Always skinned.

Size: To 110 cm and 6.8 kg (typically to 50 cm and up to 0.5 kg).

Habitat: Freshwater; prefers still water habitats such as coastal swamps, lagoons, farm dams and river backeddies. Also penetrates to remote waterways by overland migrations.

Fishery: Commenced in 1914 in Victoria with recent national catches exceeding 200 tonnes annually. The main eel species taken in Victoria and Tasmania.

Remarks: Migratory species that moves to the sea to spawn. Females reach a larger size than males with migrating adult females exceeding 48 cm and males 34 cm. Adults spend an average of 14 years in freshwater and are thought to die after spawning. Young eels (elvers) migrate back into streams. When water temperatures drop below 10°C they can hibernate without food for almost a year. Shortfin eel flesh is considered a delicacy and is ideal for smoking.

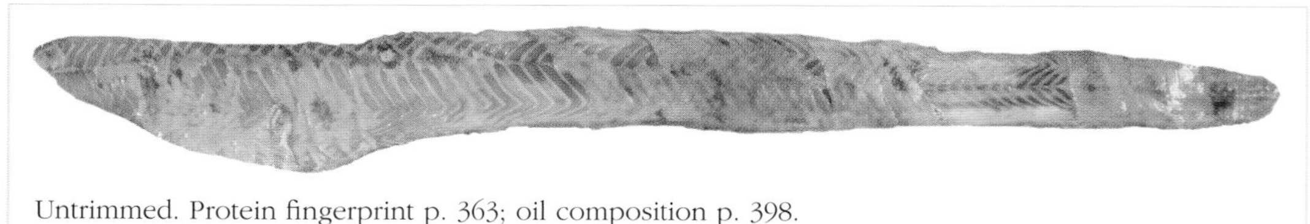

Untrimmed. Protein fingerprint p. 363; oil composition p. 398.

Emperor

Lethrinus species

Previous names: emperor bream, sweetlip, sweetlip emperor

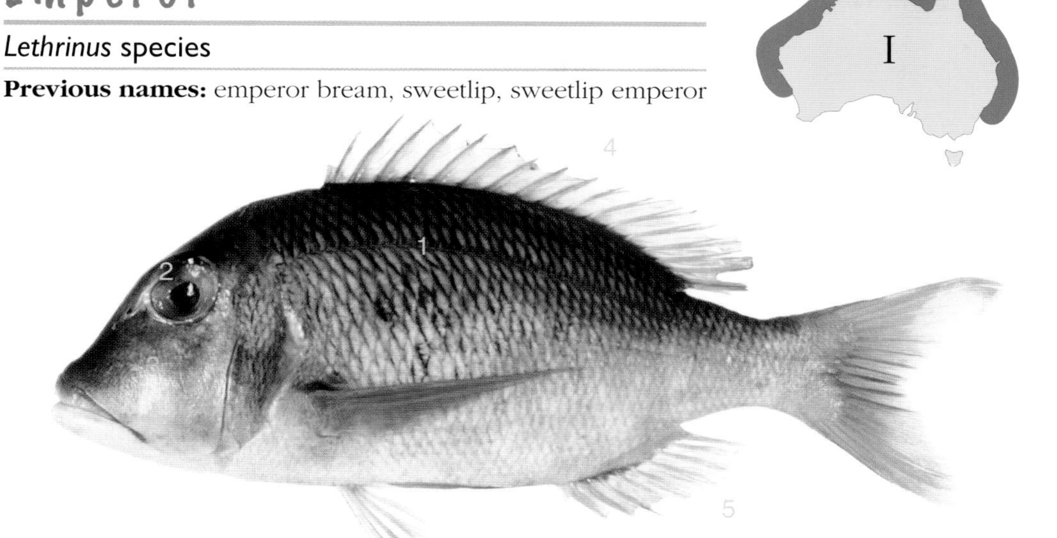

Lethrinus atkinsoni

Identifying features: 1 body moderate to deep and compressed slightly; 2 eye near top of head, snout long and pointed; 3 cheek scaleless; 4 1 continuous dorsal fin with 10 spines, 9 soft rays; 5 anal fin with 3 spines, 8–9 (usually 8) soft rays.

Comparisons: Distinctive group of perch-like fishes whose close relatives, the seabreams (subfamily Monotaxinae, p. 102), have less pointed snouts and scales on the cheek. Seaperches (*Lutjanus* species) are similar to emperors in body form but also have cheek scales.

Fillet: *L.* sp. rather deep, short to elongate, almost bottle-shaped, tapering rapidly to broad peduncle, often strongly convex above, white to pinkish or brown. Outside with intermediate or pronounced, continuous central red muscle band; relatively wide myomeres; integument usually pale. Inside flesh medium; peritoneum pale. Usually skinned, skin medium to thick, scale pockets moderate to large and defined.

Size: To about 100 cm and at least 8 kg (commonly 30–50 cm and 0.5–2.0 kg).

Habitat: Marine; broadly distributed over the continental shelf, from inshore seagrasses to coral reefs and lagoons, to depths of 220 m.

Fishery: Caught mainly by trawls, traps, and lines off northern Australia.

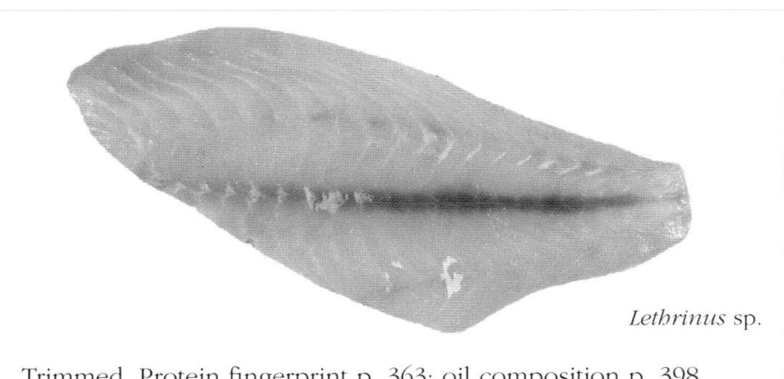

Lethrinus sp.

Trimmed. Protein fingerprint p. 363; oil composition p. 398.

Remarks: Eighteen species in Australian seas and most are marketable. All can be sold as 'emperor' but major commercial species have specific marketing names. The name 'sweetlip' has been used extensively in parts of Queensland. The flesh is considered excellent eating but a few species have a distinct taste of iodine.

Grass emperor

Lethrinus laticaudis

Previous names: blueline emperor, brown sweetlip, coral bream, emperor, grass sweetlip, snapper bream

Identifying features: 1 body very deep; 2 sides greenish or brownish, lacking bluish markings; 3 3–4 very short, blue lines extending beneath and forward of eye; 4 upper ray of pectoral fin often partly blue; 5 cheek scaleless; 6 1 continuous dorsal fin with 10 spines, 9 soft rays; 7 anal fin with 3 spines, 8 soft rays.

Comparisons: Deepest bodied and most thickly set of the emperors, it most closely resembles the spangled emperors (*L. nebulosus* and *L.* sp., p. 103). Thin bluish lines around the eyes are very short whereas the lines on spangled emperors, when they exist, extend further down the snout towards the mouth. It also lacks bluish lines along the sides.

Fillet: Deep, rather elongate, tapering rapidly near peduncle, very convex anteriorly, yellowish-white. Outside with intermediate, continuous central red muscle band; convex EL, converging HL; HS along middle of fillet; integument greyish above, whitish below. Inside flesh medium; belly flap rarely present; peritoneum silvery white. Usually skinned, skin medium, scale pockets moderate and well defined.

Size: To 70 cm and 5.6 kg (commonly 30–45 cm and 0.5–1.8 kg).

Habitat: Marine; juveniles around mangroves and seagrasses, adults mostly on coral reefs.

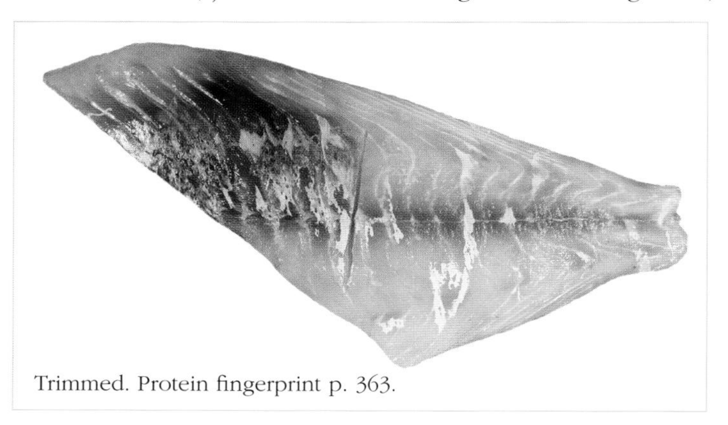

Trimmed. Protein fingerprint p. 363.

Fishery: Caught throughout northern Australia, mainly by lines, traps and trawls. Greatest quantities taken off eastern Australia and the Northern Territory.

Remarks: Popular angling species in Queensland and the Gascoyne region of Western Australia. Feeds on small crustaceans and fishes. Good foodfish, usually marketed fresh.

Longnose emperor

Lethrinus olivaceus

Previous names: emperor, longnose sweetlip

Identifying features: ① body relatively elongate; ② snout very long and pointed; ③ side greenish or olive without reddish or blue markings; ④ no scales on undersurface of pectoral-fin base; ⑤ cheek scaleless; ⑥ 1 continuous dorsal fin with 10 spines, 9 soft rays; ⑦ anal fin with 3 spines, 8 soft rays.

Comparisons: The most slender emperor species with an elongate body and pointed head. Unlike other main commercial emperors it lacks bright bluish and reddish markings. It is comparatively drab, being plain olive to purplish-grey, sometimes with darker blotches and mottling. Brown and whitish stripes down the snout are diagnostic.

Fillet: Moderately deep, rather elongate, tapering rapidly near peduncle, convex anteriorly, yellowish-white. Outside with intermediate, continuous central red muscle band; convex EL, converging HL; HS along middle of fillet; integument translucent above, silvery white below. Inside flesh medium; belly flap rarely present; peritoneum silvery white to translucent. Usually skinned, skin medium, scale pockets moderate and defined.

Size: To 100 cm and 8.2 kg (commonly 30–70 cm and 0.5–3.5 kg).

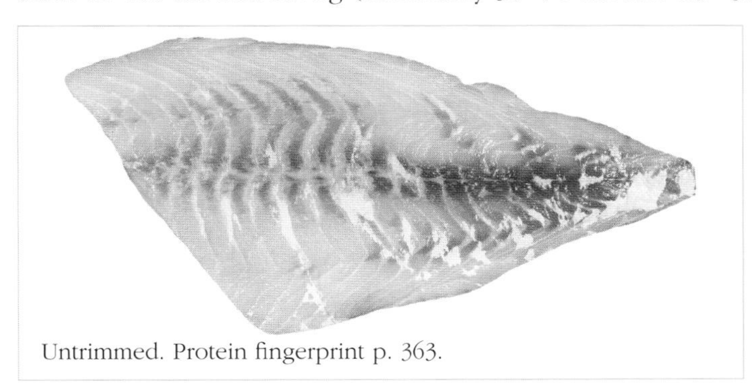

Untrimmed. Protein fingerprint p. 363.

Habitat: Marine; inshore sandy coasts, lagoons and reef slopes to depths of at least 185 m.

Fishery: Caught in small quantities mainly by lining.

Remarks: Eats fishes, octopuses and small crustaceans. Schools with fish of similar size. Excellent sport fish with high food qualities.

Redspot emperor

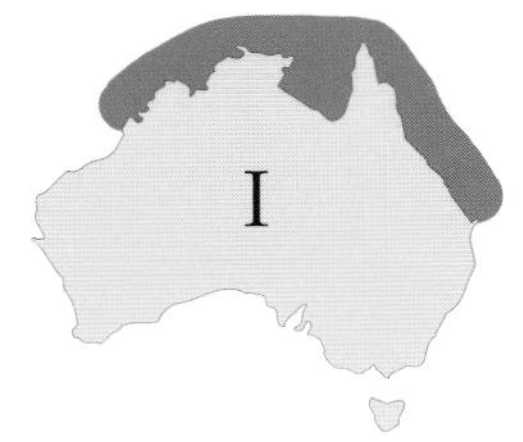

Lethrinus lentjan

Previous names: emperor, pinkear sweetlip, purplehead emperor, red-ear emperor

Identifying features: 1 body moderately deep; 2 bright red markings confined to rear edge of operculum (sometimes at pectoral-fin base); 3 mouth short; 4 sides typically yellowish or greyish with pale spots on scales above lateral line; 5 cheek scaleless; 6 1 continuous dorsal fin with 10 spines, 9 soft rays; 7 anal fin with 3 spines, 8 soft rays.

Comparisons: Several emperor species have bright red markings. The redthroat emperor (*L. miniatus*, p. 101) has more extensive reddish markings on the fins, head and body but lacks a bright red blotch along the edge of the operculum. The scarlet-cheek emperor (*L. rubriopercu latus*), which has a more elongated and darker body, has a red blotch at the top of the operculum (rather than above the pectoral fin) but lacks white-spotted shoulder scales.

Fillet: Deep, rather elongate, tapering rapidly near peduncle, convex anteriorly, yellowish-white. Outside with intermediate, continuous central red muscle band; convex EL, converging HL; HS along or just below middle of fillet; integument silvery white. Inside flesh medium; belly flap rarely present; peritoneum silvery translucent. Usually skinned, skin thick, scale pockets moderate and well defined.

Size: To 50 cm and about 1.8 kg (commonly marketed at 28–36 cm and 0.4–0.8 kg).

Habitat: Marine; continental shelf over sandy bottoms in coastal areas, deep lagoons, and near coral reefs to 50 m depth.

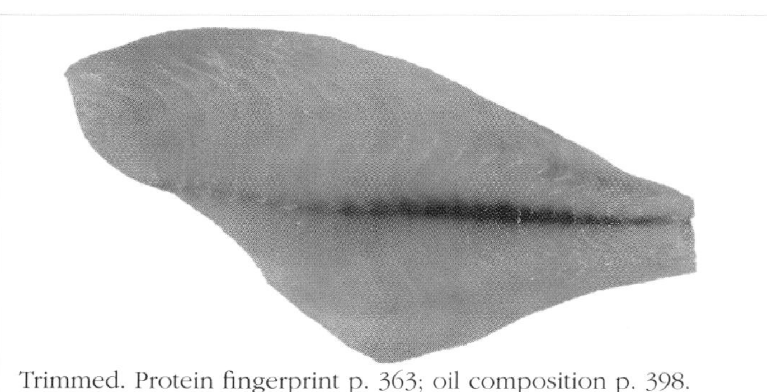

Trimmed. Protein fingerprint p. 363; oil composition p. 398.

Fishery: Dominant emperor taken in the northern demersal trawl fishery and among the two main species in Queensland's Great Barrier Reef line-fishery. Smaller quantities taken by line and trap.

Remarks: Small species marketed in quantity. The flesh is highly regarded.

Redthroat emperor

Lethrinus miniatus

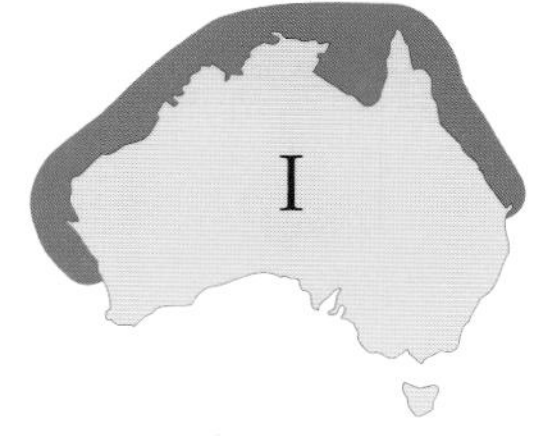

Previous names: emperor, lipper, redthroat, sweetlip emperor, tricky snapper, trumpeter

Identifying features: 1 body moderately deep; 2 bright reddish markings usually on eyeball, lips, bases of pectoral and pelvic fins, and membranes between dorsal-fin spines; 3 mouth long; 4 sides variable silvery pink to reddish-brown; 5 cheek scaleless; 6 1 continuous dorsal fin with 10 spines, 9 soft rays; 7 anal fin with 3 spines, 8 soft rays.

Comparisons: Lacks the characteristic red margin along the edge of the operculum that typifies the redspot emperor (*L. lentjan*, p. 100) and has a much longer mouth. No other Australian emperor has such an extensive array of bright red markings on the trunk and fins combined with a bright red throat.

Fillet: Moderately deep, rather elongate, upper profile almost straight, white. Outside with feeble, continuous central red muscle band; convex EL, converging HL; HS along middle of fillet; integument white. Inside flesh medium; belly flap rarely present; peritoneum silvery white to translucent. Usually skinned, skin thick, scale pockets moderate and well defined.

Size: Locally to 55 cm and 3.0 kg (commonly 30–50 cm and 0.5–2.6 kg).

Habitat: Marine; continental shelf and coral atolls in depths of 5–30 m.

Fishery: Caught mainly by traps and lines off northern Australia where it is more common off central eastern and western Australia. Also taken off Norfolk Island and by recreational fishers.

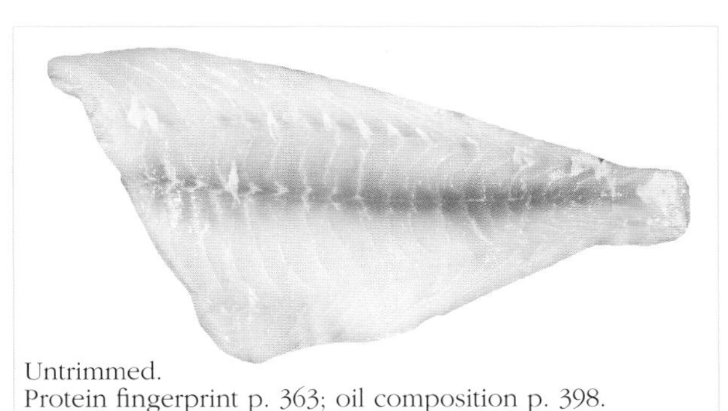

Untrimmed.
Protein fingerprint p. 363; oil composition p. 398.

Remarks: Found over coral reefs during the day, moving in small schools to deeper sand at night to feed. Exists in several colour forms that differ in the extent of red markings. Highly regarded by anglers due to its fighting qualities. The flesh has been compared in quality with coral trout (*Plectropomus* species, p. 217).

Seabream

Gymnocranius & *Monotaxis* species

Previous names: bigeye bream, coral bream, Robinson's seabream

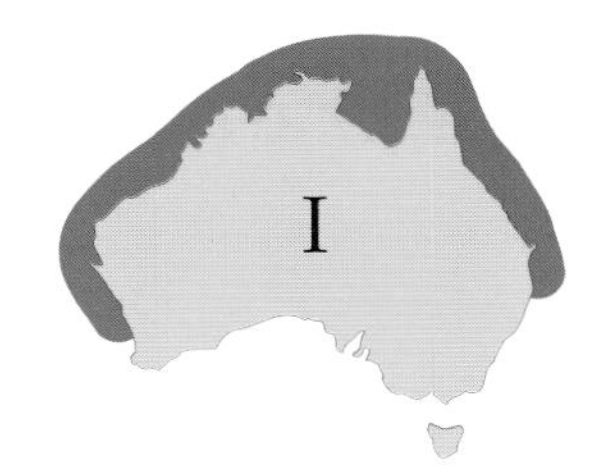

Gymnocranius grandoculis

Identifying features: ① body moderate to deep and compressed slightly; ② eye near top of head, snout relatively short and angular; ③ cheek with at least 3 rows of scales; ④ 1 continuous dorsal fin with 10 spines, 10 soft rays; ⑤ anal fin with 3 spines, 9–10 soft rays.

Comparisons: Probably more similar in appearance to breams (Sparidae) than to emperors (*Lethrinus* species) compared to which their upper body is more rounded and they have scales on the cheek (absent in emperors). No other perch-like fishes marketed have an identical combination of anal and dorsal fin counts.

Fillet: *G.* sp. moderately deep, rather elongate, upper profile almost straight, yellowish-white. Outside without red muscle band; weakly convex EL, converging HL; HS along middle of fillet; integument white, often with silvery patches. Inside flesh medium; belly flap rarely present; peritoneum white. Usually skinned, skin medium, scale pockets moderate and well defined.

Size: To 80 cm and at least 5 kg (commonly 30–55 cm and 0.8–2.4 kg).

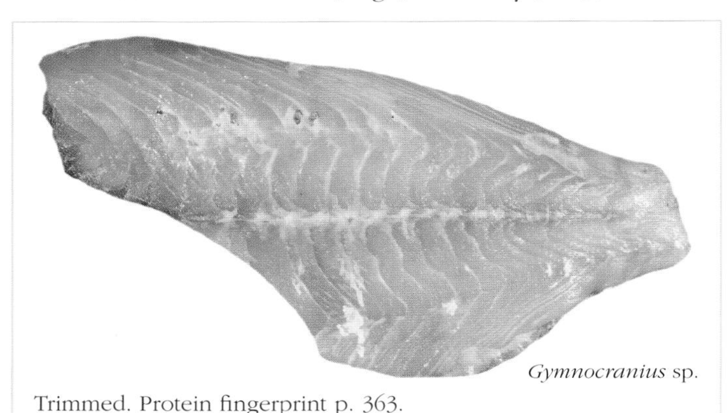

Gymnocranius sp.

Trimmed. Protein fingerprint p. 363.

Habitat: Marine; offshore continental shelf near coral and rocky reefs, mainly in depths of 10–100 m.

Fishery: Caught using lines and by trawling off northern Australia.

Remarks: Seven species of seabream occur in Australian waters and most are marketed regularly. All are good eating but can smell of iodine in some locations and seasons.

Spangled emperor

Lethrinus nebulosus & *L.* sp.

Previous names: blueline emperor, emperor, greater spangled emperor, lesser spangled emperor, nor-west snapper

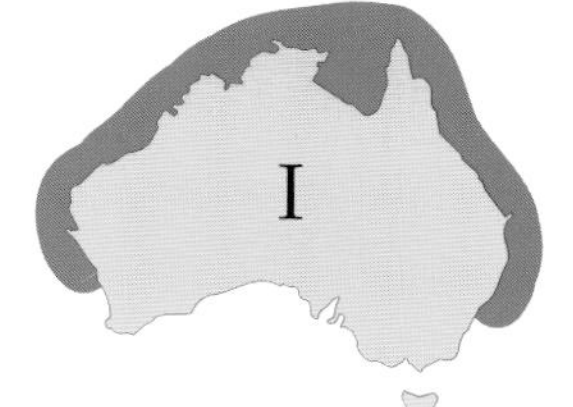

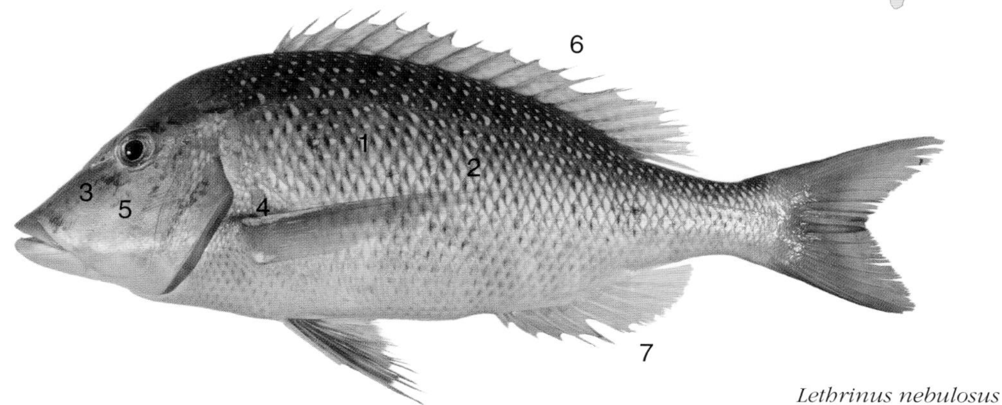

Lethrinus nebulosus

Identifying features: 1 body relatively deep; 2 sides yellowish in colour, with bluish scales forming horizontal lines (often pale); 3 spots or long bluish lines extending from eye to mouth; 4 upper ray of pectoral fin partly blue; 5 cheek scaleless; 6 1 continuous dorsal fin with 10 spines, 9 soft rays; 7 anal fin with 3 spines, 8 soft rays.

Comparisons: Consists of two commonly marketed species: spangled emperor (*L. nebulosus*) and lesser spangled emperor (*L.* sp.). The latter has bluish spots and a brown streak along the snout rather than well-defined bluish lines between the eyes and mouth. Both species resemble the grass emperor (*L. laticaudis*, p. 98) in appearance but have different head markings—lines on the snout that are either much longer or absent altogether rather than being confined to the close vicinity of the eye. The grass emperor also lacks bluish markings on the trunk.

Fillet: *L. nebulosus* moderately deep, rather elongate, upper profile convex, yellowish-white. Outside with pronounced, continuous central red muscle band; convex EL, converging HL; HS along middle of fillet; integument white. Inside flesh medium; belly flap rarely present; peritoneum silvery translucent. Usually skinned, skin thick, scale pockets moderate and defined.

Size: To 93 cm and 10 kg (commonly 30–60 cm and 0.8–40 kg).

Habitat: Marine; inshore continental shelf, from seagrasses to coral reefs and lagoons, to 75 m.

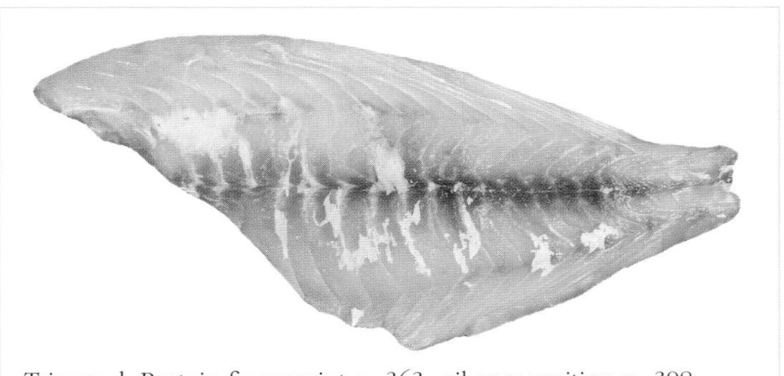

Trimmed. Protein fingerprint p. 363; oil composition p. 398.

Fishery: Caught using lines, traps and trawls off northern Australia.

Remarks: *Lethrinus* sp. has been referred to as *L. choerhynchus* or *L. punctulatus* in some recent literature. Possibly the most sought-after members of the group. Very important commercial and recreational fishes with excellent eating qualities.

Australian halibut

Psettodes erumei

Previous name: Queensland halibut

Identifying features: 1 mouth large with long sharp teeth; 2 eyes mainly on lefthand side of body (often also on right); 3 upper surface dark brown with 4 broad bands (often faint); 4 dorsal fin commencing well behind snout tip (behind level of eyes); 5 body extremely depressed; 6 caudal fin separated from dorsal and anal fins.

Comparisons: Australian halibut resembles its larger Northern Hemisphere relatives which are imported into Australia as frozen fillets. Distinctive in appearance from all other Australian flatfishes in having a greatly enlarged mouth, large teeth and relatively short dorsal fin.

Fillet: Deep, rather elongate, upper profile very convex, pale pinkish. Outside with intermediate, continuous red muscle band; convex EL, convex HL; HS along middle of fillet; integument white. Inside flesh medium; belly flap absent; peritoneum silvery translucent. Usually skinned, skin thick, scale pockets small and defined.

Size: To 64 cm and 3.2 kg (commonly 30–45 cm and 0.3–1.2 kg).

Habitat: Marine; demersal inshore on the inner continental shelf, including creeks and estuaries, on muddy bottoms to depths of at least 45 m.

Fishery: Minor commercial species taken off northern Australia mainly as bycatch by prawn trawlers.

Trimmed. Protein fingerprint p. 364.

Remarks: Adults of flatfish species usually either have both eyes on the lefthand side or righthand side of their body. This is used as an important means of distinguishing members of each family but varies more often in Australian halibut. Good quality foodfish but is often heavily parasitised.

Bay flounder

Ammotretis species

Previous names: flounder, longsnout flounder, sole, spotted flounder

Ammotretis rostratus

Identifying features: 1 mouth very small with short teeth; 2 eyes on righthand side of body; 3 snout hooked over mouth; 4 2 pelvic fins; 5 dorsal fin commencing at front tip of snout; 6 body extremely depressed; 7 caudal fin separated from dorsal and anal fins.

Comparisons: Related to the greenback flounder (*Rhombosolea tapirina*, p. 108), the only other commercial righthand flounder from temperate seas, but have a different head shape and 2 pelvic fins (rather than 1). The curved sole-like snout is unique among Australian flounders.

Fillet: *A. rostratus* moderately deep, rather elongate, upper profile slightly convex, off-white to yellowish with dark veins. Outside without red muscle band; weakly converging EL, weakly converging HL; HS along middle of fillet; integument light grey. Inside flesh fine; belly flap absent; peritoneum silvery white to translucent. Rarely skinned, skin medium, scale pockets small and defined.

Size: To 35 cm and 0.5 kg (commonly 25–33 cm and 0.2–0.4 kg).

Habitat: Marine; inshore, coastal, estuarine and inner continental shelf over mud and sand to depths of 80 m.

Fishery: Taken in moderate quantities by Danish seiners in shallow bays and by beach seines off beaches and in estuaries of southern Australia.

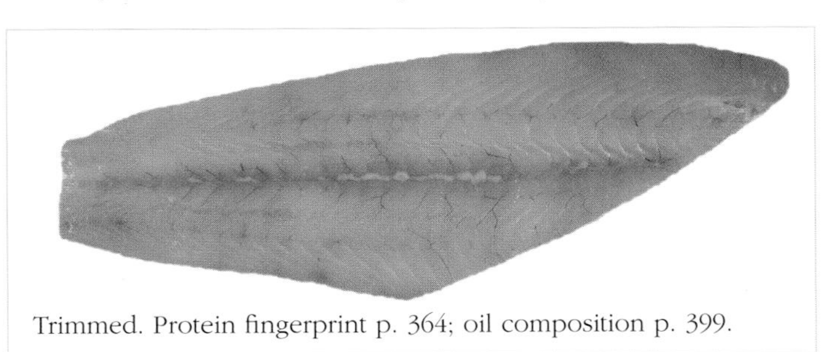

Trimmed. Protein fingerprint p. 364; oil composition p. 399.

Remarks: Two main species are marketed. The longsnout flounder (*A. rostratus*) is best distinguished from the spotted flounder (*A. lituratus*) by having fewer fin rays in the right pelvic fin (7 versus 10–13).

Flounder (page 1 of 2)

Bothidae & Pleuronectidae

Previous names: lefthand flounder, righthand flounder

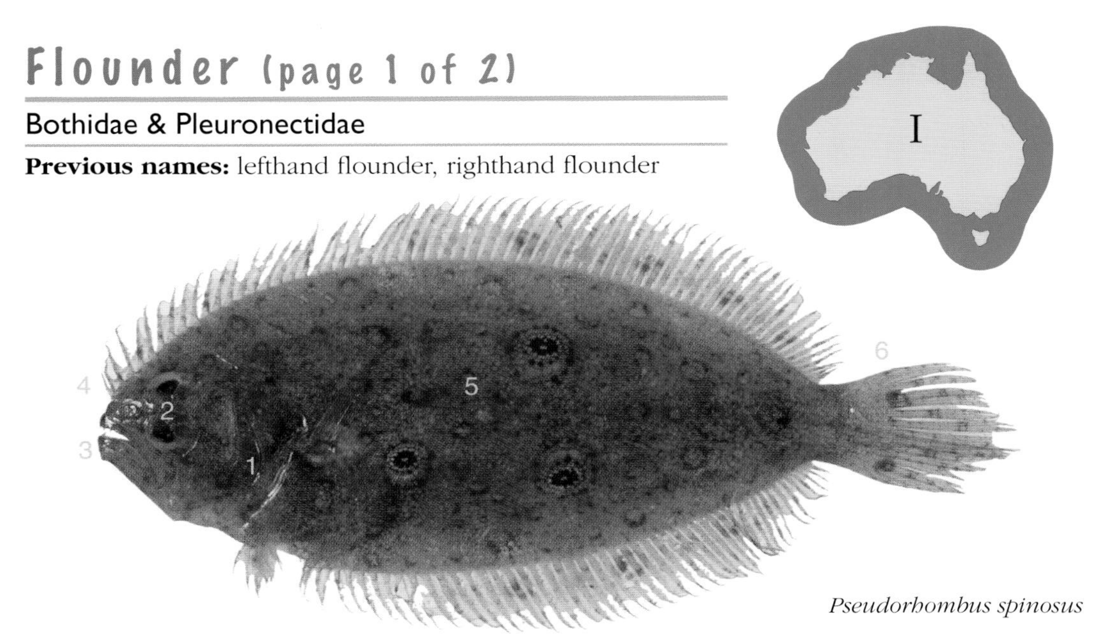

Pseudorhombus spinosus

Identifying features: ① preopercular margin evident (not concealed by thick skin); ② eyes on left or right side of body; ③ mouth small with short teeth; ④ dorsal fin commencing at eye-level or further forward; ⑤ body extremely depressed; ⑥ caudal fin always separated from dorsal and anal fins.

Comparisons: Includes two quite different families of flatfishes that are commercially important in temperate Australia. They differ from soles (Soleidae) and tongue soles (Cynoglossidae) in always having a separate tail (rather than the anal, caudal and dorsal fins mostly joined) and the edge of the preoperculum (ridge on mid-gill cover) is clearly visible (rather than concealed). The Australian halibut (*Psettodes erumei*, p. 104) has a much larger mouth and a more posteriorly located dorsal fin.

Fillet: *Pseudorhombus arsius* deep, rather elongate, upper profile very convex, greyish to white. Outside without red muscle band; convex EL, weakly converging HL; HS along middle of fillet; integument translucent. Inside flesh fine; belly flap absent; peritoneum silvery grey with melanophores. Sometimes skinned, skin medium, scale pockets small and indistinct.

Size: To at least 71 cm and 3.2 kg (commonly to 35 cm and less than 0.5 kg).

Habitat: Marine; bottom-dwelling, inshore to deep mid-continental slope from the tropics to subAntarctic seamounts.

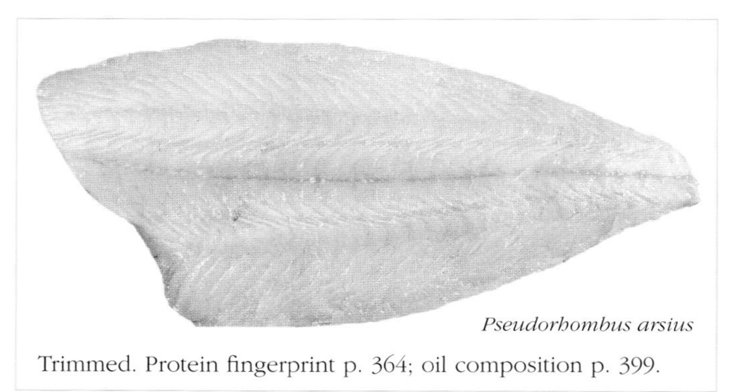

Pseudorhombus arsius

Trimmed. Protein fingerprint p. 364; oil composition p. 399.

Fishery: Bycatch throughout their range mainly from Danish seining and trawling but smaller quantities taken inshore using beach seines and gillnets. Important recreational species.

Remarks: Varied flatfish group. Also a default name for the other more frequently marketed flounders such as greenback flounder (*Rhombosolea tapirina*, p. 108).

Flounder (page 2 of 2)

Bothidae & Pleuronectidae

Pseudorhombus arsius

Remarks: Commonly called 'largetooth flounder' this species occurs on the continental shelf off all Australian states except Tasmania and Victoria in depths of 15–70 m. Compared with the following species, it has more enlarged canine teeth in the front of each jaw, a more anteriorly located dorsal-fin origin, and 2 obvious dark blotches on the upper surface. To about 50 cm and 1.2 kg. The taxonomy of *Pseudorhombus* is confused and further research is required.

Pseudorhombus jenynsii

Remarks: Commonly called 'smalltooth flounder' this species occurs along the southern continental shelf from southern Queensland to Fremantle (WA) from the coast to depths of 50 m. Compared with the preceding species, it has smaller teeth in the front of each jaw, a more posteriorly located dorsal-fin origin, and numerous, dark, gold-speckled blotches on the upper surface. Reaches at least 35 cm and about 0.7 kg. A similar flounder taken off northern Australia, and sometimes referred to as *P. jenynsii*, is probably a different species.

Neoachiropsetta milfordi

Remarks: Commonly called 'armless flounder' this species occurs widely in the subAntarctic, including off Tasmania and seamounts south to Macquarie Island, in depths of 800–1400 m. It has a variable but distinctive head shape and lacks pectoral fins. Taken as a bycatch of the Patagonian toothfish (*Dissostichus eleginoides*, p. 150) fishery. Reaches at least 71 cm and 3.2 kg.

Greenback flounder

Rhombosolea tapirina

Previous names: flounder, Melbourne flounder, southern flounder

I

Identifying features: (1) mouth very small with short teeth; (2) eyes on righthand side of body; (3) snout pointed rather than hooked over mouth; (4) 1 pelvic fin; (5) dorsal fin commencing just behind tip of snout; (6) body extremely depressed; (7) caudal fin separate from other fins.

Comparisons: Related to another group of righthand flounders, bay flounders (*Ammotretis* species, p. 105), it has a more pointed head and a single pelvic fin (rather than 2). The snout is long and fleshy but not hooked like that of bay flounders. Some other commercial flounders (*Arnoglossus* species) are a similar shape but their eyes are on the left side of the body.

Fillet: Deep, rather elongate, upper profile convex, off-white to greyish with dense, dark veins. Outside without red muscle band; parallel EL, weakly converging HL; HS along middle of fillet; integument greyish. Inside flesh fine; belly flap absent; peritoneum black. Rarely skinned, skin medium, scale pockets small and defined.

Size: To 38 cm and 0.6 kg (commonly 25–35 cm and 0.2–0.5 kg).

Habitat: Marine; bottom-dwelling in deep bays and estuaries from the coast to about 100 m.

Fishery: Caught using gillnets, spears, Danish seines and beach seines near the coast off southern Australia. Popular recreational fish. Excellent aquaculture potential.

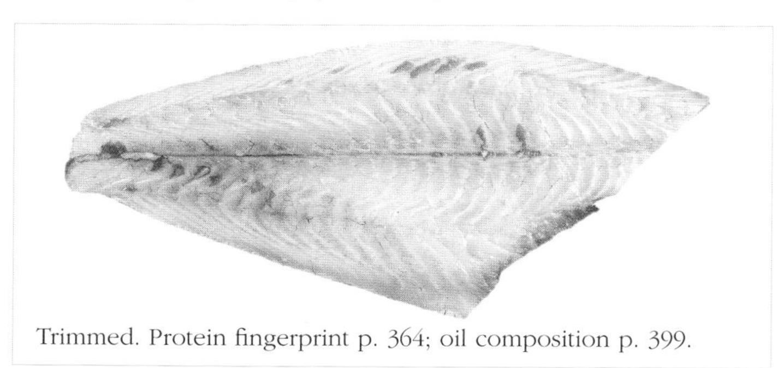

Trimmed. Protein fingerprint p. 364; oil composition p. 399.

Remarks: Only one species of *Rhombosolea* occurs in Australian waters. However, this and two other closely related species are imported from New Zealand. Among the most highly valued temperate foodfishes.

Sole

Cynoglossidae & Soleidae

Previous name: tongue sole

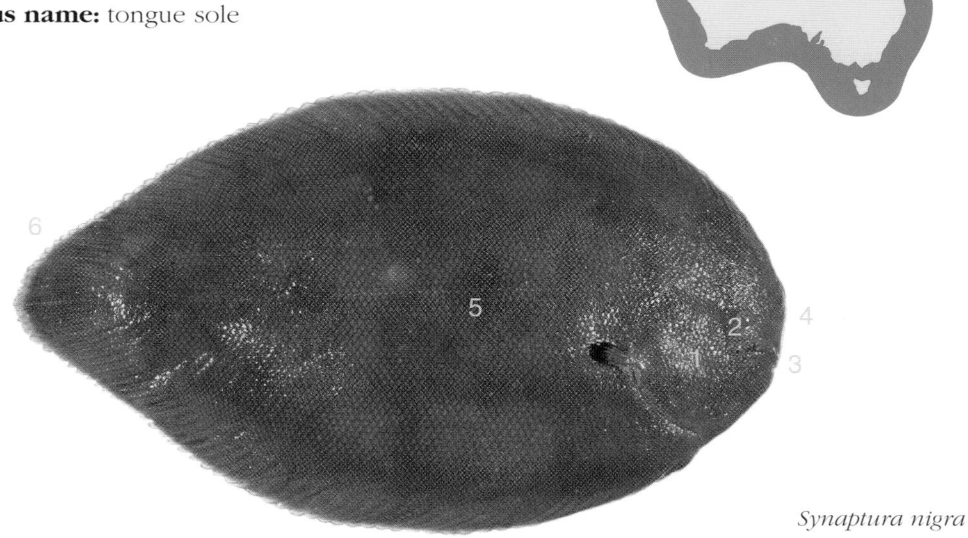

Synaptura nigra

Identifying features: ① margin of preoperculum concealed by thick skin; ② eyes on left or right side; ③ mouth small with short teeth; ④ dorsal fin commencing at eye-level or further forward; ⑤ body extremely depressed; ⑥ caudal fin mostly joined to dorsal and anal fins.

Comparisons: Includes the righthand soles (Soleidae) and lefthand tongue soles (Cynoglossidae). These differ from the flounders (Bothidae and Pleuronectidae) and halibuts (Psettodidae) in having the gill cover overlain with thick skin (preopercular margin concealed) and the anal, caudal and dorsal fins mostly joined (lacking a separate tail).

Fillet: *Synaptura nigra* deep, rather elongate, upper profile convex, pale pinkish without dark veins. Outside with pronounced, continuous red muscle band; weakly converging EL, weakly converging HL; HS along middle of fillet; integument translucent or silvery white. Inside flesh medium; belly flap absent; peritoneum silvery translucent. Usually skinned, skin thick, scale pockets small and well defined.

Size: To 35 cm and 0.5 kg (commonly 18–25 cm and 0.1–0.2 kg).

Habitat: Marine; bottom-dwelling from the coast to continental slope to at least 500 m depth.

Trimmed. Protein fingerprint p. 364.

Fishery: Bycatch of prawn trawls and beach seines and sold in small quantities.

Remarks: More than 50 sole species live in Australian seas, mostly in the tropics. Generally considered too small for local markets although this may change as fisheries move to greater use of the bycatch. Small soles are popular in Asia.

Bartail flathead

Platycephalus indicus

Previous name: flathead

Identifying features: 1 caudal fin with black and white bars and prominent central yellow blotch; 2 upper surface finely marbled with darker cross bars; 3 no greatly enlarged teeth at tip of upper jaw; 4 lower preopercular spine barely longer than upper; 5 large dark blotch inside operculum; 6 eyes their diameter or more apart; 7 no swim bladder; 8 body depressed and tapering with eyes on top of head.

Comparisons: Resembles the northern sand flathead (*P. arenarius*, p. 115) and the flagtail flathead (*P. endrachtensis*), which also have a marbled upper surface and a 'bartail'. Of these, only the flagtail flathead has a prominent yellow blotch on the caudal fin but it is located on the upper half of the fin (rather than centrally).

Fillet: Rather slender, tapering gently, pinkish with some dark veins. Outside with feeble, continuous, central red muscle band; weakly converging EL, weakly converging HL; HS along middle of fillet, dipping anteriorly; integument greyish above, silvery white below. Inside flesh medium; belly flap usually absent; peritoneum silvery translucent. Usually skinned, skin medium, scale pockets small and defined.

Size: To 100 cm and at least 4 kg (commonly 40–60 cm and 0.5–1.6 kg).

Habitat: Marine; inshore continental shelf on muddy and sandy bottoms to depths of at least 30 m. Juveniles are reported to be common in estuaries and may enter freshwater.

Fishery: Taken by handlines, seines and in small quantities as bycatch of inshore demersal trawlers.

Untrimmed. Protein fingerprint p. 364.

Remarks: Large, widely distributed flathead of inshore tropical waters. Excellent table-fish when young but tend to lack flavour when large.

Blue-spotted flathead

Platycephalus caeruleopunctatus

Previous names: longnose flathead, red-spotted flathead, sand flathead

Identifying features: 1 caudal-fin margin with 3 short, dark stripes above single dark blotch; 2 upper surface brownish with pale blue or red spots; 3 no greatly enlarged teeth at tip of upper jaw; 4 lower preopercular spine slightly longer than upper; 5 large dark blotch inside operculum; 6 several pairs of bony ridges on head behind eye; 7 no swim bladder; 8 body depressed and tapering with eyes on top of head.

Comparisons: Similar to sand flathead (*P. bassensis*, p. 117) in body form and in lacking greatly enlarged canines in the jaws, but has dark stripes and a blotch on the caudal fin (rather than 1 or 2 blotches) and several pairs of bony ridges behind the eye (rather than single pair). The blue-spotted colour pattern is distinctive but some other flatheads also have bluish spots and this can cause confusion. The southern flathead (*P. speculator*, p. 118) has smaller dark markings near the tail edge and lacks a dark blotch on the undersurface of the operculum.

Fillet: Moderately slender, pronounced taper, yellowish-white with prominent dark veins. Outside with moderate, continuous, central red muscle band; weakly converging EL, parallel HL; HS along middle of fillet, dipping anteriorly; integument greyish above, silvery white below. Inside flesh fine; belly flap usually absent; peritoneum translucent white to black. Usually skinned, skin medium, scale pockets small and defined.

Size: To at least 60 cm and 1.5 kg (commonly 35–45 cm and 0.4–0.7 kg).

Habitat: Marine; demersal on soft substrates of the mid-continental shelf mainly in 50–90 m.

Fishery: Significant bycatch of prawn and fish trawls off New South Wales. Often taken by recreational anglers offshore, including those drift-fishing for snapper (*Pagrus auratus*, p. 73).

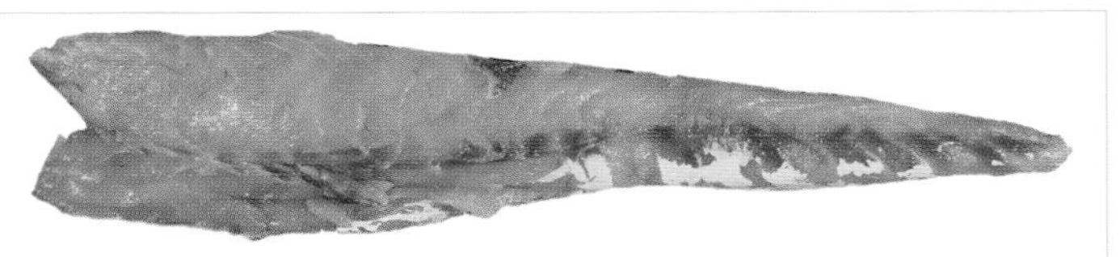

Untrimmed. Protein fingerprint p. 364; oil composition p. 399.

Remarks: The flesh has good food qualities when handled correctly.

Deepwater flathead

Neoplatycephalus conatus

Previous names: flathead, trawl flathead

Identifying features: ① caudal fin uniform greyish-green without spots; ② upper surface greyish-green without spots; ③ greatly enlarged canine teeth present at upper jaw tip; ④ teeth along side of lower jaw similar in size, not much bigger than those nearby; ⑤ outer gill arch with stumpy rakers; ⑥ large dark blotch inside operculum; ⑦ swim bladder present; ⑧ body depressed and tapering with eyes on top of head.

Comparisons: Belongs to a subgroup of flatheads with greatly enlarged canine teeth in the central upper jaw that includes those species sold as 'tiger flathead' (*N. richardsoni* and *N. aurimaculatus*, p. 119). However, unlike these species it lacks distinctive spots on the upper body and over the caudal fin and has relatively smaller, more uniformly spaced teeth in the lower jaw.

Fillet: Moderately slender, pronounced taper, yellowish-white with very few dark veins. Outside with feeble, discontinuous, central red muscle band; weakly converging EL, weakly converging HL; HS along middle of fillet, dipping anteriorly; integument silver above, white below. Inside flesh medium; belly flap usually absent; peritoneum black. Usually skinned, skin thin, scale pockets small and well defined.

Size: To 94 cm and about 4 kg (commonly 45–65 cm and 0.7–1.8 kg).

Habitat: Marine; demersal on the continental shelf and upper slope in depths from about 70–490 m.

Fishery: Dominant trawl flathead in the Great Australian Bight and replaces tiger flatheads in the western sector of the South East Fishery.

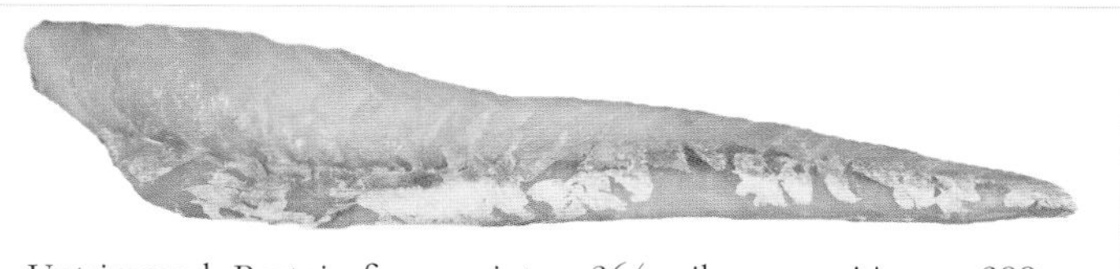

Untrimmed. Protein fingerprint p. 364; oil composition p. 399.

Remarks: Currently under-exploited but little information exists on its resource status. Flesh is of good quality when fresh.

Dusky flathead

Platycephalus fuscus

Previous names: black flathead, dusky, mud flathead

Identifying features: 1 caudal fin bluish-grey below and brown spotted above, with slightly enlarged black spot near central margin; 2 eyes very widely separated (by more than their width in adults); 3 no greatly enlarged teeth at tip of upper jaw; 4 upper and lower preopercular spines of similar length; 5 pectoral fin densely covered with small spots; 6 several bony ridges on head behind eye; 7 no swim bladder; 8 body depressed and tapering with eyes on top of head.

Comparisons: Small individuals somewhat resemble the marbled flathead (*P. marmoratus*) but their caudal fin has a bluish lower half, a black outer central spot, and finer spots on the upper half (rather than uniformly dark with a prominent white outer edge). The eyes of large dusky flathead are very widely separated.

Fillet: Moderately slender, pronounced taper, yellowish-white with dense, dark veins. Outside with feeble, continuous, central red muscle band; weakly converging EL, weakly converging HL; HS along middle of fillet, dipping anteriorly; integument translucent above, silvery white below. Inside flesh fine; belly flap usually absent; peritoneum translucent. Usually skinned, skin medium, scale pockets small and defined.

Size: To 120 cm and 15 kg (commonly 40–80 cm and 0.5–6.0 kg).

Habitat: Marine; commonly inshore in coastal bays and estuaries to depths of at least 30 m.

Fishery: Very significant species in commercial and recreational catches from New South Wales estuaries and targeted recreationally off southern Queensland. Caught mainly by gillnets and seine nets, and as a bycatch of prawn trawlers.

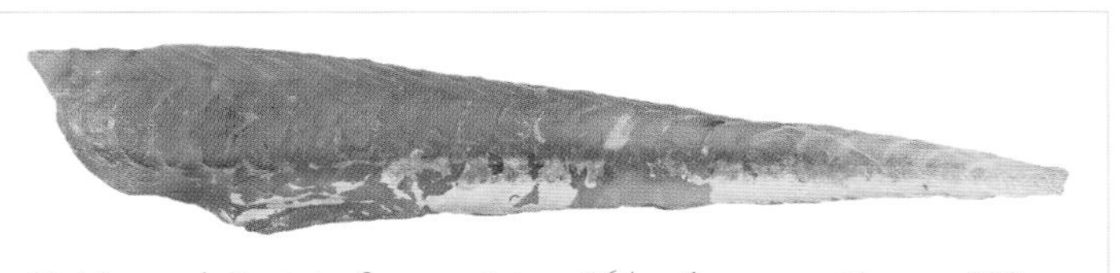

Untrimmed. Protein fingerprint p. 364; oil composition p. 399.

Remarks: Largest Australian flathead and reaches a very impressive size. An important recreational species, often taken in estuaries and near river mouths. A first-rate table-fish.

Flathead

Platycephalidae

Previous names: none

Platycephalus longispinis

Identifying features: ① body depressed and tapering; ② eyes on top of head; ③ head with prominent ridges and spines; ④ pair of sharp spines at corner of preoperculum (cheek); ⑤ 2 separate dorsal fins, the first with 6–10 thin spines; ⑥ pelvic fin bases widely separated.

Comparisons: Distinctive fishes that are unmistakable when whole in the marketplace. Their bodies are more slender than the flatfishes and the dorsal and anal fins are oriented along the midline of the body (rather than along the sides). Also, the eyes are located on the top of the head rather than one side or the other.

Fillet: *Platycephalus longispinis* moderately slender, pronounced taper, yellowish-white with some dark veins. Outside with feeble, continuous, central red muscle band; weakly converging EL, parallel HL; HS along middle of fillet, dipping anteriorly; integument grey above, silvery white below. Inside flesh fine; belly flap usually absent; peritoneum silvery translucent with melanophores. Usually skinned, skin thin, scale pockets small and defined.

Size: To about 120 cm and 15 kg (commonly 30–60 cm and 0.3–1.5 kg).

Habitat: Marine; bottom-dwellers, mostly on soft substrates of the continental shelf and upper slope to depths of about 500 m.

Fishery: Several species are commercially important with largest catches made by demersal trawlers. Inshore species targeted by recreational fishers.

Remarks: Some 40 species of flatheads occur in Australian waters. Most of the largest of these are marketed to varying extents and are popular food-fishes. The flesh is highly regarded but can be coarse. Subtleties in flavour between species has led to some of the main species being marketed separately.

Untrimmed. Protein fingerprint p. 364; oil composition p. 399.

Northern sand flathead

Platycephalus arenarius

Previous names: flathead, sand flathead

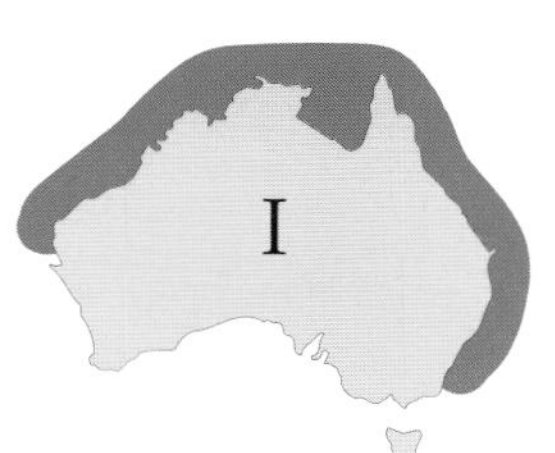

Identifying features: 1 margin of caudal-fin with 5 or 6 alternating pale and dark stripes; 2 upper surface marbled rather than spotted (without cross bars); 3 no greatly enlarged teeth at tip of upper jaw; 4 lower preopercular spine usually slightly longer than upper spine; 5 large dark blotch inside operculum; 6 eyes less than their diameter apart; 7 no swim bladder; 8 body depressed and tapering with eyes on top of head.

Comparisons: Most similar to the bartail flathead (*P. indicus*, p. 110), which also has a marbled upper surface and dark bars on the caudal fin, but the northern sand flathead lacks distinctive yellow blotch on the mid- to outer caudal fin. Instead there are oblique, brownish lines along the upper half of the tail.

Fillet: Moderately slender, pronounced taper, yellowish-white with some dark veins. Outside with feeble, continuous, central red muscle band; weakly converging EL, weakly converging HL; HS along middle of fillet, dipping anteriorly; integument with grey flecks above, silvery below. Inside flesh medium; belly flap usually absent; peritoneum blackish above, silvery below. Usually skinned, skin thin, scale pockets small and defined.

Size: To at least 45 cm and about 0.7 kg (commonly 30–42 cm and 0.3–0.6 kg).

Habitat: Marine; bottom-dwelling on soft substrates of the inner continental shelf of tropical and warm temperate Australia in depths of 5–60 m.

Fishery: Small bycatch component of inshore fisheries.

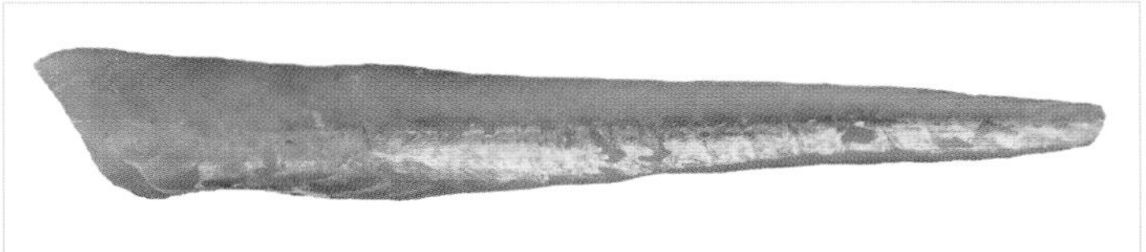

Untrimmed. Protein fingerprint p. 364; oil composition p. 399.

Remarks: The caudal fin, often with a distinct pattern or 'flag', is used in recognition of some species. Unlike other flatheads, appears to occur sporadically in schools. Provides small but tasty fillets.

Rock flathead

Platycephalus laevigatus

Previous names: grassy flathead, king flathead, smooth flathead

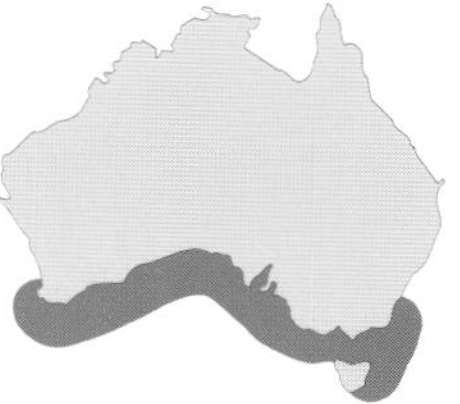

Identifying features: 1 caudal fin yellow with dense dark spots; 2 upper surface of adults marbled rather than spotted; 3 no greatly enlarged teeth at tip of upper jaw; 4 lower preopercular spine shorter than upper; 5 head narrow, deep and relatively smooth; 6 no scales on opercular flap; 7 no swim bladder; 8 body depressed and tapering with eyes on top of head.

Comparisons: Distinctive flathead with a heavily marbled upper surface. The smooth, almost rounded head with the upper preopercular spine longer than the lower spine distinguishes it from all other commercial species.

Fillet: Moderately slender, tapering gradually, greyish with dense, dark veins. Outside without central red muscle band; parallel EL, converging HL; HS along middle of fillet, dipping anteriorly; integument greyish above, silvery below. Inside flesh coarse; belly flap usually absent; peritoneum silvery white to translucent. Usually skinned, skin thick, scale pockets small and well defined.

Size: To about 60 cm and 2 kg (commonly 35–50 cm and 0.5–1.2 kg).

Habitat: Marine; coastal in shallow bays and estuaries on both soft and hard bottoms to depths of about 20 m.

Fishery: Caught in small quantities off southern Australia in beach seines and gillnets.

Remarks: Most flathead species spend time partly buried in sand and mud. However, the rock flathead has a deeper body and prefers to rest on hard, weedy bottom or within seagrass beds rather than burying. The flesh, which is possibly the highest quality of all of the flatheads, is firm and sweet.

Untrimmed. Protein fingerprint p. 364.

Sand flathead

Platycephalus bassensis

Previous names: bay flathead, common flathead, sandy flathead, slimy flathead

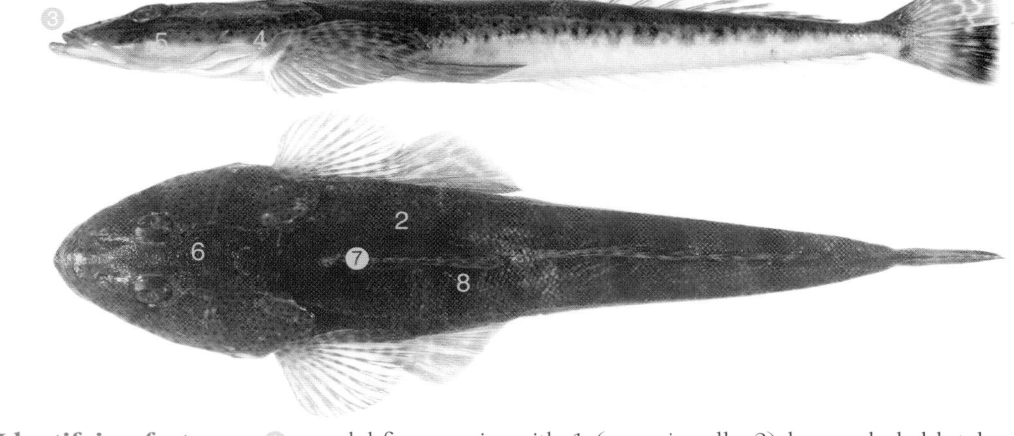

Identifying features: 1 caudal-fin margin with 1 (occasionally 2) large, dark blotches on lower half; 2 upper surface with small, widely spaced, reddish-brown flecks; 3 no greatly enlarged teeth at tip of upper jaw; 4 lower preopercular spine 1.5–2 times longer than upper; 5 dark bar evident beneath eye; 6 pair of strong bony ridges on head behind eye; 7 no swim bladder; 8 body depressed and tapering with eyes on top of head.

Comparisons: Confused with longspine flathead (*P. longispinis*, p. 114) which has a pale blue lower caudal fin (rather than grey or black) and a proportionally larger lower preopercular spine. Sand flathead and blue-spotted flathead (*P. caeruleopunctatus*, p. 111) are similar but the former has only 1–2 large, dark blotches on the lower caudal fin (rather than dark stripes above a basal blotch) and a single pair of bony ridges behind the eye (rather than several pairs). The northern sand flathead (*P. arenarius*, p. 115) is paler in colour with a barred tail.

Fillet: Moderately slender, pronounced taper, yellowish-white with some dark veins. Outside with feeble, discontinuous, central red muscle band; weakly converging EL, weakly converging HL; HS along middle of fillet, dipping anteriorly; integument with silver grey above, white below. Inside flesh fine; belly flap usually absent; peritoneum blackish above, silvery below. Usually skinned, skin thin, scale pockets small and defined.

Size: To at least 55 cm and 1.3 kg (commonly less than 40 cm and 0.5 kg).

Habitat: Marine; demersal, mostly on sand in coastal bays and estuaries, extending on to the continental shelf to depths of about 100 m.

Untrimmed. Protein fingerprint p. 364; oil composition p. 399.

Fishery: Caught in moderate quantities by trawlers and Danish seiners in open waters and by gillnet, beach seine and handline near the coast.

Remarks: Important angling fish in Victoria and Tasmania. Small but has tasty flesh.

Southern flathead

Platycephalus speculator

Previous names: Castelnau's flathead, flathead, longnose flathead, shovelnose flathead, southern blue-spotted flathead, yank flathead

Identifying features: 1 caudal-fin margin with 3–5 large black spots; 2 upper surface lacking fine dark spots; 3 no greatly enlarged teeth at tip of upper jaw; 4 lower preopercular spine slightly longer than upper; 5 no large dark blotch inside operculum; 6 pair of bony ridges on head behind eye; 7 no swim bladder; 8 body depressed and tapering with eyes on top of head.

Comparisons: Very similar to the smaller blue-spotted flathead (*P. caeruleopunctatus*, p. 111) in body form but has a single pair of bony ridges behind the eye (rather than several pairs), smaller dark spots near the tail edge (rather than a blotch and 3 short stripes), smaller spots on the pelvic fin, and no dark blotch on the undersurface of the operculum.

Fillet: Rather slender, tapering gradually, yellowish-white with dense, dark veins. Outside with feeble, continuous, central red muscle band; weakly converging EL, weakly converging HL; HS along middle of fillet, dipping anteriorly; integument marbled grey above, white below. Inside flesh medium; belly flap sometimes present; peritoneum greyish above, silvery below. Usually skinned, skin thin, scale pockets small and well defined.

Size: To 90 cm and 8 kg (commonly 35–70 cm and 0.4–2.2 kg).

Habitat: Marine; demersal in sheltered coastal bays and estuaries on sand, often near seagrass beds, to depths of about 30 m.

Fishery: Caught in beach seines, gillnets and haul nets along the southern coastline. Possibly the most commonly marketed flathead in Western Australia.

Trimmed. Protein fingerprint p. 364.

Remarks: Largest flathead along the southern coastline; was initially misidentified in eastern Bass Strait as another giant species usually from more northerly waters, the dusky flathead (*P. fuscus*, p. 113).

Tiger flathead

Neoplatycephalus aurimaculatus & *N. richardsoni*

Previous names: flathead, king flathead, trawl flathead

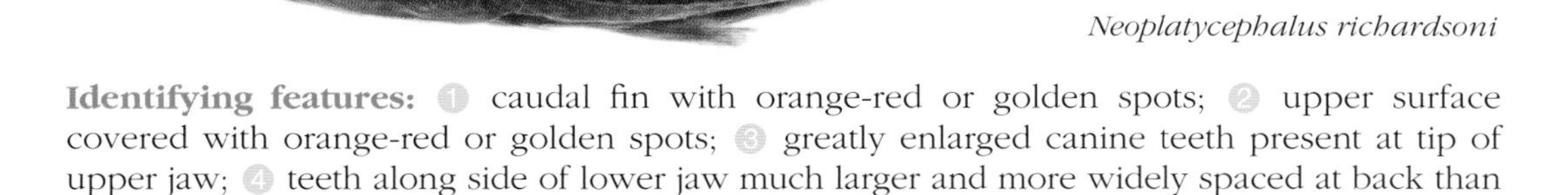

Neoplatycephalus richardsoni

Identifying features: ① caudal fin with orange-red or golden spots; ② upper surface covered with orange-red or golden spots; ③ greatly enlarged canine teeth present at tip of upper jaw; ④ teeth along side of lower jaw much larger and more widely spaced at back than towards front of jaw; ⑤ body depressed and tapering with eyes on top of head.

Comparisons: Consists of two very similar species, the true 'tiger' (*N. richardsoni*) and 'toothy' (*N. aurimaculatus*) flatheads. The toothy flathead is covered with golden spots (rather than orange to red spots), has stumpy rakers in the outer gill arch (rather than being long and slender), and lacks a swim bladder (otherwise present). Unlike their western relative the deep-water flathead (*N. conatus*, p. 112), they are spotted on the upper body and caudal fin and have larger, more widely spaced teeth in the back of the lower jaw.

Fillet: *N. richardsoni* moderately slender, pronounced taper, pinkish to yellowish-white, without prominent dark veins. Outside with feeble, discontinuous, central red muscle band; weakly converging EL, weakly converging HL; HS along middle of fillet, dipping anteriorly; integument translucent above, silvery white below. Inside flesh fine; belly flap sometimes present; peritoneum greyish with melanophores. Usually skinned, skin medium, scale pockets small and defined.

Size: To about 70 cm and possibly 3 kg (commonly 35–55 cm and 0.4–1.3 kg).

Habitat: Marine; demersal on the mid-continental shelf and upper slope in 10–400 m depth, but mainly shallower than 200 m.

Fishery: Old steam trawl fishery off southeastern Australia dating back to the early 1900s. Now mainly taken by trawl and Danish seines.

Trimmed. Protein fingerprint p. 364; oil composition p. 399.

Remarks: Tiger flathead are much more abundant and occur more offshore than toothy flathead. Tiger flathead have a swim bladder enabling them to feed in midwater. Sold whole fresh and as fillets.

Australian bass

Macquaria novemaculeata

Previous name: freshwater perch

Identifying features: ① caudal fin forked; ② forehead straight or slightly concave; ③ 12–15 rakers on lower limb of outer gill arch; ④ pair of spines on each operculum; ⑤ single dorsal fin with deep notch, 8–10 spines, 8–11 soft rays; ⑥ anal fin with 3 spines, 7–9 soft rays.

Comparisons: More slender than the golden perch (*M. ambigua*, p. 122) with a forked (rather than a rounded) caudal fin. It is almost identical to its southern estuarine relative, the estuary perch (*M. colonorum*), which has more gill rakers (14–18 rather than 12–15) and a more concave forehead above the eye. These fishes belong to a family of freshwater perches (Percichthyidae), a diverse group containing some of Australia's largest and most important commercial fishes. They have a characteristic body form and combination of fin elements.

Fillet: Moderately deep, short, tapering gently, very convex above, pinkish. Outside with intermediate, broad continuous, central red muscle band; parallel EL, weakly converging HL; HS along middle of fillet; integument translucent above, white below. Inside flesh medium; belly flap sometimes present; peritoneum silvery white. Usually skinned, skin thick, scale pockets moderate and well defined.

Size: To 58 cm and at least 3.6 kg (commonly to about 35 cm and 1 kg).

Habitat: Freshwater and estuarine; occurs in drainages and lakes of eastern Australia from Gippsland (Vic.) to the lower reaches of the Mary River system (Qld).

Trimmed. Protein fingerprint p. 364.

Fishery: Hardy aquarium fish that is suited to farming. Mainly taken in drum nets and gillnets. One of the major angling species in rivers of eastern Australia.

Remarks: Good foodfish, popular for stocking farm dams.

Barramundi

Lates calcarifer

Previous names: barra, giant perch, silver barramundi

Identifying features: 1 caudal fin rounded; 2 forehead straight or slightly concave; 3 jaw extending to end of eye or beyond; 4 young with prominent white stripe on the forehead; 5 grey green to golden brown above; 6 dorsal fins almost separate, 8–9 spines, 10–11 soft rays; 7 anal fin with 3 spines, 7–8 soft rays.

Comparisons: Belongs to the giant perch family (Centropomidae) which are similar in general appearance to freshwater perches (Percichthyidae) and freshwater grunters (Terapontidae). However, unlike commercial members of these groups, it has a tall, angular first dorsal fin that is almost separate from the second. A smaller marine relative, the sand bass (*Psammoperca waigiensis*), is dull reddish-brown above and has a shorter mouth (falling short of hind margin of the eye).

Fillet: Moderately deep, short, tapering gently, very convex above, pinkish. Outside with feeble, broad diffuse, central red muscle band; converging EL, converging HL; HS along middle of fillet; integument whitish. Inside flesh coarse; belly flap sometimes present; peritoneum silvery white to translucent. Usually skinned, skin thin, scale pockets large and defined.

Size: To 150 cm and 60 kg (commonly marketed up to 120 cm and about 0.5–10 kg).

Habitat: Freshwater; lives in rivers and creeks, but ventures downstream into estuaries and along coastal shallows to breed.

Fishery: Caught using gillnets in coastal estuaries and river mouths along northern Australian coastline. Successful aquaculture industry currently produces fingerlings and fish marketed at about 0.4–2 kg. Very popular angling species.

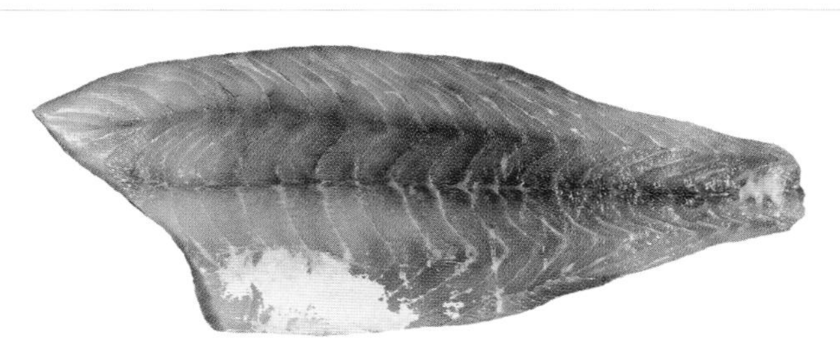

Untrimmed. Protein fingerprint p. 364; oil composition p. 399.

Remarks: Among the largest and most commercially important of our inland fishes, it has almost cult status among recreational fishers. The white flesh, which is soft and delicate, is held in great esteem.

Golden perch

Macquaria ambigua

Previous names: callop, Murray perch, white perch, yellow-belly, yellowfin perch

Identifying features: ① caudal fin rounded (not forked); ② head relatively deep and compressed, nape strongly arched; ③ prominent open pores on lower jaw; ④ filaments at tips of pelvic fins; ⑤ usually olive green grading to yellowish below; ⑥ single dorsal fin with deep notch, 8–11 spines, 11–13 soft rays; ⑦ anal fin with 3 spines, 7–10 soft rays.

Comparisons: Distinguished from the Australian bass (*M. novemaculeata*, p. 120) in having a hump on the back of adults and a rounded caudal fin. The barramundi (*Lates calcarifer*, p. 121) can also have a humped back but the golden perch's first dorsal-fin spines are uniformly taller and its pectoral-fin tips have distinctive filaments. Other giant freshwater perches, such as the Murray cod (*Maccullochella peelii*, p. 123), lack open pores on the lower jaw.

Fillet: Moderately deep, short, tapering gently, slightly bottle-shaped, very convex above, yellowish-white. Outside with intermediate, broad continuous, central red muscle band; parallel EL, weakly converging HL; HS along middle of fillet; integument silvery white. Inside flesh medium; belly flap sometimes present; peritoneum silvery white to translucent. Usually skinned, skin thick, scale pockets small and defined.

Size: To 76 cm and 9 kg (commonly to 50 cm and 4 kg). Records to 23 kg are questionable.

Habitat: Freshwater; from clear highland streams to the slow-flowing, turbid rivers and backwaters of central and eastern Australia.

Trimmed.
Protein fingerprint p. 364; oil composition p. 399.

Fishery: Historically important, although variable, fishery dating back to the early 1800s. Largest catches are made using drum nets and gillnets. Robust species with good potential for aquaculture. Valued recreationally.

Remarks: Undertakes long migrations upstream in late spring to early summer. Considered to be an excellent foodfish, with a delicate flavour. Popular for dam stocking.

Murray cod

Maccullochella species

Previous names: cod, goodoo, ponde

Maccullochella peelii

Identifying features: 1 caudal fin rounded (not forked); 2 head relatively broad, nape arched slightly; 3 outer tips of hind fins usually conspicuously white; 4 short filaments at tips of pelvic fins; 5 heavily marbled olive green to yellowish-green; 6 single dorsal fin with deep notch, 10–12 spines, 13–16 soft rays; 7 anal fin with 3 spines, 10–13 soft rays.

Comparisons: Similar to the golden perch (*Macquaria ambigua*, p. 122) but have more anal-fin rays (10–13 versus 7–10), a relatively broader snout, and lack open pores along the lower jaw. Differ from the Australian bass (*M. novemaculeata*, p. 120) in having a rounded caudal fin (rather than forked) and the barramundi (*Lates calcarifer*, p. 121) in having the last few dorsal-fin spines connected by a tall membrane.

Fillet: *Maccullochella peelii* moderately deep, short, tapering gently, weakly convex above, white. Outside with feeble, broad continuous, central red muscle band; parallel EL, parallel HL; HS along middle of fillet; integument greyish above, whitish below. Inside flesh medium; belly flap usually present; peritoneum white. Usually skinned, skin thick, scale pockets small and irregular.

Size: Probably to 180 cm and 114 kg, confirmed to 73 kg (commonly 55–65 cm and 2–5 kg).

Habitat: Freshwater; hides among fallen trees and other river debris mainly in slow-flowing, turbid rivers of the south-east. Successfully introduced to lakes and dams.

Fishery: Historical fishery dating back to the mid-1800s, with catches at lower levels in recent decades. Caught mainly in the lower reaches of the Murray (SA and Vic.) and Murrumbidgee (NSW) rivers using drum nets and gillnets.

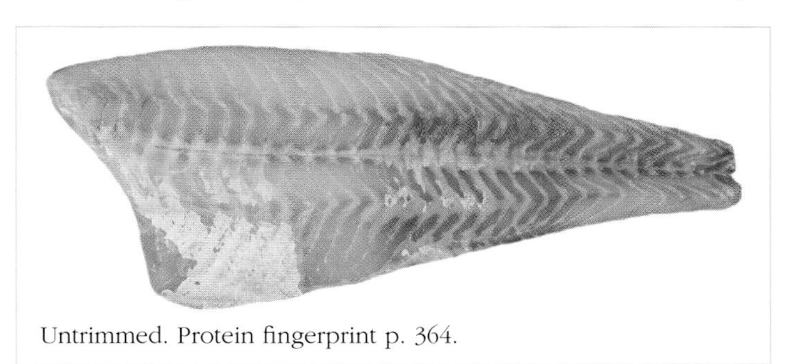

Untrimmed. Protein fingerprint p. 364.

Remarks: *M. peelii* is by far the dominant species in the fishery but two others may also form part of the catch. Most product sold whole and gutted to domestic markets where it fetches high prices, up to $23 per kg. Some live fish sales.

Redfin

Perca fluviatilis

Previous names: English perch, European perch, redfin perch

Identifying features: ① caudal fin slightly forked; ② head pointed and nape arched slightly; ③ 6 dark bands on sides; ④ black spot at end of first dorsal fin; ⑤ red to orange pelvic and anal fins; ⑥ 2 narrowly separated dorsal fins with 14–19 spines, 13–16 soft rays; ⑦ anal fin with 2 spines, 8–10 soft rays.

Comparisons: Distinctive introduced freshwater fish with typical perch-like body shape, dark bands on the body, and reddish fins. In comparison, native commercial freshwater perches are either plain-coloured or have a faint marble pattern.

Fillet: Deep, short, tapering gently, slightly bottle-shaped, weakly convex above, brownish-white. Outside with feeble, continuous, central red muscle band; convex EL, converging HL; HS along middle of fillet; integument silvery white. Inside flesh fine; belly flap sometimes present; peritoneum silvery white to translucent. Usually skinned, skin medium, scale pockets small and defined.

Size: To at least 60 cm and 4.8 kg (commonly 25–40 cm and 0.3–1.5 kg).

Habitat: Freshwater; lives in a variety of habitats including lakes, dams, swamps and slow-flowing streams. Frequently seeks cover in river weed, under rocks or among other river debris.

Fishery: Small commercial fishery based in Victoria. Regarded as an excellent angling fish, accepting a broad variety of baits.

Untrimmed. Protein fingerprint p. 364.

Remarks: Introduced to Victoria and Tasmania in the 1860s, it has dispersed widely in the Murray–Darling river system with the assistance of periodic flooding. Young make excellent aquarium specimens. The flesh is firm, white and tasty.

Eastern sea garfish

Hyporhamphus australis

Previous names: sea gar, sea garfish

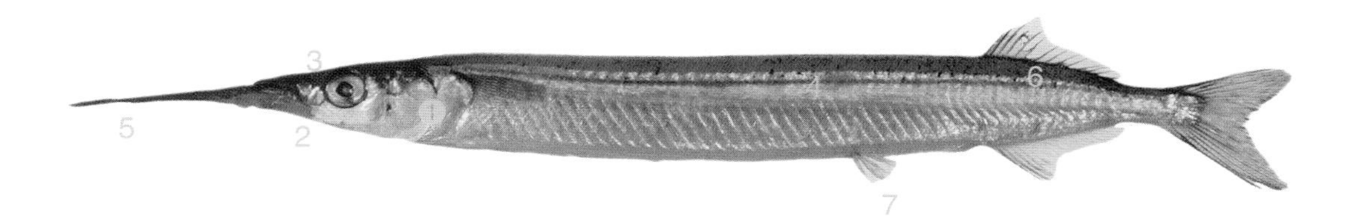

Identifying features: ① mostly 34 or more gill rakers on first arch and 27 or more on second arch; ② small sensory pore canal before eye T-shaped; ③ prominent ridge between eye and nostril; ④ silver stripe along midline; ⑤ lower jaw extended into a bill; ⑥ body elongate with dorsal and caudal fins well back on trunk; ⑦ pelvic fins well behind pectoral fins, both fins small.

Comparisons: Almost identical to its relative the southern garfish (*H. melanochir*, p. 130) but has more gill rakers (34 and 27 or more on the first and second arches respectively versus 33 and 26 or fewer). Sea garfishes differ from river garfish (*H. regularis*, p. 129) in having a branched sensory pore canal before the eye, more deciduous scales, and the pointed upper jaw is about as long as wide (rather than rounded and wider than long).

Fillet: Long, slender, taper confined to caudal peduncle, translucent grey with dark veins. Outside with very pronounced, sharp-edged, continuous, central red muscle band (often partly concealed below silver integument); parallel EL, weakly converging HL; HS along middle of fillet; integument silver, forming stripe along midline. Inside flesh fine; belly flap usually present, not lobe-like; peritoneum black. Often skinned, skin very thin, scale pockets irregular.

Size: To 45 cm and about 0.4 kg (commonly up to 33 cm and 0.1 kg).

Habitat: Marine; pelagic in coastal bays and inlets.

Fishery: Centred off New South Wales, where it is taken by beach seine and purse seine nets.

Remarks: The distributions of the eastern and southern sea garfishes overlap on the New South Wales coast and hybrid specimens of the two have been identified.

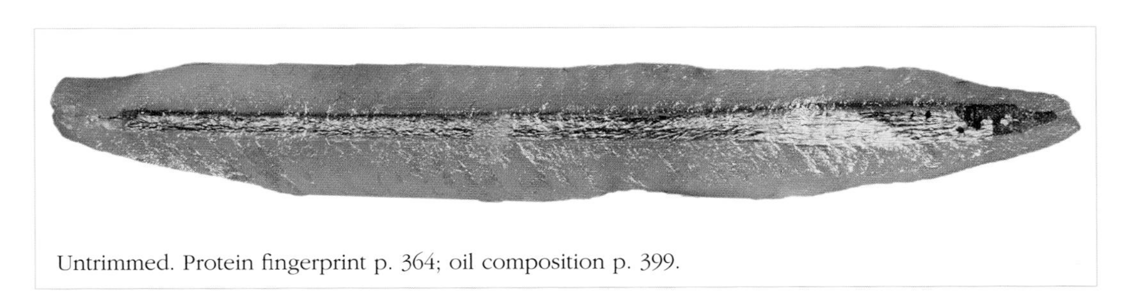

Untrimmed. Protein fingerprint p. 364; oil composition p. 399.

Garfish (page 1 of 2)

Hemiramphidae

Previous names: barred garfish, no-bill garfish, robust garfish, shortnose garfish, snubnose garfish, three-by-two garfish, tropical garfish

Arrhamphus sclerolepis

Identifying features: ① mouth small, lower jaw extended into a bill; ② body long, cylindrical to rectangular in cross-section; ③ caudal fin forked with lower lobe longer than upper; ④ dorsal and caudal fins well back on trunk; ⑤ pelvic fins well behind pectoral fins, both fins small in commercial species.

Comparisons: Very distinctive fishes. The body shape and fin positions of the longtoms (Belonidae, p. 128) and sauries (Scomberesocidae) are similar but these groups have a long bill that involves extension of both jaws (rather than just the lower jaw). The flying fishes (Exocoetidae) have greatly enlarged wing-like pectoral and pelvic fins and no long bill at the mouth tip.

Fillet: *Arrhamphus sclerolepis* long, relatively deep, taper confined to caudal peduncle, translucent grey with dark veins. Outside with very pronounced, sharp-edged, continuous, central red muscle band (sometimes partly concealed below silver integument); weakly converging EL, weakly converging HL; HS along mid- to upper fillet; integument silver, may form stripe along midline. Inside flesh fine; belly flap usually present, not lobe-like; peritoneum black. Often skinned, skin very thin, scale pockets large and defined.

Size: To 52 cm (with bill) and 0.6 kg (commonly up to 36 cm and 0.2 kg).

Habitat: Marine; pelagic, mostly near the surface, in coastal bays, rivers and estuaries.

Fishery: Caught using haul nets, ring nets and small-mesh garfish seines in both temperate and tropical regions.

Remarks: Some 18 species live in Australian seas of which three have separate marketing names. Other species shown opposite, and the three-by-two (*Hemiramphus robustus*) and black-barred (*H. far*) garfishes, are also marketed regularly. Usually sold as whole fresh fish or butterfly fillets. Also used for bait and fillets sold for sashimi. The flesh is sweet and tasty but can be bony if inadequately prepared.

Untrimmed. Protein fingerprint p. 364; oil composition p. 399.

Garfish (page 2 of 2)

Hemiramphidae

Arrhamphus sclerolepis

Remarks: Commonly called 'snubnose garfish', this species is distributed near the surface along the northern Australian coast between Shark Bay (WA) and Sydney (NSW). Best distinguished in having a comparatively short bill. Reaches about 40 cm and 0.4 kg.

Hyporhamphus quoyi

Remarks: Commonly called 'shortnose garfish', this species is distributed across northern Australia between Shark Bay (WA) and the Clarence River (NSW). Caught near the surface along the coast and marketed in Queensland throughout the year. Best distinguished in having a moderate bill that is relatively longer than the snubnose garfish (above) but shorter than other species treated here. Reaches about 38 cm and 0.2 kg.

Hyporhamphus affinis

Remarks: Commonly called 'tropical garfish', this species is distributed along the tropical coasts of Western Australia and the Northern Territory. Resembles the sea garfishes (*H. australis*, p. 125, and *H. melanochir*, p. 130) but usually has fewer anal-fin rays (17 or less rather than 18 or more). To at least 25 cm and 0.1 kg.

Longtom

Belonidae

Previous names: alligator gar, needlefish

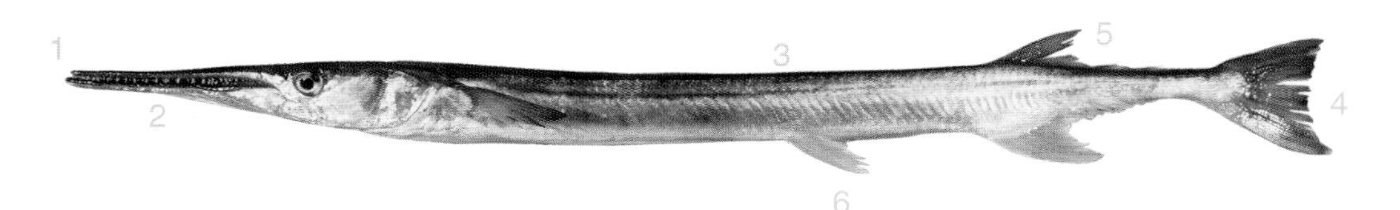

Tylosurus gavialoides

Identifying features: ① mouth large, both jaws extended into a bill; ② bill armed with needle sharp teeth; ③ body very long and slender; ④ caudal fin weakly forked, lower lobe sometimes longer than upper; ⑤ dorsal and caudal fins well back on trunk; ⑥ pelvic fins well behind pectoral fins, both fins small.

Comparisons: Very distinctive fishes. The body shape and fin positions resemble garfishes (Hemiramphidae) and sauries (Scomberesocidae) but the presence of very long upper and lower jaws armed with numerous needle-like teeth is unique in Australian seas.

Fillet: *Tylosurus gavialoides* very long, slender, gradually tapering, translucent grey with dark veins. Outside with very pronounced, sharp-edged, continuous, central red muscle band; weakly converging EL, weakly converging HL; HS along mid- to upper fillet; integument silvery white. Inside flesh fine; belly flap usually present, not lobe-like; peritoneum translucent to white with melanophores. Often skinned, skin very thin, scale pockets barely detectable.

Size: To 150 cm and at least 5.2 kg (commonly to about 100 cm and less than 2 kg).

Habitat: Marine; pelagic in coastal habitats, but also sometimes in freshwater, well offshore in the open sea or around coral atolls.

Fishery: Secondary commercial species caught in haul nets and by lining, often unintentionally by mackerel fishers.

Remarks: About a dozen species live in Australian seas. Most of these are large and several reach longer than 1 m. The most frequently marketed include the barred (*Ablennes hians*), crocodile (*T. crocodilus*) and stout (*T. gavialoides*) longtoms. Exciting recreational angling fishes with an aggressive nature; reported to have attacked wading anglers. Despite sometimes being greenish, the flesh is of above average quality.

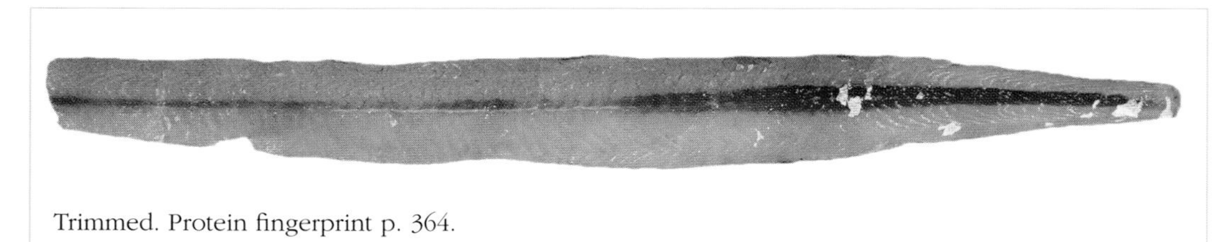

Trimmed. Protein fingerprint p. 364.

River garfish

Hyporhamphus regularis

Previous names: lakes garfish, needle gar

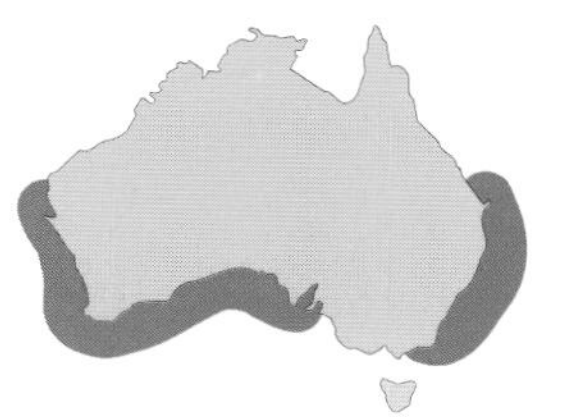

Identifying features: 1 upper jaw rounded and wider than long; 2 small sensory pore canal before eye not T-shaped; 3 prominent ridge between eye and nostril; 4 silver stripe along midline; 5 lower jaw extended into a bill; 6 body elongate with dorsal and caudal fins well back on trunk; 7 pelvic fins well behind pectoral fins, both fins small.

Comparisons: Differs from the commercially important sea garfishes (*H. australis*, p. 125, and *H. melanochir*, p. 130) in having an unbranched sensory pore canal before the eye (rather than being T-shaped) and the rounded upper jaw is distinctly wider than long (rather than pointed and about as long as wide). Its scales are better attached to the body and are often still intact when processed. Sea garfishes have comparatively delicate scales that are usually abraded away during capture and handling.

Fillet: Long, moderately deep, taper confined to caudal peduncle, translucent grey with dark veins. Outside with pronounced, sharp-edged, continuous, central red muscle band (often partly concealed below silver integument); parallel EL, weakly converging HL; HS along mid- to upper fillet; integument silver, forming stripe along midline. Inside flesh fine; belly flap usually present, not lobe-like; peritoneum silvery black. Often skinned, skin very thin, scale pockets large and defined.

Size: To 35 cm and almost 0.2 kg (commonly less than 32 cm and 0.1 kg).

Habitat: Estuarine; pelagic, lives in shallow brackish inlets and estuaries, often in the vicinity of seagrasses.

Fishery: Caught by estuarine fishers mainly using fine-mesh beach seines, and tunnel and cast nets, in depths shallower than 5 m. Shoaling fish are pinpointed by a surface trail of greenish droppings excreted during feeding.

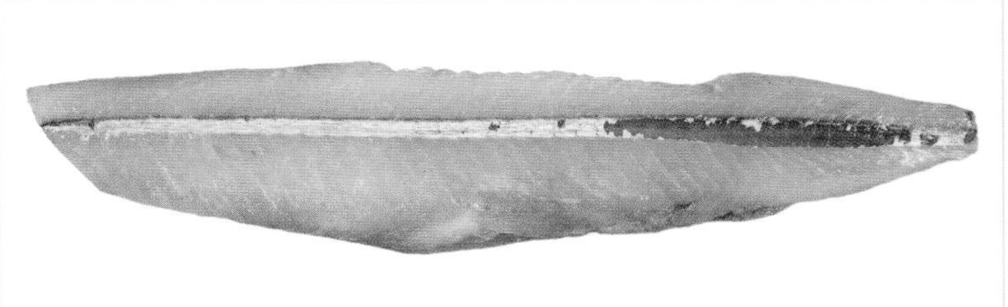

Untrimmed. Protein fingerprint p. 364; oil composition p. 399.

Remarks: Small recreational catch from line fishing and bait collection. Distinct populations occur off eastern and western Australia. It has a delicate, slightly sweet flavour and despite its small size it is well regarded as a table-fish.

Southern garfish

Hyporhamphus melanochir

Previous names: garfish, South Australian garfish, southern sea garfish

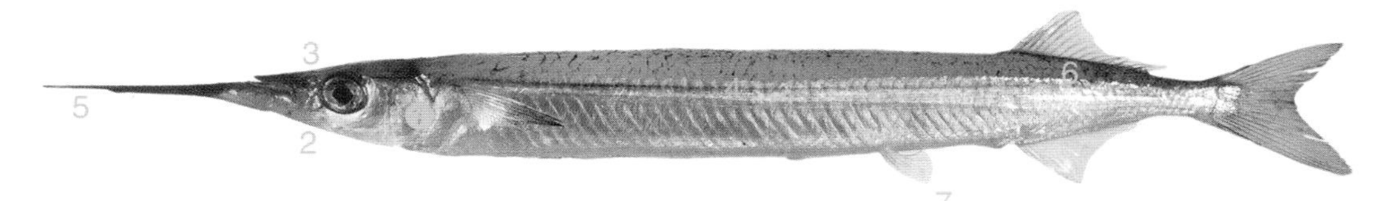

Identifying features: ① mostly 33 or less gill rakers on first arch and 26 or less on second arch; ② small sensory pore canal before eye T-shaped; ③ prominent ridge between eye and nostril; ④ silver stripe along midline; ⑤ lower jaw extended into a bill; ⑥ body elongate with dorsal and caudal fins well back on trunk; ⑦ pelvic fins well behind pectoral fins, both fins small.

Comparisons: Almost identical to the eastern sea garfish (*H. australis*, p. 125) but has fewer gill rakers (33 and 26 or fewer rakers on the first and second arches respectively versus 34 and 27 or more). They differ from river garfish (*H. regularis*, p. 129), with which they are sometimes taken, in having a T-shaped sensory pore canal before the eye, more deciduous scales, and the pointed upper jaw is about as long as wide (rather than rounded and wider than long).

Fillet: Long, moderately deep, taper confined to caudal peduncle, translucent grey with dark veins. Outside with pronounced, deep, sharp-edged, continuous, central red muscle band (often partly concealed below silver integument); parallel EL, weakly converging HL; HS along mid- to upper fillet; integument silvery grey, forming stripe along midline. Inside flesh fine; belly flap usually present, not lobe-like; peritoneum black. Often skinned, skin very thin, scale pockets irregular.

Size: To 52 cm and 0.4 kg (commonly to about 35 cm and almost 0.2 kg).

Habitat: Marine; pelagic, in coastal marine bays and inlets.

Fishery: Centred in large embayments of Victoria, the Furneaux Group, off eastern Tasmania, and the gulfs of South Australia. Also taken in large embayments and estuaries of temperate Western Australia. Caught using haul, dip and push nets, and fine-mesh garfish seines. Important recreational species.

Remarks: Largest domestic fishery is based off South Australia where garfishes are regarded among the State's premium foodfishes. The flesh is fine and delicately flavoured.

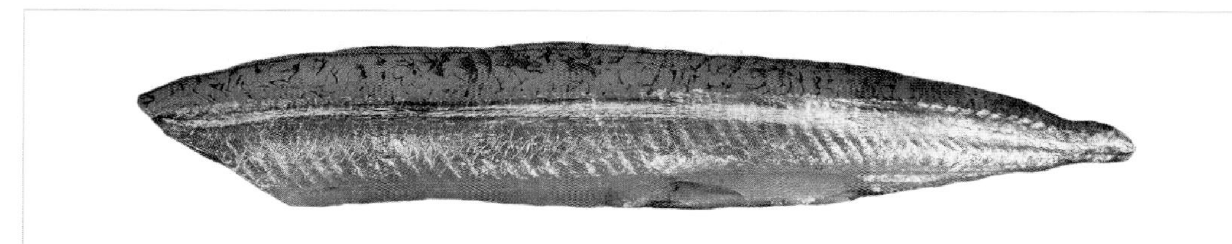

Untrimmed. Protein fingerprint p. 364; oil composition p. 399.

Barracouta

Thyrsites atun

Previous names: couta, snoek

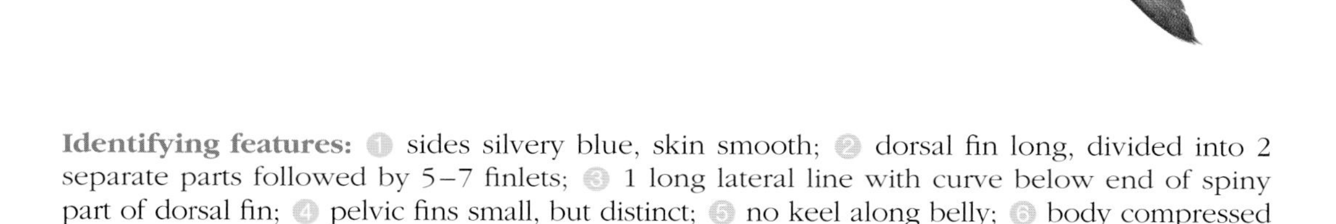

Identifying features: 1 sides silvery blue, skin smooth; 2 dorsal fin long, divided into 2 separate parts followed by 5–7 finlets; 3 1 long lateral line with curve below end of spiny part of dorsal fin; 4 pelvic fins small, but distinct; 5 no keel along belly; 6 body compressed slightly, very long; 7 head large with long fang-like teeth in upper jaw.

Comparisons: Resembles the heavier-bodied gemfish (*Rexea solandri*, p. 133) but has more finlets following both the dorsal and anal fins (5–7 versus 2) and a single lateral line. Its appearance is quite different to other commercial members of the gemfish family (Gempylidae). Should not be confused with the more tropical barracudas (Sphyraenidae, pp 201–203) which have 2 short, well separated dorsal fins.

Fillet: Rather slender, elongate, upper profile almost straight, taper slight, off-white to yellowish. Outside with feeble, continuous red muscle band; weakly converging EL, weakly converging HL; HS along middle of fillet; myomeres very narrow; integument silvery. Inside flesh medium; belly flap rarely present; peritoneum translucent white with melanophores. Often skinned, skin thin, scale pockets small and barely detectable.

Size: To 140 cm and 6 kg (commonly to about 50–100 cm and 1.5–2.5 kg).

Habitat: Marine; pelagic above the sea bed on the continental shelf, occasionally down the slope to 550 m depth. Often forming dense schools inshore near the surface.

Fishery: Dating back to the mid-1800s and caught in large quantities in the mid-1900s. Caught mainly by trolling at the surface and more recently as trawl bycatch.

Remarks: Once used widely in 'fish n chip' trade but has been replaced in recent times with other species. The flesh is somewhat soft and has a light taste.

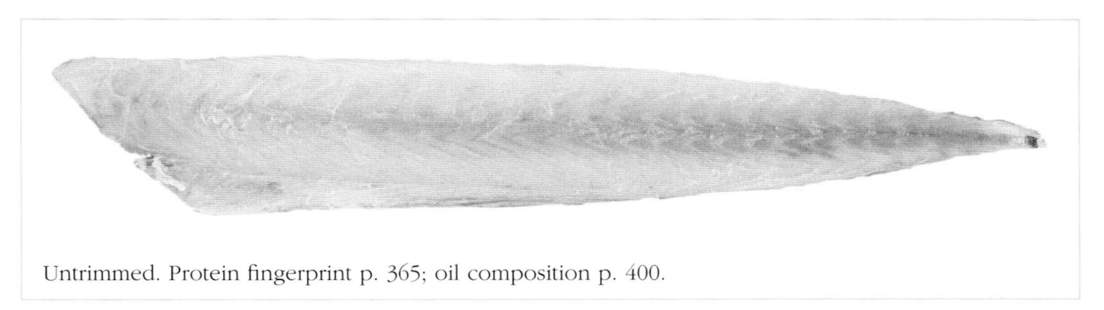

Untrimmed. Protein fingerprint p. 365; oil composition p. 400.

Escolar

Lepidocybium flavobrunneum & Ruvettus pretiosus

Previous names: black oil fish, castor oil fish, oil fish

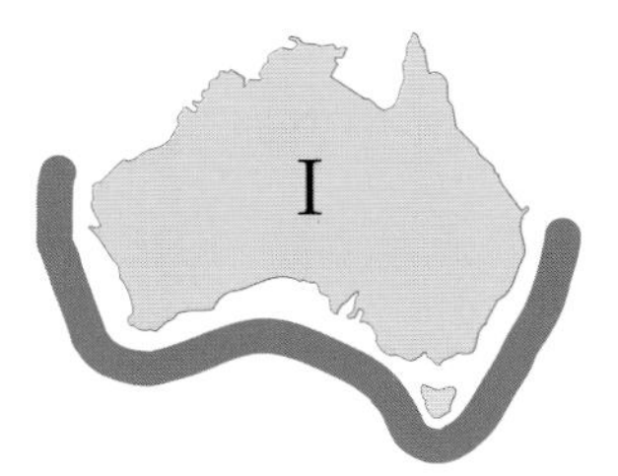

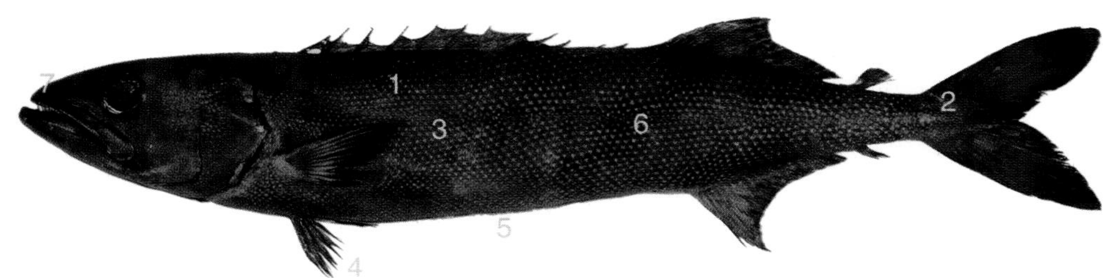

Ruvettus pretiosus

Identifying features: ① sides dark brown, grey or almost black, sometimes with rough skin; ② sometimes with prominent fleshy keel on caudal peduncle flanked by 2 smaller keels; ③ 1 long lateral line with strong curve at end of pectoral fin (may have additional curves further along body); ④ pelvic fins well developed; ⑤ distinct spiny ridge sometimes present along belly; ⑥ body solid, moderately long; ⑦ head large with long fang-like teeth in upper jaw.

Comparisons: The only commercial gemfishes (Gempylidae) to have prominent keels on the caudal peduncle and wavy lateral line (*L. flavobrunneum*) or rough spiny scales and a prominent ridge along belly (*R. pretiosus*). These characters also serve to distinguish these species from each other. Other gemfishes marketed have more silvery bodies and either lack or have very small pelvic fins.

Fillet: *R. pretiosus* moderately deep, rather elongate, upper profile almost straight, taper pronounced, milky white. Outside with intermediate, continuous red muscle band; convex EL, weakly converging HL; HS above middle of fillet anteriorly; integument white translucent. Inside flesh coarse; belly flap sometimes present; peritoneum translucent with dense melanophores. Always skinned, skin thick and rough, scale pockets small and embedded.

Size: Reported to 300 cm and probably exceeding 100 kg (commonly to about 80–150 cm and 6–20 kg).

Habitat: Marine; pelagic in the open ocean, probably concentrating around seamounts.

Fishery: Caught mainly as bycatch by tuna longliners in depths of 100–400 m. Available periodically in moderate quantities in main urban markets.

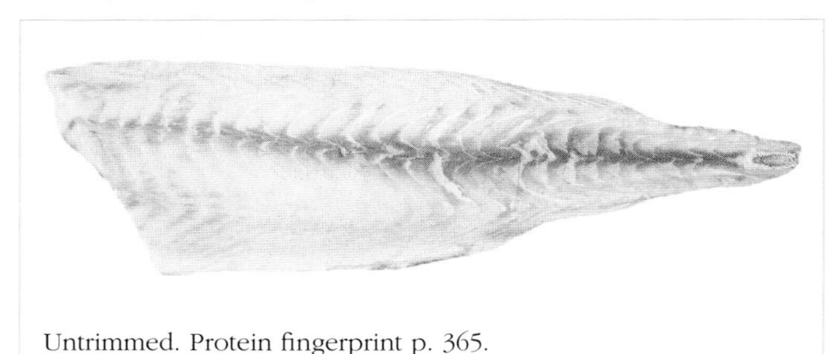

Untrimmed. Protein fingerprint p. 365.

Remarks: The names of the two species marketed as 'escolar' have been confused in domestic markets. 'Escolar' is used internationally for *L. flavobrunneum* whereas *R. pretiosus* is more widely known as 'oil fish'. Flesh, although tasty, is extremely oily and can cause diarrhoea if eaten in large quantities.

Gemfish

Rexea solandri

Previous names: hake, king couta, kingfish, silver kingfish, southern kingfish

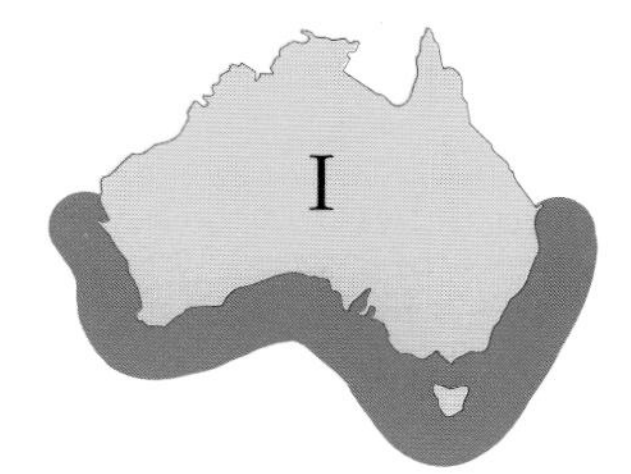

Identifying features: ① sides silvery, skin smooth; ② dorsal fin long, divided into 2 separate parts followed by 2 finlets; ③ 2 separate lateral lines; ④ pelvic fins barely visible; ⑤ no keel along belly; ⑥ body compressed slightly, moderately long; ⑦ head large with long fang-like teeth in upper jaw.

Comparisons: Resembles the barracouta (*Thyrsites atun*, p. 131) but has fewer finlets following both the dorsal and anal fins (2 versus 5–7) and 2 lateral lines (one running beneath the dorsal fins with the second arching downwards below the anterior third of the spinous part of the dorsal fin). Sometimes caught with the rarer longfin gemfish (*R. antefurcata*) from which it can be distinguished by the form of their lateral lines. The upper lateral line of the longfin gemfish finishes near the rear of the second dorsal fin (rather than forward of the middle of the fin) and the lower lateral line is straight above the anal fin (rather than wavy).

Fillet: Rather slender, elongate, upper profile almost straight, taper slight, pale pinkish. Outside with intermediate, continuous red muscle band (more pronounced posteriorly); convex EL, convex HL; HS along middle of fillet; integument silvery grey above, silvery below. Inside flesh medium; belly flap rarely present; peritoneum black. Usually skinned, skin thin, scale pockets small and barely detectable.

Size: To 120 cm and 15 kg (commonly 60–90 cm and up to 2.0–5.0 kg).

Habitat: Marine; semi-pelagic in midwater or near the bottom of the outer continental shelf and upper slope in 100–700 m depth.

Fishery: Trawl fishery originating in the 1970s and centred off New South Wales is now in decline after suffering a significant recruitment failure. A western stock is fished from west of Tasmania to Western Australia.

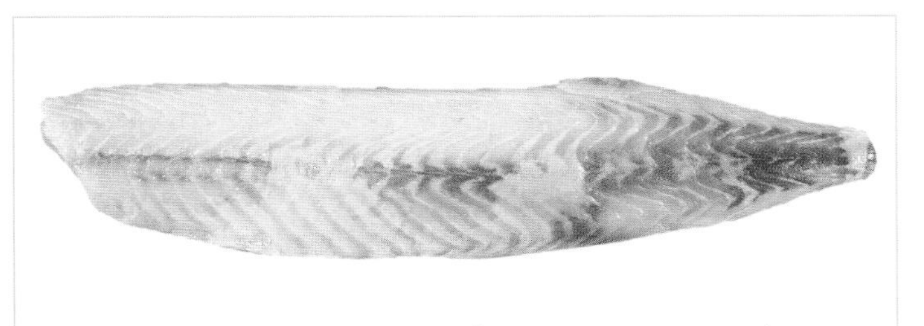

Trimmed. Protein fingerprint p. 365; oil composition p. 400.

Remarks: Sold headed and gutted or as whole fish. Retailed usually as fresh fillets. Smoked gemfish is an excellent product.

Ribbonfish

Lepidopus caudatus

Previous names: frostfish, southern frostfish

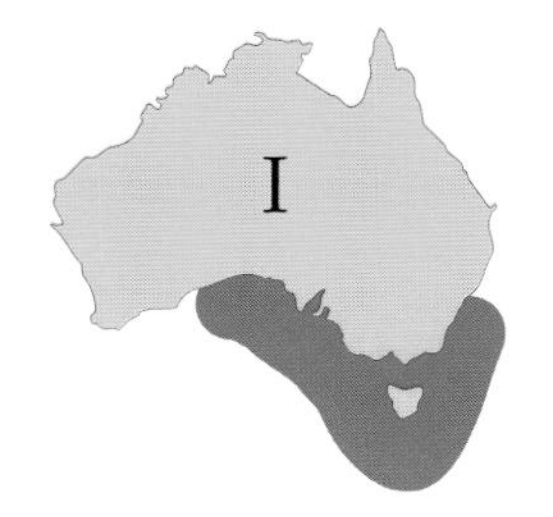

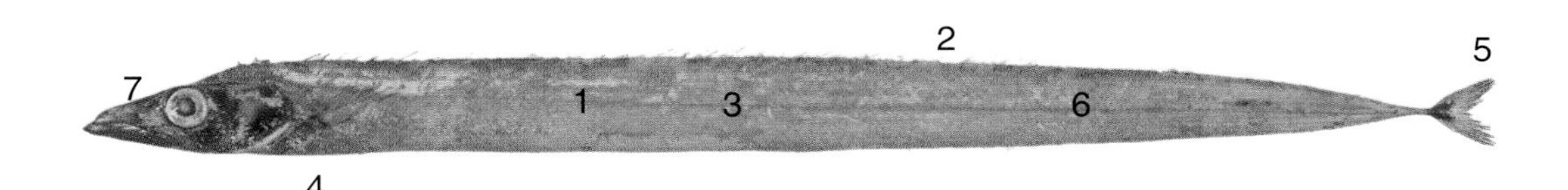

Identifying features: ① sides uniformly silvery, skin smooth; ② dorsal fin extremely long, not divided into 2 parts and lacking finlets; ③ 1 long lateral line; ④ pelvic fins minute or absent; ⑤ caudal fin comparatively small; ⑥ body extremely long, very compressed; ⑦ head small with long fang-like teeth in upper jaw.

Comparisons: Distinctive with its greatly elongated body, small head, small caudal fin, and no pelvic fins in adults. A shallow water relative, the hairtail (*Trichiurus lepturus*), which is sometimes marketed in Sydney (NSW), also has a long compressed body, small head and no pelvic fins but lacks a caudal fin (the tail tapers to a point).

Fillet: Very long, slender, barely tapering, off-white to yellowish. Outside with intermediate, discontinuous red muscle band; parallel EL, parallel HL; HS along middle of fillet; integument silvery, white streak on upper surface above midline. Inside flesh coarse; belly flap rarely present; peritoneum black. Sometimes skinned, skin thin, scale pockets absent.

Size: To about 200 cm and 8 kg (commonly to about 100–135 cm and 1–2.5 kg).

Habitat: Marine; semi-pelagic near the bottom of the continental slope mainly in 300–600 m depth off the southern coastline. Occasionally ventures onto the continental shelf and has been caught at the surface.

Fishery: Bycatch of demersal trawling and droplining operations in southeastern Australia where it is landed in small to moderate quantities locally. However, total world catches can exceed 20 000 tonnes annually.

Remarks: Trawl caught specimens may look bedraggled where the skin has been removed. However, specimens taken by line usually have undamaged skin and supply a small niche market. The flesh makes excellent eating.

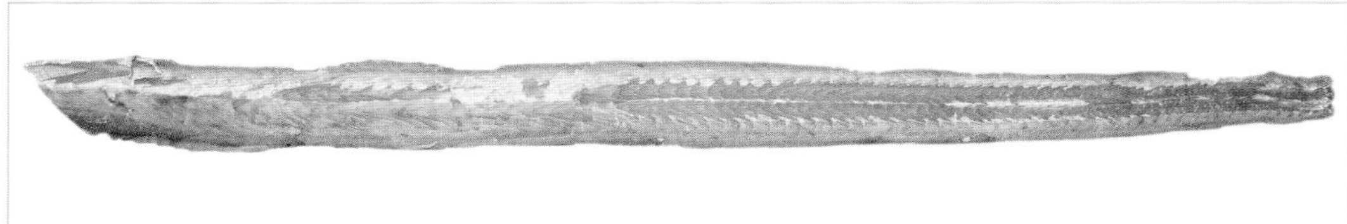

Untrimmed. Protein fingerprint p. 365.

Red mullet (page 1 of 2)

Mullidae

Previous names: barbounia, goatfish, western red mullet

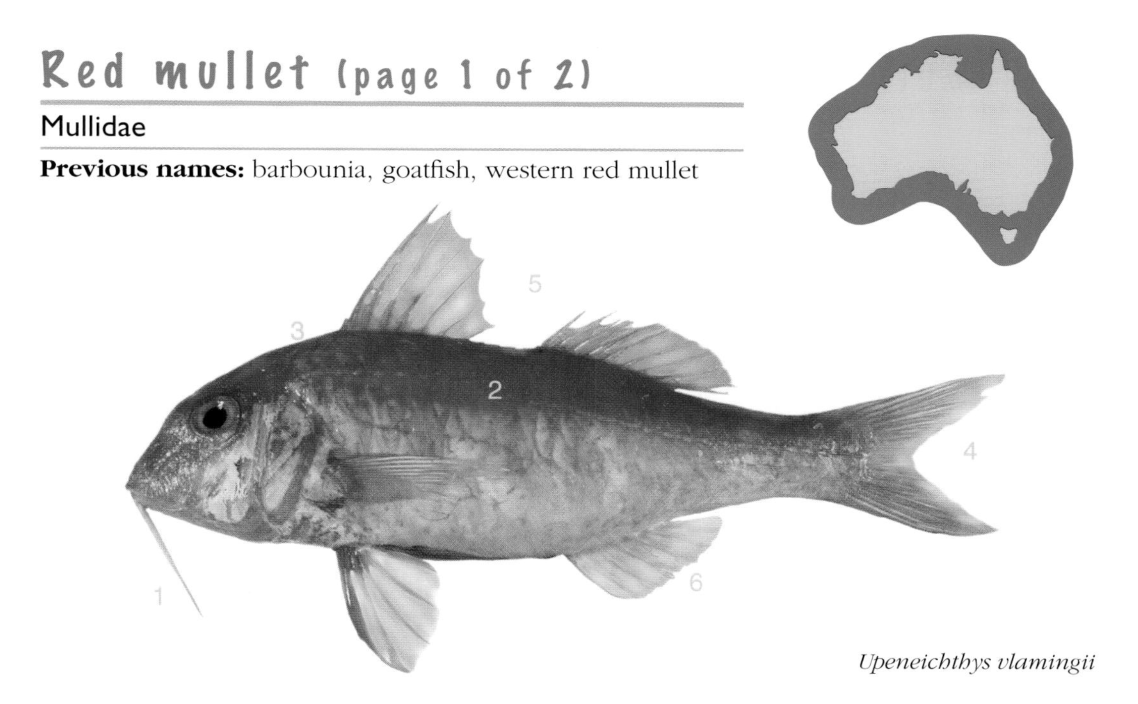

Upeneichthys vlamingii

Identifying features: ① 2 large moveable barbels on chin; ② sides often brightly coloured; ③ upper body profile more arched than lower profile; ④ caudal fin forked; ⑤ first dorsal fin with 6–8 spines, widely separated from second with 1 spine preceding 8–9 soft rays; ⑥ anal fin with 1–2 small spines, 5–8 soft rays.

Comparisons: Popular fishes in Europe where known under the same marketing name of 'red mullet' but are more widely known as 'goatfishes' (Mullidae). They have a distinctive appearance with a strongly curved upper surface and an almost straight lower surface designed for swimming near, and resting on, the bottom. Well developed barbels are located on the chin. Unrelated to the true mullets (Mugilidae), which have more streamlined bodies and swim mainly above the seabed.

Fillet: *Upeneichthys vlamingii* deep, rather elongate, tapering rapidly, pale pinkish. Outside with very broad, faint central red muscle band; parallel EL, converging HL; HS along middle of fillet; integument silvery pink. Inside flesh coarse; belly flap often present; peritoneum translucent. Usually skinned, skin medium, scale pockets large and irregular.

Size: To 50 cm and about 1.5 kg (commonly up to 35 cm and 0.8 kg).

Habitat: Marine; demersal on soft bottoms of the continental shelf to at least 200 m depth.

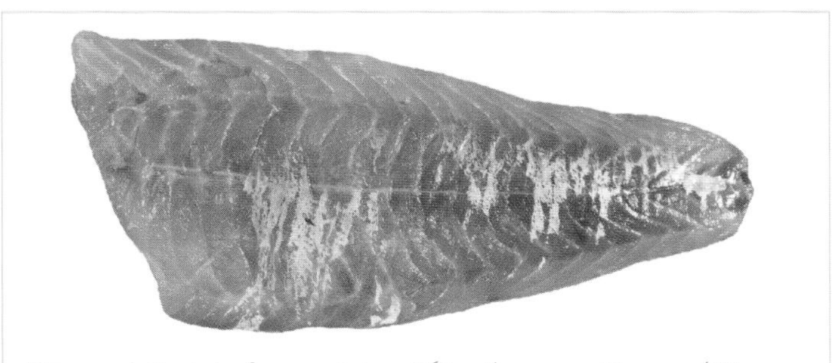

Trimmed. Protein fingerprint p. 365; oil composition p. 400.

Fishery: Caught by prawn and fish trawlers in the north. Important bycatch of the King George whiting (*Sillaginodes punctata*, p. 279) handline fishery in South Australia.

Remarks: Important food-fishes in much of Asia. Most local species have good flesh qualities but are under-rated by local consumers.

Red mullet (page 2 of 2)

Mullidae

Upeneichthys vlamingii

Remarks: Commonly called 'southern red mullet', this species is distributed south between Jurien Bay (WA) and Wilson's Promontory (Vic), including Tasmania, in depths to 80 m. Best distinguished in having relatively deep, reddish body covered in blue spots in adults. Juveniles are often almost white with a dark stripe through the eye to the upper caudal fin. Reaches at least 33 cm and 0.6 kg. Particularly popular as a foodfish in South Australia.

Upeneus tragula

Remarks: Commonly called 'bartail goatfish', this species is distributed in lagoons and near shallow reefs around northern Australia between Rottnest Island (WA) and Merimbula (NSW) in depths to 40 m. Best distinguished in having a pale (sometimes with reddish spots and blotches), slender body with a dark stripe through the eye to the mid-caudal fin. The caudal fin has black and white bars. Reaches about 30 cm and 0.4 kg.

Parupeneus indicus

Remarks: Commonly called 'yellowspot goatfish' or 'Indian goatfish', this species is distributed on the continental shelf of northern Australia from Shark Bay (WA) to Rockhampton (Qld) to at least 100 m depth. Best distinguished by the irregular yellowish blotch on the sides and a smaller black blotch on the caudal peduncle. Reaches about 50 cm and 1.1 kg.

Grunter bream

Pomadasys species

Previous names: javelin fish, smallspot javelin fish, spotted javelin fish

I

Pomadasys kaakan

Identifying features: 1 undersurface of lower jaw with a pore on each side, separated by a larger central groove; 2 sides mostly silvery (lacking vivid pattern); 3 second anal-fin spine very long and thick; 4 most of head scaly; 5 flange at base of preoperculum scaly with a serrated margin; 6 dorsal fin with 12–13 spines, 12–15 soft rays; 7 anal fin with 3 spines, 6–8 soft rays.

Comparisons: Readily identified by the presence of a pore on each side of the lower jaw followed by a central groove. Five species are known from Australian waters. They are often confused with grunters (Terapontidae, pp 140–141) and breams (Sparidae, pp 70–75). Unlike breams, they have a truncate tail (rather than being forked). Grunters, which are mostly found in freshwater, have 2 strong opercular spines (versus none or sometimes 1) and lack scales on the preopercular flange (otherwise scaly).

Fillet: *Pomadasys kaakan* moderately deep, rather elongate, tapering rapidly near peduncle, weakly convex above, white with brownish tinge. Outside with diffuse, intermediate central red muscle band; parallel EL, weakly converging HL; HS along middle of fillet; integument white. Inside flesh medium; belly flap sometimes present; peritoneum silvery white to translucent. Usually skinned, skin medium, scale pockets large and defined.

Size: To at least 70 cm and 5.1 kg (commonly to 50 cm and 2.2 kg).

Habitat: Marine; demersal, adults tend to frequent inshore waters including the lower reaches of estuaries, to a depth of about 40 m.

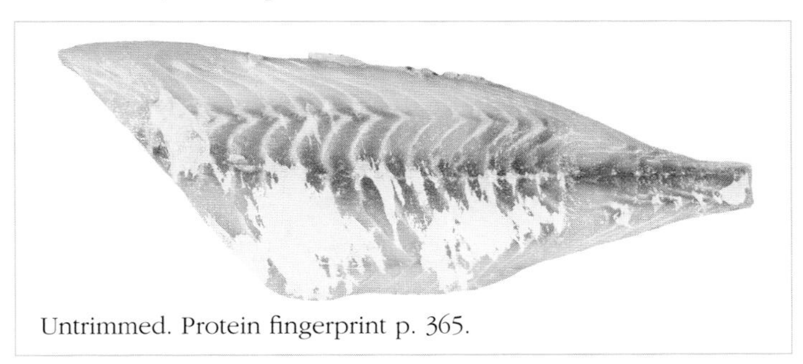

Untrimmed. Protein fingerprint p. 365.

Fishery: Caught either as bycatch of northern prawn fisheries, or targeted commercially using lines and gillnets.

Remarks: Make a characteristic grunting sound by grinding the teeth in the back of the throat when landed. Excellent foodfishes with firm flesh and a distinctive flavour.

Sweetlip bream (page 1 of 2)

Haemulidae except *Pomadasys* species

Previous names: black snapper, blackall, grey sweetlip, morwong, painted sweetlip, sand snapper, slate bream, sweetlip emperor, yellow sweetlip

Diagramma labiosum

Identifying features: ① undersurface of lower jaw with 3 pores on each side; ② sides vividly coloured or patterned (seldom silver all over); ③ second anal-fin spine long and thick; ④ most of head scaly; ⑤ flange at base of preoperculum scaly with a serrated margin; ⑥ dorsal fin with 8–14 spines, 13–26 soft rays; ⑦ anal fin with 3 spines, 6–8 soft rays.

Comparisons: Sweetlip breams, which have been given a host of confusing names, are not closely related to breams (Sparidae, p. 70–75). They are more conspicuously marked or coloured and their characteristic, thick, fleshy lips make them distinct. They also differ from the related grunter breams (*Pomadasys* species, p. 137) in having 3 pores on each side of the undersurface of the lower jaw (rather than 1, followed by a central groove) and a more striking colouration. However, these markings vary greatly with species and age.

Fillet: *Diagramma labiosum* moderately deep, rather elongate, tapering rapidly near peduncle, quite convex above, yellowish-white. Outside with continuous, intermediate central red muscle band; convex EL, converging HL; HS along middle of fillet; integument white. Inside flesh fine; belly flap sometimes present; peritoneum silvery white to translucent. Usually skinned, skin thick, scale pockets small and irregular.

Size: To 100 cm and about 12 kg (commonly to 60 cm and 3 kg).

Habitat: Marine; demersal on bottom types ranging from broken rubble on offshore reefs to inshore bays and estuaries.

Trimmed. Protein fingerprint p. 365.

Fishery: Caught as trawl bycatch, by lining on reef slopes, and by gillnets in estuaries.

Remarks: Targeted by spearfishers with local population declines noted in some regions. The flesh tends to be somewhat coarse and varies in taste from lightly flavoured to weedy.

Sweetlip bream (page 2 of 2)

Haemulidae except *Pomadasys* species

Diagramma labiosum

Remarks: Commonly called 'painted sweetlip bream', occurs along the continental shelf of northern Australia, including the outer reefs, from central New South Wales to Rottnest Island (WA). Has been called *D. pictum* in much recent literature. Best distinguished in having a relatively elongate body, and the dorsal fin barely notched with 9–10 spines. Adults are one of the few, almost plain silver-coloured sweetlip breams. Reaches 90 cm and about 6.3 kg.

Plectorhinchus flavomaculatus

Remarks: Commonly known as 'goldspot sweetlip bream', occurs on sheltered coastal reefs along northern Australia from central New South Wales to Perth (WA) in depths to about 40 m. Best distinguished in having a greyish-blue body and fins overlain with small orange spots and lines. Reaches 60 cm and about 4 kg.

Plectorhinchus polytaenia

Remarks: Commonly known as 'ribbon sweetlip bream', occurs on coastal reefs and offshore coral reefs around northern Australia from central Queensland to Onslow (WA) in depths to about 30 m. Best distinguished by the multiple, dark-edged, orange to yellow and white stripes that extend over the entire head and body. To 40 cm and about 1 kg.

Silver perch

Bidyanus bidyanus

Previous names: bidyan, black perch, bream, grunter

Identifying features: ① caudal fin forked slightly; ② 20–25 scales on midline before dorsal-fin origin; ③ end of mouth well short of eye; ④ uniformly silvery grey with dark fins; ⑤ pair of spines on each operculum; ⑥ single dorsal fin with a notch, 12–13 spines, 11–13 soft rays; ⑦ anal fin with 3 spines, 7–9 soft rays.

Comparisons: Plain-coloured freshwater fish that is quite similar to the sooty grunter (*Hephaestus fuliginosus*). It differs in having smaller scales (20–25 preceding the dorsal fin versus 10–18). Most other freshwater grunters are patterned. Its mouth is small compared to the larger freshwater perches (Percichthyidae).

Fillet: Moderately deep, rather elongate, tapering prominently, slightly convex above, white. Outside with intermediate, continuous, central red muscle band; convex EL, weakly converging HL; HS along middle of fillet; integument white. Inside flesh fine; belly flap sometimes present; peritoneum silvery white to translucent. Usually skinned, skin medium, scale pockets small and well defined.

Size: Possibly to 70 cm and 7.7 kg (commonly 35–40 cm and 0.7–1.5 kg).

Habitat: Freshwater; widespread in most of the Murray–Darling system. Prefers open, fast-moving water, particularly in rapids and on fish races.

Fishery: Wild fishery uses drum nets and gillnets. Restocking program has led to fingerlings being released in northern streams and impoundments. Resilient species, ideally suited for aquaculture and currently farmed in eastern states.

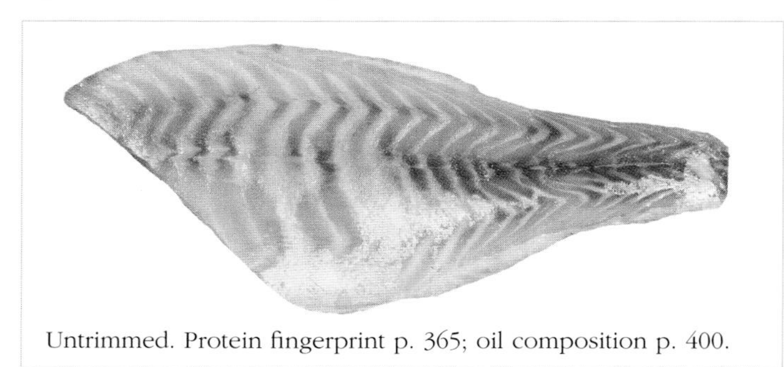

Untrimmed. Protein fingerprint p. 365; oil composition p. 400.

Remarks: Moves in schools upstream to spawn where it often becomes concentrated below dams and weirs. Together with other larger freshwater grunters, in particular the sooty grunter, likely to become more important with the expansion of fish farming. Good eating when fresh, otherwise can be dry.

Striped perch

Terapontidae

Previous names: grunter, tiger perch

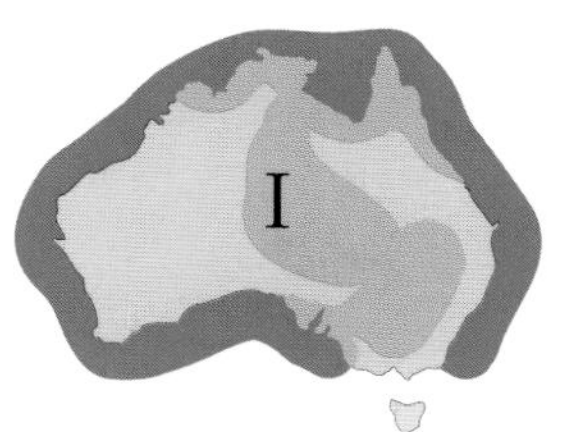

Pelates octolineatus

Identifying features: 1 caudal fin varying from forked slightly to rounded; 2 mouth small; 3 dorsal fin depressible into a groove formed by sheaths of scales; 4 pair of spines on each operculum; 5 single dorsal fin with a notch, 11–14 spines, 8–14 soft rays; 6 anal fin with 3 spines, 7–12 soft rays.

Comparisons: The true 'grunters' (Terapontidae) should not to be confused with the tropical marine 'grunter breams' (Haemulidae). Grunters, which resemble the freshwater perches (Percichthyidae) but have much smaller mouths, are particularly diverse in Australia with more than 30 species. Most occur in freshwater. The marine species, sometimes confusingly known as 'trumpeters', are silvery with dark stripes and bars.

Fillet: *Hephaestus jenkinsi* deep, short, tapering prominently, very convex above, yellowish-white. Outside with feeble, diffuse, central red muscle band; convex EL, converging HL; HS along middle of fillet; integument white. Inside flesh medium; belly flap sometimes present; peritoneum silvery white to translucent. Usually skinned, skin medium, scale pockets moderate and defined.

Size: To 45 cm and 1.1 kg (commonly 20–35 cm and 0.2–0.6 kg).

Habitat: Freshwater, estuarine, and marine in inshore coastal waters.

Fishery: Small quantities taken as bycatch using a variety of netting methods. The western sooty grunter (*H. jenkinsi*) has shown potential for aquaculture in Western Australia. Important recreational species in some areas.

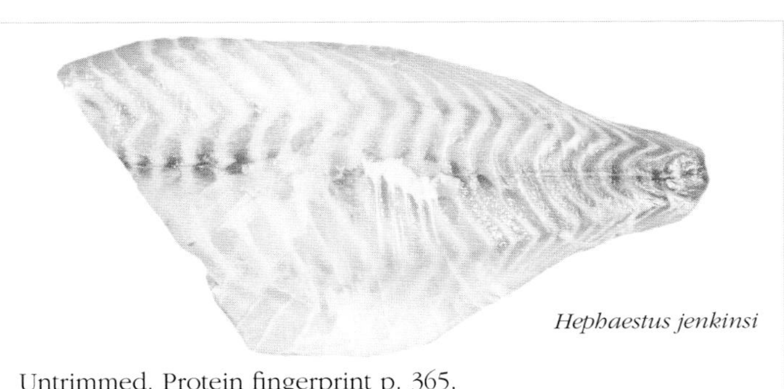

Hephaestus jenkinsi

Untrimmed. Protein fingerprint p. 365.

Remarks: Most species considered too small to be marketed. Their ability to produce loud noises from their swim bladders when captured has led to them being called 'grunters'. The flesh is reasonable but small, marine species are often used as bait.

Butterfly gurnard

Lepidotrigla species

Previous names: cocky gurnard, gurnard

Lepidotrigla vanessa

Identifying features: 1 body scales large with rough edges; 2 upper surface of pectoral fin blue edged but lacking numerous blue lines or spots; 3 lateral-line scales distinctly larger than those adjacent; 4 upper pectoral fin greatly enlarged, almost wing-like; 5 lower pectoral fin with 2–3 finger-like rays; 6 head covered in a bony casing without scales.

Comparisons: Gurnards (Triglidae) are a distinctive group of fishes with a strong bone-cased head and large, wing-like pectoral fins. About 20 Australian species of the genus *Lepidotrigla* can be marketed as 'butterfly gurnard' but many of these are small. They have much larger and rougher scales than other commercial gurnards. The main species landed is commonly called 'butterfly gurnard' (*L. vanessa*).

Fillet: *L. vanessa* moderately deep, elongate, pronounced taper, yellowish-white. Outside with feeble, continuous, central red muscle band; weakly converging EL, weakly converging HL; HS along middle of fillet; integument white. Inside flesh fine; belly flap sometimes present; peritoneum translucent. Sometimes skinned, skin thick, scale pockets small and defined.

Size: To 32 cm and 0.4 kg (commonly up to 28 cm and 0.3 kg).

Habitat: Marine; demersal on soft bottoms from inshore coastal bays to the outer continental shelf in depths to 180 m.

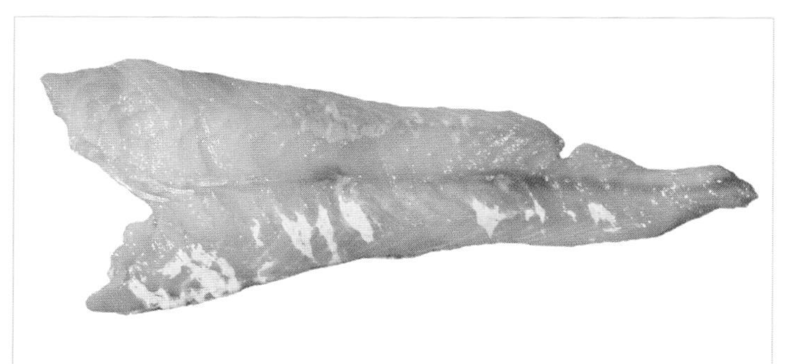

Untrimmed. Protein fingerprint p. 365; oil composition p. 400.

Fishery: Small gurnards taken as bycatch of trawl operations on continental shelf.

Remarks: Also known as 'cocky gurnard', they are often taken in greater quantity than the larger commercial members of the family. Until recently, much of the catch has been discarded. The delicate flesh is reasonably sweet and firm.

Latchet

Pterygotrigla polyommata

Previous names: gurnard, sharpbeak gurnard, spinybeak gurnard

Identifying features: ① long spines at tip of snout; ② upper surface of pectoral fin with numerous yellow spots or concentric blue and yellow lines; ③ body scales small with smooth edges; ④ lateral-line scales slightly larger than those adjacent; ⑤ upper pectoral fin greatly enlarged, almost wing-like; ⑥ lower pectoral fin with 3 finger-like rays; ⑦ head covered in a bony casing without scales.

Comparisons: Distinctive fish that can be readily distinguished from other commercial gurnards by the presence of 2 pointed spines on the snout tip.

Fillet: Moderately deep, elongate, pronounced taper, brownish-white. Outside with pronounced, continuous, central red muscle band; weakly converging EL, converging HL; HS along middle of fillet; integument silvery white. Inside flesh fine; belly flap sometimes present; peritoneum silvery white to translucent. Sometimes skinned, skin thick, scale pockets tiny and indistinct.

Size: To 62 cm and 2.6 kg (commonly 35–45 cm and 0.6–1.5 kg).

Habitat: Marine; demersal on soft bottoms of the outer continental shelf and upper slope to 450 m depth. Juveniles venture into large bays and estuaries in Tasmania.

Fishery: Large gurnard, taken mainly by demersal trawlers and Danish seiners off southern Australia, mainly in depths of 80–220 m.

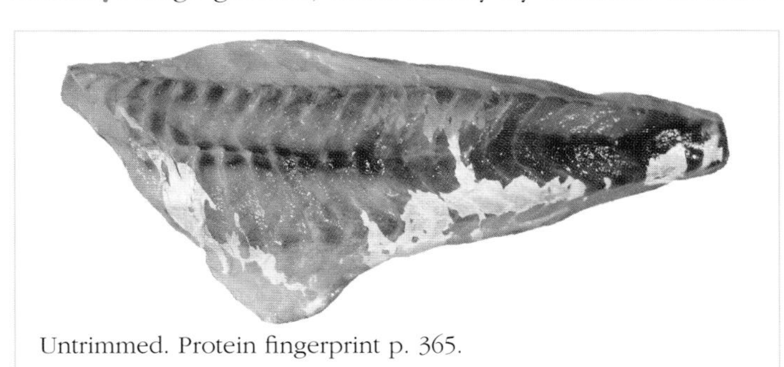

Untrimmed. Protein fingerprint p. 365.

Remarks: A second species, the painted latchet (*P. andertoni*), is taken in moderate quantities off New South Wales and probably should be sold under the same marketing name. Usually sold whole chilled, the fresh is of good quality.

Red gurnard

Chelidonichthys kumu

Previous names: flying gurnard, gurnard, kumu gurnard, kumukumu, latchet

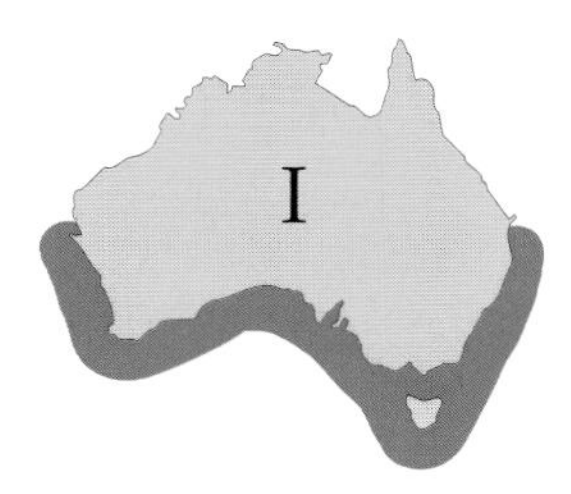

Identifying features: ① no long spines at snout tip; ② upper surface of pectoral fin with blue edge and numerous blue spots centrally; ③ body scales small with smooth edges; ④ lateral-line scales slightly larger than those adjacent; ⑤ upper pectoral fin greatly enlarged, almost wing-like; ⑥ lower pectoral fin with 3 finger-like rays; ⑦ head covered in a bony casing without scales.

Comparisons: Similar in form to the latchet (*Pterygotrigla polyommata*, p. 143) in having small, smooth-edged scales and a very large pectoral fin but lacks long spines on the snout tip and has blue spots on the mid-upper surface of the pectoral fin (rather than yellow spots or blue and yellow lines).

Fillet: Rather slender, tapering gently, yellowish-white. Outside with pronounced, continuous, central red muscle band; weakly converging EL, weakly converging HL; HS along middle of fillet; integument translucent above, silvery white below. Inside flesh fine; belly flap sometimes present; peritoneum translucent. Sometimes skinned, skin thick, scale pockets small and indistinct.

Size: To 53 cm and 2 kg (commonly up to 40 cm and 1.2 kg).

Habitat: Marine; demersal on soft bottoms, mainly on the inner continental shelf to depths of 100 m, often venturing into shallow water in large bays and estuaries.

Fishery: Caught mainly by demersal trawlers and Danish seiners off southern Australia. Catches are usually small.

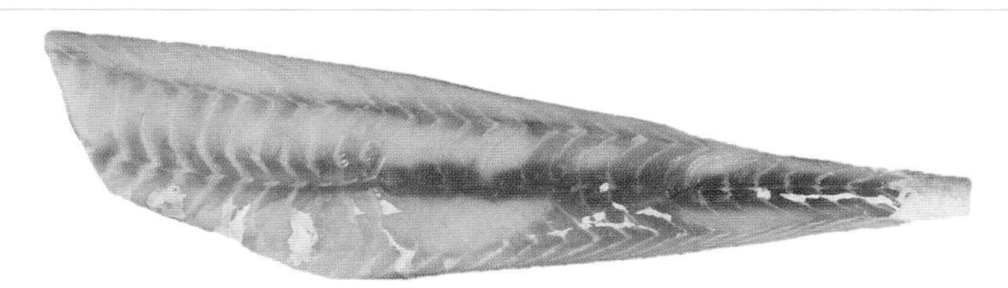

Untrimmed. Protein fingerprint p. 365.

Remarks: Should be handled carefully as the fin spines contain venom glands. Considered to be the best eating member of the group.

Anchovy

Engraulis australis

Previous names: none

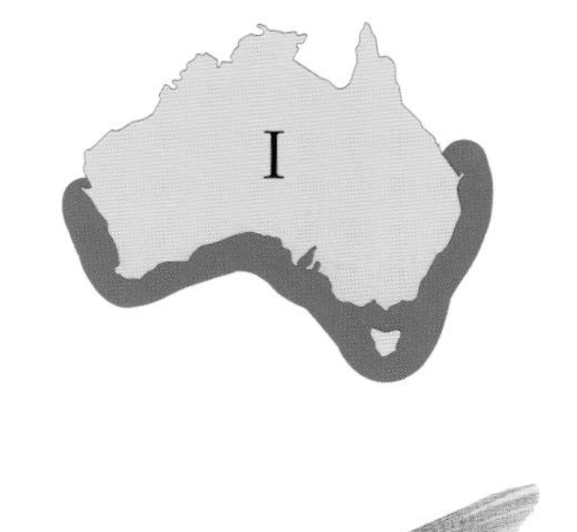

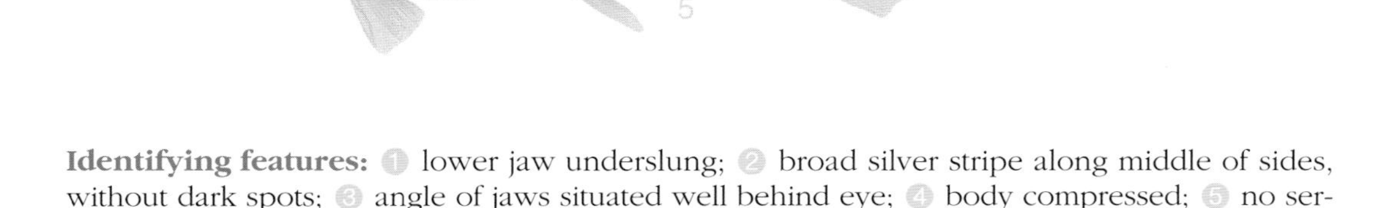

Identifying features: 1 lower jaw underslung; 2 broad silver stripe along middle of sides, without dark spots; 3 angle of jaws situated well behind eye; 4 body compressed; 5 no serrated ridge on belly; 6 1 small dorsal fin; 7 large easily removed scales.

Comparisons: Anchovies are distinctive herring-like fishes that have a long, thin, underslung lower jaw that extends back well behind the eye. Other herring-like fishes have a deep lower jaw that articulates under, or only slightly behind, the eye level. Of the 24 or so Australian species, only the main temperate species is important commercially. The anchovy's tropical relatives have more compressed bodies and scutes on the belly before the pelvic fin (otherwise absent).

Fillet: Moderately elongate, barely tapering, brownish-white. Outside with pronounced, continuous red muscle band; EL, HL and HS usually obscured by integument; integument covering most of side, translucent with melanophores above, silvery below. Inside flesh fine; belly flap indistinct; peritoneum transparent. Sometimes skinned, skin very thin, scale pockets barely detectable.

Size: To 16 cm and less than 0.1 kg (commonly to about 10 cm).

Habitat: Marine; pelagic between the surface and midwater, schooling over the continental shelf, seasonally entering deep bays and estuaries to breed.

Fishery: Caught in estuaries and nearshore waters using hoop nets, and haul and purse seines in all temperate states except South Australia.

Remarks: Three populations exist (off western, southeastern and eastern Australia) that may turn out to be separate species. More research is needed. Anchovies feed on small plankton hence the flesh is soft and oily. They are important dietary items of several large fishes and seabirds. Canned, salted, pickled or turned into fish paste. Also used for bait by both commercial and recreational fishers.

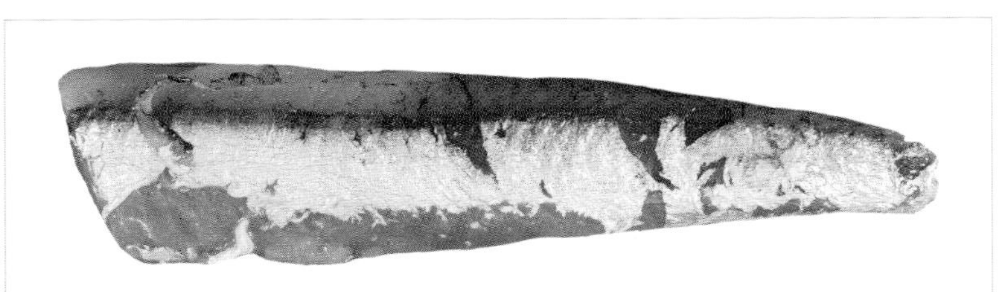

Untrimmed. Protein fingerprint p. 365.

Blue sprat

Spratelloides robustus

Previous names: none

Identifying features: ① body slightly compressed, belly rounded; ② narrow silver stripe on side not edged with dark spots; ③ dark horse-shoe marking at base of caudal fin; ④ no serrated scale ridge on belly; ⑤ 1 small dorsal fin; ⑥ upper and lower jaws extend forward equally; ⑦ large easily removed scales.

Comparisons: Belongs to the family of herrings, sardines, shads, and pilchards (Clupeidae). It is unusual within this group in having a rounded belly (rather than compressed) with smooth, unserrated scales and few anal fin rays (9–12 versus 16–29). A distinctive W-shaped scute is situated at the pelvic-fin base.

Fillet: Too small to fillet.

Size: To 10 cm and less than 0.1 kg (commonly to 8 cm). Reports to 29 cm are likely to be erroneous.

Habitat: Marine; pelagic, in huge schools in shallow inshore waters. Widespread in bays, inlets and estuaries, often in the vicinity of seagrasses.

Fishery: Caught mainly for use as bait along with anchovies, using hoop nets, haul nets and purse seines. Most abundant in Bass Strait and small quantities reach the Melbourne market as whole, fresh unprocessed fish. Also targeted off southwestern Western Australia and used locally as bait.

Remarks: Substantial resource possibly exists in southern waters but its potential remains undefined. Up to 5000 tonnes are taken annually in a fishery for the equivalent Japanese species. The flesh, which is soft and oily, is too small to fillet. Alternative preparations include pan frying and grilling of whole, headed fish. Similar species are eaten this way in Asia. Other information is provided for the protein fingerprint (p. 365).

Bony bream

Nematalosa erebi & N. vlaminghi

Previous names: none

Nematalosa erebi

Identifying features: 1 deep-bodied; 2 sides uniformly silvery; 3 no serrated ridge along midline of shoulder; 4 strong serrated ridge on belly midline; 5 1 small dorsal fin with long hind ray; 6 lower-jaw tip under upper jaw.

Comparisons: Deep-bodied shads (Clupeidae) in which the upper jaw bones form a sharp angle where they join. When viewed from the front, the centre of the upper jaw resembles an inverted V (rather than being rounded as in other commercial relatives).

Fillet: *N. erebi* moderately deep, elongate, barely tapering, pale pinkish-brown. Outside with very feeble, continuous red muscle band; parallel EL, converging HL; HS along middle of fillet; integument patchy, silver white. Inside flesh medium; belly flap usually indistinct; peritoneum transparent. Often skinned, skin medium, scale pockets large and well defined.

Size: To about 39 cm and 0.8 kg (commonly marketed at 18–25 cm and 0.1–0.2 kg).

Habitat: Freshwater; pelagic, schools in freshwater, estuaries, and coastal inlets. Also known from artesian bores. Freshwater species are most common in slow-flowing waters.

Fishery: Important inland fishery in South Australia. In other parts of Australia taken mainly as bycatch from other inshore estuarine and freshwater operations. Once discarded but now retained in increasing quantities. Targeted in estuaries and embayments of southwestern Western Australia. Often used as rocklobster bait.

Remarks: Sometimes use their soft, toothless mouths to suck up mud. Small invertebrates are filtered out using their gills. Mostly consumed by the Asian population, where these fishes are commercially important. The flesh, which needs special preparation, is extremely bony (as their name suggests) and often has a muddy taste. Potential to value add, pickled or as roll-mops.

Trimmed. Protein fingerprint p. 365; oil composition p. 400.

Pilchard

Sardinops neopilchardus

Previous name: sardine

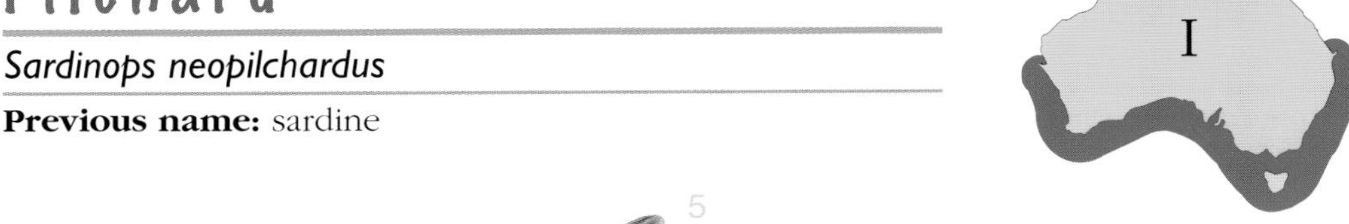

Identifying features: 1 body almost oval in cross-section; 2 narrow silver stripe along side edged below with dark blue spots; 3 no serrated ridge along midline of shoulder; 4 belly midline scales forming serrated ridge; 5 1 small dorsal fin; 6 upper and lower jaws extend forward equally; 7 large easily removed scales.

Comparisons: Members of the genus *Sardinops* are mainly referred to internationally as pilchards. They resemble the true sardine (*Sardina pilchardus*) of European waters in form but have a longer jaw (reaching about mid-eye level) and the middle rakers of the first gill arch are greatly shortened. Of other domestic commercial species, it can be distinguished from the blue sprat by its serrated belly.

Fillet: Moderately deep, barely tapering, reddish-brown. Outside with very pronounced, continuous red muscle band; weakly converging EL, weakly converging HL; HS just above middle of fillet; integument covering most of side, silver grey above, silvery white below. Inside flesh medium; belly flap indistinct; peritoneum usually evident, translucent to black. Often skinned, skin very thin, scale pockets large and irregular.

Size: To 21 cm and less than 0.1 kg (commonly to about 18 cm).

Habitat: Marine; pelagic in mid- and upper water column, mainly schooling offshore over the continental shelf to its edge, occasionally nearing the coast. Inshore schools more common off Western Australia.

Fishery: Important commercial pelagic fish caught mainly by purse seine. Smaller quantities are taken for live bait using lampara nets and occasionally beach seines. Otherwise processed for canning, as petfood or as fodder for the southern bluefin tuna (*Thunnus maccoyii*, p. 176) mariculture industry. Increasingly marketed fresh for human consumption in Western Australia.

Remarks: Sometimes referred to as *Sardinops sagax* and the correct scientific name remains a moot point. Important prey species for a diversity of large fishes, birds and marine mammals. Seasonal mass die-off has been observed in recent years which may impact greatly on their numbers. The flesh is soft and oily.

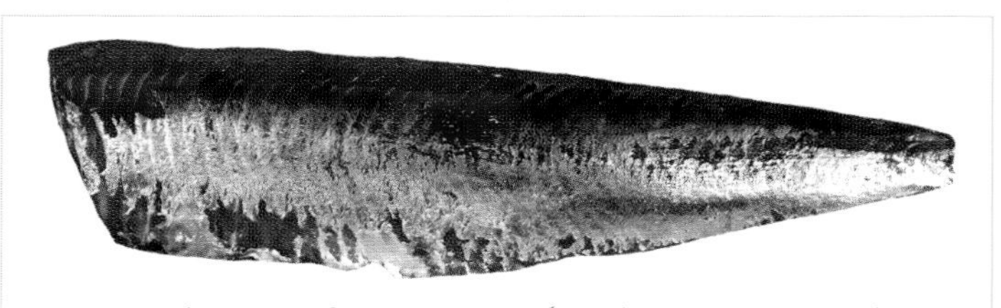

Untrimmed. Protein fingerprint p. 365; oil composition p. 400.

Sandy sprat

Hyperlophus vittatus

Previous names: glassy, white pilchard, whitebait

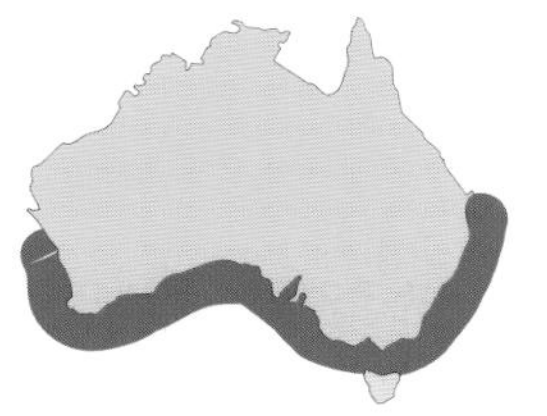

Identifying features: 1 body compressed; 2 broad silver stripe along sides lacking dark spots; 3 serrated scale ridge on midline of shoulder; 4 serrated scale ridge on belly midline; 5 1 small dorsal fin; 6 upper and lower jaws extend forward equally; 7 large, easily removed scales.

Comparisons: Distinctive herring-like fish with a serrated ridge of scales along the upper midline before the dorsal fin as well as on the belly. Its body is more compressed with a broader silver stripe than other commercial relatives. The similar sprat (*Sprattus novaehollandiae*), overlaps in distributional range but occurs further offshore. It lacks scutes on the shoulder.

Fillet: Moderately deep, tapering slightly, yellowish-white to greyish. Outside with intermediate, continuous red muscle band peppered with melanophores; weakly converging EL, weakly converging HL; HS along middle of fillet; integument silvery. Inside flesh fine; belly flap indistinct; peritoneum silvery translucent. Sometimes skinned, skin thin, scale pockets poorly defined.

Size: To 10 cm and less than 0.1 kg (commonly to about 9 cm).

Habitat: Marine; pelagic, aggregating in shallow marine bays, estuaries and off beaches in large shoals.

Fishery: Caught throughout its range in moderate quantity in small-mesh beach seines and bait nets.

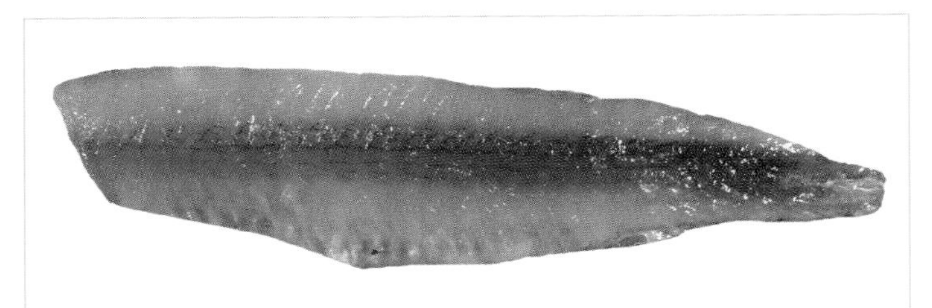

Untrimmed. Protein fingerprint p. 365; oil composition p. 400.

Remarks: Important prey species for a suite of larger fishes and seabirds. Moderate quantities packaged and frozen to be used by anglers for bait. Sold as 'whitebait', mainly in Melbourne, Sydney and Perth markets, for human consumption. Likely to be of increasing commercial value.

Patagonian toothfish

Dissostichus eleginoides

Previous names: Australian sea bass, sea bass, toothfish

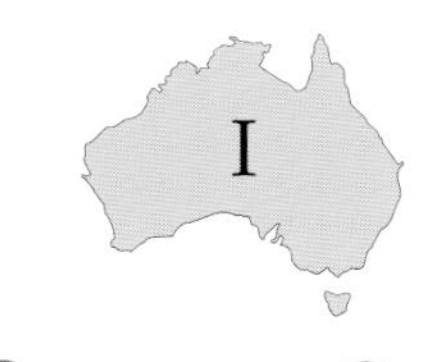

Identifying features: ① top of head with several, small, narrow scaleless patches; ② sides brownish-grey with faint darker blotches; ③ 2 separate lateral lines; ④ pelvic-fin base forward of pectoral-fin base; ⑤ single nostril on each side of head; ⑥ dorsal fin with 8–11 spines, 26–31 soft rays; ⑦ anal fin with 26–31 soft rays.

Comparisons: Large elongate, round-bodied toothfish (Nototheniidae) that is distinct from all other Australian commercial fishes in a combination of the characters above.

Fillet: Moderately deep, elongate, upper profile almost straight, taper slight, milky white. Outside with feeble, diffuse red muscle band; convex EL, converging HL; HS along middle of fillet; myomeres very narrow; integument white above, silvery below. Inside flesh fine; belly flap usually present; peritoneum translucent white with melanophores. Usually skinned, skin thick, scale pockets small and well defined.

Size: To 215 cm and 100 kg (commonly 40–65 cm and 1–3.5 kg).

Habitat: Marine; pelagic in both midwater and near the bottom of canyons in depths of 70–1500 m.

Fishery: New, strictly controlled target fishery, using demersal trawls and longlines, located on subAntarctic ridges and seamounts near Heard, Macquarie and MacDonald Islands in the remote Southern Ocean. Poaching of this resource by foreign vessels is an important jurisdictional issue.

Remarks: Incorrectly marketed under the highly confusing names 'Australian sea bass' or 'sea bass'. Toothfishes are unrelated to the sea basses (Serranidae), which include the coral trouts and rockcods (pp 213–227), and their stocks are not confined to the Australian region. Patagonian toothfish are also taken off southern South America and Kerguelen Island. Exported frozen, with a developing local market. Highly regarded, premium foodfish both locally and overseas.

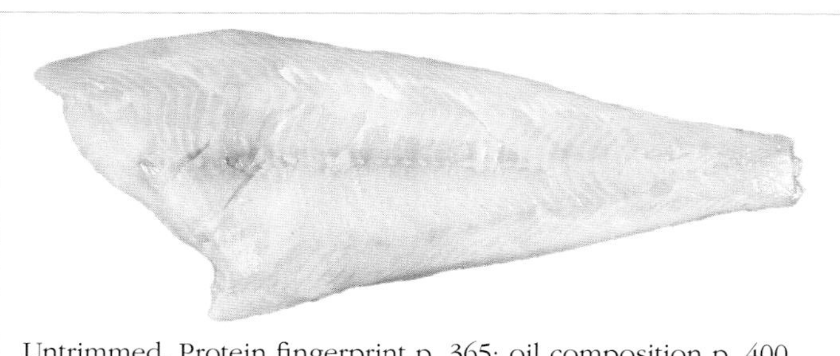

Untrimmed. Protein fingerprint p. 365; oil composition p. 400.

Black jewfish

Protonibea diacanthus

Previous names: black jew, mulloway, spotted croaker, spotted jewfish

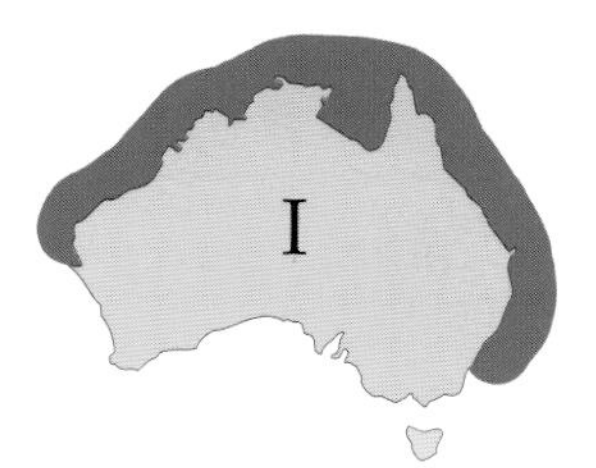

Identifying features: ① sides mostly with silvery black hue, smaller specimens have dark bands and small black spots on upper half; ② no dark spot above pectoral-fin base; ③ 3 pairs of sensory pores on chin; ④ end of caudal fin distinctly pointed; ⑤ operculum with a bony flap at top; ⑥ dorsal fin long, deeply notched with 9–10 spines in first portion, 1 spine and 22–25 soft rays in second portion; ⑦ anal fin with 2 spines, 7 soft rays.

Comparisons: Resembles the mulloway (*Argyrosomus hololepidotus*, p. 153) but has a more pointed tail and lacks a dark spot above the pectoral-fin base. The body becomes darker on death giving rise to its common name.

Fillet: Moderately elongate, narrow bottle-shaped, yellowish-white. Outside with feeble, continuous central red muscle band; convex EL, weakly converging HL; HS along middle of fillet; integument white. Inside flesh medium, with red muscle spots above midline; belly flap sometimes present; peritoneum silvery white to translucent. Usually skinned, skin thick, scale pockets very large and well defined.

Size: To 150 cm and about 40 kg (commonly to 100 cm and 8.2 kg).

Habitat: Marine; inshore in coastal bays, tidal rivers and estuaries on soft bottoms to a depth of about 60 m.

Fishery: Centred off the Northern Territory using bottom trawls and gillnets. Commonly taken in the Gulf of Carpentaria (NT and Qld).

Untrimmed. Protein fingerprint p. 366.

Remarks: Largest of the tropical jewfishes, it is extremely well-regarded as an angling fish, with renowned fighting abilities. It is mainly marketed whole or gutted and frozen, in major cities of the southern states, notably Perth. The flesh is of good eating quality.

Jewfish

Sciaenidae

Previous names: croaker, drum, jew, river jew, silver teraglin

Johnius borneensis

Identifying features: 1 body moderately elongate with silver reflections; 2 eyes near top of head; 3 bony flap at top of gill opening; 4 conspicuous sensory pores over snout tip and chin; 5 end of caudal fin slightly concave to pointed; 6 dorsal fin long, deeply notched with 9–10 spines in first portion, 1 spine and 21–44 soft rays in second portion; 7 anal fin with 2 spines, 6–12 soft rays.

Comparisons: Jewfishes have a characteristic body shape with a long, notched dorsal fin, short anal fin, and a square to pointed caudal fin. They should not be confused with the West Australian dhufish (*Glaucosoma hebraicum*, p. 200) which belongs to a group known as pearl perches (Glaucosomatidae).

Fillet: *Johnius borneensis* moderately elongate, tapering gradually, yellowish-white. Outside with intermediate, continuous central red muscle band; parallel EL, weakly converging HL; HS along or just below middle of fillet; integument silvery and white. Inside flesh medium, with faint red muscle spots above midline; belly flap often present; peritoneum translucent. Usually skinned, skin medium, scale pockets moderate and defined.

Size: To 180 cm and at least 60 kg (commonly 30–120 cm and 0.4–15 kg).

Habitat: Marine; demersal, mainly on sand and mud on the continental shelf, to at least 200 m depth. Many live inshore in coastal habitats and estuaries and some venture into freshwater.

Fishery: Generally targeted by line fishers. Several species also caught as bycatch of trawl, and gillnet fishing, mainly in tropical areas.

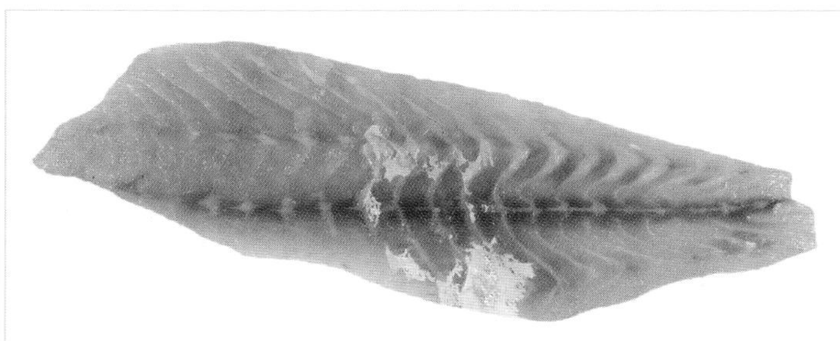

Trimmed. Protein fingerprint p. 366; oil composition p. 400.

Remarks: At least 15 local species. Many are small but some are among the region's largest fishes with considerable importance as foodfishes. Apart from the three with specific marketing names at least two others, the river jewfish (*J. borneensis*) and silver teraglin (*Otolithes ruber*), are marketed commonly.

Mulloway

Argyrosomus hololepidotus

Previous names: butterfish, jewfish, kingfish, river kingfish

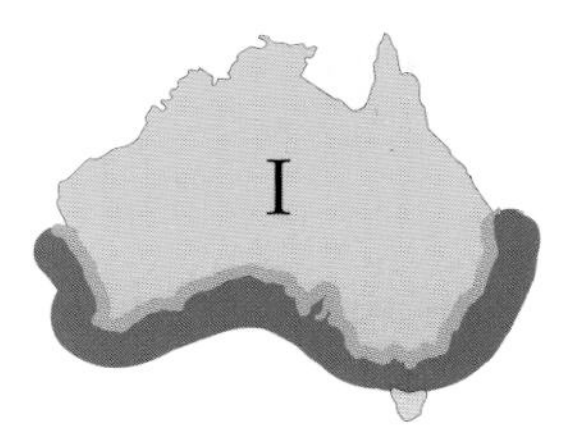

Identifying features: 1 sides mostly silvery bronze above grading to white below; 2 dark spot above pectoral-fin base; 3 3 pairs of sensory pores on chin; 4 end of caudal fin square or pointed slightly; 5 operculum with a bony flap at top, orange inside; 6 dorsal fin long, deeply notched with 10 spines in first portion, 1 spine and 26–30 soft rays in second portion; 7 anal fin with 2 spines, 7 soft rays.

Comparisons: Can be confused with the teraglin (*Atractoscion aequidens*, p. 154) but has a pointed or square caudal fin (rather than concave) and fewer anal-fin rays (7 rather than 9–10). Live mulloway have a row of pearly spots along the lateral line.

Fillet: Moderately elongate, tapering gradually, pale pinkish. Outside with pronounced, continuous central red muscle band; parallel EL, weakly converging HL; HS along middle of fillet; integument grey above, silver and white below. Inside flesh medium, with red muscle spots above midline; belly flap often present; peritoneum white translucent. Usually skinned, skin medium, scale pockets moderate and defined.

Size: To 180 cm and 61 kg (commonly to 150 cm and about 35 kg).

Habitat: Marine; demersal, mainly in coastal embayments and estuaries but also occur off ocean beaches and on inshore reefs to depths of about 100 m.

Fishery: Caught throughout range, mainly near mouths of large rivers or off surf beaches using a variety of methods that include lines, seines, gillnets, and fish trawls. Also taken as bycatch of prawn trawling on inshore and estuarine grounds.

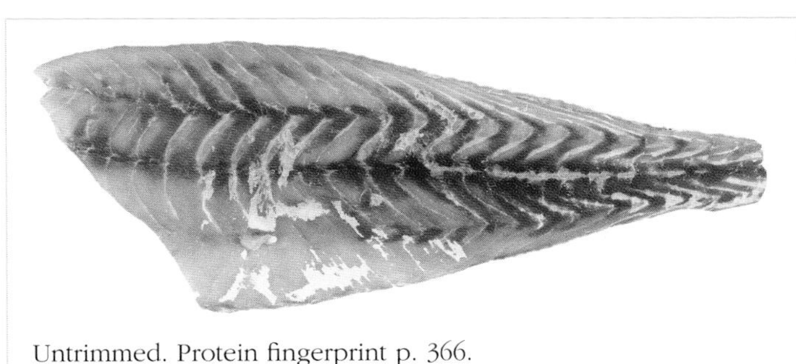

Untrimmed. Protein fingerprint p. 366.

Remarks: Recently called *A. japonicus* and further research required. Largest of Australia's jewfishes, it is regarded as a prestige angling fish. Commonly sold fresh gutted. Medium-sized fish are good eating but the flesh of small fish is often 'soapy' and that of large fish is soft and sometimes infested with worms.

Teraglin

Atractoscion aequidens

Previous names: jew, teraglin jew, trag

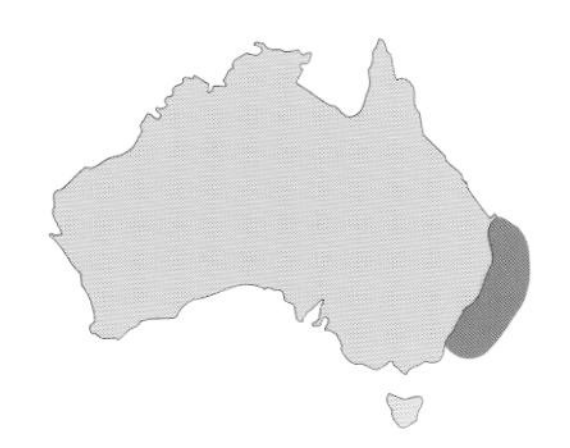

Identifying features: ① bluish to silvery with pinkish opalescence; ② dark spot above pectoral-fin base; ③ 2 pairs of minute sensory pores on chin (barely distinct); ④ hind margin of caudal fin slightly concave; ⑤ operculum with a bony flap at top, and bright yellow inside; ⑥ dorsal fin long, deeply notched with 9–10 spines in first portion, 1 spine and 28–32 soft rays in second portion; ⑦ anal fin with 2 spines, 9–10 soft rays.

Comparisons: Resembles the mulloway (*Argyrosomus hololepidotus*, p. 153) but has an indented caudal fin (rather than pointed or square) and more anal-fin rays (9–10 rather than 7). The canary yellow inner operculum and jaw edges are striking. A less important commercial species, the silver teraglin (*Otolithes ruber*), has a pointed tail and a few enlarged canines near the jaw tip.

Fillet: Moderately elongate, tapering gradually, pale pinkish. Outside with intermediate, continuous central red muscle band; converging EL, weakly converging HL; HS along middle of fillet; integument greyish above, translucent below. Inside flesh medium, with red muscle spots above midline; belly flap often present; peritoneum white translucent. Usually skinned, skin medium, scale pockets small and irregular.

Size: Supposedly to about 120 cm and 19.5 kg (commonly 40–75 cm and 0.8–3.5 kg).

Habitat: Marine; demersal, schooling mainly over gravel or broken reef in depths of 20–80 m, occurs broadly over continental shelf to at least 100 m. Juveniles occur in deep estuaries.

Fishery: Taken in small quantities by lining and trapping off New South Wales. Catches have increased recently after a period of decline. Sometimes significant in recreational catches.

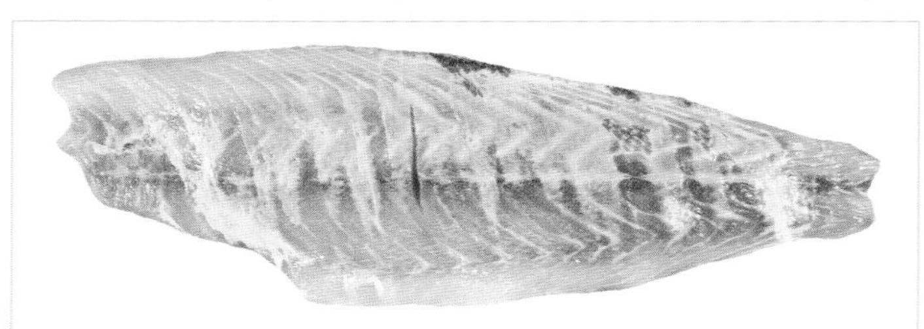

Trimmed. Protein fingerprint p. 366; oil composition p. 400.

Remarks: Occurs more offshore than the mulloway. Rises from the bottom at night to feed. Not noted to be a great fighter when caught; nevertheless, they are eagerly sought-after by recreational fishers for their excellent eating qualities. Regularly marketed in eastern Australia.

Leatherjacket (page 1 of 2)

Monacanthidae

Previous names: seine boat jacket, silver flounder

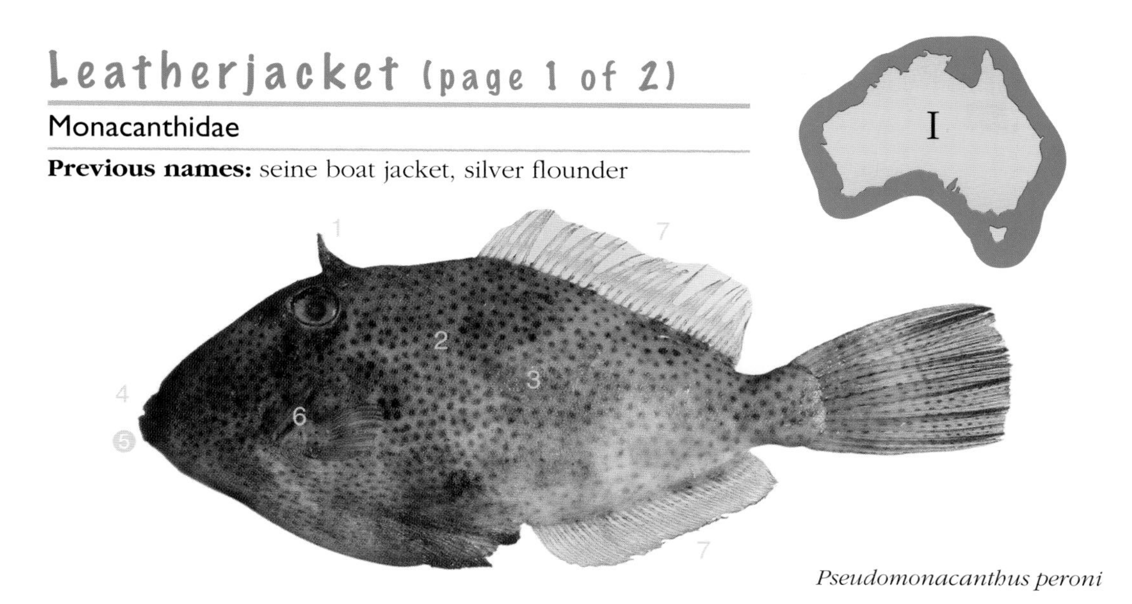

Pseudomonacanthus peroni

Identifying features: 1 first dorsal fin with 2 spines, with the first much longer and stronger; 2 body compressed; 3 skin prickly or 'furry' (without normal scales); 4 mouth very small; 5 enlarged, wide teeth, forming a 'beak'; 6 gill opening reduced to a slit above the pectoral-fin base; 7 second dorsal and anal fins almost identical in shape with unbranched soft rays.

Comparisons: Leatherjackets (Monacanthidae) are easily recognised by their velvety skin (formed of minute spiny scales), prominent first dorsal-fin spine (capable of being locked upright), small mouth and compressed body. Their tropical and less commercially important relatives, the trigger fishes (Balistidae), are similar in appearance but have 3 dorsal-fin spines (rather than 2) and hard plate-like scales covering their body.

Fillet: *Pseudomonacanthus peroni* deep, rather elongate, upper profile slightly convex, taper slight, off-white to yellowish. Outside without red muscle band; converging EL, converging HL; HS along middle of fillet; integument white. Inside flesh fine; belly flap usually absent; peritoneum silvery white to translucent. Always skinned, skin thick and spiny, scale pockets minute and embedded.

Size: To 76 cm and about 3.5 kg (commonly to about 38 cm and 0.8 kg).

Habitat: Marine; demersal and pelagic on the continental shelf and upper slope, often near reefs and sponge beds, to depths of about 200 m.

Untrimmed. Protein fingerprint p. 366; oil composition p. 401.

Fishery: Bycatch of trapping and trawling around the Australian coastline, with greatest catches made on the continental shelf. Small inshore trap fishery near Albany (WA).

Remarks: Males and females differ in colour and shape with the former being more brightly or intensely ornamented. Usually sold as fresh fillets or headed and skinned. The firm flesh varies from sweet to tasting slightly of iodine.

Leatherjacket (page 2 of 2)

Monacanthidae

Pseudomonacanthus peroni

Remarks: Commonly called 'potbelly leatherjacket', this species is distributed on deep parts of the continental shelf of both eastern and western coasts south to Byron Bay (NSW) and Perth (WA) respectively (including the Gulf of Carpentaria) probably to depths of 200 m. Best distinguished in having brown spots on the side and a large belly flap and caudal fin. To at least 38 cm and 0.7 kg.

Aluterus monoceros

Remarks: Commonly called 'unicorn leatherjacket', this species is distributed across northern Australia on the continental shelf between Perth (WA) and Sydney (NSW) to depths of about 100 m. Best distinguished in having a plain colour, almost oval shape, deep head with an extremely convex chin, and long, thin dorsal spine. Reaches at least 76 cm and 3.2 kg.

Thamnaconus degeni

Remarks: Commonly called 'Degen's leatherjacket', this species is distributed on the continental shelf between Wilsons Promontory (Vic.) and the eastern Great Australian Bight (including Tasmania) in depths of 15–70 m. Males best distinguished by the uniform bluish-green fins and bright blue lines and stripes on the sides. Females have similar fin colour but lack bluish spots. Reaches at least 28 cm and 0.3 kg.

Ocean jacket

Nelusetta ayraudi

Previous names: chinaman, chinaman leatherjacket, leatherjacket, yellow jacket

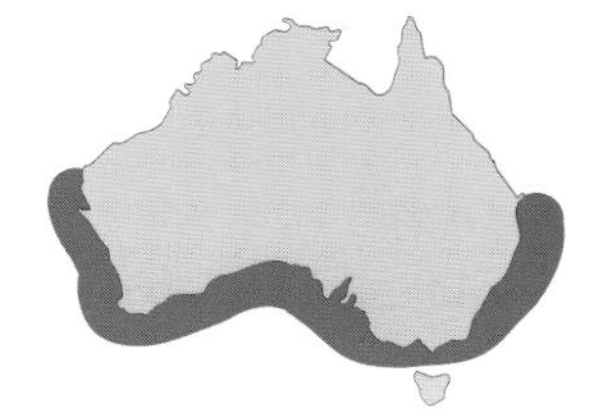

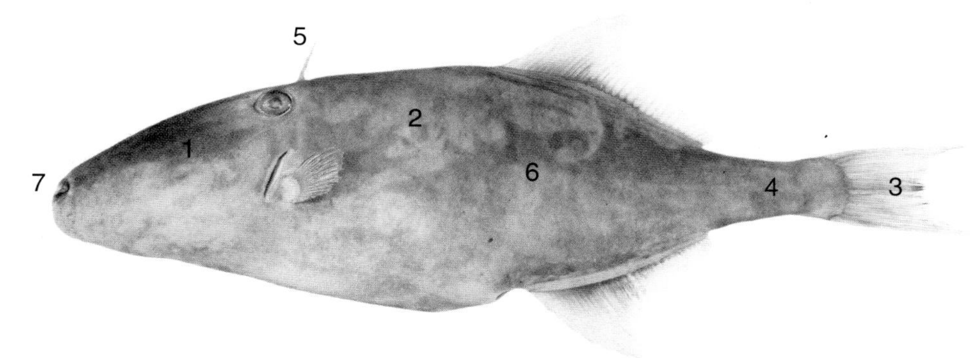

Identifying features: 1 head very long, body slender and compressed; 2 sides plain sandy brown, often with reddish blotches (may have 4 brownish stripes); 3 caudal fin uniformly yellowish-brown; 4 caudal peduncle long; 5 first dorsal-fin spine relatively short; 6 skin almost smooth (without normal scales); 7 mouth very small with beak-like teeth.

Comparisons: The comparatively plain-coloured, elongate adults, which have a long caudal peduncle and long convex snout, are distinct within the group. Young fish and females may have 4 distinct, brown stripes along the side with the upper pair coalescing in front of the eye. A few reef leatherjackets (*Meuschenia* species, p. 158) have brown stripes but have a different head shape and a less elongate caudal peduncle.

Fillet: Moderately deep, rather elongate, upper profile almost straight, taper pronounced, off-white to yellowish. Outside without red muscle band; weakly converging EL, weakly converging HL; HS along middle of fillet; integument white. Inside flesh fine; belly flap usually absent; peritoneum silvery white to translucent. Always skinned, skin very thick and spiny, scale pockets minute and embedded.

Size: To 79 cm and almost 3.5 kg (commonly to about 60 cm and 1.5 kg).

Habitat: Marine; pelagic and demersal on the continental shelf in 0–200 m depth, but adults most common on outer continental shelf. Juveniles found inshore over reef, sand and seagrass.

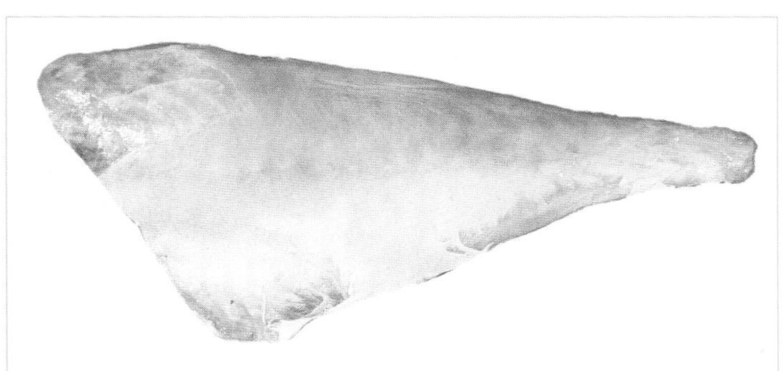

Untrimmed. Protein fingerprint p. 366; oil composition p. 401.

Fishery: Caught principally in the Great Australian Bight using traps and more recently by demersal trawlers. Once taken in quantity off New South Wales, current catches are smaller.

Remarks: Sold mostly as skinned fresh trunks in central markets of Sydney and Melbourne. The firm flesh is distinctly flavoured.

Reef leatherjacket

Meuschenia species

Previous names: silver flounder, sixspine leatherjacket

Meuschenia freycineti

Identifying features: 1 head and body compressed, moderately elongate to deep; 2 sides brightly coloured or with stripes, spots and blotches; 3 caudal fin rarely uniform in colour, often with multiple colours or black markings; 4 second dorsal fin similar in height at front and back; 5 first dorsal-fin spine relatively tall; 6 skin 'furry' (without normal scales); 7 mouth very small with beak-like teeth.

Comparisons: Among the most brightly coloured of the leatherjackets, with distinct markings on the caudal fin and males usually with spines on the caudal peduncle. The sexes differ in colour and shape.

Fillet: *Meuschenia freycineti* moderately deep, rather elongate, upper profile slightly convex, taper pronounced, pale pinkish. Outside without red muscle band; weakly converging EL, weakly converging HL; HS along middle of fillet; integument white. Inside flesh fine; belly flap usually absent; peritoneum silvery white to translucent. Always skinned, skin very thick and spiny, scale pockets minute and embedded.

Size: To 64 cm and almost 3.4 kg (commonly to about 35 cm and 0.6 kg).

Habitat: Marine; mainly inshore off temperate Australia over rocky reefs and seagrasses to depths of 30 m.

Fishery: Small specialist inshore fishery using seines and traps. Taken by line in South Australian gulfs, and as haul net bycatch over seagrass beds.

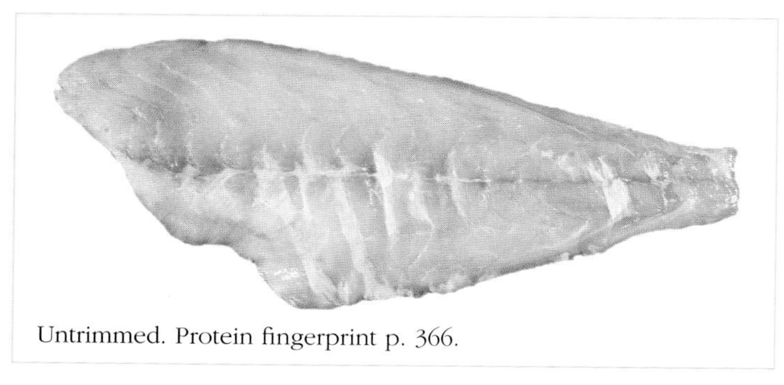

Untrimmed. Protein fingerprint p. 366.

Remarks: Four species marketed commonly. The mosaic leatherjacket (*Eubalichthys mosaicus*), caught in moderate quantities by trawlers in the Great Australian Bight, possibly should be included in this group. The flesh varies between species from delicately flavoured to strongly permeated by iodine.

Velvet leatherjacket

Parika scaber

Previous name: leatherjacket

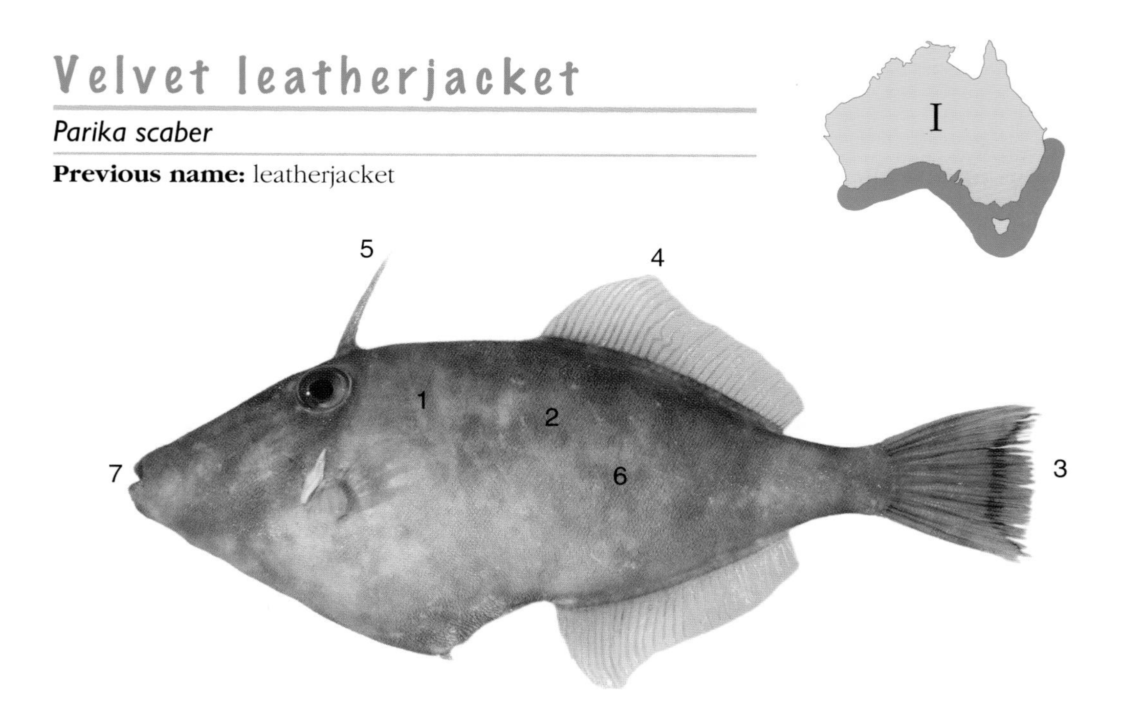

Identifying features: 1 head and body compressed, and moderately deep; 2 sides yellowish, mostly with dark blotches; 3 caudal fin yellowish-green, with black vertical line in males; 4 second dorsal fin slightly taller near front than back; 5 first dorsal-fin spine relatively tall; 6 skin 'furry' (without normal scales); 7 mouth very small with beak-like teeth.

Comparisons: Similar to other reef leatherjackets (*Meuschenia* species, p. 158) in body form but best distinguished by colour pattern (large blotches on the side and stripes down the snout are distinctive) and having slightly raised anterior rays in the second dorsal fin.

Fillet: Deep, rather elongate, upper profile slightly convex, taper pronounced, off-white to yellowish. Outside without red muscle band; converging EL, converging HL; HS along middle of fillet; integument translucent. Inside flesh fine; belly flap usually absent; peritoneum transparent. Always skinned, skin thick and spiny, scale pockets minute and embedded.

Size: To 31 cm and almost 0.5 kg (commonly to 26 cm and 0.4 kg).

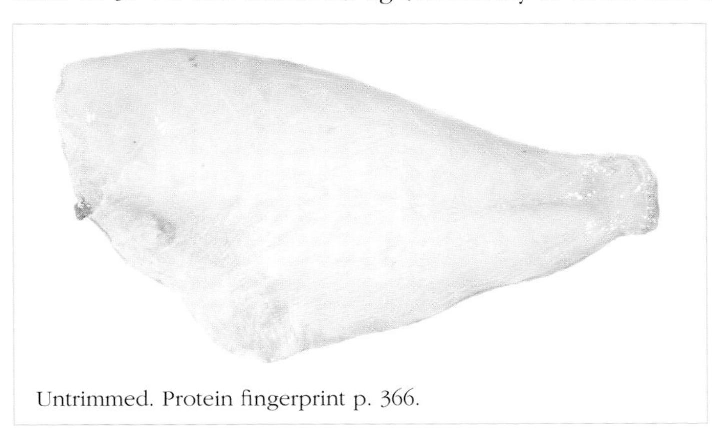

Untrimmed. Protein fingerprint p. 366.

Habitat: Marine; demersal and in midwater over hard bottoms and sponge beds in depths of 20–140 m.

Fishery: Dominant leatherjacket of the southeastern shelf trawl fishery. Aggregates, and often taken in quantity.

Remarks: The average size is small but the flesh is firm and tasty. Regarded as a nuisance fish by anglers as it will nibble away at bait.

Pink ling

Genypterus blacodes

Previous names: kingclip, ling

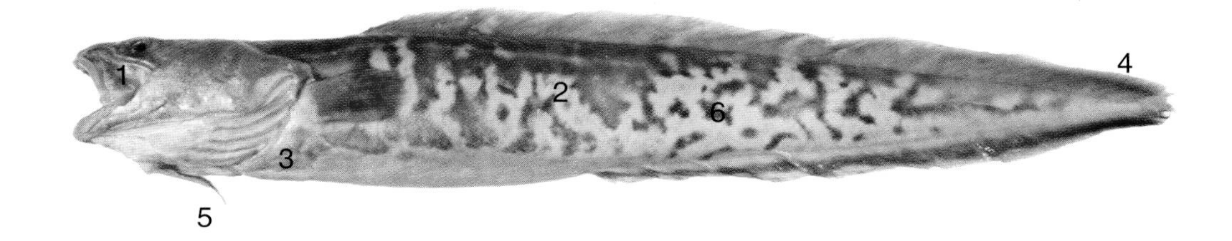

Identifying features: ① upper jaw extending to or just beyond eye; ② skin with pinkish or orange tones and mottling; ③ belly and cheek pale rather than mottled; ④ tail eel-like; ⑤ barbel-like pelvic fins originating below centre of eye; ⑥ body long, rounded.

Comparisons: Cusk-eels (Ophidiidae), such as the pink ling, are eel-like in form, and related to codfishes. Unlike eels, they have long gill openings and finger-like pelvic fins on their chin. The pink ling is similar to the more inshore dwelling rock ling (*G. tigerinus*, p. 161) but is pinkish or orange (rather than mainly greyish) with larger and less extensive mottling pattern. Its jaw is somewhat shorter than that of the rock ling.

Fillet: Long, rather slender, taper pronounced, pale pinkish to off-white. Outside with feeble, discontinuous, central red muscle band; parallel EL, weakly converging HL; HS along middle of fillet; integument patchy, translucent above, silvery white below. Inside flesh medium; belly flap usually removed, slightly lobe-like; furrow below midline anteriorly; peritoneum white. Usually skinned, skin thick, scale pockets very small and embedded.

Size: To at least 160 cm and 20 kg (commonly 50–90 cm and 0.6–4.5 kg).

Habitat: Marine; demersal on the continental shelf and upper slope in depths of 20–800 m. Seems to bury or live in holes in soft substrates.

Fishery: Caught year-round mainly by trawlers on the continental slope off southeastern Australia in depths of 300–550 m. Also taken using bottom-set longlines, mesh nets and traps and is a significant component of dropline fisheries.

Remarks: Inshore specimens tend to be orange in colour with those from the continental slope distinctively pink. The flesh has a medium texture with a mild delicate flavour.

Untrimmed. Protein fingerprint p. 366; oil composition p. 401.

Rock ling

Genypterus tigerinus

Previous name: ling

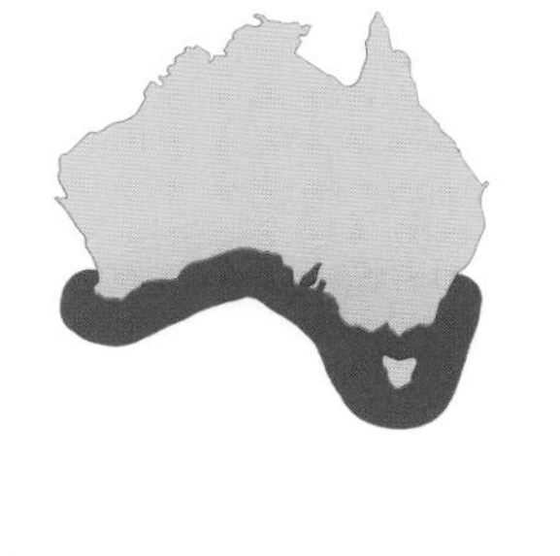

Identifying features: ① upper jaw usually extending well behind eye (when mouth closed); ② skin whitish or greyish, and with darker mottling; ③ belly and cheek mottled; ④ tail eel-like; ⑤ barbel-like pelvic fins originating below centre of eye; ⑥ body long, rounded.

Comparisons: Similar to the more offshore pink ling (*G. blacodes*, p. 160) but has a longer jaw, and is more heavily mottled without a pinkish or orange base colour.

Fillet: Long, rather slender, taper pronounced, white. Outside with feeble, continuous, central red muscle band; parallel EL, weakly converging HL; HS along middle of fillet; integument patchy, silvery white. Inside flesh fine; belly flap usually removed, slightly lobe-like; furrow below midline anteriorly; peritoneum translucent to white. Usually skinned, skin thick, scale pockets very small and embedded.

Size: To 120 cm and 9 kg (commonly about 45–75 cm and 0.7–2.2 kg).

Habitat: Marine; bottom-dwelling, mainly lives in caves and under ledges on reefs, on the inner continental shelf to depths of 60 m. Juveniles are found in estuaries on sand and among seagrasses.

Fishery: Caught in small quantities using rocklobster pots, gillnets, and Danish seines.

Remarks: Divers searching for rocklobsters under ledges are occasionally confronted head on by this ling often mistaking it for the conger eel (*Conger verreauxi*, p. 94). The white chin with barbels is typical. Marketed fresh, whole gutted or as fillets. The flesh is highly regarded by consumers in southern states with the bulk of the catch being sold in the Melbourne fish market.

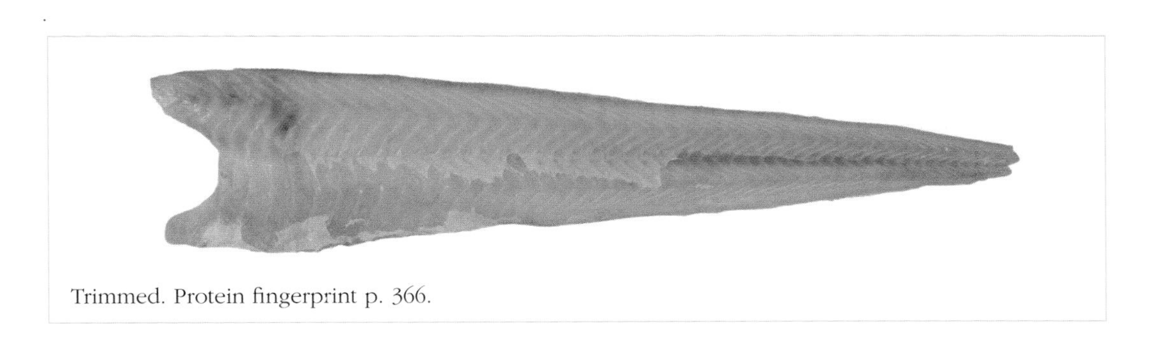

Trimmed. Protein fingerprint p. 366.

Tusk

Dannevigia tusca

Previous names: Australian tusk, tuskfish

Identifying features: ① upper jaw extending to or just beyond eye; ② skin pinkish-brown on back, not mottled; ③ belly greyish or pale; ④ tail eel-like; ⑤ barbel-like pelvic fins originating below hind margin of eye; ⑥ body long, rounded.

Comparisons: Differs from the lings (*Genypterus* species, pp 160–161) in being plain-coloured (rather than mottled), and with a more stumpy body and posteriorly positioned pelvic fin. Juveniles sometimes have a few bands on the posterior part of the body. Several similar cusk-eels, which are caught as bycatch in the deepwater prawn and scampi grounds on the continental slope off tropical Australia, are rarely landed.

Fillet: Long, relatively deep, taper pronounced, white. Outside with feeble, discontinuous, central red muscle band; convex EL, weakly converging HL; HS drops below middle of fillet; integument patchy, white above, silvery white below. Inside flesh medium; belly flap sometimes removed, not lobe-like; furrow below midline anteriorly; peritoneum silvery white. Usually skinned, skin thick, scale pockets very small and defined.

Size: To at least 61 cm and 2.7 kg (commonly to 50 cm and 1.4 kg).

Habitat: Marine; bottom-dwelling on the outer continental shelf and upper slope in depths of 140–360 m.

Fishery: Caught in moderate quantities as bycatch of the Great Australian Bight trawl fishery where it seems to replace the pink ling (*G. blacodes*, p. 160) as the dominant cusk-eel.

Untrimmed. Protein fingerprint p. 366; oil composition p. 401.

Remarks: Landings are likely to increase with possible expansion of the trawl fishing effort in the Great Australian Bight, but the size of the resource is unknown. The flesh has been compared to European cod in quality, but with a slightly richer flavour.

Albacore

Thunnus alalunga

Previous name: albacore tuna

Identifying features: ① very long pectoral fin, reaching back to finlets; ② white outer edge on caudal fin; ③ 25–31 gill rakers on first gill arch; ④ second dorsal fin distinctly lower than first dorsal fin; ⑤ small scales on posterior half of body; ⑥ upper surface dark blue without dark spots or striped pattern; ⑦ upper surface of tongue with 2 longitudinal ridges; ⑧ pronounced, thickened scale patch near pectoral-fin base; ⑨ large fleshy keel on caudal peduncle flanked by 2 smaller keels.

Comparisons: Distinctive tuna with an exceptionally long pectoral fin. Like the southern bluefin tuna (*T. maccoyii*, p. 176) and the bigeye tuna (*T. obesus*, p. 164), the lower surface of its liver is striated with the central lobe longer than the left or right lobes. The southern bluefin tuna has more rakers on the first gill arch (31–40 versus 25–31). The bigeye tuna has a heavier body which is thickest below the middle of the first dorsal fin (rather than at the end of the fin), yellow anal finlets with dark tips (rather than uniformly dark) and lacks a pale outer edge of the caudal fin.

Fillet: Moderately deep, rather short, upper profile convex, bottle-shaped, pale pinkish. Outside with pronounced, continuous red muscle band (evident on inside); parallel EL, converging HL; HS along middle of fillet; integument blackish above, silvery white below. Inside flesh coarse; belly flap usually present; peritoneum translucent. Sometimes skinned, skin medium, scale pockets small and defined.

Size: To 127 cm (fork length) and about 55 kg (commonly 50–90 cm and 3–22 kg).

Habitat: Marine; pelagic in open ocean in both temperate and tropical regions, but usually in water masses exceeding 13°C.

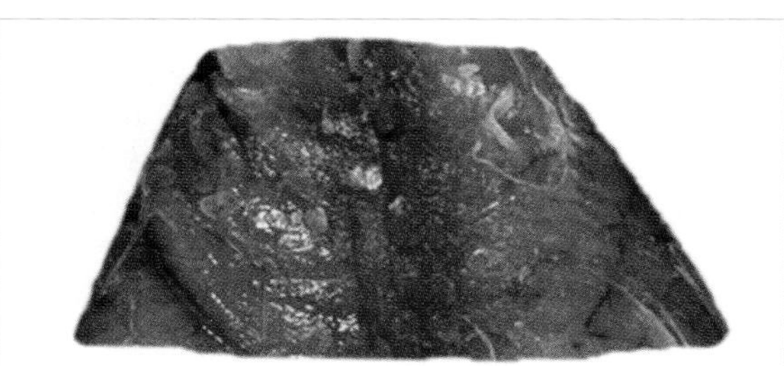
Steak. Protein fingerprint p. 366; oil composition p. 401.

Fishery: Caught throughout the year as bycatch of fisheries for larger tunas. Mainly by longline but also overseas by live-bait fishing, trolling and purse seining. Significant recreational catch.

Remarks: Well regarded, with dry white flesh when cooked, very high in calories and ideally suited to canning. Often referred to as 'chicken of the sea'.

Bigeye tuna

Thunnus obesus

Previous names: none

Identifying features: 1 moderate pectoral fin, not reaching to end of second dorsal-fin base; 2 no white outer edge on caudal fin; 3 23–31 gill rakers on first gill arch; 4 second dorsal fin slightly shorter than first dorsal fin; 5 small scales on posterior half of body; 6 upper surface dark blue without dark spots or striped pattern; 7 upper surface of tongue with 2 longitudinal ridges; 8 pronounced thickened scale patch near pectoral-fin base; 9 large fleshy keel on caudal peduncle flanked by 2 smaller keels.

Comparisons: Heavy bodied tuna with a relatively large eye. The lower surface of its liver is striated with the central lobe longer than the left or right lobes like the albacore (*T. alalunga*, p. 163) and southern bluefin tuna (*T. maccoyii*, p. 176). The albacore has a less robust body which is thickest below the end of the first dorsal fin (rather than at the middle of the fin), dark anal finlets (rather than yellow with dark tips) and a pale outer edge of the caudal fin (rather than plain). The southern bluefin tuna has more rakers on the first gill arch (31–40 versus 23–31). The second dorsal and anal fins of the bigeye tuna are never greatly extended, unlike in yellowfin tuna (*T. albacares*, p. 180), with which it is often caught.

Fillet: Moderately deep, rather short, upper profile convex, bottle-shaped, reddish. Outside with pronounced, continuous red muscle band (evident on inside); HS along middle of fillet. Inside flesh medium; belly flap usually present.

Size: To 236 cm (fork length) and 197 kg (commonly to about 180 cm and 100 kg).

Habitat: Marine; pelagic in open ocean in warm temperate and tropical regions to depths of about 850 m.

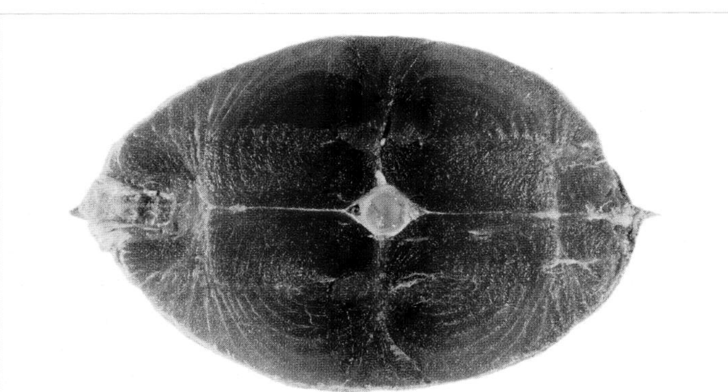

Cutlet. Protein fingerprint p. 366; oil composition p. 401.

Fishery: Targeted domestically off southeastern and southwestern Australia using longlines, with additional Japanese catches off Western Australia. Smaller quantities taken by pole-and-line, and handline, off Queensland.

Remarks: Highly regarded for sashimi, only exceeded in price on Japanese markets by the bluefin tunas. Fillet information incomplete.

Blue mackerel

Scomber australasicus

Previous names: Pacific mackerel, slimy mackerel

Identifying features: ① dorsal fins widely separated; ② diagonal lines above lateral line, narrow broken bars on sides; ③ 5 finlets behind the dorsal and anal fins; ④ clear fatty eyelid at front and back of eye; ⑤ body covered in small scales; ⑥ no pronounced scale patch near pectoral-fin base; ⑦ 2 small fleshy keels on caudal peduncle.

Comparisons: Small, slender mackerel with narrow bars on sides. Unlike larger relatives, the Spanish mackerels (Scomberomorini), bonitos (Sardini) and tunas (Thunnini), it lacks a large central keel between the 2 small keels at the base of the caudal fin and has very widely separated dorsal fins. Distinguished from the jack mackerel (*Trachurus declivis*, p. 267), with which it often schools, by its wavy colour pattern and the absence of scutes along the sides.

Fillet: Moderately deep, elongate, upper profile slightly convex, taper gradual, reddish-brown. Outside with broad, very pronounced, continuous red muscle band; weakly converging EL, weakly converging HL; HS along middle of fillet; integument silvery grey above, more whitish below. Inside flesh coarse; belly flap sometimes present; peritoneum translucent. Sometimes skinned, skin thin, scale pockets small and indistinct.

Size: To at least 50 cm (fork length) and at least 1.5 kg (commonly 20–35 cm and 0.2–0.7 kg).

Habitat: Marine; pelagic near the surface around continental margins and islands primarily in temperate seas but wanders into tropical regions.

Fishery: Under-utilised locally, where it is caught mainly as bycatch of the jack mackerel fishery off southeastern Australia. Significant target species of purse seine fishery off New South Wales. Elsewhere annual catches of this genus may exceed 2 million tonnes. Pier-based recreational fishery in temperate Western Australia.

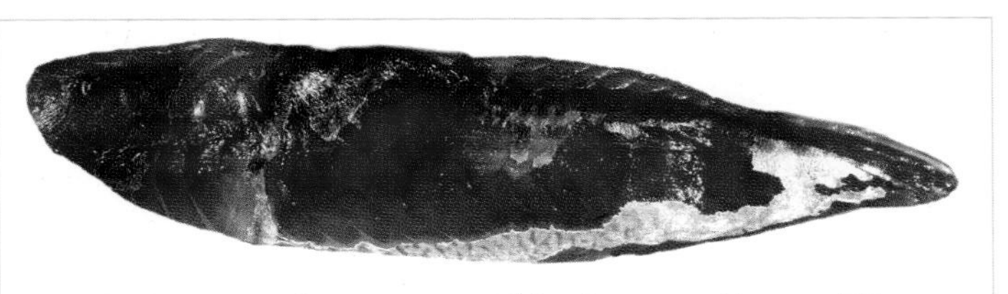

Trimmed. Protein fingerprint p. 366; oil composition p. 401.

Remarks: Important foodfish of larger mackerels, tunas, sharks, seals and dolphins. It has potential for canning but is also sold fresh, frozen, smoked and salted. The flesh is of reasonable quality when fresh.

Bonito

Sarda species

Previous names: Australian bonito, horse mackerel, oriental bonito

Sarda australis

Identifying features: ① dorsal fins narrowly separated; ② horizontal stripes present both above and below lateral line; ③ first dorsal fin with 17–19 spines; ④ small scales over posterior half of body; ⑤ upper surface of tongue lacking longitudinal ridges; ⑥ pronounced, thickened scale patch near pectoral-fin base; ⑦ large fleshy keel on caudal peduncle flanked by 2 smaller keels.

Comparisons: Two very similar species. The Australian bonito (*S. australis*) has 19–21 gill rakers on the first gill arch (rather than 8–13 in the oriental bonito, *S. orientalis*) and the stripes on the back are almost horizontal (rather than quite oblique). Unlike the frigate mackerel (*Auxis thazard*, p. 167), mackerel tuna (*Euthynnus affinis*, p. 171), and skipjack tuna (*Katsuwonus pelamis*, p. 174), bonitos have stripes both above and below the lateral line and lack longitudinal ridges on the tongue.

Fillet: *S. australis* moderately deep, rather elongate, almost oval, upper profile strongly convex, reddish-brown. Outside with pronounced, continuous red muscle band (evident on inside); parallel EL, weakly converging HL; HS along middle of fillet; integument translucent. Inside flesh coarse; belly flap usually present; peritoneum translucent. Sometimes skinned, skin medium, scale pockets absent.

Size: To about 102 cm (fork length) and about 11 kg (commonly to 35–55 cm and 1–3 kg).

Habitat: Marine; pelagic over the continental shelf, often shoaling with other tunas near the coast, in warm temperate and tropical parts of the Indo–Pacific.

Fishery: Small targeted fishery using lines and purse seines; mostly sold in Sydney. The oriental bonito is occasionally trolled off Western Australia, where it is also an important recreational target.

Untrimmed. Protein fingerprint p. 366; oil composition p. 401.

Remarks: Considered to be excellent bait for larger reef and gamefish. The pale reddish flesh, which is soft with a delicate flavour, is ideal for canning, but has received mixed reviews as a fresh foodfish.

Frigate mackerel

Auxis thazard

Previous names: frigate tuna, leadenall

Identifying features: 1 dorsal fins widely separated; 2 diagonal wavy lines confined to back above lateral line; 3 no dark spots below pectoral fin; 4 10–12 spines in first dorsal fin; 5 no scales on posterior half of body (except on lateral line); 6 upper surface of tongue with 2 longitudinal ridges; 7 pronounced, thickened scale patch near pectoral-fin base; 8 large fleshy keel on caudal peduncle flanked by 2 smaller keels.

Comparisons: Small tuna with a distinctive, dark wavy pattern of lines on its back and very widely spaced first and second dorsal fins. The dorsal fins are very close together—often almost touching at their bases—in all other members of this family with a corselet of scales on the body behind the head; this includes the true bonitos (*Sarda* species, p. 166), mackerel tuna (*Euthynnus affinis*, p. 171), and skipjack tuna (*Katsuwonus pelamis*, p. 174). The more slender bullet tuna (*Auxis rochei*), which has fewer and broader lines on the back, is less commonly marketed.

Fillet: Moderately deep, rather elongate, upper profile weakly convex, reddish-brown. Outside with intermediate, continuous red muscle band; converging EL, converging HL; HS above middle of fillet; integument grey translucent. Inside flesh medium, darker than outside; belly flap usually present; peritoneum translucent. Sometimes skinned, skin thin, scale pockets absent behind corselet.

Size: To 58 cm (fork length) and about 3.6 kg (commonly 25–40 cm and 0.3–1.3 kg).

Habitat: Marine; pelagic, widespread in tropical and temperate latitudes over the continental shelf and in the nearby open ocean.

Fishery: Locally caught incidentally, mainly by pole-and-line and purse seine, although taken overseas by a variety of tangle nets, seines and trawls.

Trimmed. Protein fingerprint p. 366.

Remarks: Often creates problems for purse seine vessels targeting large tunas as it gets meshed in the webbing. Considered to be a good foodfish but the flesh deteriorates rapidly. Best canned, dried or smoked.

Grey mackerel

Scomberomorus semifasciatus

Previous names: broad-barred mackerel, Spanish mackerel, tiger mackerel

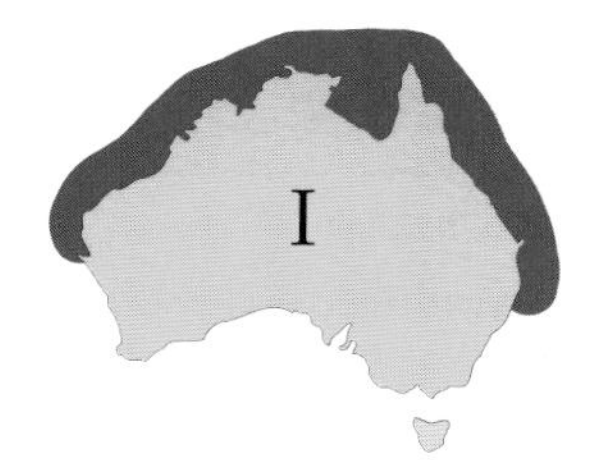

Identifying features: 1 dorsal fins narrowly separated; 2 dark vertical bars on sides; 3 black blotch at front of first dorsal fin; 4 lateral line dipping sharply below end of second dorsal fin; 5 13–15 spines in first dorsal fin; 6 enlarged teeth, triangular or knife-like; 7 no pronounced scale patch near pectoral-fin base; 8 large fleshy keel on caudal peduncle flanked by 2 smaller keels.

Comparisons: Grey mackerel, which have dark bars along the side (often faint or absent in adults), are similar to the largest mackerel species, the common Spanish mackerel (*S. commerson*, p. 177). However, in the grey mackerel these bars are broader and the front of the first dorsal fin is jet black (rather than bluish). Some other related species also have a dark anterior dorsal fin but have spots (rather than plain or with bars).

Fillet: Moderately deep, rather elongate, upper profile slightly convex, taper gradual, off-white to brownish. Outside with very pronounced, continuous red muscle band (also evident on inside); parallel EL, weakly converging HL; HS along middle of fillet; integument silvery. Inside flesh medium; belly flap sometimes present; peritoneum silvery white to translucent. Usually skinned, skin medium, scale pockets small and indistinct.

Size: To 120 cm (fork length) and 10 kg (commonly up to 90 cm and averaging 1.3–2.7 kg).

Habitat: Marine; pelagic over the continental shelf off northern Australia.

Fishery: Among the four main commercial mackerels contributing to one of Queensland's main fin-fisheries. Adults caught mainly by surround nets and gillnets and also by trolling. Young taken in estuaries and near the coast in set nets and on lines. Important target of mackerel fishers in Western Australia and the Northern Territory.

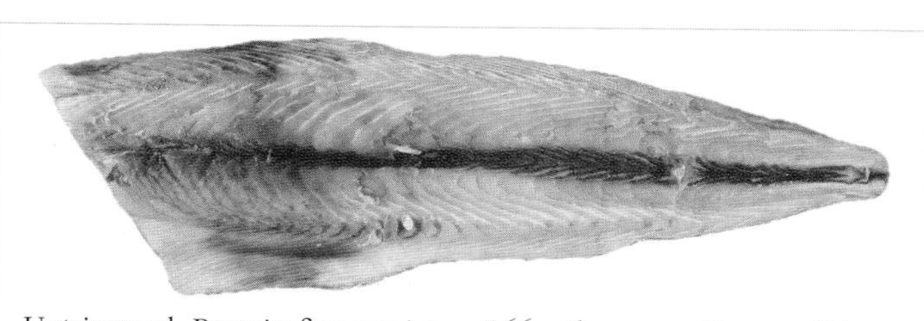

Untrimmed. Protein fingerprint p. 366; oil composition p. 401.

Remarks: Sold as whole, fresh fillets; the flesh is high grade.

Longtail tuna

Thunnus tonggol

Previous names: bluefin tuna, northern bluefin tuna, tuna

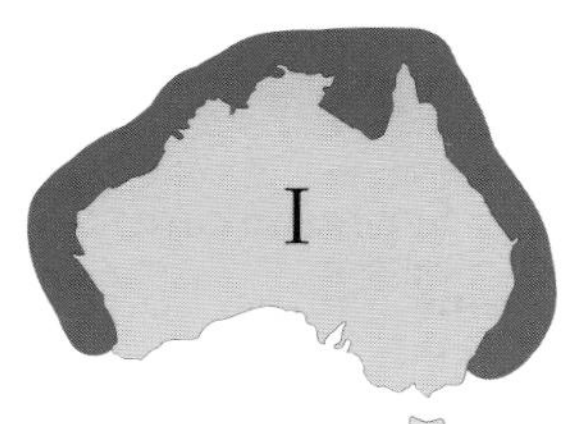

Identifying features: 1 moderate pectoral fins, reaching close to origin of second dorsal fin; 2 tail noticeably elongate and slender behind anal fin; 3 19–27 gill rakers on first gill arch; 4 second dorsal fin not greatly extended in adults; 5 small scales on posterior half of body; 6 upper surface dark blue without dark spots or striped pattern; 7 top of tongue with 2 longitudinal ridges; 8 pronounced, thickened scale patch near pectoral-fin base; 9 large fleshy keel on caudal peduncle flanked by 2 smaller keels.

Comparisons: The only other Australian tuna, apart from the yellowfin tuna (*T. albacares*, p. 180), having the lower surface of its liver not striated and the right lobe much longer than either the left or centre lobes. However, it differs from the yellowfin in having fewer rakers on the first gill arch (19–27 versus 26–34), shorter second dorsal and anal fins, and its corselet extends well past the tip of the pectoral fin. The slender body, with an abnormally long and slender tail, is unique within the large tuna species.

Fillet: Moderately deep, rather elongate, upper and lower profiles convex, reddish-brown. Outside with very pronounced, continuous red muscle band (evident on inside); convex EL, converging HL; HS along middle of fillet; integument blackish above, translucent below. Inside flesh medium; belly flap usually present; peritoneum white. Usually skinned, skin thick, scale pockets small and defined.

Size: To at least 136 cm (fork length) and 36 kg (commonly 80–95 cm and 10–15 kg).

Habitat: Marine; pelagic in surface waters over the continental shelf and in the open ocean. Mainly in warm temperate and tropical regions of the Indo–Pacific.

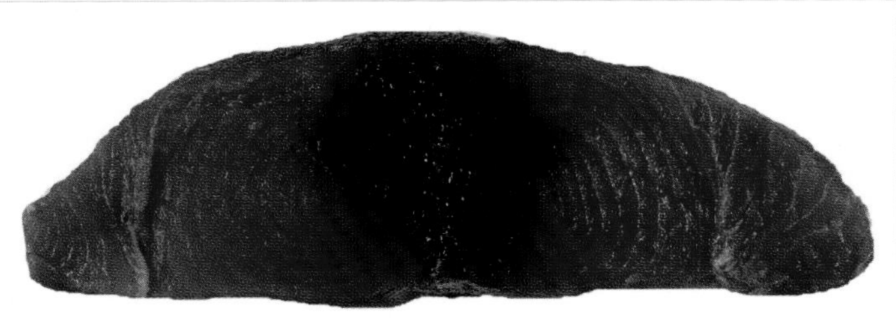

Cutlet. Protein fingerprint p. 366; oil composition p. 401.

Fishery: Locally mainly by trolling and as longline bycatch off tropical eastern Australia.

Remarks: Highly regarded recreational fish capable of rapid acceleration. Secondary commercial tuna used for local sashimi market. Flesh is very dark.

Mackerel

Tribes Scomberomorini & Scombrini

Previous names: mackerel tuna, Spanish mackerel

Rastrelliger kanagurta

Identifying features: ① body streamlined, usually slender; ② head mostly short; ③ no pronounced scale patch on sides above and behind pectoral-fin base; ④ no cartilaginous ridges on tongue; ⑤ teeth typically enlarged, compressed, triangular or knife-like; ⑥ caudal peduncle with 2 or 3 fleshy keels on each side; ⑦ 2 dorsal fins (depressible into grooves) followed by 5–12 finlets; ⑧ pectoral fin placed high on body.

Comparisons: Species that may be marketed as 'mackerel' include members of two major subgroups of the family Scombridae, the tribes Scomberomorini (Spanish mackerels) and Scombrini (true mackerels). The appearance of these fishes, with their characteristic body shape, short fins, finlets on the caudal peduncle, and thinly forked or moon-shaped caudal fins, makes them distinct from other commercial fish families. Other major subgroups of the family, the tribes Thunnini (tunas) and Sardini (bonitos), are represented by more robust species with a well-defined scale patch behind the head. The border between this patch, known as a corselet, and the scaleless posterior part of the body is usually well defined. Another group, the butterfly mackerels (*Gasterochisma* species), have more compressed bodies covered in large scales, and the pelvic fins can be depressed into a deep groove.

Fillet: See fillet descriptions of the mackerels included here.

Size: To 210 cm (fork length) and at least 83 kg (commonly to 125 cm and 15 kg).

Habitat: Marine; pelagic, mainly in surface waters over the continental shelf but some venturing into the open ocean. Most species confined to tropical seas with only one species, the blue mackerel (*Scomber australasicus*, p. 165), regularly found in southern waters.

Fishery: Important commercial fishes in Australia in both market price and tonnage. Caught mainly by trolling but also taken by baited line by both commercial and recreational fishers. Some species appear to be under-utilised.

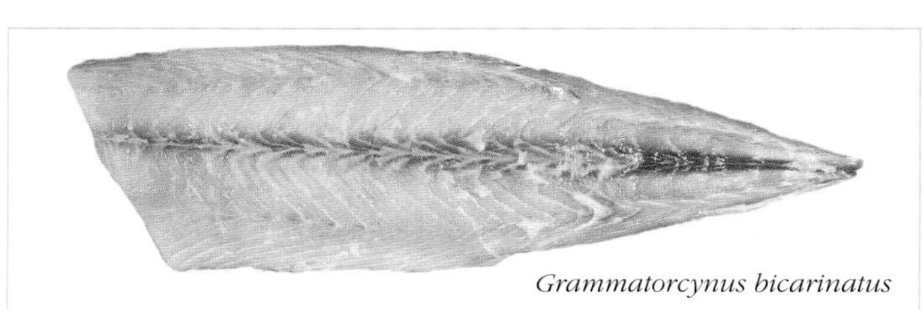

Grammatorcynus bicarinatus

Untrimmed. Protein fingerprint p. 366; oil composition p. 401.

Remarks: Highly regarded commercial and recreational fishes, mostly in the north. Less valued as sashimi than tunas, usually sold whole fresh or as fillets. Flesh spoils quickly if handled poorly.

Mackerel tuna

Euthynnus affinis

Previous names: bonito, jack mackerel, kawa kawa, little tuna, tuny

Identifying features: ① dorsal fins narrowly separated; ② diagonal stripes confined to oval area above lateral line; ③ up to 5 dark spots below pectoral fin; ④ 14–16 spines in first dorsal fin; ⑤ no scales on posterior half of body; ⑥ upper surface of tongue with 2 longitudinal ridges; ⑦ pronounced, thickened scale patch near pectoral-fin base; ⑧ large fleshy keel on caudal peduncle flanked by 2 smaller keels.

Comparisons: Small tuna with a distinctive, dark-striped pattern on its back and usually a few dark spots below the pectoral fin. True bonitos (*Sarda* species, p. 166), frigate mackerel (*Auxis thazard*, p. 167) and skipjack tuna (*Katsuwonus pelamis*, p. 174) also have striped patterns on their body. However, bonitos lack longitudinal ridges on the tongue, the skipjack tuna has stripes on the belly (rather than on the back), and the frigate mackerel has widely separated dorsal fins. Large tunas (*Thunnus* species) have scales on the posterior half of the body which are lacking in mackerel tuna.

Fillet: Moderately deep, rather short, upper profile slightly convex, dark red to brownish. Outside with broad, very pronounced, continuous red muscle band (evident on inside); convex EL, converging HL; HS along middle of fillet; integument white. Inside flesh fine; belly flap usually present; peritoneum transparent. Sometimes skinned, skin thin, scale pockets absent or variable.

Size: To 100 cm (fork length) and 19 kg (commonly to 60 cm and about 3 kg).

Habitat: Marine; pelagic over the continental shelf, often near the coast, in both warm temperate and tropical regions of the Indo–Pacific.

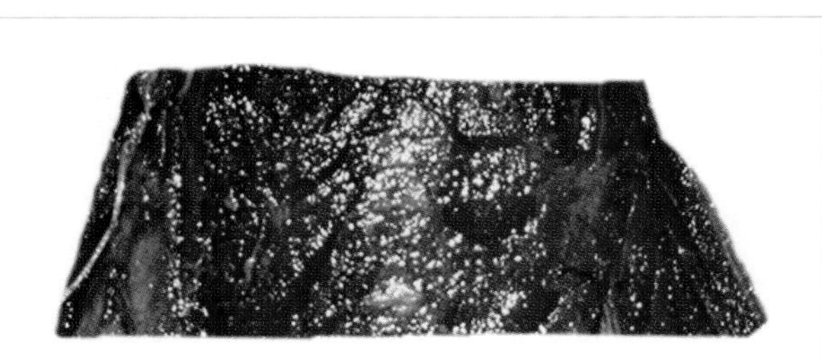

Steak. Protein fingerprint p. 366; oil composition p. 401.

Fishery: Caught mainly by trolling off Queensland and northern New South Wales while targeting mackerel and larger tunas. Popular light tackle gamefish.

Remarks: Apparently, the dark flesh is best steamed after briefly immersing it in hot water. It deteriorates rapidly if not handled correctly.

School mackerel

Scomberomorus queenslandicus

Previous names: doggie mackerel, mackerel

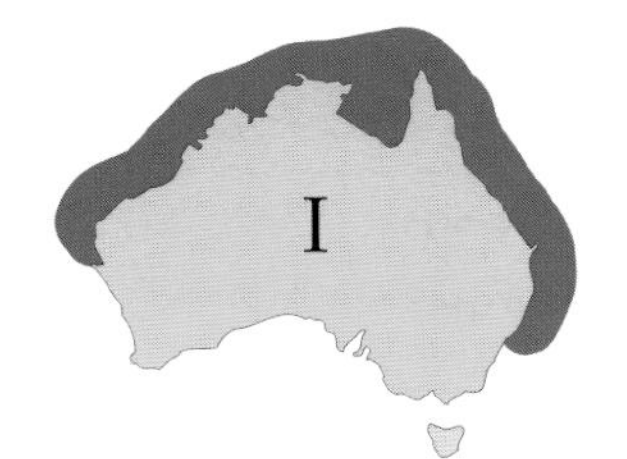

Identifying features: ① dorsal fins narrowly separated; ② indistinct greyish spots on sides behind first dorsal fin (extending further forward in smaller specimens); ③ black blotch at front of first dorsal fin; ④ lateral line descending gently below second dorsal fin; ⑤ 16–18 spines in first dorsal fin; ⑥ enlarged teeth, triangular or knife-like; ⑦ no pronounced scale patch near pectoral-fin base; ⑧ large fleshy keel on caudal peduncle flanked by 2 smaller keels.

Comparisons: Resembles the grey mackerel (*S. semifasciatus*, p. 168) in body form and with the first half of the first dorsal fin black. However, the school mackerel has distinctive large greyish-yellow blotches on the hind half of the body (rather than plain or with dark vertical bars). The spotted mackerel (*S. munroi*, p. 178) is also spotted but has spots extending well forward onto the anterior half of the body and has a uniformly bluish-black, spinous dorsal fin.

Fillet: Moderately deep, elongate, upper profile weakly convex, taper gradual, off-white to yellowish. Outside with pronounced, continuous red muscle band (also evident on inside); parallel EL, weakly converging HL; HS along middle of fillet; integument silvery. Inside flesh medium; belly flap sometimes present; peritoneum translucent. Usually skinned, skin medium, scale pockets undetectable.

Size: To 100 cm (fork length) and 8 kg (commonly 50–80 cm and averaging 1.7–4.5 kg).

Habitat: Marine; surface pelagic, schooling species moving into inshore bays and estuaries during spring and winter. Thought to migrate to the northern gulfs.

Fishery: Among Queensland's four main commercial mackerels contributing to a fishery which exceeds 1000 tonnes annually. Adults caught by trolling by both commercial and recreational fishers.

Trimmed. Protein fingerprint p. 366; oil composition p. 401.

Remarks: School mackerel are strong swimmers capable of rapid acceleration. Retailed as whole fresh fillets. Like other large mackerels, such as the grey mackerel, the fillets have densely-packed myomeres and are considered to be of high grade.

Shark mackerel

Grammatorcynus bicarinatus

Previous name: salmon mackerel

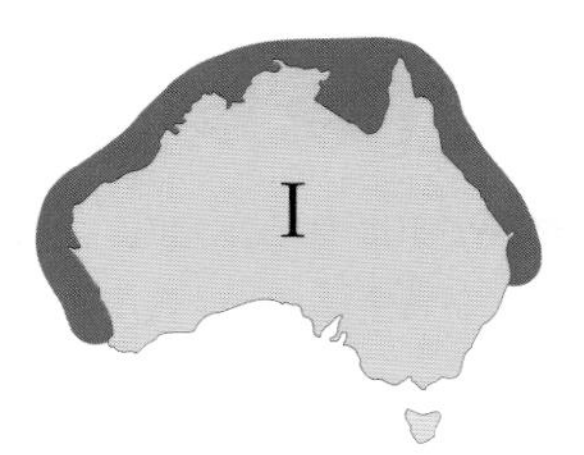

Identifying features: 1 dorsal fins narrowly separated; 2 no dark lines on sides; 3 2 lateral lines branching off below front of first dorsal fin; 4 6–7 finlets behind dorsal and anal fins; 5 body covered in small scales; 6 no pronounced, thickened scale patch near pectoral-fin base; 7 large fleshy keel on caudal peduncle flanked by 2 smaller keels.

Comparisons: Resembles the Spanish mackerels (*Scomberomorus* species) in body form but has a unique lateral-line configuration. The lateral line consists of a long branch that follows the profile of the dorsal surface and a second branch that descends vertically from the upper branch below the front of the dorsal fin and then more or less follows the ventral profile. Other mackerels have only a single lateral line. A close relative, the smaller scad mackerel (*G. bilineatus*), has a larger eye, more gill rakers (more than 18 rather than 14–15), and lacks the scattering of dark spots along the side.

Fillet: Moderately deep, elongate, upper profile slightly convex, taper gradual, pale pink to reddish. Outside with pronounced, continuous red muscle band (also evident on inside); convex EL, weakly converging HL; HS along middle of fillet; integument translucent. Inside flesh medium; belly flap sometimes present; peritoneum silvery white to translucent. Usually skinned, skin medium, scale pockets small and defined.

Size: To 110 cm (fork length) and 13.5 kg (commonly 50–85 cm and averaging 2–6 kg).

Habitat: Marine; inshore surface pelagic, forming dense schools. Concentrates in channels draining reef lagoons and migrates over reef flats on the rising tide to feed on congregating baitfishes.

Untrimmed. Protein fingerprint p. 366.

Fishery: Trolled in moderate quantities off Queensland and caught recreationally off Western Australia.

Remarks: Highly regarded table-fish but has a slight smell of ammonia similar to shark flesh.

Skipjack tuna

Katsuwonus pelamis

Previous names: oceanic bonito, skipjack, striped tuna, stripey

I

Identifying features: 1 dorsal fins narrowly separated; 2 3–5 horizontal stripes on belly below lateral line; 3 no dark spots below pectoral fin; 4 14–16 spines in first dorsal fin; 5 no scales on posterior half of body (except on lateral line); 6 upper surface of tongue with 2 longitudinal ridges; 7 pronounced, thickened scale patch near pectoral-fin base; 8 large fleshy keel on caudal peduncle flanked by 2 smaller keels.

Comparisons: Small, stocky tuna easily identified by dark horizontal stripes confined to the lower half and sides. Of these, the lowermost stripes are often most distinct against a silvery white belly. The closely related mackerel tuna (*Euthynnus affinis*, p. 171) and frigate mackerel (*Auxis thazard*, p. 167) have horizontal or diagonal stripes only on their backs. True bonitos (*Sarda* species, p. 166) can have stripes on the belly but lack longitudinal ridges on the tongue.

Fillet: Moderately deep, rather short, upper profile slightly convex, dark brown. Outside with broad, pronounced, continuous red muscle band (pronounced on inside); convex EL, converging HL; HS above middle of fillet; integument white. Inside flesh medium; belly flap sometimes removed; peritoneum translucent to white. Usually skinned, skin medium, scale pockets absent beyond corselet.

Size: To 108 cm (fork length) and 34.5 kg (commonly to 75 cm and about 10 kg).

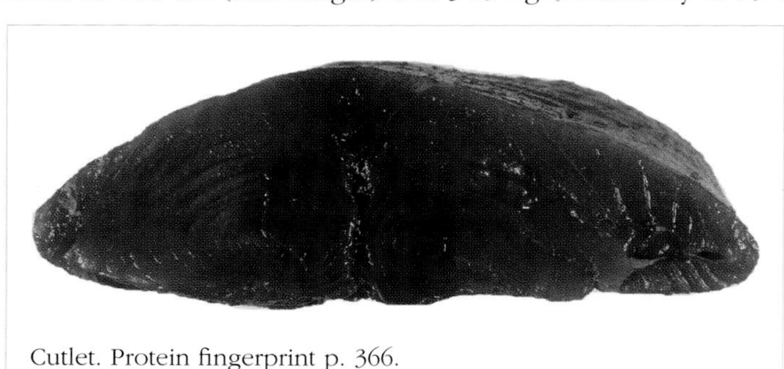

Cutlet. Protein fingerprint p. 366.

Habitat: Marine; pelagic, mostly near the surface of all temperate and tropical seas.

Fishery: Caught mainly by trolling, pole-and-lining, and purse seining. Smaller incidental longline catches.

Remarks: Undertakes long migrations in huge shoals. Usually canned but also marketed fresh or frozen.

Slender tuna

Allothunnus fallai

Previous names: none

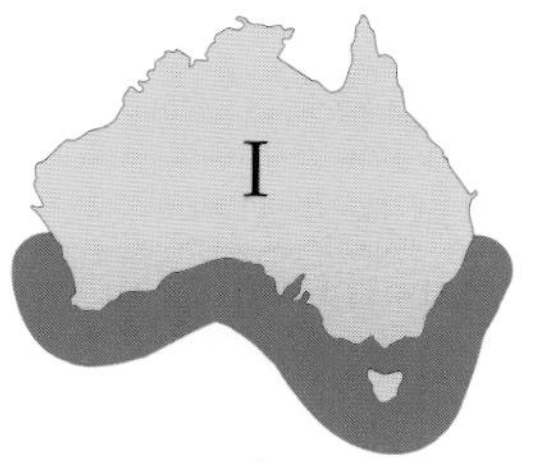

Identifying features: ① very short pectoral fin, not reaching middle of first dorsal fin; ② body relatively slender; ③ 70–80 gill rakers on first gill arch; ④ tiny teeth in jaws; ⑤ upper surface dark blue in colour without dark spots or striped pattern; ⑥ top of tongue lacking longitudinal ridges; ⑦ pronounced, thickened scale patch near pectoral-fin base; ⑧ large fleshy keel on caudal peduncle flanked by 2 smaller keels.

Comparisons: Shares with true tunas (*Thunnus* species) the lack of a pronounced colour pattern of dark spots or stripes. However, its body is more slender, it has many more gill rakers (exceeding 70 on the first arch versus less than 30), and its teeth are smaller and more numerous (more than 40 on each side of the upper jaw rather than less than 30).

Fillet: Deep, rather elongate, upper profile convex, almost bottle-shaped, off-white to yellowish. Outside with pronounced, continuous red muscle band (very evident on inside); parallel EL, converging HL; HS along middle of fillet; integument silvery or translucent. Inside flesh fine; belly flap sometimes present; peritoneum silvery white to translucent. Usually skinned, skin medium, scale pockets small and indistinct.

Size: To 96 cm (fork length) and less than 10 kg (commonly to 85 cm and about 8 kg).

Habitat: Marine; pelagic, probably widely distributed in temperate seas of the Southern Hemisphere. May occur slightly deeper than other tunas.

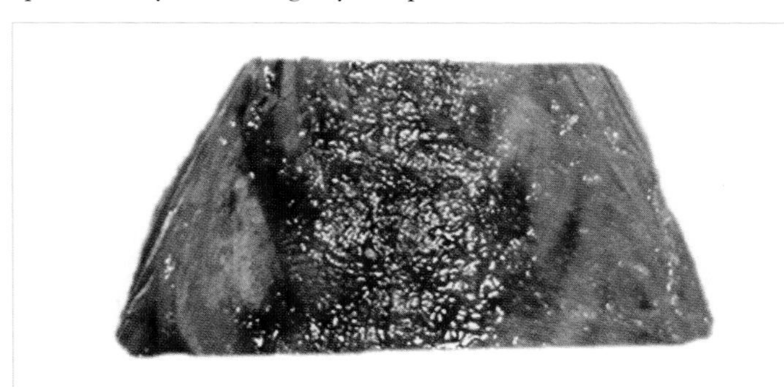

Steak. Protein fingerprint p. 366.

Fishery: No specific fishery. Landings mainly as incidental bycatch of jack mackerel (*Trachurus* species, p. 267) and other tuna fisheries.

Remarks: Considered to feed primarily on plankton making it less likely to be caught by longline. Substantial catches have been taken by purse seining. Presently a low value species but the oily, pale flesh is considered fine eating.

Southern bluefin tuna

Thunnus maccoyii

Previous names: bluefin, bluefin tuna, tuna

Identifying features: ① short pectoral fins, not reaching back to origin of second dorsal fin; ② no white outer edge on caudal fin; ③ 31–40 gill rakers on first gill arch; ④ second dorsal fin barely taller than first dorsal fin in adults; ⑤ small scales on posterior half of body; ⑥ upper surface dark blue without dark spots or striped pattern; ⑦ top of tongue with 2 longitudinal ridges; ⑧ pronounced, thickened scale patch near pectoral-fin base; ⑨ large fleshy keel on caudal peduncle flanked by 2 smaller keels.

Comparisons: Like the albacore (*T. alalunga*, p. 163) and bigeye tuna (*T. obesus*, p. 164), the lower surface of its liver is striated with the central lobe longer than the left or right lobes but differs in having more rakers on the first gill arch (31–40 versus 23–31). The albacore also has a much longer pectoral fin and dark anal finlets (rather than yellow with dark tips). Unlike the yellowfin tuna (*T. albacares*, p. 180), with which its primary range overlaps slightly in the south-east, its second dorsal and anal fins are never greatly extended.

Fillet: Moderately deep, rather elongate, upper profile convex, somewhat bottle-shaped, pinkish to reddish-brown. Outside with pronounced, continuous red muscle band (pronounced on inside); converging EL, converging HL; HS along middle of fillet; integument blackish above, whitish below. Inside flesh medium; belly flap usually removed; peritoneum translucent to whitish. Usually skinned, skin thick, scale pockets small and poorly defined.

Size: To 225 cm (fork length) and possibly 200 kg (commonly to 180 cm and 100 kg).

Cutlet. Protein fingerprint p. 366; oil composition p. 401.

Habitat: Marine; pelagic in the open ocean, mainly in cool temperate waters.

Fishery: Caught locally by pole-and-line, purse seining and drift longlining in the Great Australian Bight and off Tasmania. Farmed in sea cages off South Australia.

Remarks: One of Australia's premium seafood species. Largely air freighted fresh chilled to Japan where it can fetch very high prices for use as sashimi.

Spanish mackerel

Scomberomorus commerson

Previous name: snook

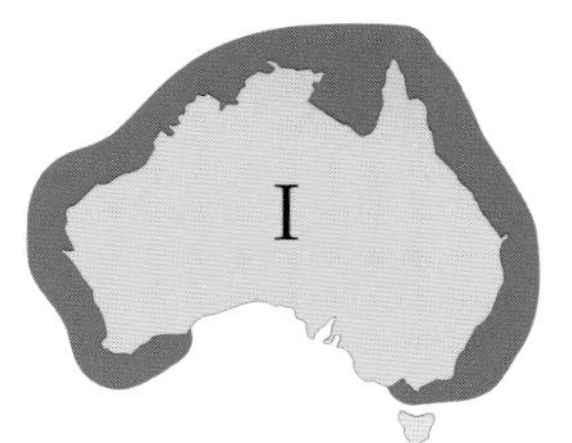

Identifying features: ① dorsal fins narrowly separated; ② dark wavy vertical bars on sides, often faint but most pronounced on lower half; ③ no black blotches on dorsal fins; ④ lateral line wavy posteriorly, descending to below the midline behind second dorsal fin; ⑤ 15–18 spines in first dorsal fin; ⑥ enlarged teeth, triangular or knife-like; ⑦ no pronounced scale patch near pectoral-fin base; ⑧ large fleshy keel on caudal peduncle flanked by 2 smaller keels.

Comparisons: Resembles the grey mackerel (*S. semifasciatus*, p. 168), which also has dark bars along the side in young (often faint or absent in adults), but the spinous part of the dorsal fin is uniform in colour (bluish rather than black initially followed by white). Another barred species, the wahoo (*Acanthocybium solandri*), is more slender with a longer snout, has more dorsal fin spines (23–27 versus 15–18 spines), and its lateral line drops below the middle of the spinous dorsal fin (rather than behind the second dorsal fin).

Fillet: Moderately deep, elongate, upper profile slightly convex, taper gradual, off-white to yellowish. Outside with very pronounced, continuous red muscle band (also evident on inside); parallel EL, weakly converging HL; HS along middle of fillet; integument translucent. Inside flesh medium; belly flap sometimes present; peritoneum transparent. Usually skinned, skin medium, scale pockets small and indistinct.

Size: To at least 200 cm (fork length) and 50 kg (commonly 55–125 cm and averaging 2–15 kg).

Habitat: Marine; pelagic near the surface along continental margins and near islands. Thought to be both resident and capable of undertaking extended longshore migrations.

Fishery: The most sought-after commercial mackerel, contributing about 700 tonnes annually to one of Queensland's main fin-fisheries. Catches are seasonal with the bulk made in late winter and spring. Adults caught mainly by trolling. Also a very important commercial and recreational species in Western Australia.

Remarks: Locally sold mainly whole fresh or as fillets, but elsewhere also salted or dried. The flesh is considered to be of high quality and is reputed to have good keeping qualities.

Untrimmed. Protein fingerprint p. 366; oil composition p. 401.

Spotted mackerel

Scomberomorus munroi

Previous names: spottie

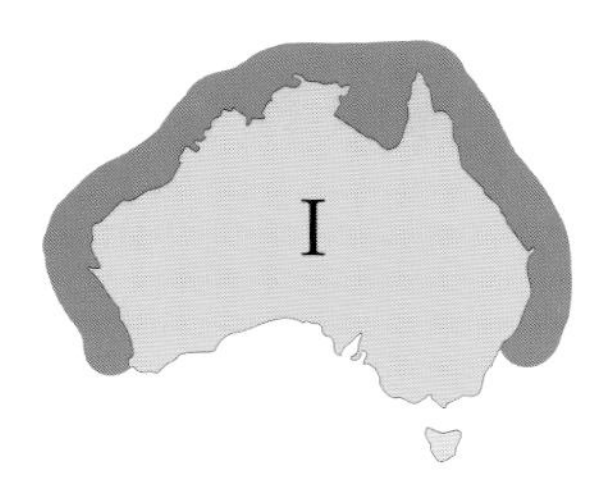

Identifying features: ① dorsal fins narrowly separated; ② dark, eye-sized spots along midline posterior to pectoral fin; ③ most of first dorsal fin black; ④ lateral line wavy but not descending substantially; ⑤ 20–22 spines in first dorsal fin; ⑥ enlarged teeth, triangular or knife-like; ⑦ no pronounced scale patch near pectoral-fin base; ⑧ large fleshy keel on caudal peduncle flanked by 2 smaller keels.

Comparisons: Members of the genus *Scomberomorus*, known generally as Spanish mackerels, differ from other mackerels and tunas in having large, sharp teeth in the jaws, a central keel on the caudal peduncle, a single lateral line, and an indistinct corselet of scales on the side. The school mackerel (*S. queenslandicus*, p. 172) is also spotted but has larger spots confined to the posterior half of the body and only the front half of the spinous dorsal fin is black (rather than all of the fin).

Fillet: Moderately deep, elongate, upper profile slightly convex, taper gradual, pale pinkish inside, more whitish outside. Outside with very pronounced, continuous red muscle band (also evident on inside); parallel EL, weakly converging HL; HS along middle of fillet; integument silvery grey above, silvery white below. Inside flesh medium; belly flap sometimes present; peritoneum silvery white to translucent. Usually skinned, skin medium, scale pockets small and indistinct.

Size: To 104 cm (fork length) and about 10 kg (commonly 50–80 cm and averaging 1.6–4 kg).

Habitat: Marine; surface pelagic, schooling species moving into inshore bays and estuaries, mostly during summer and autumn.

Trimmed. Protein fingerprint p. 366; oil composition p. 401.

Fishery: A major commercial mackerel caught by trolling and by line fishing in the vicinity of reefs.

Remarks: Occasionally ventures as far south as Fremantle in Western Australia, where it has been caught recreationally. Sold as whole fresh fillets, the flesh is high grade.

Tuna

Tribes Sardini and Thunnini

Previous name: leaping bonito

Cybiosarda elegans

Identifying features: 1 body streamlined, usually fusiform; 2 head rather large; 3 pronounced, raised scale patch on sides above and behind pectoral-fin base; 4 cartilaginous ridges on tongue of many species; 5 teeth relatively small, not compressed, conical; 6 caudal peduncle with 3 fleshy keels on each side; 7 2 dorsal fins (depressible into grooves) followed by 5–12 finlets; 8 pectoral fins placed high on body.

Comparisons: Species that may be marketed as 'tuna' include members of two major subgroups of the family Scombridae, the tribes Thunnini (tunas) and Sardini (bonitos). The appearance of these fishes, with their robust fusiform body shape, corselet of scales on the body behind the head, short fins, finlets on the caudal peduncle, and thin moon-shaped caudal fins, makes them distinct from other commercial fish families. Other major subgroups of the family, the tribes Scomberomorini (Spanish mackerels) and Scombrini (true mackerels), are represented by more slender species that lack a well-defined scale patch behind the head. Butterfly mackerels (*Gasterochisma* species), have more compressed bodies covered in large scales, and the pelvic fins can be depressed into a deep groove. True tunas have a pair of longitudinal ridges on the tongue which are lacking in bonitos.

Fillet: See fillet descriptions of the tunas included here.

Size: To at least 225 cm (fork length) and 200 kg (commonly to about 120 cm and 50 kg).

Habitat: Marine; pelagic, mainly near surface and in midwater over the continental shelf and in the open ocean in both temperate and tropical seas. Most school and some species enter coastal waters seasonally.

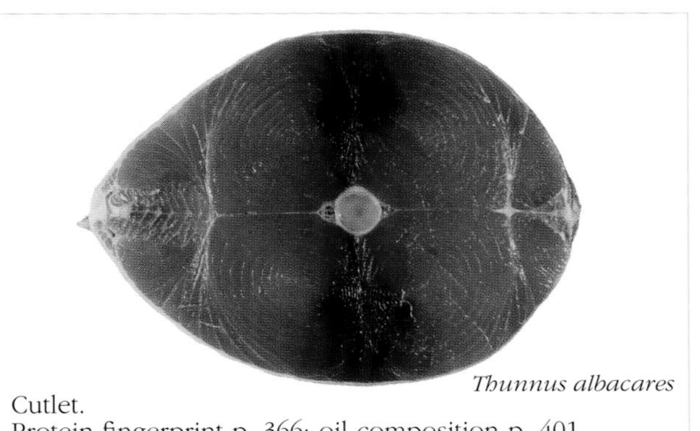

Thunnus albacares

Cutlet.

Protein fingerprint p. 366; oil composition p. 401.

Fishery: Among the most important commercial finfishes due to their high average value. Caught by longlining, purse seining and trolling by both commercial and recreational fishers.

Remarks: Highly regarded foodfishes, particularly for sashimi market, but also sold as steaks and cutlets. Flesh is mostly dark and oily.

Yellowfin tuna

Thunnus albacares

Previous names: none

Identifying features: ① moderate to long pectoral fins, sometimes reaching just beyond origin of second dorsal fin; ② tail not excessively slender behind anal fin; ③ 26–34 gill rakers on first gill arch; ④ second dorsal fin greatly extended in adults; ⑤ small scales on posterior half of body; ⑥ upper surface dark blue without dark spots or striped pattern; ⑦ top of tongue with 2 longitudinal ridges; ⑧ pronounced, thickened scale patch near pectoral-fin base; ⑨ large fleshy keel on caudal peduncle flanked by 2 smaller keels.

Comparisons: Like the longtail tuna (*T. tonggol*, p. 169), the lower surface of its liver is not striated and the right lobe is much longer than either the left or centre lobes. However, it differs in having more rakers on the first gill arch (26–34 versus 19–27), much longer second dorsal and anal fins, and a heavier-bodied tail. The southern bluefin tuna (*T. maccoyii*, p. 176), has a striated liver and shorter second dorsal and anal fins.

Fillet: Moderately deep, rather short, upper profile convex, bottle-shaped, reddish. Outside with pronounced, continuous red muscle band (evident on inside); HS along middle of fillet. Inside flesh medium; belly flap usually present.

Size: To 208 cm (fork length) and at least 176 kg (commonly 50–190 cm and 4–100 kg).

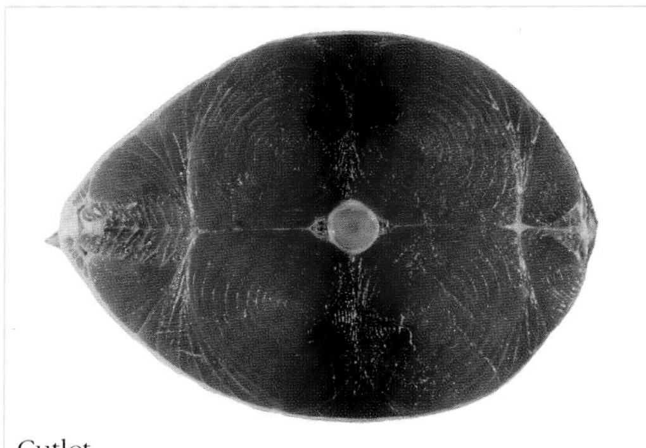

Cutlet.
Protein fingerprint p. 366; oil composition p. 401.

Habitat: Marine; pelagic in the open ocean, mainly in warm temperate and tropical waters.

Fishery: Caught mainly using drifting longlines off eastern and western Australia by local vessels. More broad-based Japanese catch off northern Western Australia and near margins of the Exclusive Economic Zone.

Remarks: Extremely important and widely resourced tropical tuna. Air freighted fresh chilled to Japan but increasing quantities now absorbed by Australian sashimi market. Fillet information incomplete.

Milkfish

Chanos chanos

Previous name: salmon herring

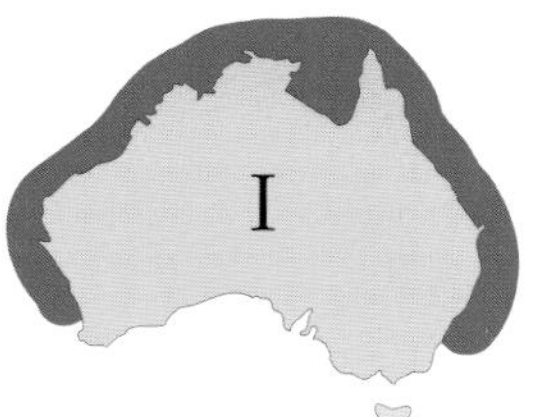

Identifying features: ① body silvery; ② mouth very small; ③ eye covered with gelatinous membrane; ④ large, deeply forked caudal fin; ⑤ single, short-based dorsal fin; ⑥ no adipose fin on dorsal surface of caudal peduncle; ⑦ pelvic fin well behind pectoral fins; ⑧ lateral line present on body.

Comparisons: Superficially resembles the herrings (clupeoids), to which it is only distantly related, but differs in having a much larger caudal fin, a continuous lateral line (usually absent in herrings), and always lacking scutes on the belly. The giant herring (*Elops hawaiiensis*), which is a type of 'tenpounder' rather than a true herring, has a more elongate body and a much larger mouth.

Fillet: Moderately elongate, barely tapering, translucent grey. Outside with pronounced, wide, continuous red muscle band; weakly converging EL, weakly converging HL; HS along middle of fillet; integument patchy, silvery grey to black above, white below. Inside flesh fine; belly flap distinct, not lobe-like; peritoneum black. Sometimes skinned, skin thick, scale pockets moderate-sized and defined.

Size: To at least 120 cm and about 16 kg (commonly 70–80 cm and 6–10 kg).

Habitat: Marine; pelagic in the open sea, migrating inshore to estuaries and bays to spawn. Produces copious quantities of eggs and develops inshore from ribbon-like larvae.

Fishery: Important foodfish in south-east Asia, with much of the production coming from aquaculture. Only recently marketed in quantity locally due to increasing demand. Caught mainly by seine and gillnet.

Trimmed. Protein fingerprint p. 367.

Remarks: Powerful sport fish capable of fast swimming and rapid acceleration. It is a specialised feeder with sand worms, algae, and nutrified mud being among the main dietary components. The flesh, which is laced with small bones, is rather bland and flavourless.

Moonfish

Lampris guttatus & L. immaculatus

Previous name: opah

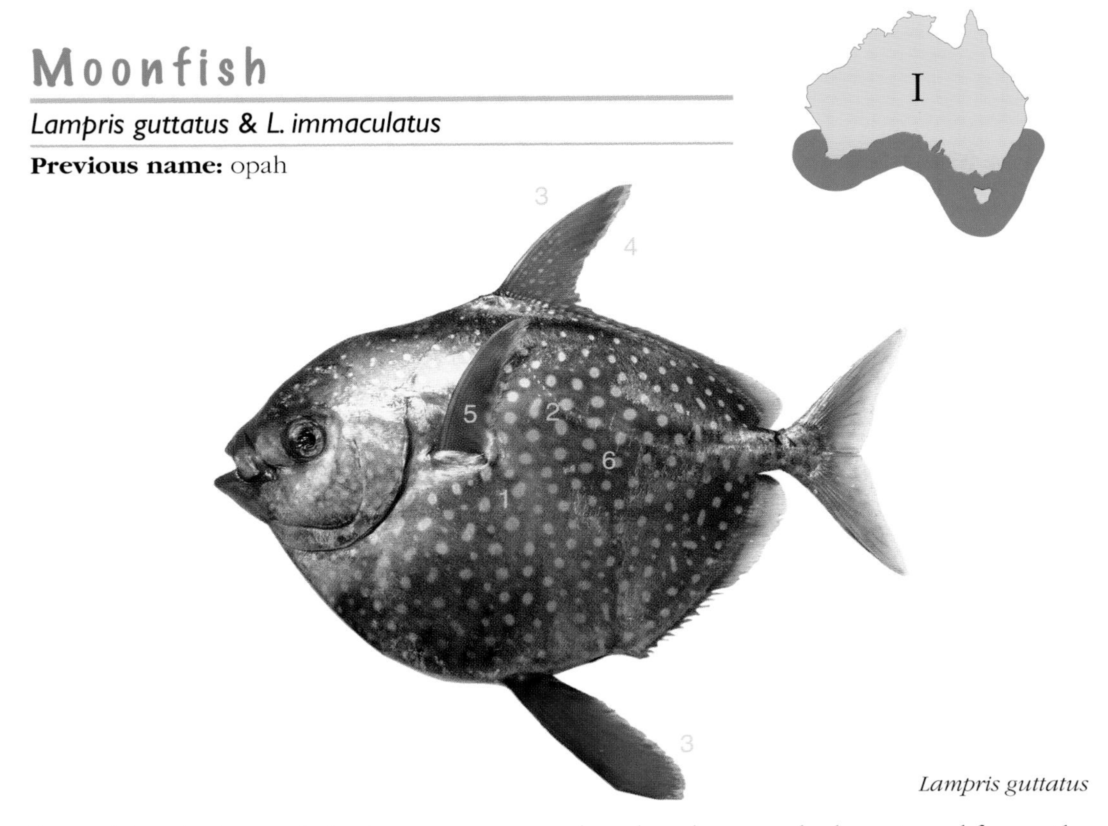

Lampris guttatus

Identifying features: (1) body very compressed, and oval to round when viewed from side; (2) body silvery blue (often covered with large pale spots); (3) fins bright red; (4) dorsal fin long-based, with front part much taller; (5) pectoral fin high on body, capable of moving up and down (rather than forward and backward); (6) scales minute.

Comparisons: The striking appearance of moonfishes makes them unique within fishes. Two species occur locally, although the spotted moonfish (*L. guttatus*) is marketed most frequently. It differs from the southern moonfish (*L. immaculatus*) in having a deeper, more oval body and is spotted (rather than plain-coloured).

Fillet: *L. guttatus* very deep, short, V-shaped; profile tapering sharply below, convex above, orange. Outside with narrow, pronounced, continuous, central red muscle band; extremely convex EL, converging HL; HS well above middle of fillet; integument translucent. Inside flesh fine; belly flap mostly absent; peritoneum transparent. Always skinned, usually sold as cutlets.

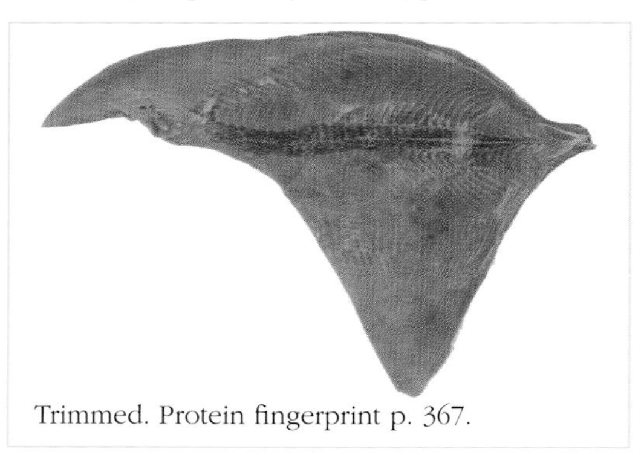

Trimmed. Protein fingerprint p. 367.

Size: Reported to attain more than 180 cm and 270 kg (commonly landed at 80–150 cm and 20–50 kg).

Habitat: Marine; pelagic vagrant in upper water masses of the open ocean, rarely venturing inshore.

Fishery: Caught mainly as bycatch of tuna longliners and more recently by pelagic trawls.

Remarks: Thought to eat squid, jellyfishes, and small fishes and crustaceans. Very highly regarded foodfishes in some regions.

Banded morwong

Cheilodactylus spectabilis

Previous name: brown-banded morwong

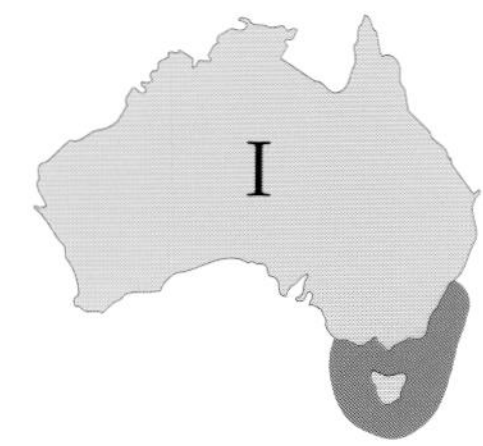

Identifying features: ① sides reddish-brown, colour uniform or broken up by 7 pale bands; ② anal fin with 3 spines, 8 soft rays; ③ 48–54 lateral-line scales; ④ caudal fin lacking a broad dark margin; ⑤ dorsal fin with 17–18 spines, 26–27 soft rays; ⑥ lower pectoral-fin rays only slightly longer than those above and partly detached; ⑦ lips thick and rubbery.

Comparisons: Most similar to the red morwong (*C. fuscus*, p. 187) but usually has a more pronounced banded pattern, no dark fin margins, fewer segmented dorsal-fin rays (less than 28 versus more than 29), and fewer scales in the lateral line (less than 55 versus more than 59). Other banded species of morwong are occasionally marketed but these either have fewer bands or are covered with spots.

Fillet: Moderately deep, rather elongate, convex above, off-white to yellowish. Outside with broad, intermediate, diffuse central red muscle band; top of shoulder with small red muscle patch; convex EL, converging HL; HS along middle of fillet; integument silvery white. Inside flesh medium; belly flap usually removed; peritoneum black. Usually skinned, skin thick, scale pockets moderate and defined.

Size: To 70 cm and about 4 kg (commonly 40–55 cm and 1–1.8 kg).

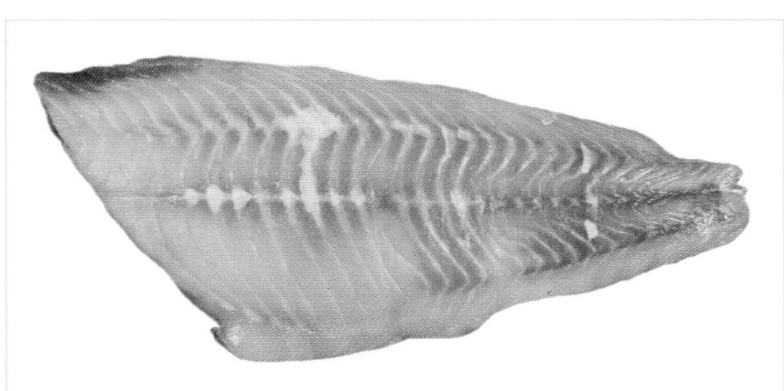

Untrimmed. Protein fingerprint p. 367.

Habitat: Marine; demersal, mainly lives in kelp, near caves and under ledges on shallow, coastal, rocky reefs.

Fishery: Caught mainly in Tasmania using gillnets, for live foodfish markets of Sydney and Melbourne.

Remarks: Rarely takes a hook so is of little value to anglers. The flesh has a delicate flavour but can be dry.

Blue morwong

Nemadactylus valenciennesi

Previous names: morwong, queen snapper

Identifying features: 1 sides silvery blue with yellowish tinge; 2 anal fin with 3 spines, 17–19 soft rays; 3 64–68 lateral-line scales; 4 no dark band across nape; 5 bright blue and yellow lines around eyes; 6 dorsal fin with 16–17 spines, 30–31 soft rays; 7 1 pectoral-fin ray very much longer than those above and detached; 8 lips thick and rubbery.

Comparisons: Distinctive morwong with striking colour pattern of blue and yellow lines and stripes on the head and body. Other bluish and greyish morwongs lack these yellowish markings. Most similar to its eastern relative, the grey morwong (*N. douglasii*, p. 185), which was also formerly known as 'blue morwong'. In addition to colour differences, the latter has fewer anal-fin soft rays (16–17 versus 17–19) and fewer lateral-line scales (less than 59 versus more than 63).

Fillet: Moderately deep, rather elongate, convex above, off-white to yellowish. Outside with broad, intermediate, diffuse central red muscle band; top of shoulder with large red muscle patch; convex EL, converging HL; HS along middle of fillet; integument white. Inside flesh medium; belly flap usually removed; peritoneum black. Usually skinned, skin medium, scale pockets moderate and well defined.

Size: To 100 cm and 12 kg (commonly 45–80 cm and 1.3–5.7 kg).

Habitat: Marine; demersal on the continental shelf, living over reefs and sponge gardens, from inshore to depths of more than 100 m.

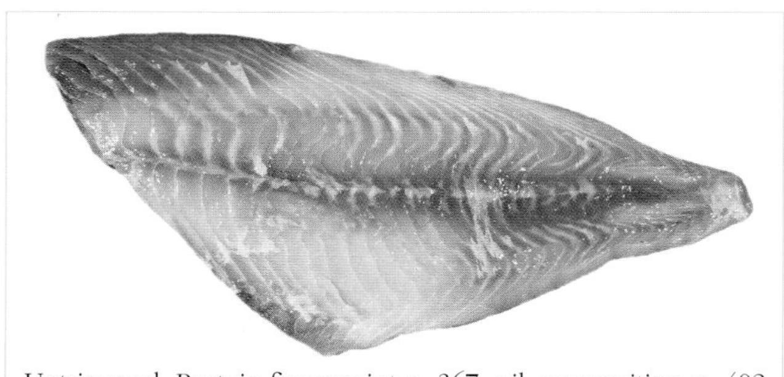

Untrimmed. Protein fingerprint p. 367; oil composition p. 402.

Fishery: Caught mainly using gillnets, lines or traps off South Australia and Western Australia. More recently in moderate quantities as bycatch of the Great Australian Bight demersal trawl fishery.

Remarks: Highly regarded foodfish in southern states, where it is the target of recreational fishers. The flesh is whitish with mild flavour.

Grey morwong

Nemadactylus douglasii

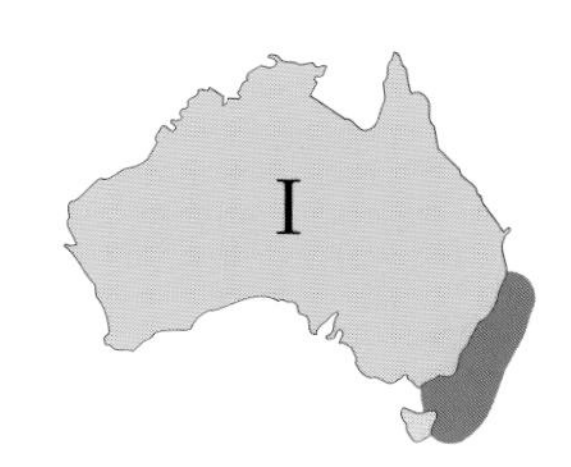

Previous names: blue morwong, butterfish, douglas morwong, morwong, rubberlip morwong

Identifying features: 1 sides uniform silvery to pale greyish-blue; 2 anal fin with 3 spines, 16–17 soft rays; 3 53–58 lateral-line scales; 4 no dark saddle across nape; 5 no yellow or blue lines around eyes; 6 dorsal fin with 17–18 spines, 27–28 soft rays; 7 1 pectoral-fin ray very much longer than those above and detached; 8 lips thick and rubbery.

Comparisons: Closely related to the main commercial morwong (*N. macropterus*, p. 186) but is more bluish, lacks a dark saddle on the nape, and has more anal-fin soft rays (16–17 versus 14–15). A western relative, the blue morwong (*N. valenciennesi*, p. 184), is more strongly patterned with blue overlain with yellow lines.

Fillet: Moderately deep, rather elongate, convex above, off-white to pinkish. Outside with broad, intermediate, diffuse central red muscle band; top of shoulder with large red muscle patch; convex EL, converging HL; HS along middle of fillet; integument white. Inside flesh medium; belly flap usually removed; peritoneum black. Often skinned, skin medium, scale pockets moderate and defined.

Size: To 80 cm and almost 6 kg (commonly 40–60 cm and 0.8–3.0 kg).

Habitat: Marine; demersal, aggregating mainly on coastal and offshore reefs on the continental shelf to at least 100 m depth.

Trimmed. Protein fingerprint p. 367; oil composition p. 402.

Fishery: Caught mainly using traps off New South Wales. Also taken by handline, bottom-set longline and as bycatch of the south-east demersal trawl fishery. Very significant recreational and charter-boat catch.

Remarks: Sometimes incorrectly marketed as 'morwong'; the flesh has been described as succulent.

Morwong

Nemadactylus macropterus & N. sp.

Previous names: butterfish, jackass morwong, king morwong, perch, rubberlip, seabream, silver perch, terakihi

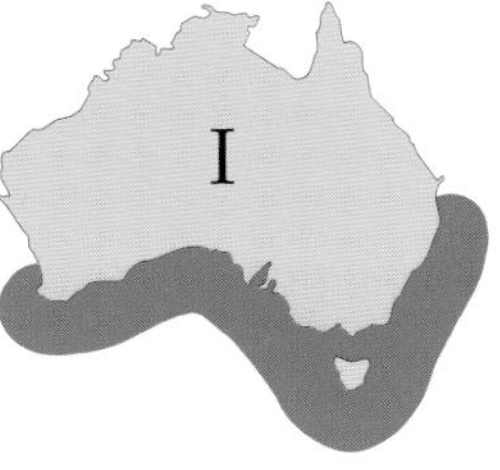

Nemadactylus macropterus

Identifying features: ❶ sides uniform silvery grey above, and whitish below; ❷ anal fin with 3 spines, 12–15 soft rays; ❸ 59–60 lateral-line scales; ❹ prominent dark band across nape; ❺ no yellow or blue lines around eyes; ❻ dorsal fin with 17–18 spines, 25–28 soft rays; ❼ 1 pectoral-fin ray very much longer than those above and detached; ❽ lips thick, rubbery.

Comparisons: Two similar species that resemble the grey morwong (*N. douglasii*, p. 185). They are more greyish-white (rather than bluish), have a dark saddle on the nape, and fewer anal-fin soft rays (15 or less versus 16–17). The blue morwong (*N. valenciennesi*, p. 184), has a more pronounced colour pattern of blue and yellow lines. Compared with the jackass morwong (*N. macropterus*), the king morwong (*N.* sp.) has an additional dark band on the outer upper half of the pectoral fin and fewer anal-fin rays (12 versus 14–15).

Fillet: *N. macropterus* moderately deep, rather elongate, convex above, off-white to pinkish. Outside with feeble, continuous central red muscle band; top of shoulder with large red muscle patch; convex EL, converging HL; HS along middle of fillet; integument white. Inside flesh medium; belly flap usually removed; peritoneum black. Often skinned, skin medium, scale pockets moderate and defined.

Size: To 70 cm and 4.5 kg (commonly 40–60 cm and 0.9–3 kg).

Habitat: Marine; demersal on the continental shelf and upper slope to at least 360 m depth. Smaller fishes occur near the coasts of Victoria and Tasmania.

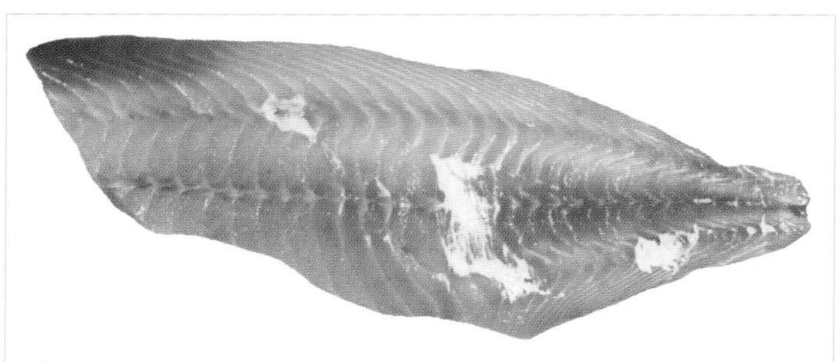

Trimmed. Protein fingerprint p. 367; oil composition p. 402.

Fishery: Includes the main commercial morwong. Taken by demersal trawl off south-eastern Australia and minor catches in the Great Australian Bight. Also taken in traps and by droplines off New South Wales.

Remarks: Small quantities taken by anglers. Firm, white, mildly flavoured flesh.

Red morwong

Cheilodactylus fuscus

Previous name: mowie

Identifying features: 1 sides almost uniform reddish-brown above, whitish below (not banded anteriorly); 2 anal fin with 3 spines, 8 soft rays; 3 60–64 lateral-line scales; 4 caudal fin with a broad, dark margin; 5 dorsal fin with 17 spines, 30–34 soft rays; 6 lower pectoral-fin rays slightly longer than those above and partly detached; 7 lips thick and rubbery.

Comparisons: Most similar to the banded morwong (*C. spectabilis*, p. 183) which also has a reddish rather than silvery or bluish back. However, it is usually plain-coloured (sometimes with bands confined to the caudal peduncle), has dark fin margins (rather than being plain), more segmented dorsal-fin rays (more than 29 versus fewer than 29), and more scales in the lateral line (more than 59 versus fewer than 55). Other banded species of morwong are occasionally marketed but these either have fewer bands or are covered with spots.

Fillet: Moderately deep, rather elongate, convex above, off-white to pinkish. Outside with intermediate, continuous central red muscle band; top of shoulder with large red muscle patch; convex EL, converging HL; HS along middle of fillet; integument translucent above, white below. Inside flesh medium; belly flap usually removed; peritoneum black. Usually skinned, skin medium, scale pockets small and defined.

Size: To 65 cm and 3 kg (commonly 35–45 cm and 0.5–0.9 kg).

Habitat: Marine; demersal, lives inshore on shallow, coastal, rocky reefs. Often occurs in congregations.

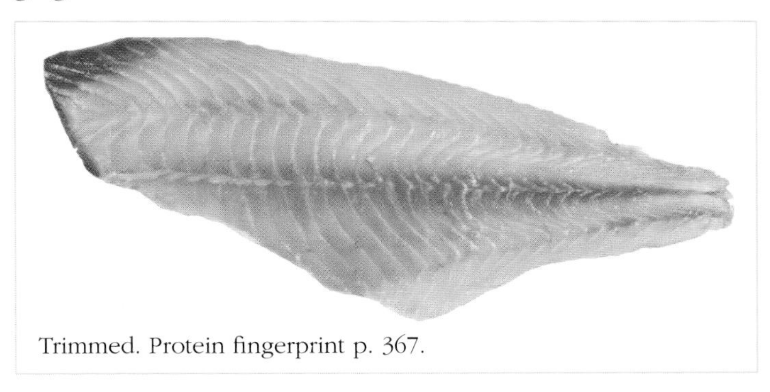

Trimmed. Protein fingerprint p. 367.

Fishery: Caught using traps, gillnets and lines, mainly for the live foodfish markets of Sydney.

Remarks: Taken by anglers and spearfishers in moderate quantities. The flesh, which has a delicate flavour and is somewhat softer than most other morwongs, is highly regarded.

Diamondscale mullet

Liza vaigiensis

Previous names: largescale mullet, mullet

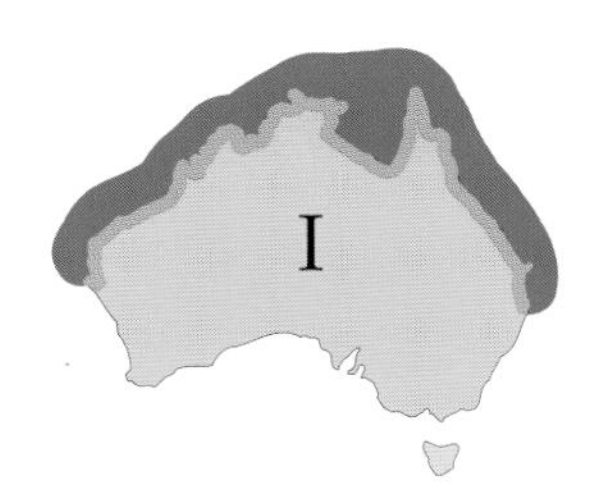

Identifying features: 1 pectoral fin black; 2 caudal-fin margin almost straight when fin stretched; 3 body silver olive with dark-edged scales; 4 first dorsal fin with 4 spines; 5 dorsal fins widely separated; 6 pelvic fin well forward of first dorsal fin; 7 anal fin with 3 spines, 8 soft rays.

Comparisons: Readily recognisable both in the catch or when viewed from the surface due to its large, sharply defined, black-edged scales and striking jet black pectoral fin. No other Australian mullet has a black pectoral fin and a truncate caudal fin. The caudal fin of most mullet species is forked or at least has a slightly concave hind margin.

Fillet: Moderately deep, rather elongate, gently tapering to very deep caudal peduncle, pale pinkish. Outside with pronounced, discontinuous central red muscle band; convex EL, weakly converging HL; HS along middle of fillet; integument patchy, white. Inside flesh fine; belly flap sometimes removed; peritoneum black. Usually skinned, skin medium, scale pockets large and defined.

Size: Possibly to 70 cm and 4.6 kg (commonly 30–50 cm and 0.4–1.5 kg).

Habitat: Marine; pelagic, often in very shallow water, near the surface along the tropical coastline. Also common offshore around islands and coral reefs.

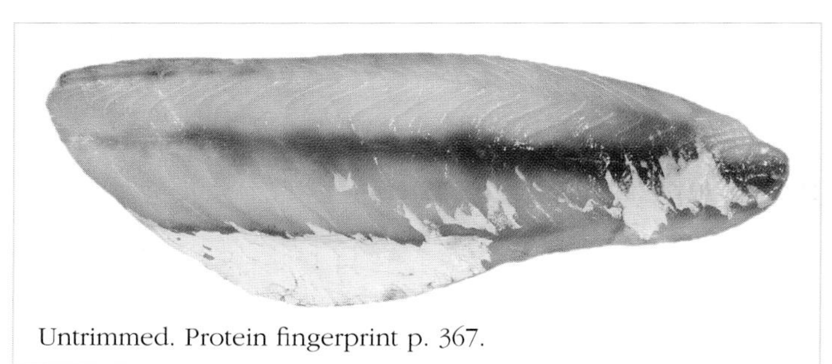

Untrimmed. Protein fingerprint p. 367.

Fishery: Caught mainly using beach seines, mostly off the Queensland coast.

Remarks: One of the largest mullet species; because of its bulk, even small fish provide substantial fillets. The flesh is rather oily and makes ideal marlin bait. Highly regarded table-fish.

Mullet (page 1 of 2)

Mugilidae

Previous names: bluetail mullet, fantail mullet, flat-tail mullet, flicker mullet, jumping mullet, lano mullet, sand mullet, tallegalane, tygum mullet, yelloweye mullet

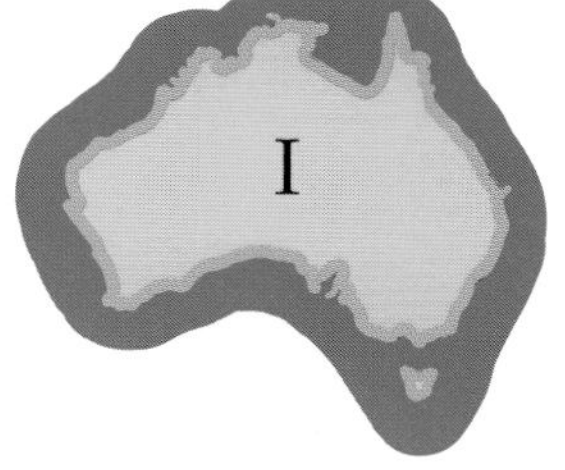

Aldrichetta forsteri

Identifying features: ① body silvery; ② first dorsal fin with 4 spines; ③ dorsal fins widely separated; ④ mouth small, teeth minute when present; ⑤ gill rakers long; ⑥ lateral line faint or absent; ⑦ pelvic fin well forward of first dorsal fin; ⑧ anal fin with 3 spines, 8–13 soft rays.

Comparisons: Distinctive fishes that most closely resemble small baitfishes known as 'hardyheads' or 'prettyfishes' (Atherinidae). Mullets have a silver body, which is covered in rather large, clear scales, and is robust anteriorly, usually becoming more compressed towards the tail. The head shape, fin sizes and positions are also important defining characters. The barracudas or pikes (*Sphyraena* species, pp 201–203) have similar fin arrangements but are more slender and have larger heads with long, sharp teeth.

Fillet: *Aldrichetta forsteri* rather slender, gently tapering to deep caudal peduncle, off-white to yellowish. Outside with broad, pronounced, continuous central red muscle band; parallel EL, converging HL; HS along middle of fillet; integument blackish above, silvery white below. Inside medium; belly flap sometimes removed; peritoneum white with melanophores. Usually skinned, skin medium, scale pockets moderate and defined.

Size: To at least 80 cm and 8 kg (marketed usually from 25–50 cm and 0.3–1.5 kg).

Habitat: Marine; pelagic, schools near the surface, mostly in estuaries and near the coast, over soft bottoms. Some migrate to headwaters of tidal rivers and others venture into freshwater.

Fishery: Caught near the coast and in estuaries using mainly haul and beach seines, gillnets, and tunnel, pound and ring nets. Caught seasonally or year round depending on species.

Remarks: Traditionally important foodfishes in Australia. Some 16 or so species occur here and at least eight are frequently marketed. The two most commercially important species have separate marketing names but all may be marketed as 'mullet'. The flesh varies in quality between species and most need to be eaten fresh. Consequently, they are usually sold fresh, either as chilled fillets or whole.

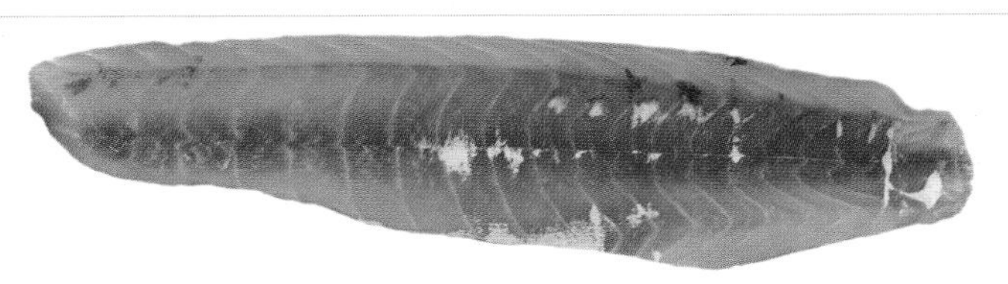

Trimmed. Protein fingerprint p. 367; oil composition p. 402.

Mullet (page 2 of 2)

Mugilidae

Aldrichetta forsteri

Remarks: Commonly called 'yelloweye mullet', this species is distributed around the southern coast between Newcastle (NSW) and Kalbarri (WA), including Tasmania. Best distinguished by body shape, having 12–13 soft rays in the anal fin, and lacking transparent eyelids and dark spots at the pectoral-fin base. Reaches at least 50 cm and 1.2 kg.

Liza argentea

Remarks: Commonly called 'flat-tail mullet', this species is distributed around the southern coast between Cooktown (Qld) and Kalbarri (WA), including the northern coast of Tasmania. Best distinguished by body shape, 10 soft rays in the anal fin, a dark spot at the pectoral-fin base, and lacking transparent eyelids. To 45 cm (rarely exceeding 30 cm) and 1.0 kg.

Valamugil seheli

Remarks: Commonly called 'bluetail mullet', this species is distributed around the coast of northern Australia between Noosa (Qld) and Exmouth Gulf (WA). Best distinguished by body shape, 9 soft rays in the anal fin, bluish caudal fin, a long, yellow pectoral fin, a dark spot at the pectoral-fin base, and lacking transparent eyelids. Reaches about 70 cm and 4.8 kg.

Sea mullet

Mugil cephalus

Previous names: bully mullet, hardgut mullet, hardgut river mullet, mangrove mullet, mullet, poddy mullet, river mullet

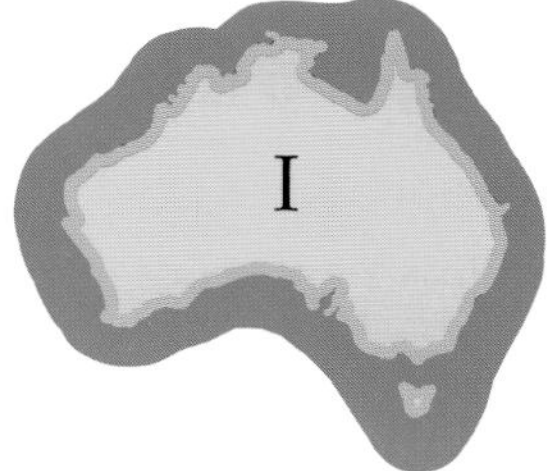

Identifying features: 1 transparent fatty eyelid covering most of eye; 2 caudal-fin margin forked; 3 scales not dark-edged; 4 first dorsal fin with 4 spines; 5 dorsal fins widely separated; 6 pelvic fin well forward of first dorsal fin; 7 anal fin with 3 spines, 8 soft rays.

Comparisons: Unlike other commercial mullets, the sea mullet has a characteristic fatty eye lid extending over most of the front and back of the eye to reach the pupil. It is probably occasionally confused with a less common relative, Broussonnet's sea mullet (*M. broussonnetii*), which has more anal-fin soft rays (9 rather than 8).

Fillet: Moderately deep, rather elongate, gently tapering to very deep caudal peduncle, pink to reddish-brown. Outside with broad, intermediate, diffuse central red muscle band; convex EL, converging HL; HS along middle of fillet; integument patchy, white. Inside flesh medium; belly flap sometimes removed; peritoneum translucent to white. Usually skinned, skin thin, scale pockets moderate and defined.

Size: To at least 80 cm and 8 kg (commonly 30–45 cm and 0.5–1.5 kg).

Habitat: Marine; pelagic in coastal marine habitats and estuaries and occasionally venturing into freshwater. Worldwide, mainly in warm temperate and tropical regions.

Fishery: Major commercial mullet taken off beaches using set and surround nets, mainly in Queensland, New South Wales and Western Australia. Also sought-after by recreational fishers mostly using beach seines and gillnets in states where these are allowed. Only occasionally taken by anglers.

Untrimmed. Protein fingerprint p. 367; oil composition p. 402.

Remarks: Largest Australian mullet. Highly regarded table-fish, usually marketed whole or as fillets. Roe is exported.

Coral perch (page 1 of 2)

Scorpaenidae

Previous name: ocean perch

Scorpaena cardinalis

Identifying features: 1 pronounced bony ridge (mostly spiny) on cheek beneath the eye; 2 head spiny with 1–2 spines on operculum; 3 caudal fin rounded or truncate (not forked); 4 1 dorsal fin (sometimes with notch), with 11–17 strong spines, 8–17 soft rays; 5 anal fin with 1–3 strong spines, 3–9 soft rays; 6 body moderate or short, usually slightly compressed; 7 pectoral fins large, often with thickened rays.

Comparisons: Coral perches, also known as scorpionfishes, belong to a group known as the 'mail-cheeked' fishes. They all have a solid bone beneath the eye, often appearing as a ridge, and the group includes the gurnards and flatheads. The body form is distinctive.

Fillet: *Scorpaena cardinalis* moderately deep, short, pronounced taper, convex above, greyish-white. Outside without central red muscle band; weakly converging EL, weakly converging HL; HS below middle of fillet; integument greyish or translucent. Inside flesh fine; belly flap absent; peritoneum silvery to translucent. Usually skinned, skin thick, scale pockets small, well defined.

Size: To about 50 cm and 2.0 kg (adults usually 20–35 cm and 0.1–0.8 kg).

Habitat: Marine; occur in a variety of habitats from inshore in estuaries to the deep continental slope in 1000 m depth or more. Most live around hard bottoms or those rich in large invertebrates; others have adapted to life on sandy bottoms or among seagrasses.

Trimmed. Protein fingerprint p. 368.

Fishery: Caught mostly as bycatch of trawl fisheries or to a lesser extent by lining.

Remarks: About 100 species occur locally, of which several are important aquarium fishes, but only a dozen or so are marketed. They have venomous spines in the dorsal, anal and pelvic fins which are capable of inflicting a painful wound.

Coral perch (page 2 of 2)

Scorpaenidae

Scorpaena cardinalis

Remarks: Commonly called 'red rock cod', this species is distributed on the inner continental shelf off eastern Australia from Noosa Head (Qld) south to at least Jervis Bay (NSW) in depths to over 100 m. However, the taxonomy is confused and more than one species may be caught in this area. Best distinguished in having a marbled reddish body, a dorsal fin with 12 spines and 9–10 soft rays, a black blotch over the last outer membranes of the spinous dorsal fin, and usually with pronounced reddish spots on the lower head, belly and anal fin. Reaches at least 45 cm and 1.4 kg.

Neosebastes thetidis

Remarks: Commonly called 'thetis fish', this species is distributed on the outer continental shelf off southern Australia between Newcastle (NSW) and Rottnest Island (WA), including Tasmania, in depths of 20–240 m. Best distinguished in having a pale, stumpy body with large dark blotches on the back, a dorsal fin with 13 spines and 8–9 soft rays, a bony ridge over each eye, and no deep groove on the nape. Reaches about 35 cm and almost 0.8 kg.

Trachyscorpia capensis

Remarks: Commonly called 'cape scorpionfish', this species is distributed on the outer continental shelf off southern Australia between Lakes Entrance (Vic.) and Margaret River (WA), including Tasmania, in depths of 530–870 m. Best distinguished in having a uniform pinkish-red body, a dorsal fin with 13 spines and 9 soft rays, no bony ridge over each eye, and a long black blotch along the margin of the spinous dorsal fin. Reaches at least 40 cm and 1.2 kg.

Ocean perch

Helicolenus barathri & H. percoides

Previous names: coral cod, coral perch, red gurnard perch, red perch, red rock perch, sea perch

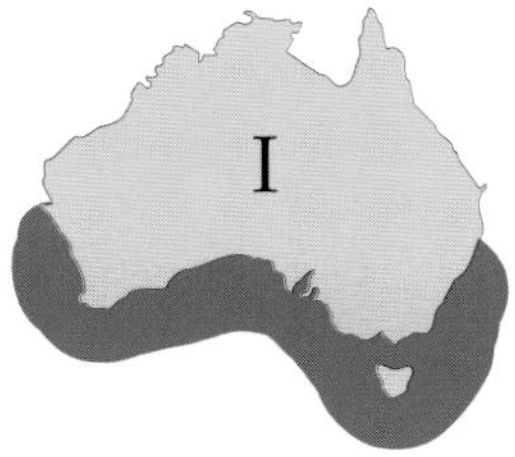

Helicolenus barathri

Identifying features: 1 body reddish with greenish or reddish flecks or bands; 2 smooth, bony ridge on cheek beneath eye; 3 no scales between eyes; 4 caudal fin truncate or forked slightly; 5 dorsal fin with 12 strong spines, 11–14 soft rays, last spine distinctly shorter than first soft ray; 6 anal fin with 3 strong spines, 5 soft rays; 7 pectoral fins large.

Comparisons: Most easily confused with the deepwater scorpionfishes (*Trachyscorpia* species) but the latter have only 9 anal-fin soft rays (rather than 11 or more), obvious scale pockets between the eyes (rather than scaleless), and the bony ridge under the eyes is raised upward with a spiny edge. Two main species are landed. The reef ocean perch (*H. percoides*) has more pronounced banding on the body and fewer dorsal-fin soft rays (mostly 11–12 versus mostly 13–14) than the ocean perch (*H. barathri*).

Fillet: *H. barathri* moderately deep, short, pronounced taper, slightly convex above, yellowish-white. Outside with feeble, discontinuous, central red muscle band; weakly converging EL, converging HL; HS below middle of fillet; integument patchy, silvery white. Inside flesh medium; belly flap usually absent; peritoneum black. Usually skinned, skin thin, scale pockets moderate and well defined.

Size: To 47 cm and 1.8 kg (commonly less than 35 cm and 0.8 kg).

Habitat: Marine; demersal on the shelf and slope off southern Australia in 10–800 m depth.

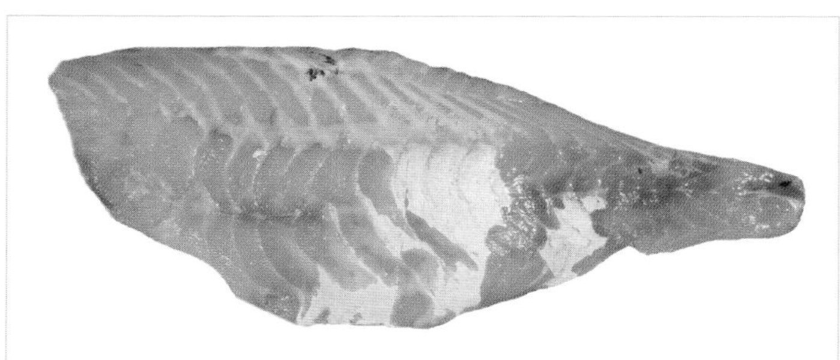

Trimmed. Protein fingerprint p. 368; oil composition p. 402.

Fishery: Includes two South East Trawl Fishery quota species partitioned by depth, with the reef ocean perch taken more inshore on the continental shelf.

Remarks: Taxonomically confused and more species of the genus may occur in our seas. The flesh is of reasonable quality.

Black oreo

Allocyttus niger & *A. verrucosus*

Previous names: black dory, black oreo dory, warty oreo

Allocyttus niger

Identifying features: 1 predorsal profile straight or slightly concave; 2 1–2 rows of bony 'plates' or protuberances on sides (often present as scar in large adults); 3 skin rough, scales hard to dislodge; 4 second spine of dorsal fin much thicker and taller than first; 5 eye very large; 6 dark, stumpy body with a small caudal fin and narrow caudal peduncle; 7 mouth can be extended forward.

Comparisons: Two similar species. The main commercial species, the black oreo (*A. niger*), differs from the warty oreo (*A. verrucosus*) in having a larger pelvic fin (reaching the anal fin rather than falling short of the anal fin), a single row of scaly protuberances on the belly (rather than 2 rows of bony 'plates'), and relatively taller first dorsal-fin spines. The spikey oreo (*Neocyttus rhomboidalis*, p. 197) has a more concave nape with proportionally much larger spines in the dorsal and anal fins and lacks protuberances or 'plates' on the belly.

Fillet: *A. niger* moderately deep, bottle-shaped, tapering sharply, very convex above, yellowish-white. Outside without central red muscle band; weakly converging EL, converging HL; HS well below middle of fillet, EL closer to HS than dorsal margin; integument translucent. Inside flesh fine; belly flap always absent; peritoneum black. Always skinned, skin thick, scale pockets small and well defined.

Size: To 49 cm and almost 2 kg (commonly 25–35 cm and 0.4–1.0 kg).

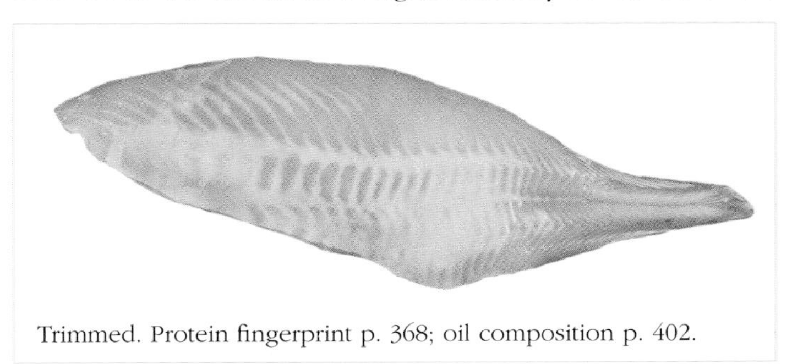

Trimmed. Protein fingerprint p. 368; oil composition p. 402.

Habitat: Marine; demersal on the slope in 340–1630 m depth (usually 750–1200 m), often on hard bottoms and over seamounts.

Fishery: Trawled often in large quantities, mainly south of Bass Strait.

Remarks: The flesh is firm with a delicate taste.

Smooth oreo

Pseudocyttus maculatus

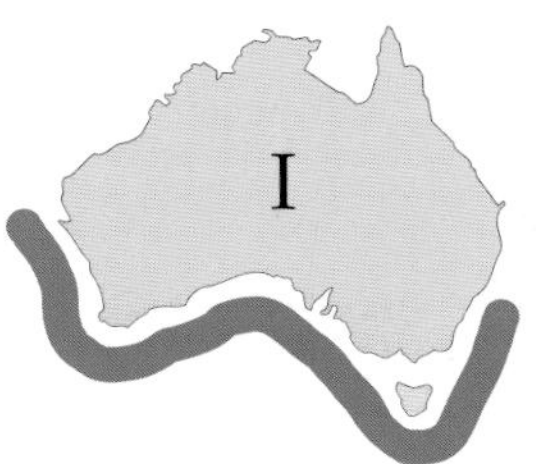

Previous names: smooth dory, smooth oreo dory, spotted dory, spotted oreo

Identifying features: ❶ predorsal profile almost straight behind eye; ❷ second spine of dorsal fin small and only slightly shorter than first; ❸ no bony 'plates' or protuberances on sides; ❹ skin smooth, scales rub off easily; ❺ eye very large; ❻ dark, oval body with a small caudal fin and narrow caudal peduncle; ❼ mouth can be extended forward.

Comparisons: Oreos (Oreosomatidae) differ from dories (Zeidae) in having dark (rather than silvery) bodies, rougher skin, relatively larger eyes, and are more heavily built. The smooth oreo is somewhat atypical in having removable scales, smaller spines and a more elongate body.

Fillet: Moderately deep, bottle-shaped, tapering sharply, very convex above, white. Outside without central red muscle band; parallel EL, converging HL; HS well below middle of fillet, EL closer to HS than dorsal margin; integument translucent. Inside flesh fine; belly flap always absent; peritoneum grey or black. Always skinned, skin medium, scale pockets small and irregular.

Size: To 68 cm and at least 4.5 kg (commonly 35–45 cm and 1–2 kg).

Habitat: Marine; demersal on the mid-continental slope in depths of 650–1500 m (usually 850–1100 m), mostly near deep underwater pinnacles.

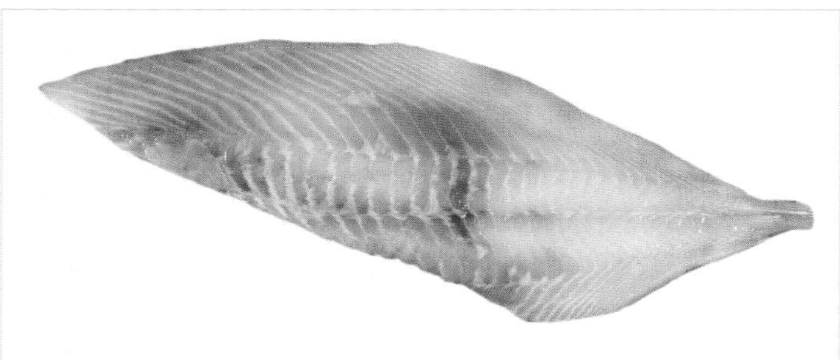

Trimmed. Protein fingerprint p. 368; oil composition p. 402.

Fishery: Often trawled in large quantities off southern Australia, mainly near Tasmania or on seamounts to the south. Once discarded, oreos are now almost fully utilised.

Remarks: Considered to be the premium foodfish of the group; the flesh has a medium texture and a mild but distinctive taste.

Spikey oreo

Neocyttus rhomboidalis

Previous names: deepwater dory, oreo dory, oxeye oreo, spikey dory, spikey oreo dory

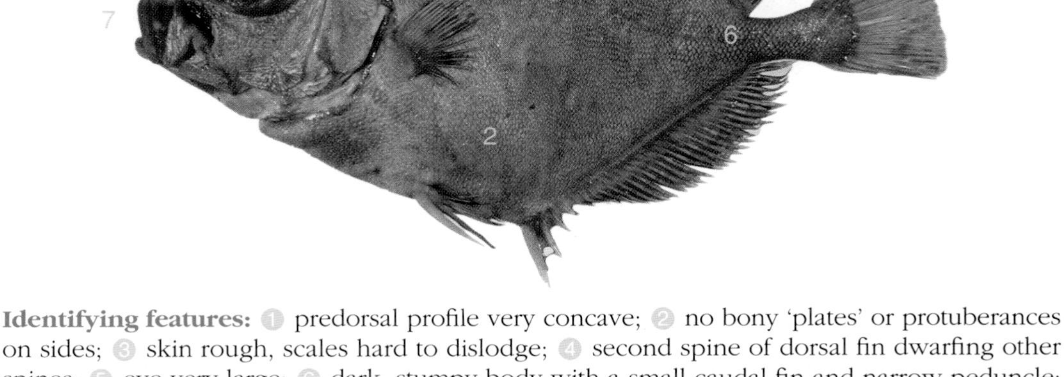

Identifying features: 1 predorsal profile very concave; 2 no bony 'plates' or protuberances on sides; 3 skin rough, scales hard to dislodge; 4 second spine of dorsal fin dwarfing other spines; 5 eye very large; 6 dark, stumpy body with a small caudal fin and narrow peduncle; 7 mouth can be extended forward.

Comparisons: Most similar to black oreos (*Allocyttus niger* and *A. verrucosus*, p. 195) but has a more concave nape with proportionally much larger spines in the dorsal and anal fins and lacks 'plates' or protuberances on the belly. Unlike smooth oreo (*Pseudocyttus maculatus*, p. 196), it has a more angular body shape, spiny (rather than smooth) skin, and larger fin spines.

Fillet: Moderately deep, bottle-shaped, tapering sharply, very convex above, white. Outside without central red muscle band; weakly converging EL, converging HL; HS well below middle of fillet, EL closer to HS than dorsal margin; integument translucent. Inside flesh fine; belly flap absent; peritoneum black. Always skinned, skin thick, scale pockets small and well defined.

Size: To at least 47 cm and about 2 kg (commonly 30–40 cm and 0.8–1.5 kg).

Trimmed. Protein fingerprint p. 368; oil composition p. 402.

Habitat: Marine; demersal, aggregates on the slope in 200–1240 m depth (usually 450–800 m), usually over hard bottoms.

Fishery: Trawled, often in large quantities, mainly near Tasmania or on seamounts to the south.

Remarks: Firm, almost rubbery, flesh with delicate taste.

Pearl perch (page 1 of 2)

Glaucosoma species

Previous names: northern dhufish, northern pearl perch, threadfin pearl perch

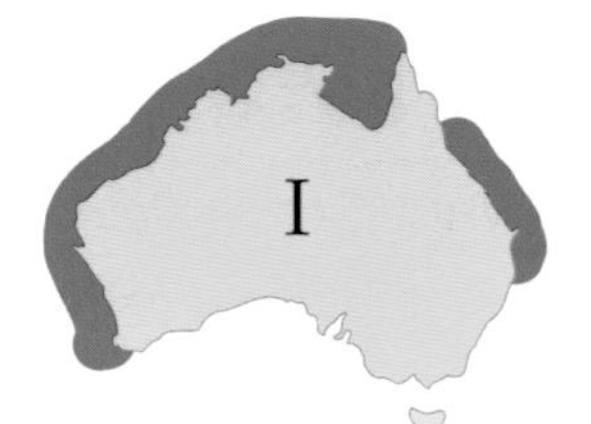

Glaucosoma scapulare

Identifying features: 1 body silvery, deep and slightly compressed; 2 lateral line almost straight and extending onto tail; 3 dorsal-fin and anal-fin spines gradually increasing in size; 4 upper-jaw bone scaly; 5 hind margin of caudal fin almost square or tips with trailing filaments; 6 dorsal fin with 8 spines, 11–14 soft rays; 7 anal fin with 3 spines, 9–12 soft rays.

Comparisons: Pearl perches differ from members of other Australian warm temperate and tropical perch-like fishes in the combination of the above characters. Unlike the three species sold under this name, the related West Australian dhufish (*Glaucosoma hebraicum*, p. 200), lacks a flattened bone at the top of the opercular margin and has pale gill rakers and peritoneum.

Fillet: *G. scapulare* deep, short, tapering prominently, slightly convex above, yellowish-white. Outside with feeble, continuous, central red muscle band; convex EL, converging HL; HS along middle of fillet; integument white. Inside flesh medium; belly flap sometimes present; peritoneum black. Usually skinned, skin medium, scale pockets moderate and well defined.

Size: To about 70 cm and about 7 kg (commonly to 50 cm and 2.5 kg).

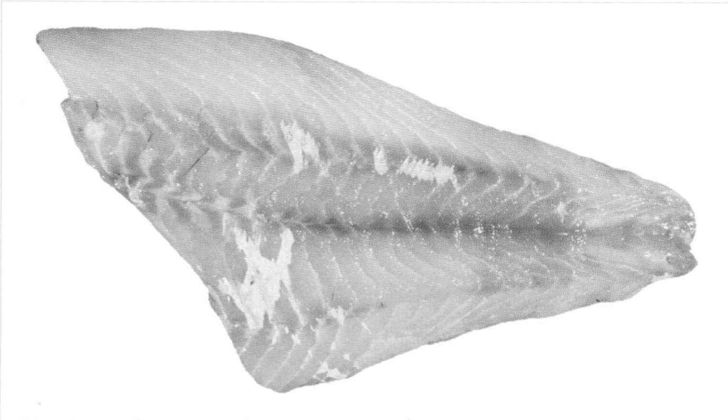

Untrimmed. Protein fingerprint p. 368; oil composition p. 402.

Habitat: Marine; demersal near rocky habitats of the continental shelf.

Fishery: Caught mainly using lines although small catches of tropical species are taken by trawlers.

Remarks: Superior table-fishes with white flesh and excellent flavour and texture. The common pearl perch (*G. scapulare*), is the east coast equivalent of the much revered 'dhufish'.

Pearl perch (page 2 of 2)

Glaucosoma species

Glaucosoma scapulare

Remarks: Commonly called 'pearl perch', this species is distributed off eastern Australia between Rockhampton (Qld) and Port Jackson (NSW) in depths to 150 m. Best distinguished in having a dark, flattened, bony plate at the top of the operculum and faint narrow lines along its silvery body. Reaches about 70 cm and at least 7.3 kg.

Glaucosoma magnificum

Remarks: Commonly called 'threadfin pearl perch', this species is widely distributed off northern Australia between Exmouth Gulf (WA) and Shelbourne Bay (Qld) in depths to at least 100 m. Best distinguished in lacking a flattened bony plate at the top of the operculum, and having 3 dark bands on the head and long filamentous caudal-fin rays. Reaches at least 32 cm and 0.8 kg.

Glaucosoma buergeri

Remarks: Commonly called 'northern pearl perch', this species is widely distributed off south-east Asia and locally off Western Australia between Shark Bay and the northern Kimberley in depths to at least 100 m. Best distinguished in lacking a flattened bony plate at the top of the operculum, and having a dark band through the eye, and black gill rakers and body cavity membrane. Reaches at least 45 cm and 2.5 kg.

West Australian dhufish

Glaucosoma hebraicum

Previous names: dhufish, jewfish, WA pearl perch

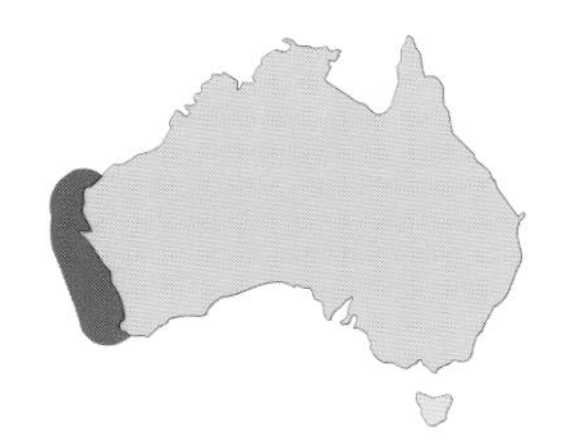

Identifying features: 1 no shield-like bone at top of operculum; 2 dark curved bar through eye (often faint in mature females); 3 gill rakers and body cavity membrane pale; 4 body silvery with dark stripes (sometimes faint); 5 dorsal-fin and anal-fin spines gradually increasing in size; 6 hind margin of caudal fin almost square; 7 dorsal fin with 8 spines, 11 soft rays; 8 anal fin with 3 spines, 9 soft rays.

Comparisons: Distinguished from the other main commercial member of the family, the pearl perch (*G. scapulare*, p. 199), by the absence of a flattened, pearly bone (covered with a black membrane) near the top of the operculum. The northern pearl perch (*G. buergeri*, p. 199) has black gill rakers and body cavity membrane (rather than pale).

Fillet: Deep, short, tapering prominently, convex above, yellowish-white. Outside with moderate, broad continuous, central red muscle band; convex EL, weakly converging HL; HS along or slightly below middle of fillet; integument white. Inside flesh medium; belly flap sometimes present; peritoneum silvery white to translucent. Usually skinned, skin medium, scale pockets large and well defined.

Size: To 122 cm and possibly 32 kg (commonly to 80 cm and up to 8 kg).

Untrimmed. Protein fingerprint p. 368; oil composition p. 402.

Habitat: Marine; demersal on rocky bottoms on the inner shelf to at least 120 m depth.

Fishery: Off Western Australia using mainly handlines and in smaller quantities by gillnets, spears and longlines.

Remarks: One of Australia's most highly regarded foodfishes. Highly sought-after, with the recreational catch possibly exceeding the commercial catch. Marketed fresh.

Pike

Sphyraena novaehollandiae

Previous names: seapike, shortfin pike, snook

Identifying features: ① single, long gill raker on outer gill arch (located at angle); ② no black spot at base of each pectoral fin; ③ more than 100 scales in lateral line (usually 125–135); ④ body very slender; ⑤ sides silvery, lacking faint horizontal stripes; ⑥ mouth large with long canine teeth; ⑦ dorsal fins widely separated; ⑧ first dorsal fin with 6 spines, second with 1 spine, 9 soft rays.

Comparisons: Occasionally confused with another southern fish, the smaller longfin pike (*Dinolestes lewini*), which is similar in appearance but is shorter-bodied, with a longer anal-fin base and a brighter yellow tail. The other (tropical) pikes, the striped seapikes (*Sphyraena* species, pp 202–203), are very similar externally but the pike lacks bars and stripes, has relatively small scales (well over 100 in the lateral line) and only a single raker in the outer gill arch (rather than none or 2–3).

Fillet: Long, slender, barely tapering, off-white to yellowish. Outside with pronounced, continuous central red muscle band; parallel EL, parallel HL; HS along middle of fillet, dipping anteriorly; myomeres small, integument greenish-grey above, silvery white below. Inside flesh coarse; belly flap sometimes removed; peritoneum translucent. Usually skinned, skin medium, scale pockets small and well defined.

Size: To 110 cm and about 5.3 kg (commonly up to 80 cm and 2 kg).

Habitat: Marine; pelagic in small groups close to the surface, mainly near reefs and over seagrass beds in shallow water. Also aggregates near deeper reefs.

Fishery: Caught mainly using haul nets and gillnets off southern Australia in winter and spring as bycatch of fisheries targeting southern garfish (*Hyporhamphus melanochir*, p. 130) and school whiting (*Sillago* species, pp 281–282) over seagrass beds. Also caught trolling. Caught by recreational fishers both angling and trolling, using lures, small fish or squid as bait.

Remarks: High quality table-fish with firm and tasty flesh. Has excellent smoking qualities.

Untrimmed. Protein fingerprint p. 368; oil composition p. 403.

Striped seapike (page 1 of 2)

Sphyraena species

Previous names: barracuda, pickhandle, seapike

Sphyraena obtusata

Identifying features: 1 body very slender; 2 sides silvery, often with faint horizontal stripes; 3 mouth large with long canine teeth; 4 lower jaw protruding further forward than upper jaw; 5 gill rakers mostly reduced to spiny plates (sometimes 1–3 near angle); 6 dorsal fins widely separated; 7 first dorsal with 5–6 spines, second with 1–2 spines, 8–9 soft rays.

Comparisons: Australian fishes sold under the names of 'pike' or 'seapike' (Sphyraenidae) should not be confused with the unrelated 'freshwater pikes' (Esocidae) of the Northern Hemisphere. In many ways they resemble elongate mullets (Mugilidae) but have larger heads and longer, sharper teeth. Nine species occur locally and most of these reach markets in varying quantities. Despite the marketing name, not all species are striped.

Fillet: *S. obtusata* long, moderately slender, barely tapering, pale pinkish. Outside with pronounced, continuous central red muscle band; parallel EL, weakly converging HL; HS along middle of fillet; myomeres moderate, integument silvery grey above, silvery white below. Inside flesh medium; belly flap sometimes removed; peritoneum silvery white to translucent. Usually skinned, skin medium, scale pockets small and indistinct.

Size: To 240 cm and more than 45 kg (commonly less than 100 cm and 5 kg).

Habitat: Marine; pelagic and demersal, usually in schools but sometimes alone, near coral reefs on the continental shelf.

Fishery: Taken mainly as bycatch of a range of line and net fisheries around the entire continent. Trawled in small quantities in the tropics.

Remarks: Group of important foodfishes throughout the Asia-Pacific region. The flesh is mostly soft and delicately flavoured but may be greyish and rather strongly scented. Very large fish can contain ciguatoxin and should not be eaten.

Trimmed. Protein fingerprint p. 368.

Striped seapike (page 2 of 2)

Sphyraena species

Sphyraena obtusata

Remarks: Commonly called 'striped seapike', this species is distributed around northern Australia from Sydney (NSW) to Albany (WA). Best distinguished in having 2 gill rakers on the outer arch, the first dorsal fin originating forward of the pectoral-fin tips, a short anal-fin base, a brownish stripe along entire length, and a yellowish caudal fin. Reaches about 55 cm and less than 1 kg.

Sphyraena barracuda

Remarks: Commonly called 'barracuda', this species is distributed across northern Australia from southern Queensland to Albany (WA). Best distinguished in having no gill rakers on the outer arch, a low lateral-line scale count (less than 90), upright teeth, faint bars on back, and whitish fin tips. Reaches 240 cm and 45 kg.

Sphyraena jello

Remarks: Commonly called 'slender seapike', this species is widely distributed off northern Australia. Distinguished in having no gill rakers on the outer arch, high lateral-line scale count (more than 100), large teeth in the lower jaw sloping backwards, V-shaped bars on back, and the posterior fins without white tips. Reaches at least 165 cm and possibly 20 kg.

Ray's bream (page 1 of 2)

Bramidae

Previous names: castagnole, bigscale pomfret, flathead pomfret, golden pomfret, southern Ray's bream

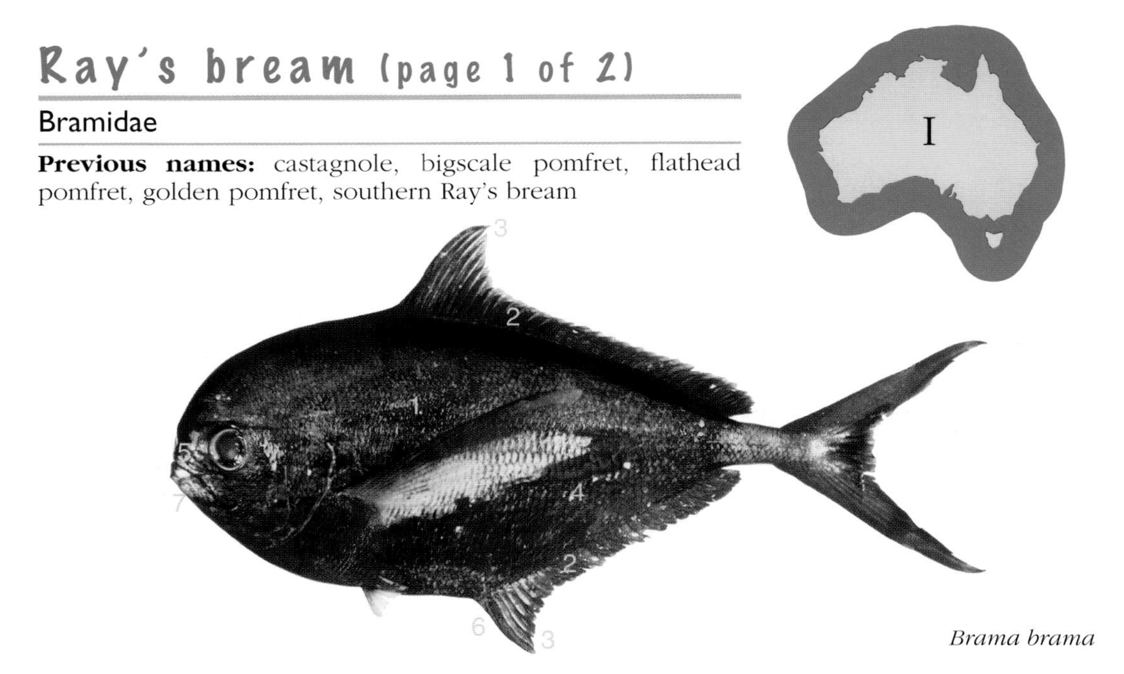

Brama brama

Identifying features: 1 sides usually silvery or silvery brown; 2 dorsal and anal fins covered in scales and fixed in position; 3 leading fin-rays flexible, elongated; 4 body scales vary greatly in shape and size; 5 snout short; 6 no anal-fin spines; 7 lower jaw strong.

Comparisons: Distinctive oceanic fishes with each of the species recognisable by slightly different body shapes. Their body is deeper and more compressed than tunas and other pelagic fishes with which they are caught. The are readily identified from the inshore perches by their long, scaly dorsal and anal fins that lack hard spines. Some trevallas (Centrolophidae) have these features but their bodies are soft and flabby rather than firm.

Fillet: *Brama brama* moderately deep, rather elongate, tapering sharply, slightly convex above, yellowish-white. Outside with continuous, feeble central red muscle band; convex EL, converging HL; HS along middle of fillet; myomeres dense, integument pale grey above, white below. Inside flesh fine; belly flap sometimes present; peritoneum white. Usually skinned, skin medium, scale pockets tall and defined.

Size: To at least 110 cm (fork length) and about 45 kg (commonly 30–50 cm and 0.5–2 kg).

Habitat: Marine; pelagic in all oceans adjacent to Australia and extending south to the extremities of the Exclusive Economic Zone.

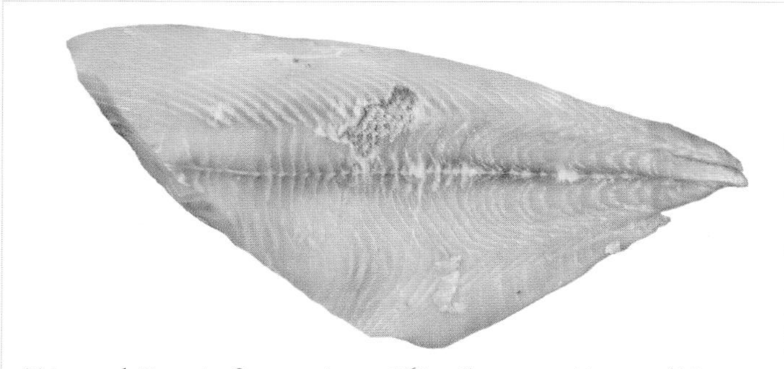

Trimmed. Protein fingerprint p. 368; oil composition p. 403.

Fishery: Caught locally off southern Australia as bycatch of longline fisheries for tunas and more recently by pelagic trawl. Probably under-utilised and expansion likely.

Remarks: About four species are landed. Extremely popular with those fortunate enough to have tried the white, oily flesh. Considered by some to taste like chicken.

Ray's bream (page 2 of 2)

Bramidae

Brama brama

Remarks: Commonly called 'Ray's bream', this species is widely distributed throughout oceans of the Southern Hemisphere below 30°S. A very similar form in the Northern Hemisphere appears to be another species. Best distinguished by its mostly silvery, strongly compressed body, humped head, and high gill raker count (15–18 on outer arch). To at least 60 cm and about 2.5 kg.

Taractichthys longipinnis

Remarks: Commonly called 'bigscale pomfret', this species is widely distributed in open oceans, but is most abundant in temperate latitudes. Best distinguished in having a deep, stocky, almost oval body, long anterior rays in the dorsal and anal fins, and a yellowish outer caudal-fin margin. Reaches well over 120 cm and possibly 45 kg.

Xenobrama microlepis

Remarks: Commonly called 'golden pomfret', this species is widely distributed in the South Pacific and Indian Oceans, mainly between 38–55°S off Australia. Best distinguished by its silvery to golden brown colouration, robust elongate body, relatively small scales (lateral-line scales 83–95), and low gill raker count (10–12 on outer arch). Reaches at least 60 cm and almost 3 kg.

Rabbitfish

Siganus species

Previous names: black trevally, happy moments, spinefoot

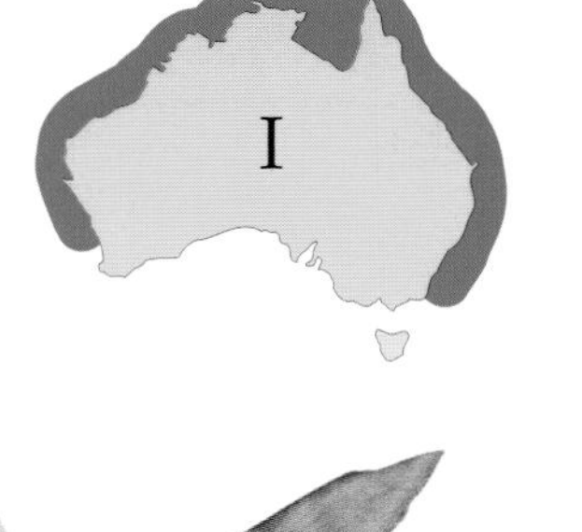

Siganus nebulosus

Identifying features: ① pelvic fin with 1 spine at each end, 3 soft rays between; ② anal fin with 7 spines, 9 soft rays; ③ dorsal fin with 13 spines (plus 1 anterior embedded spine), 10 soft rays; ④ dorsal-fin spines and anal-fin spines alternating sequentially between the left and right side of their respective midline; ⑤ mouth small; ⑥ skin smooth, scales minute.

Comparisons: Primarily tropical species with ovate compressed bodies. Each has high numbers of dorsal-fin (13) and anal-fin (7) spines, and a unique pelvic fin with a spine at each end. The 16 or so Australian species are all very similar and difficult to distinguish. Colour patterns can be useful but many species change colour considerably after capture.

Fillet: *S. nebulosus* moderately deep, rather elongate, upper profile convex, taper slight, pale pinkish. Outside with pronounced, continuous red muscle band; convex EL, converging HL; HS above middle of fillet anteriorly; integument white. Inside flesh fine; belly cavity enormous, flap usually absent; peritoneum transparent. Usually skinned, skin very thick, scale pockets minute and embedded.

Size: To 55 cm and 1.3 kg (commonly to about 25 cm and 0.3 kg).

Habitat: Marine; shallow water (less than 50 m depth), mostly among weed, algae or seagrass but sometimes on coral reefs. Juveniles occasionally enter rivers and creeks.

Fishery: Taken as trawl bycatch and by seine and tunnel nets in northern Australian bays and estuaries. Sometimes called 'bait-thieves' by recreational fishers. Good aquaculture potential.

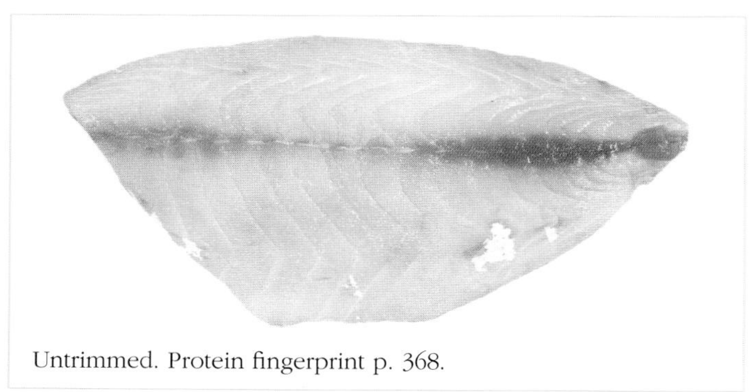

Untrimmed. Protein fingerprint p. 368.

Remarks: Previously marketed as 'black trevally' although unrelated to the true trevallies (Carangidae). Passionately disliked by anyone unfortunate enough to be stung by the venomous spines. However, the firm, flavoursome flesh makes good eating. Marketed in small quantities in Australia but very popular in many parts of Asia.

Alfonsino

Beryx splendens

Previous name: alfonsin

Identifying features: ① dorsal fin with 4 spines, 13–15 soft rays; ② body elongate; ③ dorsal-fin base shorter than anal-fin base; ④ spine present below nostril; ⑤ anal fin with 4 spines, 25–32 soft rays.

Comparisons: Most similar to imperador (*B. decadactylus*, p. 209) but has a shallower body and fewer dorsal-fin soft rays (13–15 versus 16–20). Distinguished from remaining redfishes and the closely related roughies (Trachichthyidae) by having only 4 dorsal-fin spines. The roughies also have obvious scutes along the belly.

Fillet: Moderately deep, short, tapering sharply, slightly convex above, pale pinkish. Outside without central red muscle band; parallel EL, converging HL; HS along middle of fillet, EL much closer to HS than dorsal margin; integument pink above, pinkish to white below. Inside flesh medium; belly flap sometimes present; peritoneum black. Usually skinned, skin thin, scale pockets moderate and indistinct.

Size: To at least 49 cm and 1.7 kg (commonly 30–45 cm and 0.3–1.0 kg).

Habitat: Marine; demersal on the continental slope and seamounts in depths of 210–680 m (typically 450–650 m). Elsewhere shown to school near the bottom during the day and move up the water column (as much as 200 m off the bottom) at night.

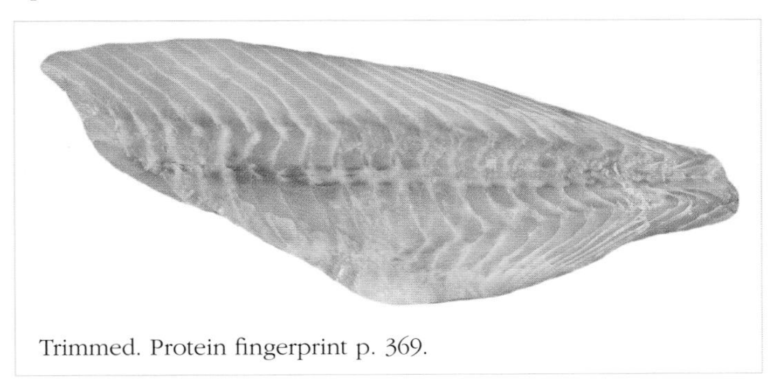

Trimmed. Protein fingerprint p. 369.

Fishery: Taken in small quantities by demersal trawl, dropline and longline through much of its range. Sold fresh, either whole or as skinned fillets.

Remarks: Considered good eating and increasing in popularity. The flesh is white when cooked, moist and reasonably soft.

Bight redfish

Centroberyx gerrardi

Previous names: nannygai, redfish, red snapper

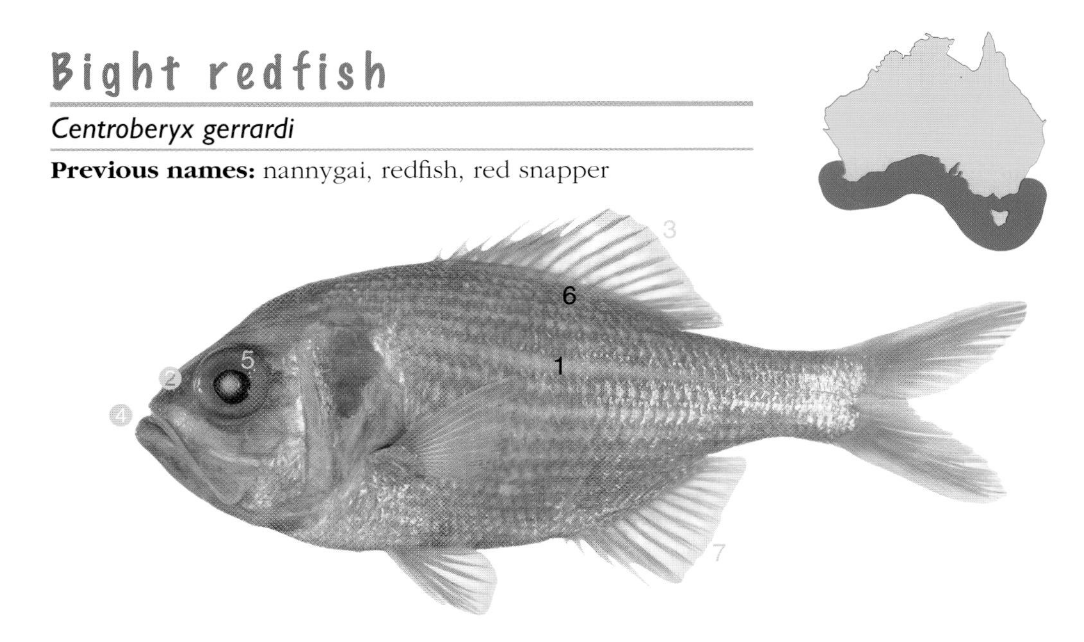

Identifying features: ① indistinct white stripe along lateral line; ② nostrils separated by about 3 times diameter of anterior nostril; ③ dorsal fin with 6 spines, 13 soft rays; ④ a few teeth at tip of jaws larger than those adjacent; ⑤ eye rim red; ⑥ dorsal-fin base longer than anal-fin base; ⑦ anal fin with 4 spines, 12–13 soft rays.

Comparisons: The only redfish with a white stripe along the lateral line. Also distinguished from other redfishes by having a combination of widely separated nostrils, 6 dorsal-fin spines and enlarged teeth at the jaw tips. Lacks the belly scutes of roughies (Trachichthyidae).

Fillet: Very deep, short, tapering sharply, very convex above, pale pinkish. Outside with weak, discontinuous central red muscle band; parallel EL, converging HL; HS below middle of fillet, EL closer to HS than dorsal margin; integument pinkish to white. Inside flesh fine; belly flap sometimes present; peritoneum grey or black. Usually skinned, skin thin, scale pockets large and indistinct.

Size: To 66 cm and 4.6 kg (commonly 30–50 cm and 0.4–2.2 kg).

Habitat: Marine; on and near rocky reefs on the continental shelf and upper slope in depths of 5–300 m (typically 80–210 m). Inshore specimens often solitary or in pairs.

Untrimmed. Protein fingerprint p. 369; oil composition p. 403.

Fishery: Trawled in the Great Australian Bight and landed in medium quantities. Also taken commercially and recreationally off southern Western Australia using demersal gillnets and lines.

Remarks: Only juveniles have been recorded east of Port Phillip Bay (Vic.) and from Tasmania. Probably a slow growing, long-lived species. The flesh is good eating, particularly when fresh.

Imperador

Beryx decadactylus

Previous names: none

Identifying features: 1 dorsal fin with 4 spines, 16–20 soft rays; 2 body deep; 3 dorsal-fin base shorter than anal-fin base; 4 spine present below nostril; 5 anal fin with 4 spines, 26–30 soft rays.

Comparisons: Closely related to alfonsino (*B. splendens*, p. 207) but has a much deeper body and more dorsal-fin soft rays (16–20 versus 13–15). Has fewer dorsal-fin spines than the remaining redfishes (4 versus 6–7) and a shorter dorsal-fin base. The similar-looking roughies (Trachichthyidae) also have more dorsal-fin spines (6–9) and obvious scutes along the belly.

Fillet: Very deep, short, tapering sharply, slightly convex above, yellowish-white. Outside without central red muscle band; parallel EL, converging HL; HS along middle of fillet, EL much closer to HS than dorsal margin; integument pinkish-white or orange. Inside flesh medium; belly flap rarely present; peritoneum black. Usually skinned, skin medium, scale pockets moderate and defined.

Size: To at least 60 cm and about 4 kg (commonly 30–50 cm and 0.4–2.4 kg).

Habitat: Marine; demersal on the upper continental slope and southern seamounts in depths of 300–530 m (more typically 400–500 m).

Trimmed. Protein fingerprint p. 369.

Fishery: Taken in small quantities by demersal trawl and longline. Sold fresh, either whole or as skinned fillets.

Remarks: Often confused with alfonsino and incorrectly sold under that name. The flesh of the two is similar.

Redfish

Centroberyx affinis

Previous names: nannygai, red snapper

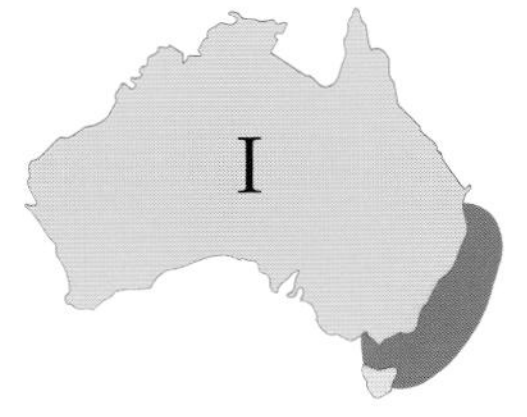

Identifying features: 1 dorsal fin with 7 spines, 11–12 soft rays; 2 eye rim mostly red in colour; 3 caudal fin without black or reddish tips; 4 nostrils large, almost touching one another; 5 dorsal-fin base longer than anal-fin base; 6 anal fin with 4 spines, 11–12 soft rays.

Comparisons: Unique among redfishes in having 7 spines in the dorsal fin. Lacks the belly scutes typical of the similar-looking roughies (Trachichthyidae).

Fillet: Deep, short, tapering sharply, very convex above, pale pinkish. Outside with feeble, continuous central red muscle band; parallel EL, converging HL; HS below middle of fillet, EL closer to HS than dorsal margin; integument pinkish to orange. Inside flesh medium; belly flap rarely present; peritoneum greyish-white or black. Usually skinned, skin medium, scale pockets large and defined.

Size: To 51 cm and 2 kg (commonly 20–30 cm and 0.2–0.4 kg).

Habitat: Marine; benthopelagic in depths of 5–450 m (typically 100–250 m). Feed in mid-water but adults school near rocky reefs and muddy bottoms. Juveniles enter deeper coastal bays.

Fishery: Trawled mainly off New South Wales between Sydney and Eden but also taken off Victoria and occasionally Tasmania. Smaller catches taken by Danish seines and in fish traps. Caught by recreational anglers fishing near deep reefs.

Trimmed. Protein fingerprint p. 369; oil composition p. 403.

Remarks: Feeds on small fishes, crustaceans and molluscs. Frequently sold fresh (usually as skinned fillets) in Sydney and Melbourne. The flesh is soft with a delicate flavour.

Swallowtail

Centroberyx lineatus

Previous names: none

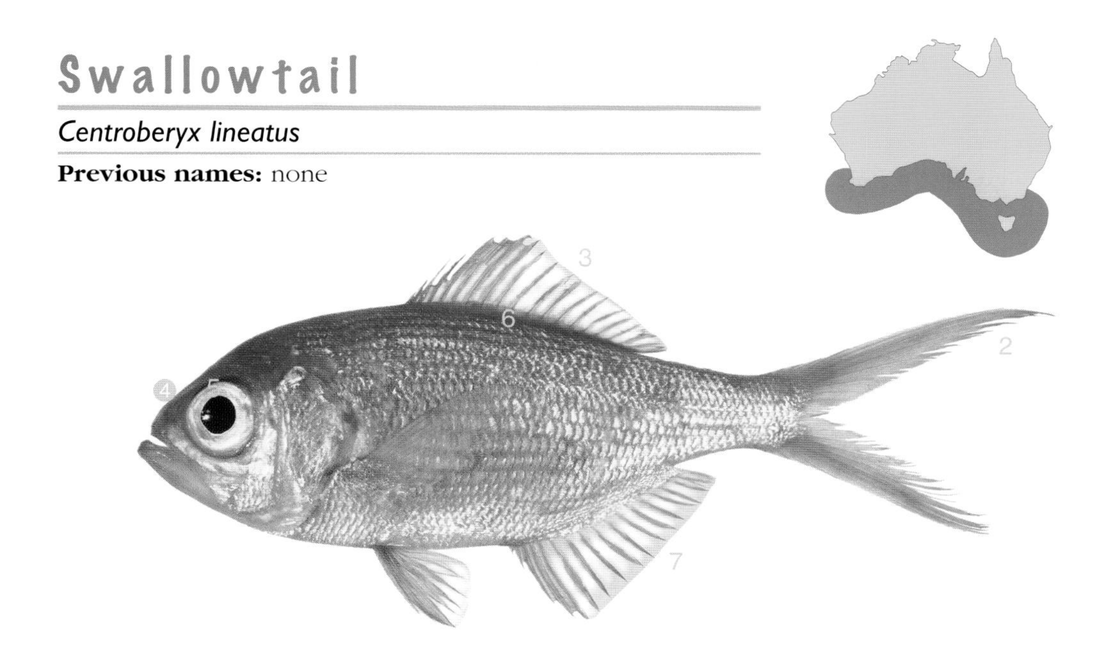

Identifying features: ① body elongate; ② caudal fin deeply forked, elongate, tips distinctly extended; ③ dorsal fin with 6 spines, 14–15 (usually 14) soft rays; ④ nostrils separated by about their diameter; ⑤ eye rim red; ⑥ dorsal-fin base longer than anal-fin base; ⑦ anal fin with 4 spines, 13–14 (usually 14) soft rays.

Comparisons: Distinguished from other redfishes by having a rather elongate body and caudal fin, and 6 dorsal-fin spines. The closely related roughies (Trachichthyidae) have obvious belly scutes.

Fillet: Deep, short, tapering sharply, slightly convex above, pale pinkish. Outside with feeble, continuous central red muscle band; convex or parallel EL, converging HL; HS near middle of fillet, EL closer to HS than dorsal margin; integument orange to pinkish-white. Inside flesh fine; belly flap rarely present; peritoneum greyish-white or black. Usually skinned, skin thin, scale pockets large and defined.

Size: To 43 cm and 0.9 kg (commonly 25–35 cm and 0.2–0.7 kg).

Habitat: Marine; forms schools near rocky reefs on the continental shelf and upper continental slope in depths of 10–300 m (usually 80–120 m). Individuals occasionally school with the similar redfish (*C. affinis*, p. 210).

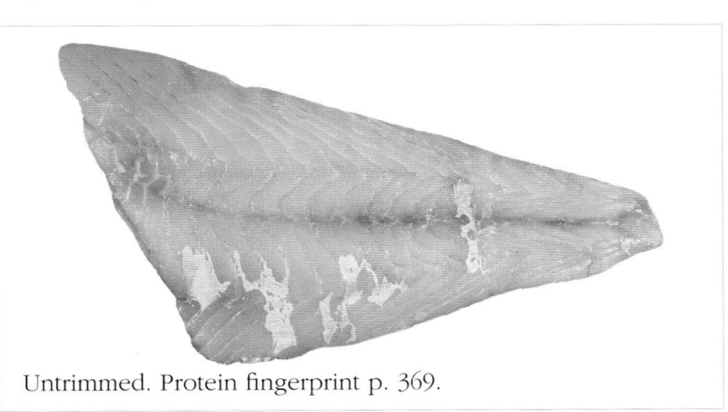

Untrimmed. Protein fingerprint p. 369.

Fishery: Taken as trawl bycatch through most of its range and using demersal gillnets off Western Australia. Sometimes taken in traps. Marketed fresh locally.

Remarks: Attractive species with good market appeal. The flesh is soft like other redfishes but is penetrated by the end of the swim bladder.

Yelloweye redfish

Centroberyx australis

Previous name: yelloweye red snapper

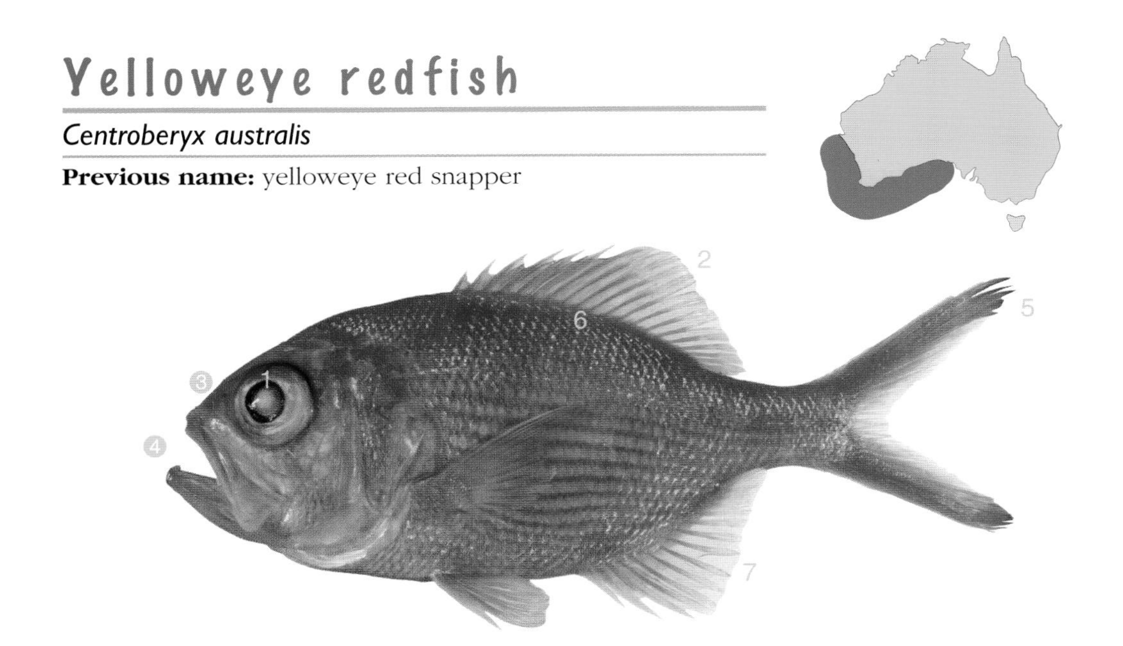

Identifying features: 1 eye rim yellow; 2 dorsal fin with 6 spines, 13–14 soft rays; 3 nostrils separated by about diameter of the anterior nostril; 4 all teeth at tip of jaws of equal size; 5 caudal-fin tips usually darker than rest of fin; 6 dorsal-fin base longer than anal-fin base; 7 anal fin with 4 spines, 12–13 soft rays.

Comparisons: Previously confused with redfish (*C. affinis*, p. 210) but has fewer dorsal-fin spines (6 versus 7) and a yellow (rather than red) eye rim. Also closely related to Bight redfish (*C. gerrardi*, p. 208) but lacks a white stripe along the lateral line and has the nostrils placed close together. A similar but undescribed *Centroberyx* species, the smalleye redfish, has a relatively small yellow-rimmed eye. Distinguished from roughies (Trachichthyidae) by the lack of obvious scutes along the belly.

Fillet: Very deep, short, tapering sharply, very convex above, yellowish-white. Outside with very feeble, diffuse central red muscle band; parallel EL, converging HL; HS along middle of fillet, EL closer to HS than dorsal margin; integument pinkish-white to orange. Inside flesh medium; belly flap sometimes present; peritoneum black. Usually skinned, skin medium, scale pockets large and defined.

Size: To 51 cm and about 2.2 kg (commonly 30–40 cm and 0.4–1.0 kg).

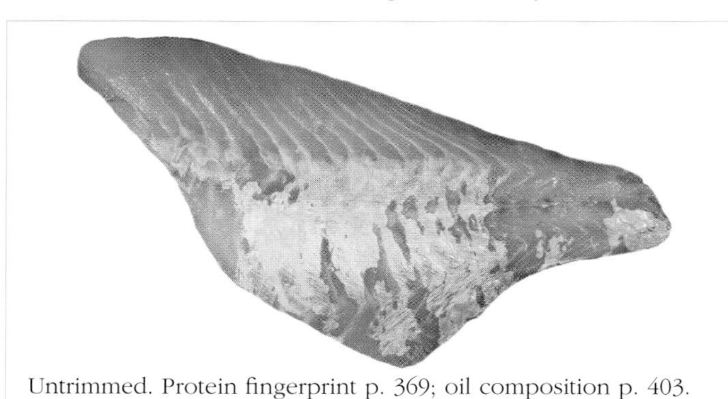

Untrimmed. Protein fingerprint p. 369; oil composition p. 403.

Habitat: Marine; often forms schools off the bottom in depths of 80–560 m (typically 200–300 m).

Fishery: Trawled off southern Western Australia and sometimes off western South Australia. Fresh product is sold locally or exported to Sydney and Melbourne.

Remarks: Considered good eating, similar to redfish.

Bar rockcod

Epinephelus ergastularius & *E. septemfasciatus*

Previous names: barcod, rockcod

Epinephelus ergastularius

Identifying features: 1 sides with 8–9 dark bands (faint in large individuals); 2 1–4 moderate spines on lower margin of preoperculum; 3 edge of groove of upper jaw black in colour; 4 caudal fin truncate or rounded; 5 operculum angular posteriorly, with 3 spines; 6 single dorsal fin with 11 strong spines, 13–15 soft rays; 7 anal fin with 3 strong spines, 9–10 soft rays.

Comparisons: Represents at least two Australian species of 'rockcods' or 'groupers' (*Epinephelus* species), each having prominent vertical bands which are more pronounced in small specimens. A western species, *E. septemfasciatus*, has a more rounded caudal fin than *E. ergastularius*, the common form off New South Wales. Two species of bar rockcod appear to be landed off eastern Australia. Although very similar in appearance, these can be separated by protein fingerprinting. More taxonomic work is needed on this group to determine their identity. Some other tropical rockcods are similarly banded but their bands differ in position, size or number and they usually have fewer soft rays in the anal fin (8 rather than 9).

Fillet: Not available.

Size: Eastern forms possibly to about 160 cm and 70 kg (commonly 60–100 cm and 4–20 kg); western form slightly smaller.

Habitat: Marine; demersal on both hard and soft bottoms of the continental shelf and slope. Juveniles inshore, with adults to depths of 370 m.

Fishery: Caught commercially in small quantities mainly using droplines. Juveniles taken as bycatch of traps and prawn trawls. *E. ergastularius* is the dominant species of the northern New South Wales line fishery. It is marketed regularly but in small quantities at the Sydney Fish Market.

Remarks: Good table-fishes, with white flesh and firm texture. The flesh of large adults is coarser than that of smaller fish. Other information is provided for protein fingerprints (p. 369).

Barramundi cod

Cromileptes altivelis

Previous name: humpback grouper, humpback rockcod

Identifying features: ❶ body pale with fine black spots; ❷ pronounced hump on back behind head; ❸ operculum angular posteriorly, with 3 spines; ❹ dorsal fin continuous, spinous part indistinct; ❺ caudal fin rounded; ❻ mouth and head comparatively small; ❼ dorsal fin with 10 strong spines, 17–19 soft rays; ❽ anal fin with 3 strong spines, 10 soft rays.

Comparisons: Immediately distinguishable from other rockcods in having a barramundi-like body shape with a hump on the back behind the head, no canine teeth in lower jaw, and a slit-like (rather than circular) posterior nostril. However, it differs from the unrelated barramundi (*Lates calcarifer*, p. 121) in having a dense coverage of black spots.

Fillet: Moderately deep, rather elongate, taper pronounced, slightly convex above, yellowish-white. Outside with feeble, continuous, central red muscle band; parallel EL, weakly converging HL; HS slightly below middle of fillet; integument translucent. Inside flesh medium; belly flap sometimes present; peritoneum silvery white to translucent. Usually skinned, skin thick, scale pockets small and defined.

Size: To about 70 cm and 4.8 kg (commonly 40–60 cm and 1.0–2.8 kg).

Habitat: Marine; demersal and cave-dwelling, mainly on coral reefs and offshore atolls to depths of at least 40 m.

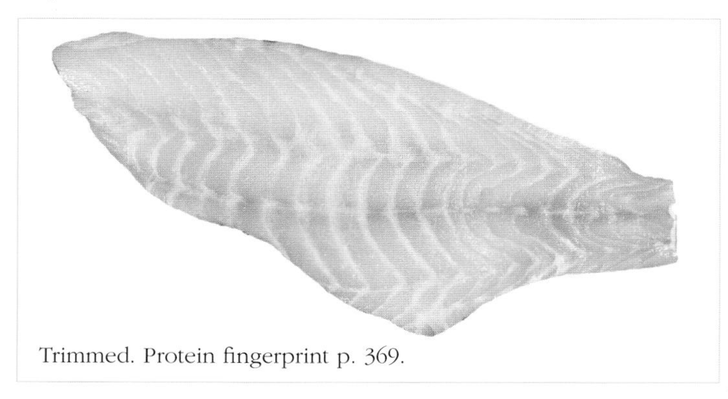
Trimmed. Protein fingerprint p. 369.

Fishery: Caught commercially in small quantities mainly using traps, bottom-set lines and by demersal fish trawlers. Sought-after by recreational fishers and divers.

Remarks: Distinctive species. Attractive as an aquarium fish but also highly regarded as a table-fish. The delicate, white flesh has excellent flavour and texture.

Blacktip rockcod

Epinephelus fasciatus

Previous names: footballer cod, rockcod

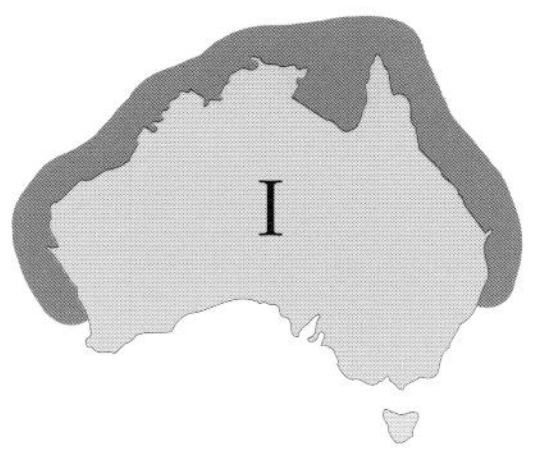

Identifying features: ① sides reddish, often with faint bands; ② membranes of spiny part of dorsal fin usually black (sometimes dark red); ③ rim of eye usually black; ④ caudal fin rounded slightly; ⑤ operculum angular posteriorly, with 3 spines; ⑥ single dorsal fin with 11 strong spines, 15–17 soft rays; ⑦ anal fin with 3 strong spines, 7–8 soft rays.

Comparisons: Brightly coloured rockcod identifiable from its relatives in having a faintly banded reddish body (which lacks bluish or black spots), a black or dark red tip on the dorsal fin and usually a black rim around the eye. Most other rockcods have a darker body colour with more pronounced bands and spots.

Fillet: Moderately deep, rather elongate, slightly tapering, slightly convex above, yellowish-white. Outside with feeble, continuous, central red muscle band; parallel EL, converging HL; HS slightly below middle of fillet; integument translucent. Inside flesh medium; belly flap sometimes present; peritoneum silvery white to translucent. Usually skinned, skin medium, scale pockets small and well defined.

Size: To about 50 cm and 2.3 kg (commonly 30–40 cm and 0.5–1.2 kg).

Habitat: Marine; demersal, mainly on shallow tropical reefs of the continental shelf but occurs in depths to 160 m.

Trimmed. Protein fingerprint p. 369.

Fishery: Caught commercially in moderate quantities mainly using lines, traps and demersal fish trawlers.

Remarks: One of the most widely distributed and abundant members of the genus and may consist of more than one species. Despite its small size, it is considered to be one of Australia's best eating fishes with distinctively flavoured, firm, white flesh.

Coral cod

Cephalopholis species

Previous names: blue-spotted cod, tomato cod

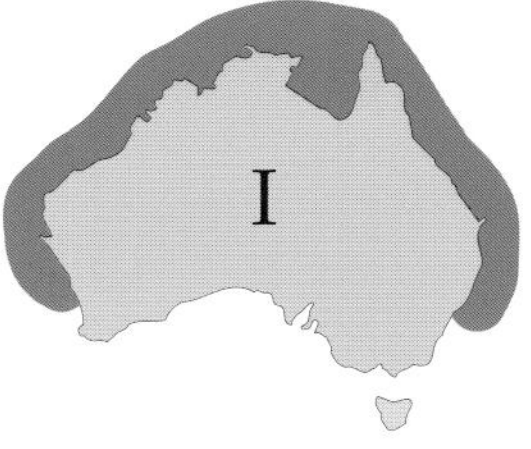

Cephalopholis cyanostigma

Identifying features: ① body relatively deep and compressed; ② angle of preoperculum finely serrated; ③ usually brightly coloured; ④ operculum angular posteriorly, with 3 spines; ⑤ dorsal fin continuous with spinous part distinct; ⑥ caudal fin rounded; ⑦ dorsal fin with 9 slender spines, 13–17 soft rays; ⑧ anal fin with 3 slender spines, 8–9 soft rays.

Comparisons: At least 11 species can be marketed as 'coral cod'. These relatively deep-bodied and compressed fishes are brilliantly coloured with reddish or brownish bodies and bright spots and bands. Differ from the similarly named coral trouts (*Plectropomus* and *Variola* species, pp 217–218), in having a rounded caudal fin (rather than almost truncate or lunar-shaped) and more dorsal-fin spines (9 rather than 7–8 in most species). The only rockcod with 9 dorsal-fin spines (*Aethaloperca rogaa*) is plain-coloured greyish to black.

Fillet: *C. cyanostigma* moderately deep, rather elongate, tapering prominently, almost straight above, white. Outside with feeble, discontinuous, central red muscle band; weakly converging EL, weakly converging HL; HS slightly below middle of fillet; integument silvery white. Inside flesh medium; belly flap sometimes present; peritoneum silvery to translucent. Usually skinned, skin very thick, scale pockets small and defined.

Size: To about 50 cm and possibly 2.2 kg (adults usually 30–40 cm and 0.5–1.2 kg).

Habitat: Marine; demersal, mainly in shallow coralline to deep inter-reefal habitats on the shelf, mostly shallower than 50 m depth. Some species occur among seagrasses.

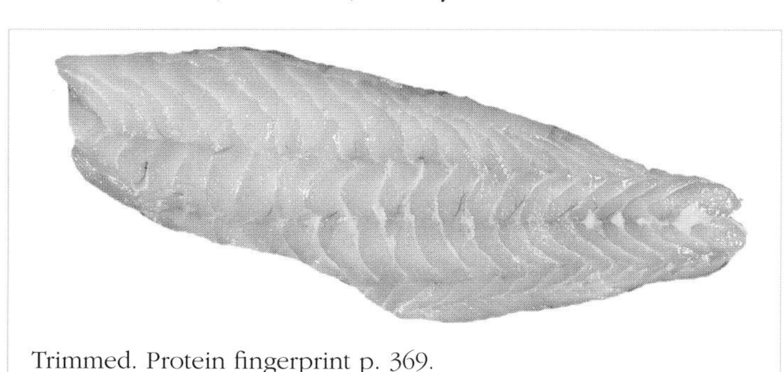

Trimmed. Protein fingerprint p. 369.

Fishery: Caught opportunistically mainly using traps, lines and by prawn and demersal fish trawlers.

Remarks: Excellent table-fishes with all species having white flesh and excellent flavour and texture.

Coral trout (page 1 of 2)

Plectropomus & *Variola* species

Previous names: Chinese footballer, coronation trout, footballer cod, lunartail rockcod

Plectropomus leopardus

Identifying features: 1 operculum mildly angular posteriorly, with 3 spines; 2 large, forward pointing spines on lower margin of preoperculum; 3 dorsal fin continuous with spinous part distinct; 4 caudal fin lunate to square (not rounded or forked); 5 mouth and head large; 6 dorsal fin with 7–9 strong spines, 10–15 soft rays; 7 anal fin with 3 strong spines, 8 soft rays.

Comparisons: At least four species are sold under the collective marketing name 'coral trout'. These species, which are adorned with bluish spots, are very similar in body shape and colour. Differ from rockcods in having either fewer dorsal-fin spines (7–8 in the *Plectropomus* species versus 9–11), or have a more convex caudal-fin margin with greatly elongated soft rays near the end of the dorsal and anal fins (as in *Variola* species). The smaller, but similarly coloured coral cods (*Cephalopholis* species, p. 216), have a rounded caudal fin and more dorsal-fin spines (9) than all species other than the coronation trouts (*Variola* species).

Fillet: *P. leopardus* moderately deep, rather elongate, tapering gradually, slightly convex above, white. Outside with feeble, discontinuous, central red muscle band; parallel EL, weakly converging HL; HS slightly below middle of fillet; integument translucent above, white below. Inside flesh medium; belly flap sometimes present; peritoneum silvery white to translucent. Usually skinned, skin thick, scale pockets small and indistinct.

Size: To about 120 cm and possibly 21 kg (adults usually 35–80 cm and 0.8–9.0 kg).

Habitat: Marine; demersal, mainly in shallow to mid-depth coralline and tropical reef habitats over the continental shelf to about 100 m depth.

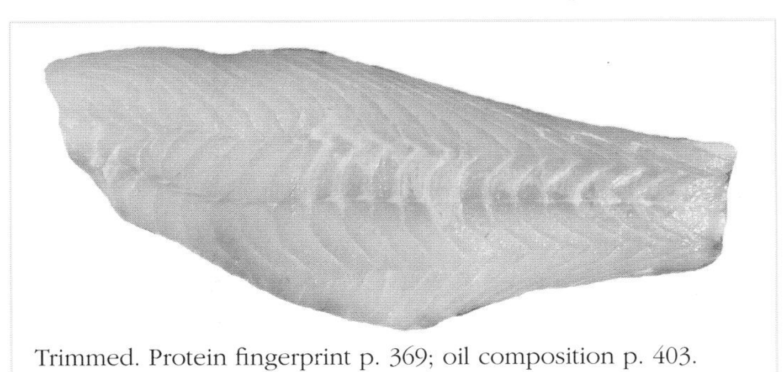

Trimmed. Protein fingerprint p. 369; oil composition p. 403.

Fishery: Caught using lines, nets and traps with smaller catches of soft-bottom species taken by prawn and demersal fish trawlers. Sought-after recreational species.

Remarks: One of Australia's premium table-fish groups with all species having white flesh and excellent flavour and texture.

Coral trout (page 2 of 2)

Plectropomus & *Variola* species

Plectropomus leopardus

Remarks: Commonly called 'common coral trout', this species is distributed across northern Australia from southern Queensland to Dongara (WA) in depths to 100 m. Best distinguished from other coral trouts in being covered in more numerous, smaller bluish spots. Reaches about 80 cm and about 9.0 kg.

Plectropomus maculatus

Remarks: Commonly called 'barcheek coral trout', this species is distributed across northern Australia from southern Queensland to Shark Bay (WA) in depths to 50 m. Best distinguished from other coral trouts in being covered in less numerous, larger bluish spots and in having short bluish lines (rather than spots) on the upper head. Reaches about 125 cm and about 25 kg.

Variola louti

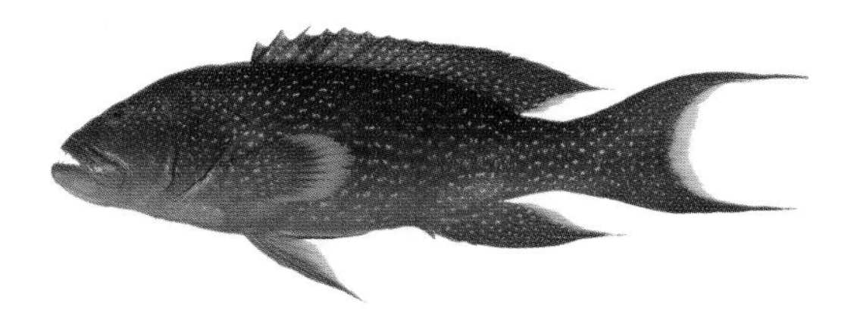

Remarks: Commonly called 'coronation trout', this species is distributed across northern Australia from southern Queensland to Shark Bay (WA) in depths to 240 m. Best distinguished from the true coral trouts (*Plectropomus* species) in having a lunate tail and long fin rays near the end of the dorsal and anal fins. To at least 81 cm and 4.5 kg.

Estuary rockcod

Epinephelus coioides

Previous names: estuary cod, orange-spotted cod, rockcod

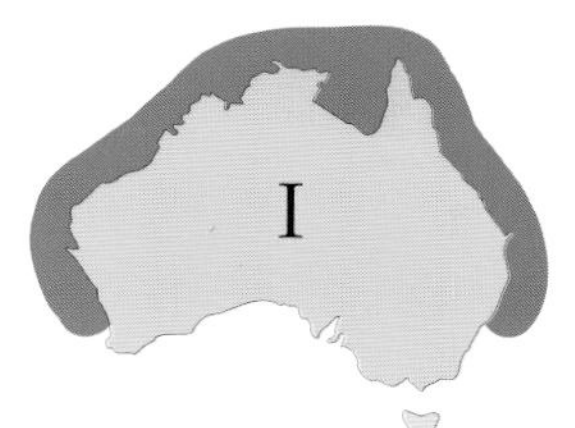

Identifying features: ① sides with dense coverage of similar-sized, orange or reddish-brown spots; ② no pale spots or blotches on body; ③ 2–3 rows of teeth on sides of lower jaw; ④ caudal fin rounded; ⑤ operculum angular posteriorly, with 3 spines; ⑥ single dorsal fin with 11 strong spines, 13–16 soft rays; ⑦ anal fin with 3 strong spines, 8 soft rays.

Comparisons: Largest of a group of dark-spotted rockcods. Most often confused with another major commercial species, the malabar grouper (*E. malabaricus*), but has larger, orange to brownish (rather than almost black) spots and no pale blotches on body, and 2–3 (rather than 4–5) rows of teeth on the mid-sides of the lower jaw. The smaller greasy rockcod (*E. tauvina*) is similar but can be distinguished from young estuary rockcod by a black blotch at the base of the last four dorsal-fin spines. Most other spotted rockcods have denser spotting.

Fillet: Moderately deep, rather elongate, tapering prominently, almost straight above, brownish-white. Outside with intermediate, continuous, central red muscle band; parallel EL, weakly converging HL; HS slightly along middle of fillet; integument white. Inside flesh coarse; belly flap sometimes present; peritoneum silvery white to translucent. Usually skinned, skin very thick, scale pockets moderate and defined.

Size: To possibly 180 cm and 100 kg (commonly marketed at 40–120 cm and 1–25 kg).

Habitat: Marine; demersal, usually alone or in small groups, covering a variety of habitats from lower rivers and estuaries to offshore reefs. Juveniles inshore, with adults occurring deeper, to 100 m or more.

Fishery: Caught sporadically, mainly using traps, seines, bottom-set lines and by demersal fish trawlers. Sought-after recreational fish.

Remarks: Maximum size is unknown due to confusion with other large spotted rockcods and the gigantic Queensland groper (*E. lanceolatus*). Small fish to about 8 kg are well regarded as table-fish but the flesh is coarse in large individuals.

Untrimmed.
Protein fingerprint p. 369; oil composition p. 403.

Hapuku

Polyprion americanus & P. oxygeneios

Previous names: bass groper, groper, hapuka, hapuka cod, longnose hapuku, New Zealand groper

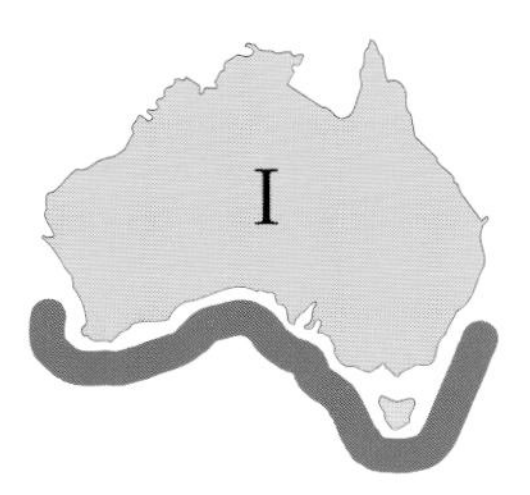

Polyprion oxygeneios

Identifying features: 1 operculum broadly angular posteriorly, with obvious ridge ending in strong spine; 2 uniformly greyish or greyish-pink; 3 dorsal fin continuous with spinous part distinct; 4 caudal fin truncate; 5 mouth and head large; 6 dorsal fin with 11–12 strong spines, 12–13 soft rays; 7 anal fin with 3 strong spines, 8–10 soft rays.

Comparisons: Temperate, rockcod-like fishes that can be distinguished from their distant relatives by their plain-coloured body, small pectoral fins, truncate caudal fin, and prominent horizontal ridge on the operculum that ends in a spine. The true rockcods have bright or more complex colour patterns.

Fillet: *P. oxygeneios* moderately deep, somewhat elongate, tapering gently, weakly convex above, brownish-white. Outside with pronounced, broad continuous, central red muscle band; parallel EL, weakly converging HL; HS slightly along middle of fillet; integument white. Inside flesh coarse; belly flap sometimes present; peritoneum silvery white to translucent. Usually skinned, skin thick, scale pockets small and well defined.

Size: To 180 cm and 70 kg (commonly marketed at 80–120 cm and 7–30 kg).

Habitat: Marine; mainly demersal over deep coastal reefs and canyons of the continental slope to 450 m depth. Large adults may live beneath flotsam in the open ocean.

Untrimmed. Protein fingerprint p. 369; oil composition p. 403.

Fishery: Caught as secondary catch of dropline fisheries off temperate Australia. Also taken by trawlers.

Remarks: Among the most highly priced fishes marketed in temperate Australia, with firm, flavoursome flesh. Landed in small quantities.

Honeycomb rockcod

Epinephelus quoyanus

Previous names: longfin cod, longfin rockcod, rockcod, spotted groper, wirenet cod

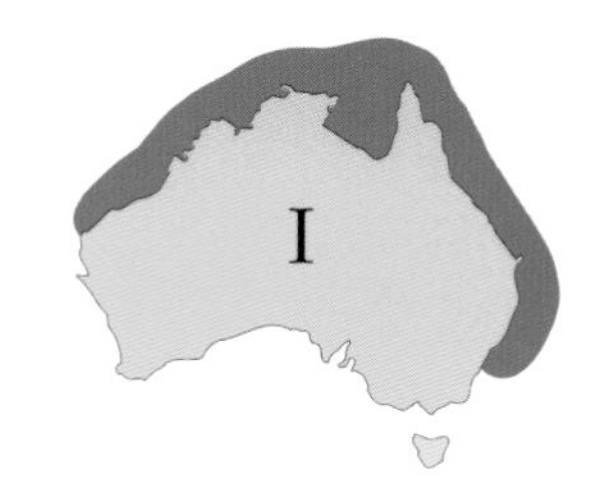

Identifying features: ① body and fins whitish with dense coverage of large, round to hexagonal, brownish-black spots forming a 'honeycomb' pattern; ② 2 oblique dark bars on breast; ③ large dark blotch at base of outsized pectoral fin; ④ caudal fin rounded; ⑤ operculum angular posteriorly, with 3 spines; ⑥ single dorsal fin with 11 strong spines, 16–18 soft rays; ⑦ anal fin with 3 strong spines, 8 soft rays.

Comparisons: Small rockcod with distinctive honeycomb pattern of dark almost hexagonal spots separated by thin pale lines. It differs from other rockcods with a honeycomb pattern, such as the common birdwire rockcod (*E. merra*, p. 225), in having an exceptionally large pectoral fin with a dark bar at its base, and prominent bars on the breast just forward of this fin (rather than large spots).

Fillet: Moderately deep, rather elongate, tapering prominently, slightly convex above, pale pinkish. Outside with feeble, continuous, central red muscle band; parallel EL, weakly converging HL; HS slightly below middle of fillet; integument translucent. Inside flesh medium; belly flap sometimes present; peritoneum silvery white to translucent. Usually skinned, skin very thick, scale pockets small and defined.

Size: To about 55 cm and at least 2.3 kg (commonly marketed at 30–40 cm and 0.4–1.2 kg).

Trimmed. Protein fingerprint p. 369.

Habitat: Marine; demersal, living mainly among broken coral and rubble in very shallow, silty water on reefs.

Fishery: Caught incidentally using traps, beach seines, and handlines along the tropical coastal margin.

Remarks: Only average as a table-fish with tough and somewhat bland flesh.

Longfin perch

Caprodon longimanus

Previous name: pink maomao

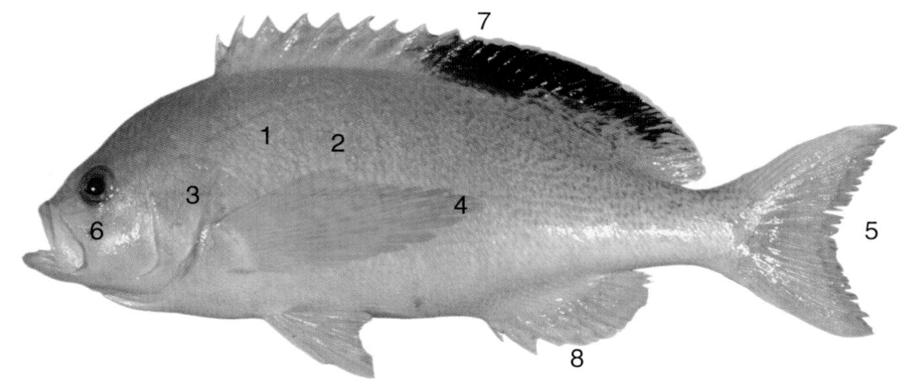

Identifying features: 1 sides reddish or pink, mature males yellowish with huge black blotch on dorsal fin; 2 body deep behind head, compressed; 3 operculum not angular posteriorly, with 3 spines; 4 pectoral fin somewhat pointed; 5 caudal fin with weak fork; 6 mouth and head relatively short; 7 dorsal fin with 10 spines, 19–20 soft rays; 8 anal fin with 3 strong spines, 8 soft rays.

Comparisons: Temperate relative of the rockcods. The body shape and colouration, including the presence of a dark blotch on the dorsal fin of males, makes it easily identifiable from fishes sold as 'rockcod'. Other smaller perches occurring locally are rarely marketed because of their small average size.

Fillet: Moderately deep, rather elongate, tapering prominently, slightly convex above, pale pinkish. Outside with feeble, continuous, central red muscle band; parallel EL, weakly converging HL; HS slightly below middle of fillet; integument translucent above, white below. Inside flesh medium; belly flap sometimes present; peritoneum silvery white to translucent. Usually skinned, skin medium, scale pockets medium and indistinct.

Size: To at least 50 cm and 1.6 kg (commonly marketed at 30–40 cm and 0.5–1.0 kg).

Habitat: Marine; often occurs in huge schools over reefs of the inner continental shelf of eastern Australia to 80 m depth.

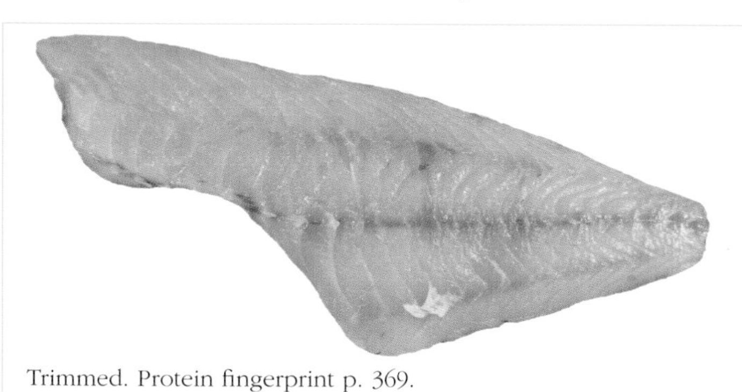

Trimmed. Protein fingerprint p. 369.

Fishery: Caught in moderate quantities by lining and trapping off New South Wales, with occasional catches by trawlers over hard ground.

Remarks: Good, medium to high priced foodfish with increasing domestic landings due to improved fishing practices.

Maori rockcod

Epinephelus undulatostriatus

Previous names: Maori cod, rockcod

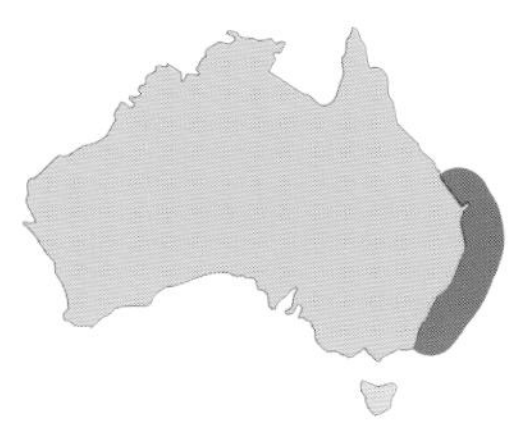

Identifying features: ① body with wavy orange brown lines; ② head densely covered with small orange brown spots; ③ margins of hind fins yellow; ④ caudal fin rounded; ⑤ operculum angular posteriorly, with 3 spines; ⑥ single dorsal fin with 11 strong spines, 15–17 soft rays; ⑦ anal fin with 3 strong spines, 8 soft rays.

Comparisons: Distinct among rockcods in having narrow, wavy lines along the sides and spots on the head. The specklefin rockcod (*E. ongus*, p. 225) is similar but has a dark grey or black (rather than yellow) edge above the upper lip (resembling a moustache) and lacks yellow margins along the dorsal, anal and caudal fins.

Fillet: Moderately deep, rather elongate, tapering prominently, slightly convex above, pale pinkish. Outside with feeble, continuous, central red muscle band; parallel EL, weakly converging HL; HS slightly below middle of fillet; integument silvery white. Inside flesh medium; belly flap sometimes present; peritoneum silvery translucent. Usually skinned, skin thick, scale pockets small and indistinct.

Size: To at least 65 cm and about 5.5 kg (commonly marketed at 30–50 cm and 0.5–1.5 kg).

Habitat: Marine; demersal, from coastal reefs to the Great Barrier Reef, to at least 80 m depth.

Trimmed. Protein fingerprint p. 369.

Fishery: Caught in small quantities off eastern Australia mainly using traps and lines.

Remarks: Actively sought by anglers for its fighting abilities. Reported to swim well above the bottom to ambush prey and to take floating bait. Commonly marketed in Brisbane. Very good table-fish with firm, flavoursome flesh.

Rockcod (page 1 of 2)

Aethaloperca, Anyperodon & Epinephelus species

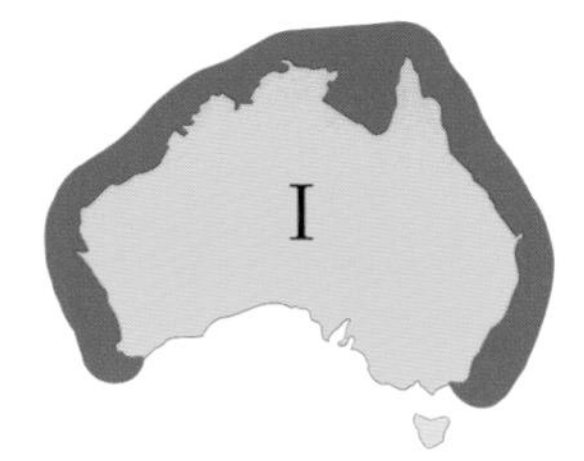

Previous names: chinaman cod, flowery cod, greasy cod, groper, grouper, potato cod, redflush rockcod, reef cod, slimy cod, spotted cod, threeline rockcod, whiteline rockcod

Epinephelus morrhua

Identifying features: 1 operculum angular posteriorly, with 3 spines; 2 no large, forward-pointing spines on lower margin of preoperculum; 3 dorsal fin continuous with spinous part distinct; 4 caudal fin lunate to rounded (not forked); 5 mouth and head large; 6 dorsal fin with 9–11 strong spines, 13–18 soft rays; 7 anal fin with 3 strong spines, 7–10 soft rays.

Comparisons: The 35 or so species sold as 'rockcod' are also confusingly referred to as 'cod' but are unrelated to the true cods (Gadiformes). Rockcods are essentially similar to each other in shape but usually differ subtly in colour. Nevertheless, identification can be difficult even for a specialist. They differ from related fishes marketed as 'coral trout' (p. 217) in having 9–11 dorsal-fin spines (versus 7–8 in the *Plectropomus* species), and a less convex caudal-fin margin without the hind soft rays of the dorsal and anal fins greatly elongated (as in *Variola* species).

Fillet: *E. morrhua* moderately deep, rather elongate, tapering prominently, distinctly convex above, yellowish-white to pinkish. Outside with intermediate, continuous, central red muscle band; parallel EL, weakly converging HL; HS along middle of fillet; integument white to translucent. Inside flesh coarse; belly flap sometimes present; peritoneum silvery white to translucent. Usually skinned, skin thick, scale pockets small and defined.

Size: To about 300 cm and 400 kg (adults usually 30–120 cm and 0.5–25 kg).

Habitat: Marine; demersal in coralline and rocky reef habitats over the continental shelf and upper slope to depths of about 400 m.

Trimmed. Protein fingerprint p. 369; oil composition p. 403.

Fishery: Caught mainly using lines, nets and traps although considerable catches of some tropical species are taken by trawlers.

Remarks: Highly regarded table-fishes with white flesh and excellent flavour and texture. Several species are large and eight of the most valuable of these have separate marketing names.

Rockcod (page 2 of 2)

Aethaloperca, Anyperodon & *Epinephelus* species

Epinephelus ongus

Remarks: Commonly called 'specklefin rockcod', this cryptic species is distributed off shallow reefs of northern Australia between Mooloolaba (Qld) and the Rowley Shoals (WA) in depths to at least 25 m. Best distinguished in having wavy lines and pale blotches along the sides with a dark edge above the upper lip resembling a moustache but no yellow fin tips. Reaches about 35 cm and about 0.9 kg.

Epinephelus merra

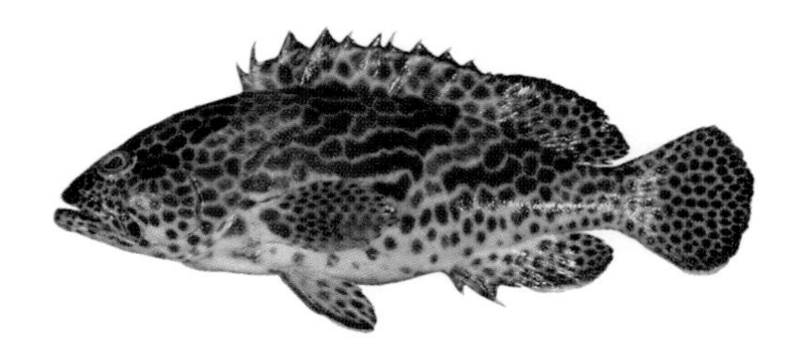

Remarks: Commonly called 'birdwire rockcod', this species is distributed in very shallow, sheltered water on reefs of northern Australia between Lord Howe Island (NSW) and the Dampier Archipelago (WA) rarely exceeding depths of 20 m. Best distinguished in having a distinctive honeycomb pattern of brownish spots and lines separated by narrower pale lines, distinctly darker and smaller spots on the head than the body, and lacking prominent bars on the breast just forward of the pectoral fin (with a few large spots instead). Reaches about 35 cm and about 0.9 kg.

Epinephelus heniochus

Remarks: Commonly called 'threeline rockcod', this species is distributed over soft bottoms of the mid-continental shelf of northern Australia between southern Queensland and the North West Shelf (WA) in depths of 40–230 m. Best distinguished from other rockcods in having a uniform brownish-pink body with a few faint lines radiating from the back of the eye. Reaches at least 43 cm and about 1.3 kg.

White-spotted rockcod

Epinephelus multinotatus

Previous names: rankin cod, rockcod, white-blotched rockcod

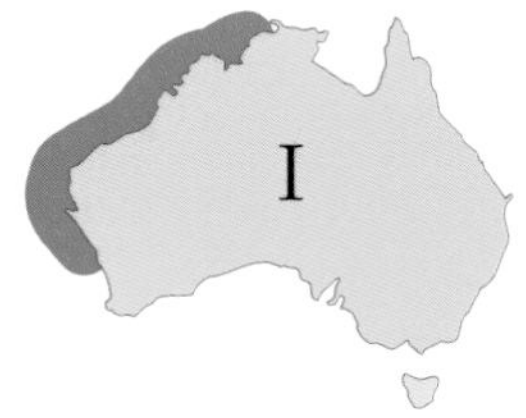

Identifying features: ❶ body purplish-grey with moderate coverage of variable white blotches; ❷ blotches less contrasted on fins than on body; ❸ rather deep bodied; ❹ caudal fin truncate; ❺ operculum angular posteriorly, with 3 spines; ❻ single dorsal fin with 11 strong spines, 15–17 soft rays; ❼ anal fin with 3 strong spines, 8–9 soft rays.

Comparisons: Distinctive, dark-bodied rockcod rather densely covered with white, irregularly shaped blotches but lacking dark or brightly coloured spots. It is among those species that have a straight-edged caudal fin.

Fillet: Moderately deep, rather elongate, tapering prominently, slightly convex above, brownish-white. Outside with pronounced, continuous, central red muscle band; parallel EL, converging HL; HS along middle of fillet; integument white. Inside flesh coarse; belly flap sometimes present; peritoneum silvery white to translucent. Usually skinned, skin thick, scale pockets small and irregular.

Size: To 80 cm and about 9 kg (commonly marketed at 50–70 cm and 1.6–5 kg).

Habitat: Marine; demersal on both soft and hard bottoms of the continental shelf of northern Australia. Juveniles occur on coral reefs, but adults live in deeper water to at least 90 m depth.

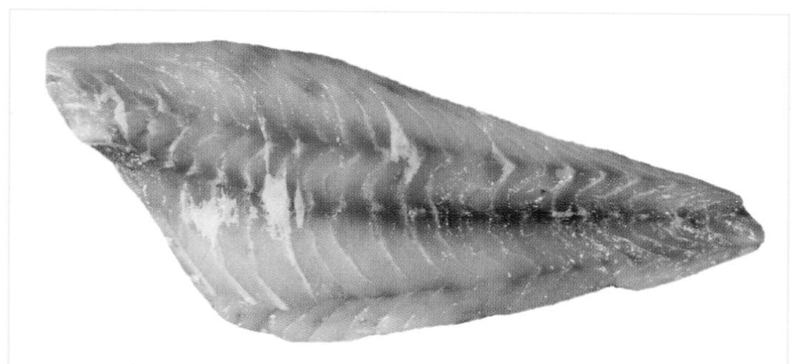

Untrimmed. Protein fingerprint p. 369; oil composition p. 403.

Fishery: Caught mainly off northwestern Australia using traps, hook-and-line and bottom trawl nets.

Remarks: Adults eat small fishes and crabs. Young are thought to mimic damselfishes in order to get closer to their prey. Most of the catch is sold in Perth and Sydney. Good table-fish with firm, flavour-some flesh.

Yellow-spotted rockcod

Epinephelus areolatus

Previous name: rockcod

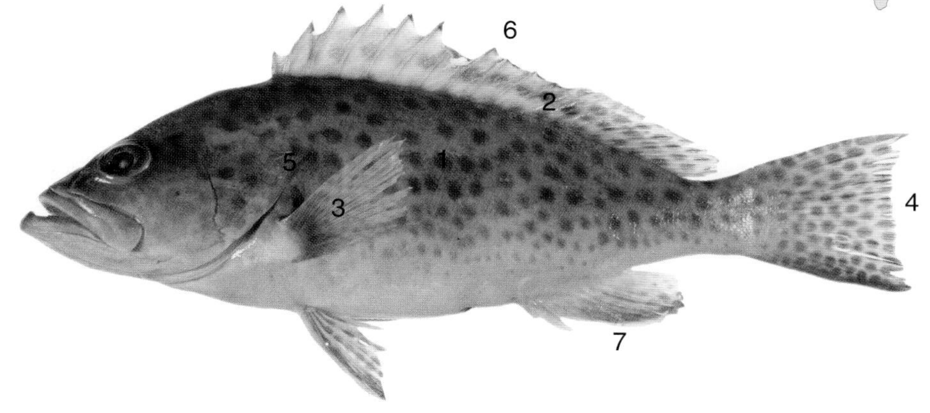

Identifying features: ① body and fins whitish with dense coverage of large, yellowish-brown spots; ② spots in rows on dorsal fin; ③ small spots on pectoral fin confined to rays; ④ caudal fin truncate, usually with a white margin; ⑤ operculum angular posteriorly, with 3 spines; ⑥ single dorsal fin with 11 strong spines, 15–17 soft rays; ⑦ anal fin with 3 strong spines, 7–8 soft rays.

Comparisons: Another rockcod with a honeycomb pattern. However, its large spots tend to be yellowish or golden rather than dark like most of its relatives. Also its truncate caudal fin with a pale margin is present in very few members of the genus.

Fillet: Moderately deep, rather elongate, tapering gently, convex above, yellowish-white. Outside with intermediate, continuous, central red muscle band; parallel EL, converging HL; HS just below middle of fillet; integument silvery white. Inside flesh medium; belly flap sometimes present; peritoneum possibly silvery white. Usually skinned, skin thick, scale pockets small and irregular.

Size: To 40 cm and about 1.2 kg (commonly marketed at 30–35 cm and 0.4–0.8 kg).

Habitat: Marine; demersal on a variety of bottom types of the tropical continental shelf in depths to 200 m. Generally found in turbid water.

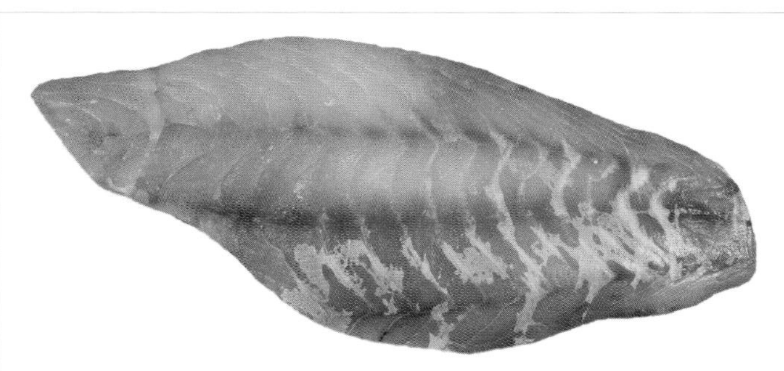

Trimmed. Protein fingerprint p. 369; oil composition p. 403.

Fishery: Caught inshore in far northern waters using traps, handlines and by prawn trawlers.

Remarks: Being landed in increasing quantities off Western Australia and the Northern Territory and sold in Perth. Good table-fish with firm, flavoursome flesh.

Darwin's roughy

Gephyroberyx darwinii

Previous names: none

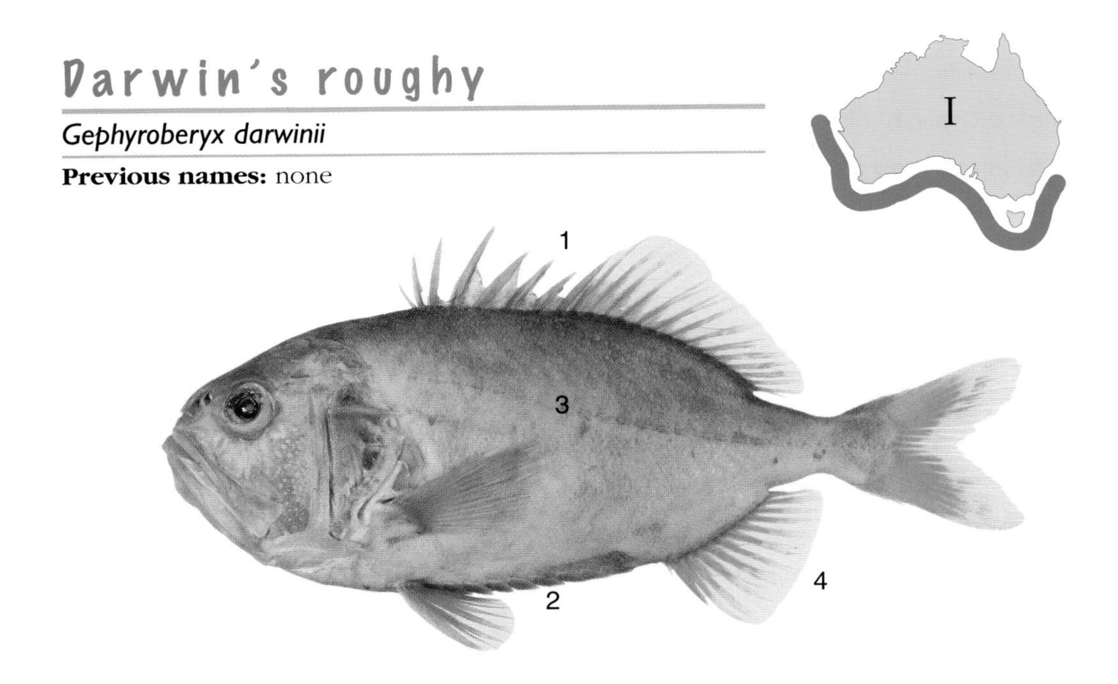

Identifying features: 1 dorsal fin with 8 (occasionally 9) spines (middle spines longest), 13 soft rays; 2 10–12 strong scutes along belly; 3 body pale pink (fading to orange after death); 4 anal fin with 3 spines and 11 soft rays.

Comparisons: Most similar to the orange roughy (*Hoplostethus atlanticus*, p. 229) but has 8–9 spines in the dorsal fin (rather than 6) and 10–12 strong scutes along the belly (rather than 19–25 weak scutes). Also differs from another commercially caught roughy, the giant sawbelly (*H. gigas*), which has 5–7 (usually 6) dorsal-fin spines the last of which is longest. Distinguished from similar groups such as the redfishes (Berycidae) by the presence of obvious scutes along the belly.

Fillet: Deep, rather elongate, tapering sharply, very convex above, milky white. Outside with feeble, continuous central red muscle band; converging EL, converging HL; HS below middle of fillet, EL closer to HS than dorsal margin; integument silvery grey above, silver below. Inside flesh fine; belly flap usually absent; peritoneum black. Always skinned, skin medium, scale pockets moderate-sized and indistinct.

Size: To 50 cm and 2.5 kg (commonly 30–40 cm and 0.5–1.2 kg).

Habitat: Marine; demersal on the mid-slope in depths of 640–820 m and probably shallower.

Trimmed. Protein fingerprint p. 369.

Fishery: Taken by demersal trawl mainly in the Great Australian Bight and occasionally off New South Wales.

Remarks: Distribution may extend further north off both coasts. Increasing in popularity, particularly in Western Australia and South Australia. The flesh is moist and a little firmer than orange roughy.

Orange roughy

Hoplostethus atlanticus

Previous names: deepsea perch, sea perch

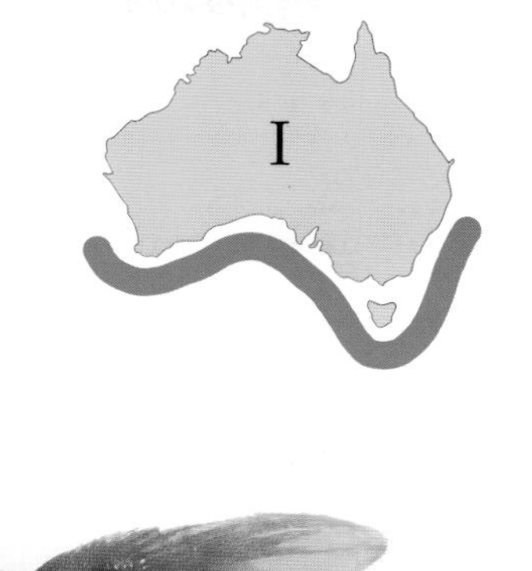

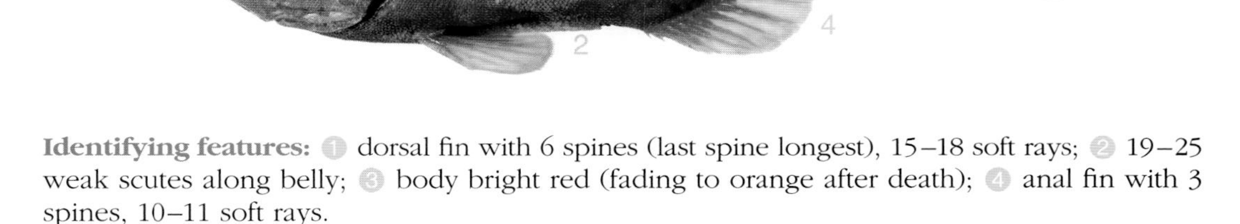

Identifying features: 1 dorsal fin with 6 spines (last spine longest), 15–18 soft rays; 2 19–25 weak scutes along belly; 3 body bright red (fading to orange after death); 4 anal fin with 3 spines, 10–11 soft rays.

Comparisons: Differs from other roughies such as Darwin's roughy (*Gephyroberyx darwinii*, p. 228) and giant sawbelly (*H. gigas*) by having 19–25 weak scutes along the belly (other species have fewer, stronger scutes). Also has only 6 dorsal-fin spines (8–9 in Darwin's roughy). Roughies are distinguished from similar groups by the obvious scutes along the belly.

Fillet: Deep, rather elongate, tapering sharply, slightly convex above, milky white. Outside with feeble, continuous central red muscle band; converging EL, converging HL; HS below middle of fillet, EL closer to HS than dorsal margin; integument silvery grey, sometimes with traces of orange skin. Inside flesh fine; belly flap usually absent; peritoneum black. Always skinned, skin medium, scale pockets small and defined.

Size: To at least 60 cm and 3.5 kg (commonly 35–45 cm and 0.8–1.5 kg).

Habitat: Marine; forms demersal schools on the mid-slope and seamounts in depths of 500–1400 m (more typically 750–1050 m).

Fishery: Targeted by demersal trawlers off the southern coastline, including nearby seamounts, as far north as Perth (WA) and Sydney (NSW). Spawning aggregations have been targeted off Tasmania since the late 1980s. Numbers declined soon after exploitation commenced and the fishery is strictly managed to protect remaining stocks.

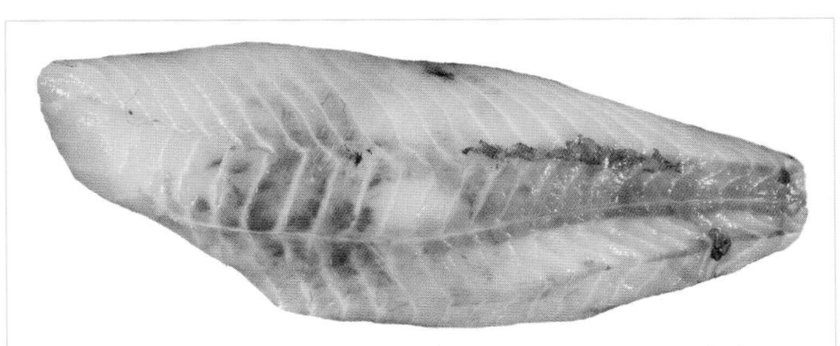

Trimmed. Protein fingerprint p. 369; oil composition p. 404.

Remarks: Sold after deep-skinning to remove both the skin and a layer of oil. The oil has been used as a lubricant and in the cosmetic industry. Black specimens are occasionally caught but appear to be very rare. A very popular foodfish with soft, moist, white flesh and a mild taste. Good export market.

Atlantic salmon

Salmo salar

Previous name: salmon

Identifying features: ① jaws rather short, reaching to underneath eye; ② anal-fin base short (9–10 rays), caudal peduncle long; ③ adipose fin present; ④ scattering of black spots (often faint) on body; ⑤ dorsal fin centrally located, with short base; ⑥ pectoral fins placed low on side; ⑦ pelvic fins well behind pectoral fins.

Comparisons: Similar in appearance to the brown trout (*S. trutta*, p. 231) with which it is sometimes confused. Like the Atlantic salmon, the brown trout also lacks a spotted caudal fin but has a larger mouth, shorter caudal peduncle, and its spots are often reddish or dark with a pale halo (versus uniformly black). It also has sharp teeth at the head of the vomer (front of the roof of the mouth) which are weak or absent in the Atlantic salmon. Other salmonids marketed in Australia have spotted caudal fins.

Fillet: Deep, elongate, slightly tapering, orange. Outside with pronounced, continuous red muscle band; convex EL, converging HL; HS along middle of fillet; integument patchy, greyish above, white below. Inside flesh medium; belly flap distinct, not protruding; peritoneum transparent. Usually skinned, skin medium, scale pockets small and irregular.

Size: To 150 cm and 38 kg (commonly harvested at about 65–75 cm fork length and 4–6 kg).

Habitat: Sea-going native of streams, rivers and lakes draining into the North Atlantic Ocean. Introduced for aquaculture but escapees remain coastal.

Fishery: Farmed in sea cages in the bays of southeastern Tasmania since the mid-1980s, with recent interest in South Australia. Farm escapees taken recreationally.

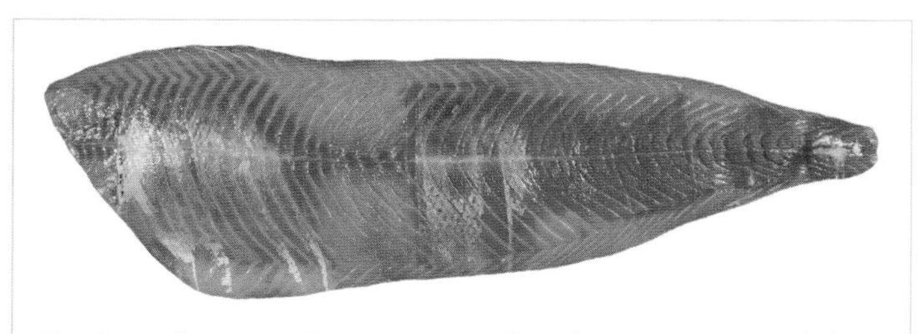

Untrimmed. Protein fingerprint p. 369; oil composition p. 404.

Remarks: Highest valued salmonid, forming an important aquaculture industry in Tasmania. The flesh is distinctive. Marketed fresh, whole, as cutlets and fillets, and smoked.

Trout

Oncorhynchus mykiss & *Salmo trutta*

Previous names: brown trout, ocean trout, rainbow trout, sea trout

Oncorhynchus mykiss

Identifying features: 1 jaws large, reaching past eye level; 2 anal-fin base short (8–12 soft rays), caudal peduncle short; 3 adipose fin present; 4 small reddish or black spots on body; 5 dorsal fin centrally located, with short base; 6 pectoral fins placed low on side; 7 pelvic fin well behind pectoral fins.

Comparisons: The rainbow trout (*O. mykiss*) differs from the brown trout (*S. trutta*) in having numerous prominent spots on its caudal fin and often a pinkish stripe along its midline. Both smaller than salmon farmed in Australia. The brown trout and the Atlantic salmon (*S. salar*, p. 230) are easily confused. However, the brown trout has a larger mouth, shorter caudal peduncle and more pronounced pigmentation.

Fillet: *S. trutta* deep, elongate, slightly tapering, flesh variable orange to white. Outside with pronounced, continuous red muscle band; convex EL, weakly converging HL; HS along middle of fillet; integument white. Inside flesh medium; belly flap distinct, not protruding; peritoneum transparent. Usually skinned, skin thick, scale pockets small and well defined.

Size: Locally to 90 cm and 14 kg (harvested at about 30–45 cm and 0.5–1.0 kg).

Habitat: Sea-going natives of streams, rivers and lakes draining into the North Atlantic (*S. trutta*) and eastern Pacific (*O. mykiss*) Oceans. Locally, prefer lakes and gravelly streams with a moderate to fast flow. Also found in the sea.

Fishery: Among the most important recreational angling fishes of southern Australia. Commercial component reared in ponds and cages.

Remarks: The chinook salmon (*O. tschawytscha*), once released in Victoria, resembles the rainbow trout in having a spotted caudal fin but its anal fin is much longer-based, with more rays (13–19). Trout flesh tends to be richer in colour, usually orange or pink, when caught in some lakes or the sea. Similarly, fish from rivers can have pale, somewhat softer flesh.

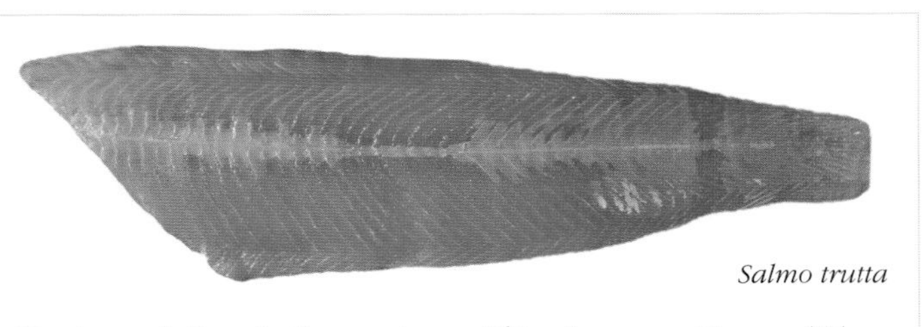

Salmo trutta

Untrimmed. Protein fingerprint p. 369; oil composition p. 404.

Whitebait

Lovettia sealii & *Galaxias* species

Previous names: galaxias, jollytail, native trout

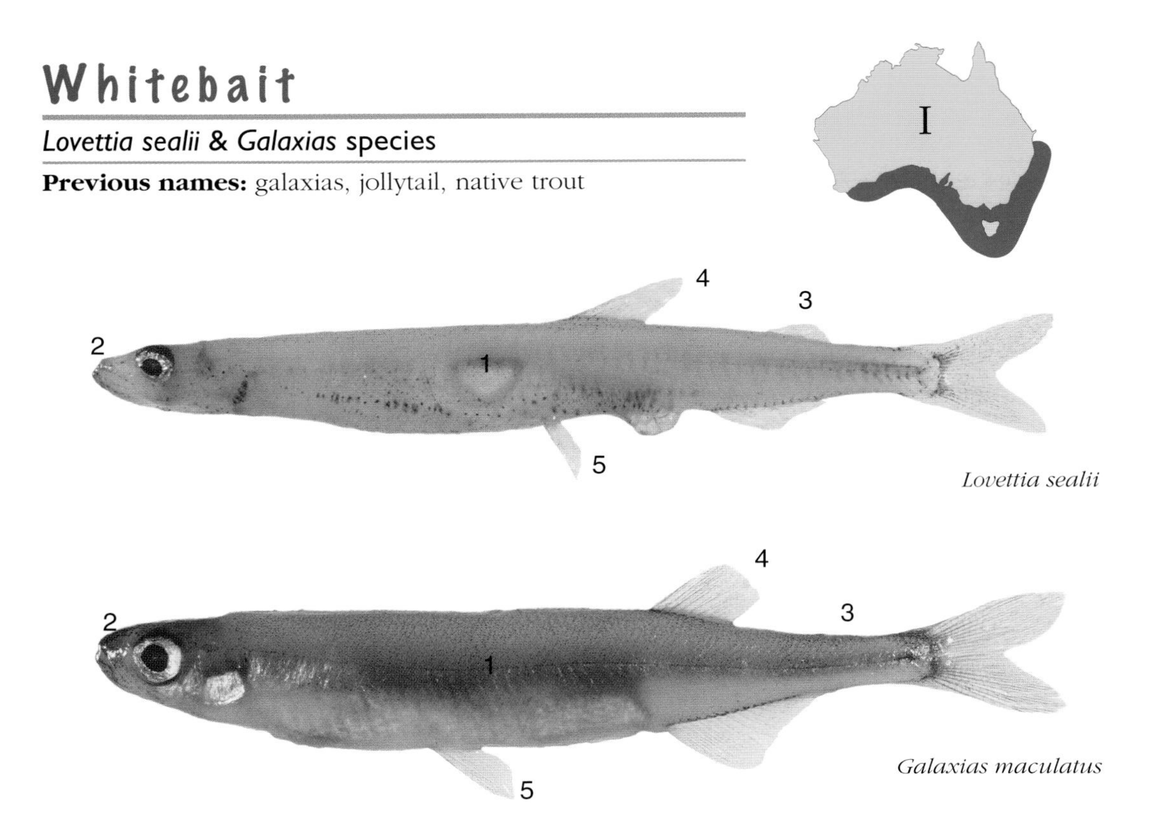

Lovettia sealii

Galaxias maculatus

Identifying features: (1) body small, elongate, scaleless, largely translucent; (2) snout usually long, pointed and flattened; (3) adipose fin present or absent; (4) dorsal-fin with short base, located well forward of, or partially over, anal fin; (5) pelvic fin well behind pectoral fins.

Comparisons: Young native trout (*Galaxias* species) resemble adult Tasmanian whitebait (*L. sealii*) but lack an adipose fin and the dorsal fin is situated above the anal fin. Both species have a distinctive pattern of black spots and are readily separated from each other in the catch. Imported Asian whitebait (Salangidae) are very similar in appearance but have an adipose fin, an extremely depressed head, and the dorsal fin located partly over the anal fin.

Fillet: Too small to fillet, and marketed whole.

Size: Adult native trout to 20 cm and Tasmanian whitebait to 7.7 cm, but commonly to about 6 cm.

Habitat: Sea-going; migrating into rivers in huge shoals to breed.

Fishery: Confined to Tasmania where they are taken during the riverine spawning migration in scoop and fyke nets. Catches exceeded 500 tonnes in the 1940s but the fishery subsequently collapsed and has been closed since 1974. Stocks have recovered slightly and, if recovery is sustained, could support a small, controlled fishery in the future.

Remarks: Only small quantities of young migrating native trout are caught with Tasmanian whitebait during the annual spawning run. A product imported from New Zealand and marketed as 'whitebait' is exclusively young native trout. Other information is provided for protein fingerprints (p. 369).

Crimson seaperch

Lutjanus erythropterus

Previous names: crimson snapper, red snapper, saddletail seaperch, scarlet snapper, seaperch

Identifying features: ① mouth relatively small, maxilla length much less than distance between bases of last dorsal-fin and anal-fin rays; ② dorsal profile near eye slightly convex; ③ body of adults uniformly pink, crimson or red, with a faint dark saddle over the upper caudal peduncle; ④ body deep and compressed; ⑤ dorsal fin with 11 spines, 12–14 soft rays; ⑥ dorsal-fin and anal-fin bases, especially soft portions, with covering of scales.

Comparisons: Most similar to saddletail seaperch (*L. malabaricus*, p. 242) but with a smaller mouth and a convex dorsal profile near the eyes (rather than straight or slightly concave). The darktail seaperch (*L. lemniscatus*, p. 244) is similar but has 10 (rather than 11) dorsal-fin spines and a dusky brown or black caudal fin. Distinguished from most other *Lutjanus* species by the lack of spots or stripes (in adult specimens) and from other tropical snappers such as king snapper (*Pristipomoides filamentosus*, p. 237) by the relatively deep body.

Fillet: Deep, rather elongate, tapering sharply, upper profile strongly convex, pale pinkish. Outside with pronounced, continuous, central red muscle band; convex EL, converging HL; HS along middle of fillet; integument translucent above, pinkish or silvery white below. Inside flesh medium; belly flap mostly absent; peritoneum silvery white to transparent. Usually skinned, skin medium, scale pockets small and defined.

Size: To at least 65 cm and about 3.8 kg (commonly to 60 cm and 3 kg).

Habitat: Marine; often around reefs but also on trawling grounds to depths of about 100 m.

Untrimmed.
Protein fingerprint p. 370; oil composition p. 404.

Fishery: Taken by trawls, lines and in traps in northern Australia between Shark Bay (WA) and Brisbane (Qld). High-quality product landed in Darwin is often exported to other capital cities and elsewhere.

Remarks: Flesh differs from the saddletail seaperch and appeals to different ethnic markets. Cooked flesh is moderately firm and white.

Goldband snapper

Pristipomoides multidens & *P. typus*

Previous names: goldband jobfish, jobfish, sharptooth jobfish, sharptooth snapper

I

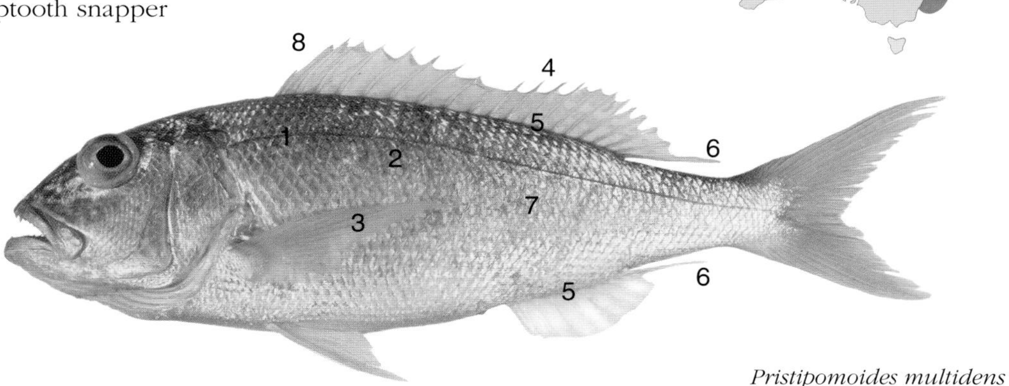

Pristipomoides multidens

Identifying features: (1) scales relatively large, 48–52 in lateral line; (2) body usually yellow, but sometimes rosy red, often with stripes and spots; (3) pectoral fins longer than snout length; (4) dorsal fin not deeply notched between spinous and soft portions; (5) dorsal-fin and anal-fin bases without a covering of scales; (6) last ray of dorsal and anal fins longer than those preceding; (7) body elongate and slightly compressed; (8) dorsal fin with 10 spines, 11–12 soft rays.

Comparisons: The goldband snapper (*P. multidens*) has transverse yellow bars on top of the head whereas the sharptooth snapper (*P. typus*) has longitudinal yellow stripes. Both are distinguished from the similar king snapper (*P. filamentosus*, p. 237) in having larger scales (48–52 versus 60–65 in lateral line) and usually a yellow body, from ruby snappers (*Etelis* species, p. 241) in lacking a notch between the spinous and soft ray portions of the dorsal fin and from the green jobfish (*Aprion virescens*, p. 235) in having longer pectoral fins.

Fillet: *P. multidens* deep, rather elongate, tapering sharply, upper profile moderately convex, pinkish. Outside with intermediate to pronounced, continuous, central red muscle band; parallel EL, weakly converging HL; HS along middle of fillet; integument white. Inside flesh medium; belly flap absent; peritoneum silvery white to translucent. Usually skinned, skin medium, scale pockets large and defined.

Size: To 90 cm and at least 6 kg (commonly to 65 cm and almost 3 kg).

Habitat: Marine; mostly over rocky bottoms of the continental shelf in depths of 40–200 m.

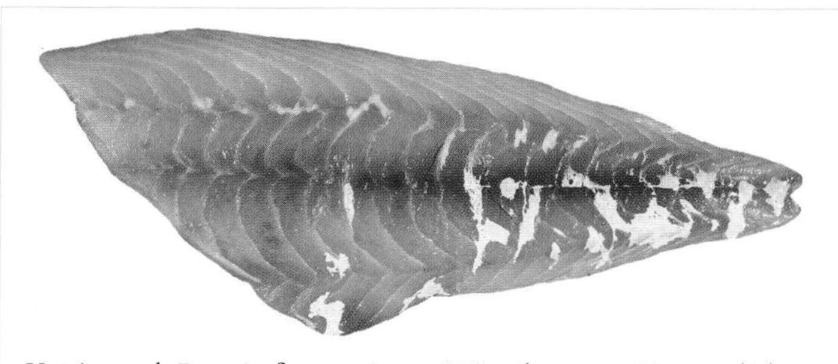

Untrimmed. Protein fingerprint p. 370; oil composition p. 404.

Fishery: Mostly off Western Australia and the Northern Territory but occasionally taken commercially and by recreational anglers off Queensland. Main catching methods are demersal longlines, traps and trawls.

Remarks: Very good eating. Firm, flaky flesh with a delicate flavour.

Green jobfish

Aprion virescens

Previous name: jobfish

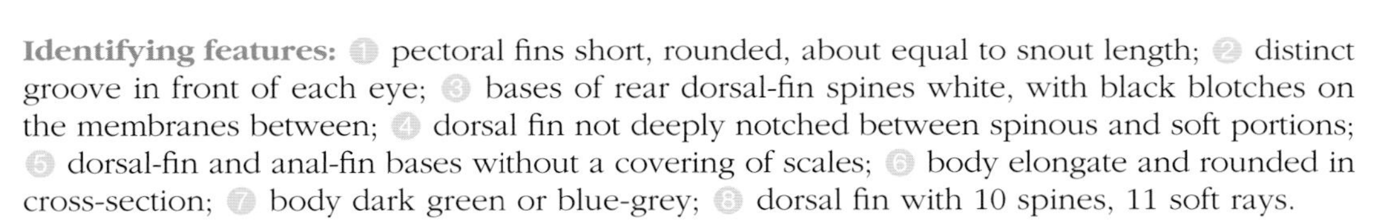

Identifying features: 1 pectoral fins short, rounded, about equal to snout length; 2 distinct groove in front of each eye; 3 bases of rear dorsal-fin spines white, with black blotches on the membranes between; 4 dorsal fin not deeply notched between spinous and soft portions; 5 dorsal-fin and anal-fin bases without a covering of scales; 6 body elongate and rounded in cross-section; 7 body dark green or blue-grey; 8 dorsal fin with 10 spines, 11 soft rays.

Comparisons: Distinctive species with an almost cylindrical dark green or blue grey body and short pectoral fins. Elongate relatives, the king snapper (*Pristipomoides filamentosus*, p. 237) and ruby snappers (*Etelis* species, p. 241), differ slightly in body shape and are uniformly pinkish, reddish or brownish.

Fillet: Moderately deep, elongate, tapering gently, upper profile weakly convex, pale pinkish. Outside with intermediate, continuous, central red muscle band; convex EL, converging HL; HS along middle of fillet; integument greyish-white above, white below. Inside flesh fine; belly flap present, not protruding; peritoneum silvery white to translucent. Usually skinned, skin fine, scale pockets large and well defined.

Size: To about 110 cm and 15.7 kg (commonly to 90 cm and 10 kg).

Habitat: Marine; usually near rock or coral reefs from near the surface inshore to depths of about 100 m. Sometimes forms small schools although larger individuals tend to be solitary.

Fishery: Caught using various line-fishing methods (most commonly droplines) and occasionally as a trawl bycatch. Taken commercially and recreationally throughout much of northern Australia.

Untrimmed. Protein fingerprint p. 370.

Remarks: Strong fighters and a tasty prize when landed. However, large specimens have been implicated in cases of ciguatera poisoning. Like most members of this group, the flesh is firm, flaky and white.

Hussar

Lutjanus adetii

Previous names: none

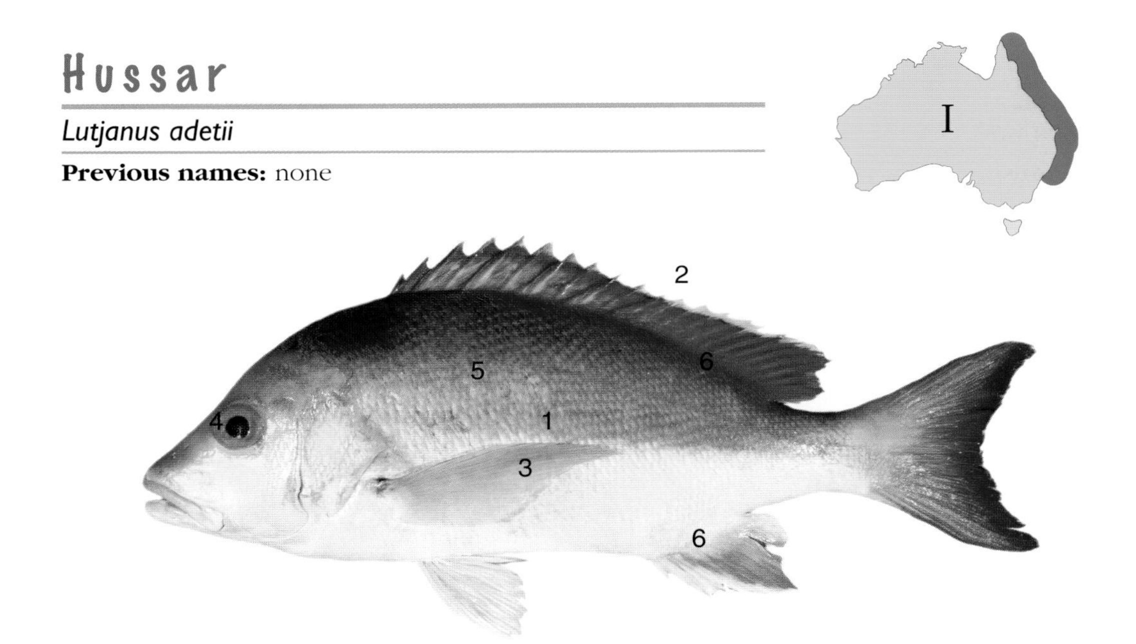

Identifying features: 1 single, broad golden-brown or yellow stripe along middle of sides, slightly narrower than eye diameter; 2 dorsal fin with 11 spines, 14 soft rays; 3 pectoral fin with 17 rays; 4 eye surrounded by yellow or orange; 5 body deep and laterally compressed; 6 dorsal-fin and anal-fin bases, especially soft portions, with covering of scales.

Comparisons: Readily identified by the broad golden-brown or yellow stripe along the sides of the body. The brownband seaperch (*L. vitta*, p. 244) has a similar but narrower stripe, 10 (rather than 11) dorsal-fin spines and 15–16 (rather than 17) pectoral-fin rays.

Fillet: Deep, rather elongate, tapering sharply, upper profile moderately convex, pale pinkish. Outside with feeble, discontinuous, central red muscle band; convex EL, weakly converging HL; HS along middle of fillet; integument silvery white. Inside flesh medium; belly flap mostly absent; peritoneum silvery white to translucent. Usually skinned, skin medium, scale pockets moderate-sized and variably defined.

Size: To 50 cm and about 1.6 kg (commonly to 35 cm and 0.7 kg).

Habitat: Marine; coastal waters on rock or coral reefs or over rubble bottoms to depths of about 20 m. Forms large schools.

Untrimmed. Protein fingerprint p. 370; oil composition p. 390.

Fishery: Regularly marketed in Queensland after being caught on handlines or as trawl bycatch.

Remarks: Considered a nuisance by some recreational fishers for taking bait intended for more favoured species such as the red emperor (*L. sebae*, p. 240). Usually marketed fresh. The firm, white flesh has a delicate taste.

King snapper

Pristipomoides filamentosus

Previous names: rosy job, rosy jobfish

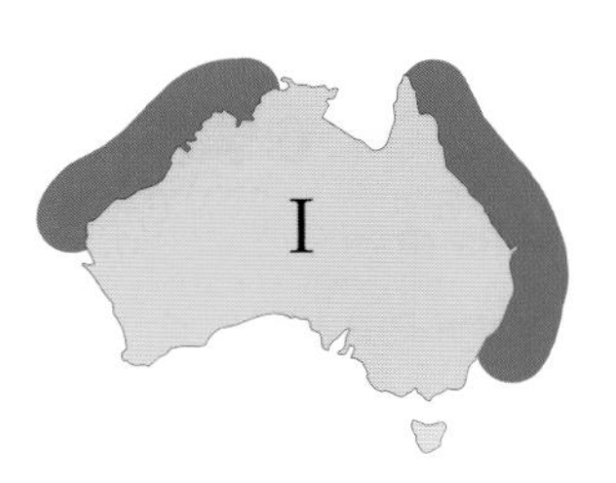

Identifying features: ① scales relatively small, 60–65 in lateral line; ② sides of body uniformly pink, crimson, red or brown, without stripes; ③ pectoral fins longer than snout length; ④ dorsal fin not deeply notched between spinous and soft portions; ⑤ last ray of dorsal and anal fins longer than those preceding; ⑥ dorsal-fin and anal-fin bases without a covering of scales; ⑦ body elongate and slightly compressed; ⑧ dorsal fin with 10 spines, 12 soft rays.

Comparisons: Similar elongate shape to goldband snappers (*P. multidens* and *P. typus*, pp 234) but usually pink or red in body colour (rather than yellow) and with more scales in the lateral line (60–65 versus 48–52). Has a longer pectoral fin than the green jobfish (*Aprion virescens*, p. 235) and lacks the dorsal-fin notch of ruby snappers (*Etelis* species, p. 241).

Fillet: Moderately deep, rather elongate, tapering gently, upper profile slightly convex, pale pinkish. Outside with pronounced, continuous, central red muscle band; convex EL, converging HL; HS along middle of fillet; integument white. Inside flesh medium; belly flap often present, not protruding; peritoneum silvery white to translucent. Usually skinned, skin medium, scale pockets large and irregular.

Size: To about 90 cm and 6.5 kg (commonly to 60 cm and 2.2 kg).

Habitat: Marine; over and near rocky and coralline reefs of the outer continental shelf and upper slope in depths of 90–360 m.

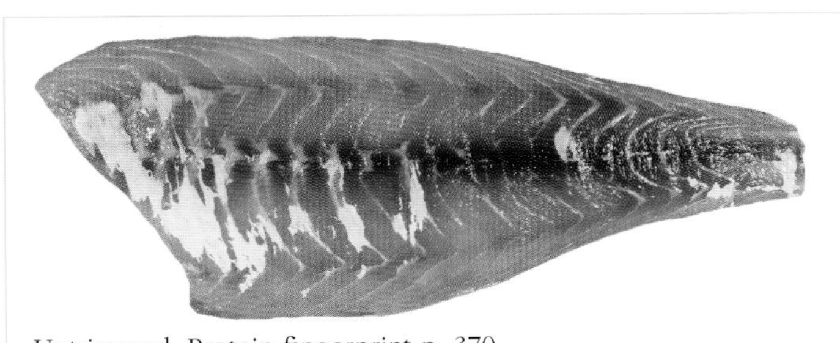

Untrimmed. Protein fingerprint p. 370.

Fishery: Caught using droplines and traps, mostly off Queensland and northern New South Wales but also off the Northern Territory and Western Australia.

Remarks: Excellent table-fish that commands a high price. The flesh is firm, flaky and tasty.

Mangrove jack

Lutjanus argentimaculatus

Previous names: red bream, seaperch

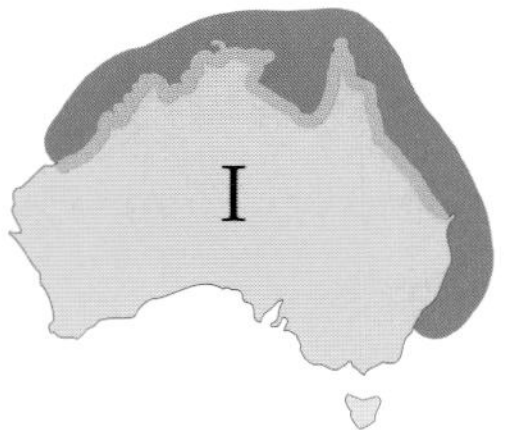

Identifying features: 1 back and sides of adults green-brown or red, each scale with a darker centre (body of juveniles with a series of thin white bars); 2 greatly enlarged canine teeth at front of jaws; 3 dorsal fin with 10 spines, 13–14 soft rays; 4 preopercular notch poorly developed; 5 body deep and moderately compressed; 6 dorsal-fin and anal-fin bases, especially soft portions, with covering of scales.

Comparisons: Sometimes confused with at least two local seaperches. The fingermark seaperch (*L. johnii*, p. 244) is usually paler with a large black spot on the upper side mainly above the lateral line and the red bass (*L. bohar*), which is considered poisonous (ciguatera), has a distinctive groove between the eye and the nostril on each side of the head.

Fillet: Deep, rather elongate, tapering sharply, upper profile moderately convex, pinkish. Outside with pronounced, continuous, central red muscle band; convex EL, converging HL; HS along middle of fillet; integument translucent above, white below. Inside flesh medium; belly flap sometimes present; peritoneum silvery white to translucent. Usually skinned, skin thick, scale pockets large and defined.

Size: Possibly to 120 cm and 16 kg (commonly to 75 cm and at least 5 kg).

Habitat: Marine; however, juveniles commonly enter mangrove estuaries and rivers to the extent of the tidal influence. Adults prefer sheltered inshore coral reefs, especially in areas of heavy silting, or offshore trawling grounds to depths of 120 m.

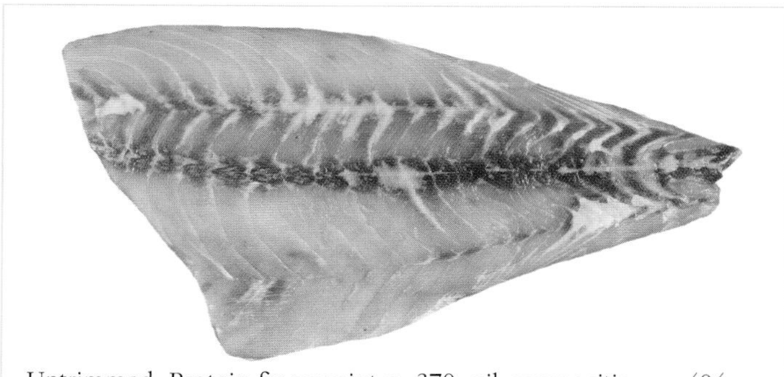

Untrimmed. Protein fingerprint p. 370; oil composition p. 404.

Fishery: Fished throughout its range using handlines, traps and gillnets and occasionally taken in trawls. Very popular target of recreational anglers.

Remarks: Immense fighting ability. Excellent foodfish but not seen in large quantities. The flesh is sweet and flaky but can be dry in large specimens.

Moses seaperch

Lutjanus russelli & *L.* sp.

Previous names: fingermark bream, Moses perch, red bream, Russell snapper, seaperch

Lutjanus russelli

Identifying features: ① body white, pink or dusky brown with a black spot on upper side (partly covering the lateral line) and often with a series of narrow golden-brown or black stripes along the sides; ② caudal fin white, pink, red or dusky brown, never yellow; ③ dorsal fin with 10 spines, 14 soft rays; ④ body deep and compressed; ⑤ dorsal-fin and anal-fin bases, especially soft portions, with covering of scales.

Comparisons: Eastern Moses seaperch (*L. russelli*) differs from western Moses seaperch (*L.* sp.) in having yellow pectoral, pelvic and anal fins (rather than pink or translucent). Both species differ from the blackspot seaperch (*L. fulviflamma*) in never having a yellow caudal fin and in having 14 (rather than usually 13) dorsal-fin soft rays.

Fillet: *L. russelli* deep, rather elongate, tapering sharply, upper profile moderately convex, pale pinkish. Outside with feeble, continuous, central red muscle band; convex EL, converging HL; HS along middle of fillet; integument translucent silvery white. Inside flesh medium; belly flap sometimes present; peritoneum silvery white to translucent. Usually skinned, skin medium, scale pockets moderate-sized and defined.

Size: To 53 cm and 2 kg (commonly to 45 cm and 1.2 kg).

Habitat: Marine; occur in estuaries and bays and on coastal and offshore reefs to depths of at least 80 m.

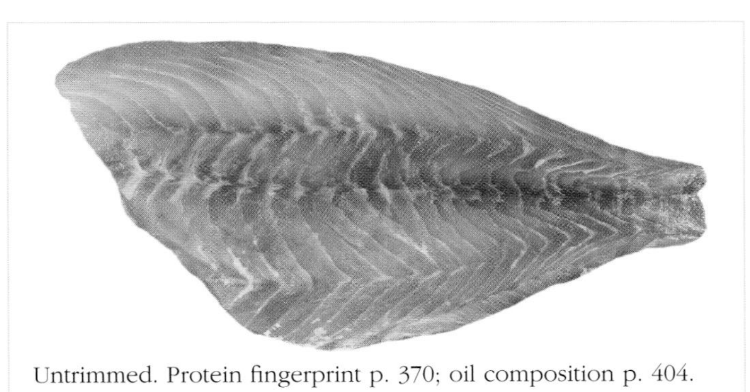

Untrimmed. Protein fingerprint p. 370; oil composition p. 404.

Fishery: Commonly caught and sold throughout its range. Capture methods include droplines, trolled lures, traps and demersal trawls.

Remarks: Further work is required to determine which of these two species is the true *L. russelli* as both have long been lumped under the one name. The firm, tasty flesh cooks to white flakes.

Red emperor

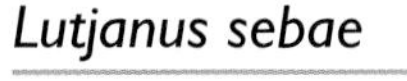

Lutjanus sebae

Previous names: none

Identifying features: ① dorsal fin with 11 spines, 16 (rarely 15) soft rays; ② body red, pink or white with 1 broad darker bar along the head and 2 across the body (often faint in large adults); ③ very pronounced preopercular notch; ④ body very deep, compressed; ⑤ dorsal-fin and anal-fin bases, especially soft portions, with covering of scales.

Comparisons: Distinctive seaperch with 3 broad dark bars on the head and body (may be faint in adults), 15–16 dorsal-fin soft rays, a very pronounced preopercular notch and a very deep body. Unrelated to the true emperors (*Lethrinus* species) which have fewer dorsal-fin soft rays (9–10) and lack scales on the cheek.

Fillet: Deep, rather elongate, tapering sharply, upper profile strongly convex, yellowish-white to pinkish. Outside with very pronounced, continuous, central red muscle band; almost parallel EL, converging HL; HS along or just below middle of fillet; integument white. Inside flesh medium; belly flap mostly absent; peritoneum translucent. Usually skinned, skin thick, scale pockets moderate-sized and well defined.

Size: Exceeding 100 cm and 22 kg (commonly to 70 cm and about 6 kg).

Habitat: Marine; often in the vicinity of coral reefs but also over soft bottoms to depths of about 100 m.

Fishery: Taken by demersal and semi-pelagic trawls, traps, droplines, handlines or longlines off Western Australia, the Northern Territory and Queensland. Highly esteemed by recreational fishers.

Remarks: Lesser seaperches are sometimes incorrectly sold as 'red emperor'. One of the best known and most popular foodfishes in Australia. The flesh is excellent to eat, firm and flaky with a delicate flavour.

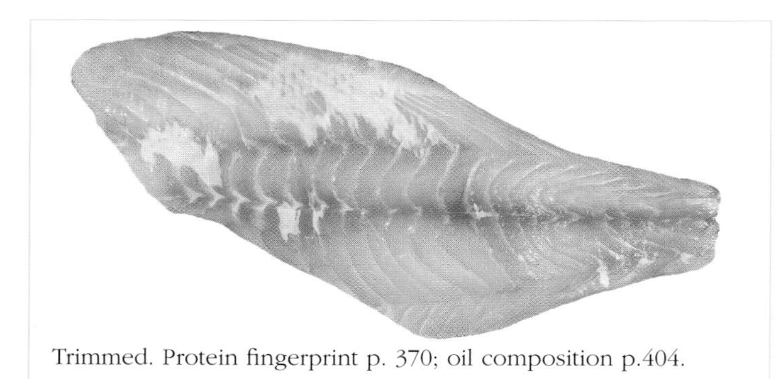

Trimmed. Protein fingerprint p. 370; oil composition p.404.

Ruby snapper

Etelis species

Previous name: flame snapper

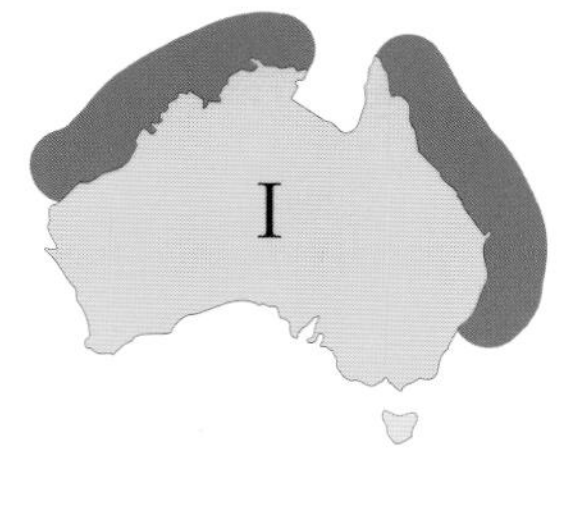

Etelis coruscans

Identifying features: ① dorsal fin deeply notched between spinous and soft portions, and with 10 spines, 11 soft rays; ② maxilla scaled; ③ upper sides of body uniformly pink or red; ④ body elongate and somewhat fusiform; ⑤ dorsal-fin and anal-fin bases without a covering of scales.

Comparisons: Distinguished from other members of this group by the deeply notched dorsal fin. The king snapper (*Pristipomoides filamentosus*, p. 237) is similar in colour but has a proportionally more rounded head and shorter caudal-fin lobes. Other similar tropical snappers lack scales on the maxilla. The seaperches (*Lutjanus* species) are deeper bodied and have scales on the dorsal-fin and anal-fin bases.

Fillet: *E. coruscans* moderately deep, rather elongate, tapering gently, upper profile almost straight, pale pinkish. Outside with intermediate, continuous, central red muscle band; parallel EL, weakly converging HL; HS along middle of fillet; integument white. Inside flesh medium; belly flap usually absent; peritoneum white. Usually skinned, skin very thick, scale pockets large and indistinct.

Size: To at least 100 cm and at least 8 kg (commonly to 55 cm and about 1.7 kg).

Habitat: Marine; on rocky bottoms of the outer continental shelf and upper continental slope in depths of 90–300 m.

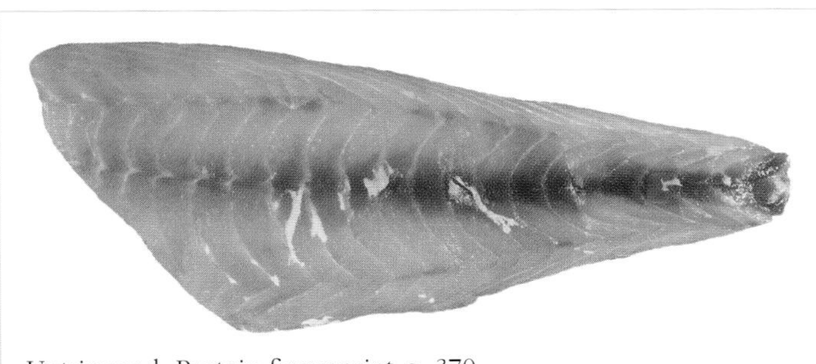

Untrimmed. Protein fingerprint p. 370.

Fishery: Taken by droplines, longlines or demersal trawls throughout their range.

Remarks: Attractive fishes which have only recently been introduced to Australian markets but have gained rapid acceptance. The flesh, which is of exceptional quality, is retailed as fresh or frozen fillets.

Saddletail seaperch

Lutjanus malabaricus

Previous names: red bass, red jew, red snapper, ruby emperor, saddletail, scarlet seaperch, seaperch

Identifying features: ① mouth relatively large, maxilla length about equal to distance between bases of last dorsal-fin and anal-fin rays; ② dorsal profile near eye straight or slightly concave; ③ body of adults uniformly pink, crimson or red, with a dark saddle over the upper caudal peduncle; ④ body deep and compressed; ⑤ dorsal fin with 11 spines, 12–14 soft rays; ⑥ dorsal-fin and anal-fin bases, especially soft portions, with covering of scales.

Comparisons: Often confused with crimson seaperch (*L. erythropterus*, p. 233) but has a larger mouth, a straight or concave (rather than convex) dorsal profile near the eyes and a broader space between the eye and the upper jaw. The dark saddle on the caudal peduncle is characteristic. Lacks the body stripes or spots of most other seaperches and has a deeper body than tropical snappers, including the ruby snappers (*Etelis* species, p. 241).

Fillet: Moderately deep, rather elongate, tapering sharply, upper profile strongly convex, yellowish-white. Outside with pronounced, continuous, central red muscle band; almost convex EL, converging HL; HS along or just below middle of fillet; integument white. Inside flesh medium; belly flap mostly absent; peritoneum translucent. Usually skinned, skin thick, scale pockets large and well defined.

Size: To about 100 cm and 13.6 kg (commonly to 70 cm and at least 5 kg).

Habitat: Marine; coastal and offshore reefs and over soft bottoms to depths of 100 m. Often schools with crimson seaperch.

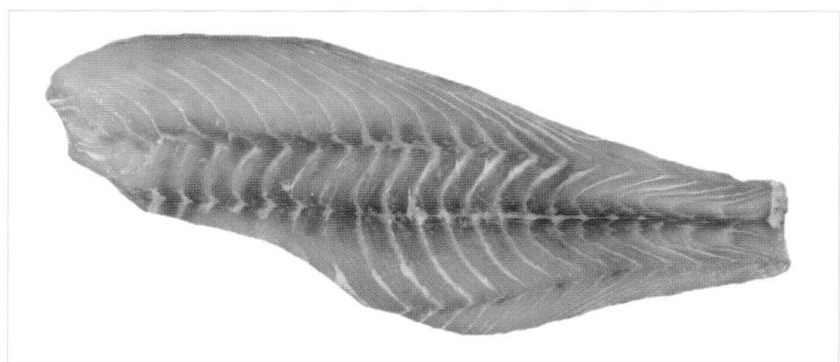

Trimmed. Protein fingerprint p. 370; oil composition p. 404.

Fishery: Throughout most of northern Australia where it is taken by traps, demersal and semi-pelagic trawls, droplines, handlines or longlines. Targeted by recreational anglers throughout its range.

Remarks: Sometimes erroneously marketed as 'red emperor'. The flesh is flaky and excellent to eat.

Seaperch (page 1 of 2)

Lutjanus species

Previous name: snapper

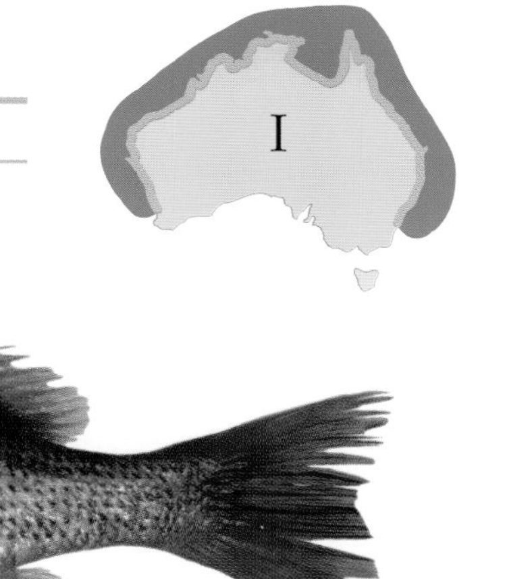

Lutjanus johnii

Identifying features: ① dorsal-fin and anal-fin bases, especially soft portions, with covering of scales; ② small to large fishes, body usually deep and compressed; ③ dorsal fin with 10–12 spines, 12–16 soft rays; ④ lateral-line scales 42–51; ⑤ mouth medium, protractile; ⑥ notch present on lower margin of preoperculum, variable in development.

Comparisons: Twenty six species of these perch-like fishes occur in Australian waters and most are harvested. Distinguished from other snappers (Lutjanidae) included here by being deeper-bodied and in having a covering of scales on the bases of the soft dorsal and anal fins. Similar to emperors and seabreams (Lethrinidae) but have more dorsal-fin soft rays (12–16 versus 9–10). Also have scales on the cheek which are absent in emperors (*Lethrinus* species).

Fillet: *L. lemniscatus* moderately deep, rather elongate, tapering sharply, upper profile strongly convex, pale pinkish. Outside with intermediate, continuous, central red muscle band; almost convex EL, converging HL; HS along middle of fillet; integument translucent above, silvery white below. Inside flesh medium; belly flap sometimes present; peritoneum silvery white to translucent. Usually skinned, skin medium, scale pockets small and defined.

Size: To 120 cm and at least 22 kg (commonly to 60 cm and 4 kg).

Habitat: Marine; often associated with inshore reefs but some species inhabit trawling grounds to depths of at least 200 m.

Fishery: Caught commercially and recreationally; main methods are handlines, demersal longlines, traps and trawls.

Lutjanus lemniscatus

Trimmed.
Protein fingerprint p. 370; oil composition p. 404.

Remarks: Seaperches (*Lutjanus* species) and other members of the family Lutjanidae are sometimes called 'snappers'. In Australia, this causes confusion with a well known bream, *Pagrus auratus*, that is marketed as 'snapper' (p. 73). Some larger species, such as the red bass (*L. bohar*), can cause ciguatera poisoning but generally seaperches are considered excellent eating.

Seaperch (page 2 of 2)

Lutjanus species

Lutjanus johnii

Remarks: Commonly called 'fingermark seaperch', or 'golden snapper' this species is distributed in far northern Australia between about Cape Leveque (WA) and Cooktown (Qld) in depths to 80 m. Best distinguished in having a generally pale golden body and a large black spot on the upper side. Reaches at least 90 cm and 12.4 kg.

Lutjanus lemniscatus

Remarks: Commonly called 'darktail seaperch', this species has a tropical distribution between Carnarvon (WA) and Cooktown (Qld) in depths to 80 m. As its name suggests, it is best distinguished in having a dusky brown or black caudal fin. It also has 10 dorsal-fin spines and a poorly developed preopercular notch. Reaches 65 cm and about 4.5 kg.

Lutjanus vitta

Remarks: Commonly called 'brownband seaperch', this northern species is distributed between North West Cape (WA) and Hervey Bay (Qld) in depths of 10–40 m. Best distinguished in having a narrow dark stripe along the middle of the sides, 10 dorsal-fin spines and 15–16 pectoral-fin rays. Reaches 40 cm and 1.0 kg.

Stripey seaperch

Lutjanus carponotatus

Previous names: seaperch, Spanish flag, striped seaperch, stripey

Identifying features: 1 sides of body yellow or blue-grey with 8–9 orange or yellow stripes; 2 black spot on upper corner of pectoral-fin base; 3 no black spot on upper side; 4 dorsal fin with 10 spines, 14–16 (usually 15) soft rays; 5 dorsal-fin and anal-fin bases, especially soft portions, with covering of scales.

Comparisons: Distinguished from other seaperches in having stripes along the sides, a black spot on the pectoral-fin base and 14–16 dorsal-fin soft rays. The fiveline seaperch (*L. quinquelineatus*) has fewer blue stripes and lacks the spot on the pectoral-fin base. Deeper bodied than tropical snappers such as goldband snappers (*Pristipomoides multidens* and *P. typus*, p. 234).

Fillet: Deep, short, tapering sharply, upper profile strongly convex, pinkish. Outside with intermediate, continuous, central red muscle band; convex EL, converging HL; HS along middle of fillet; integument translucent above, white below. Inside flesh medium; belly flap sometimes present, peritoneum silvery white to translucent. Usually skinned, skin thick, scale pockets moderate and well defined.

Untrimmed.
Protein fingerprint p. 370; oil composition p. 404.

Size: To at least 40 cm and 1.0 kg (commonly to 35 cm and 0.7 kg).

Habitat: Marine; often forms small schools near inshore coral reefs and on soft bottoms to depths of about 80 m.

Fishery: Often taken by linefishers and as a trawl bycatch throughout much of northern Australia. Popular angling fish.

Remarks: In plentiful supply in the markets and sells well. The firm, flaky flesh is delicately flavoured.

Silver biddy

Gerreidae

Previous names: roach, silverbelly

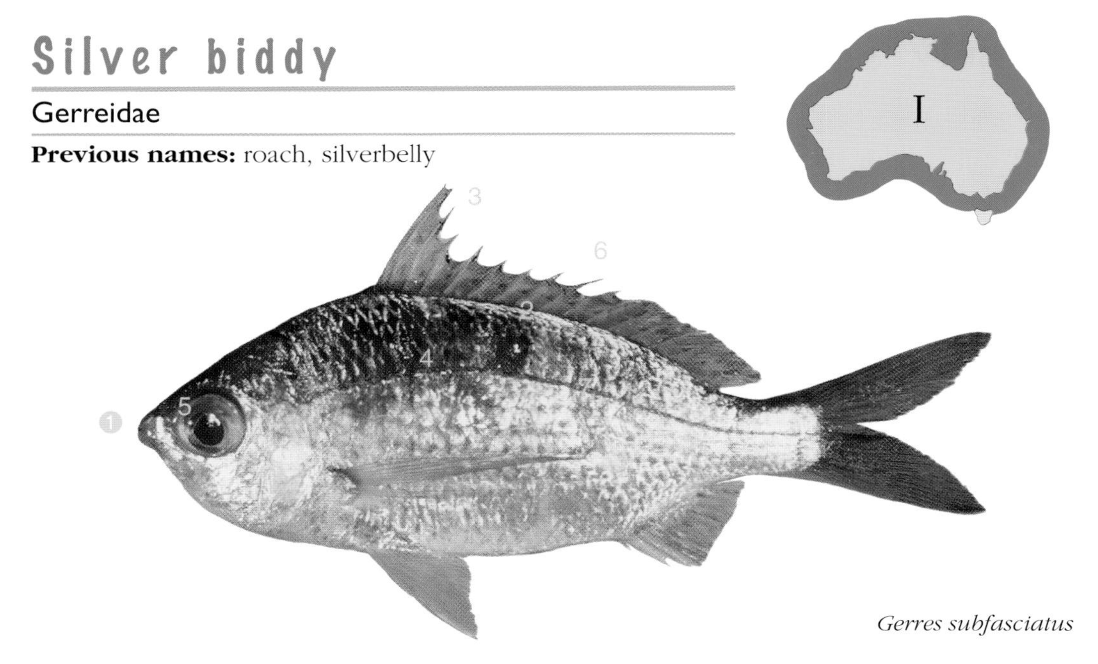

Gerres subfasciatus

Identifying features: ① mouth protrusible, opening downward to form a tube; ② dorsal-fin base long; ③ anterior dorsal-fin spines usually much longer than those following; ④ body usually silver, often with faint bars or spots; ⑤ eye large and very close to mouth; ⑥ dorsal fin with 9–10 spines, 9–17 soft rays; ⑦ body compressed.

Comparisons: Superficially similar to breams (Sparidae) but with a protrusible mouth, a more elongate body and a larger eye. The 18 or so Australian species are distinguished from each other by colour patterns, body depth, and fin-element counts and lengths.

Fillet: *Gerres subfasciatus* deep, rather short, tapering rapidly, weakly convex above, greyish with dense dark veins. Outside with continuous, intermediate central red muscle band; weakly converging EL, converging HL; HS along middle of fillet; integument silvery grey above, more whitish below. Inside flesh medium; belly flap usually present; peritoneum translucent. Sometimes skinned, skin thin, scale pockets moderate and irregular.

Size: Commercial species to 26 cm and 0.2 kg (commonly to 18 cm and 0.1 kg).

Habitat: Marine; coastal and estuarine waters to depths of 100 m (typically less than 40 m). Form large schools over sand, mud or seagrass beds.

Fishery: Taken by seines, cast nets, tunnel nets and gillnets and as bycatch of shallow trawl fisheries. Frequently marketed fresh, mainly in Brisbane and Sydney and sometimes in Perth.

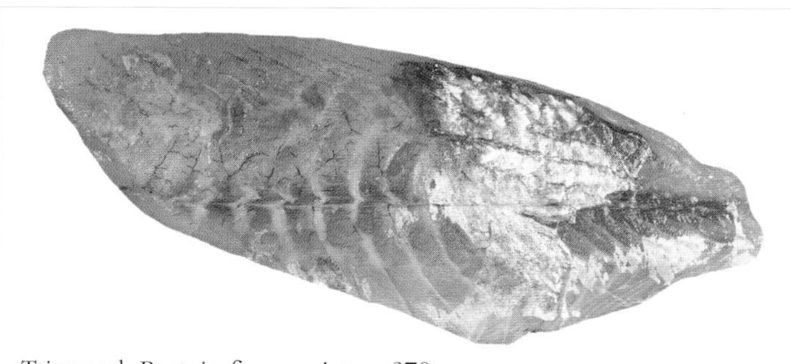

Trimmed. Protein fingerprint p. 370.

Remarks: Among the smallest commercial fishes in Australia and many silver biddy species are too small to be marketed. The flesh is firm and delicately flavoured when fresh and well suited to pan-frying.

Stargazer

Uranoscopidae

Previous name: monkfish

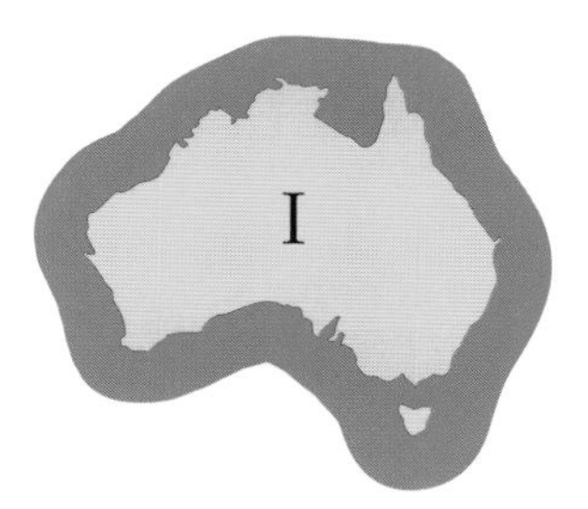

Kathetostoma canaster

Identifying features: ① head large, cube-like and bony; ② mouth cavernous, directed upwards and armed with large, sharp teeth; ③ eyes on top of head; ④ spine present above pectoral fin; ⑤ body very stocky anteriorly, tapering posteriorly.

Comparisons: Unique, bulldog-like head shape, unlikely to be confused with the members of any other seafood group. Over 20 species in Australian waters, most with distinctive colour patterns and head sculpturing.

Fillet: *Kathetostoma canaster* moderately deep, elongate, upper profile almost straight, taper slight, pale pinkish. Outside with feeble, diffuse red muscle band centred well above midline; weakly converging EL, weakly converging HL; HS along middle of fillet; integument translucent or silvery white. Inside flesh fine; belly flap usually absent; peritoneum transparent. Usually skinned, skin very thick, without scales.

Size: To at least 75 cm and 8 kg (commonly to 50 cm and about 6 kg).

Habitat: Marine; bury in sand or mud on the continental shelf and slope in water depths of about 5–900 m.

Fishery: Common trawl bycatch on the outer continental shelf and upper continental slope in southeastern Australia and the Great Australian Bight. Marketed regularly but in small quantities. Occasionally taken by recreational anglers.

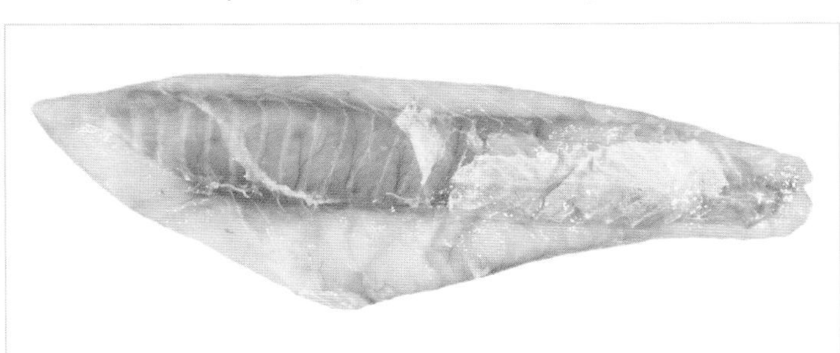

Trimmed. Protein fingerprint p. 370.

Remarks: To be handled with care due to strong jaws, sharp teeth and a strong, venomous spine above the pectoral fin. Many species are too small to be of commercial value but larger specimens provide good eating. The flesh is firm and has been sold illegally in Europe as 'lobster tails'.

Threadfin bream (page 1 of 2)

Nemipteridae

Previous names: butterfly bream, monocle bream, rosy threadfin bream

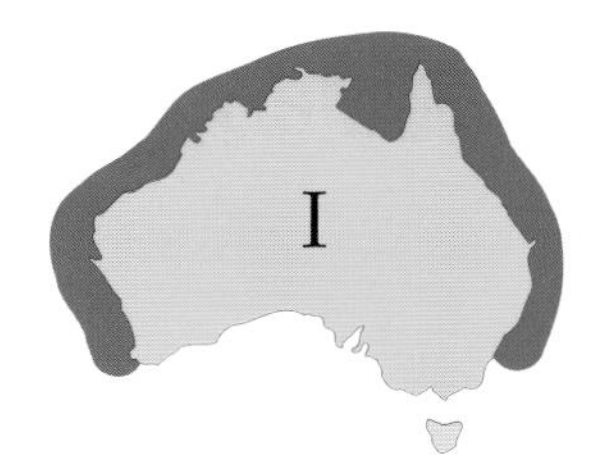

Nemipterus furcosus

Identifying features: 1 anal fin with 3 spines and 7–8 soft rays; 2 long-based dorsal fin with 10 spines and 9–10 soft rays; 3 body elongate and compressed; 4 usually brightly coloured with stripes and/or spots on the body and fins.

Comparisons: Thirty or so Australian species, distinguished from each other using colour patterns, length of caudal-fin filaments and development of a spine below the eye. Other brightly coloured tropical seafood species such as emperors (Lethrinidae) and seaperches (Lutjanidae) are deeper bodied and have more anal-fin soft rays (8–10 in emperors and 7–11 in seaperches versus 7–8 in threadfin breams).

Fillet: *Nemipterus furcosus* moderately deep, elongate, tapering evenly, weakly convex above, yellowish-white. Outside with continuous, pronounced central red muscle band; parallel EL, converging HL; HS along middle of fillet; integument silvery white. Inside flesh medium; belly flap rarely present; peritoneum silvery white to translucent. Usually skinned, skin medium, scale pockets large and well defined.

Size: To at least 44 cm and about 1.2 kg (commonly to 30 cm and 0.4 kg).

Habitat: Marine; demersal in depths of about 10–400 m (usually less than 100 m). Found on sand or mud bottoms among sponges and soft corals or near coral reefs.

Fishery: Heavily exploited by foreign trawlers from 1973–91 but now taken only in small quantities off northwestern Australia. Bycatch of otter trawling or caught in small-mesh nets. Often used as bait by recreational anglers.

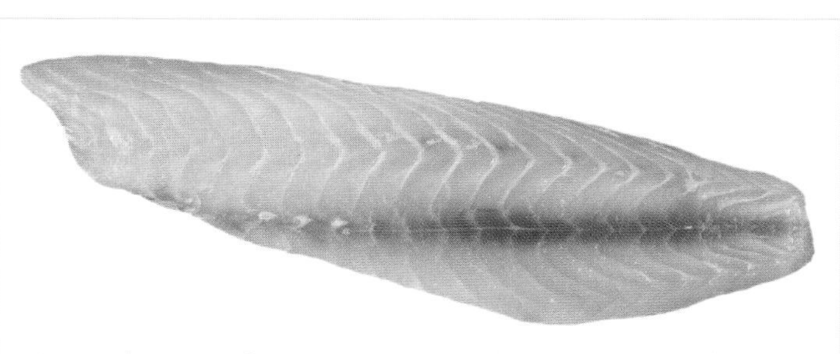

Trimmed. Protein fingerprint p. 370; oil composition p. 405.

Remarks: Popular in some parts of Asia where pan-fried fish are consumed with scales and bones intact. The flesh is firm and flaky if well handled, otherwise soft.

Threadfin bream (page 2 of 2)

Nemipteridae

Nemipterus furcosus

Remarks: Commonly called 'rosy threadfin bream', this tropical species is distributed between Dampier (WA) and Hervey Bay (Qld) in depths of 10–110 m. Best distinguished by being relatively drab in colour (pink with faint darker bars dorsally and silvery white ventrally). Reaches 35 cm and 0.7 kg.

Nemipterus peronii

Remarks: Commonly called 'notched threadfin bream', this species is distributed throughout northern Australia between North West Cape (WA) and Brisbane (Qld) in depths of 10–100 m. Best distinguished by deep notches in the membrane between the dorsal-fin spines and a yellow tip to the upper lobe of the caudal fin. Reaches 30 cm and 0.4 kg.

Scolopsis taeniopterus

Remarks: Commonly called 'redspot monocle bream', this species has a tropical distribution between Shark Bay (WA) and Mackay (Qld) in depths of 10–50 m. Best distinguished by a small red spot on the upper pectoral-fin base, blue and yellow stripes in front of the eye, and a spine immediately below the eye. To 23 cm and 0.2 kg.

Blue threadfin

Eleutheronema tetradactylum

Previous names: blue salmon, Cooktown salmon, threadfin

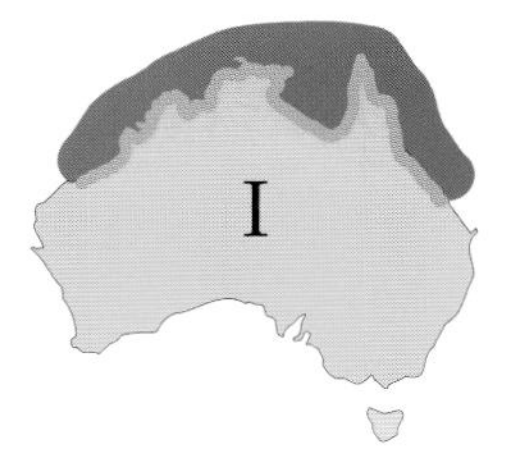

Identifying features: 1 3–4 filaments at bottom of each pectoral fin, longest barely extending past pelvic fin; 2 anal fin with 3 spines, 13–17 soft rays; 3 mouth behind tip of snout; 4 2 widely separated dorsal fins; 5 pectoral fins inserted very low on body; 6 caudal fin deeply forked.

Comparisons: Threadfins are unrelated to true salmons (Salmonidae). They are distinctive in having up to 16 unconnected filaments at the base of each pectoral fin, clear fatty tissue over the eye, and a blunt, rounded snout. Their mouth is under the head rather than at its tip. The blue threadfin differs from the other main commercial species, the king threadfin (*Polydactylus sheridani*, p. 251), in having fewer and shorter pectoral-fin filaments (usually 4 on each side versus usually 5), a proportionally larger eye, and more anal-fin soft rays (13–17 versus 9–11).

Fillet: Moderately deep, elongate, gently tapering to deep caudal peduncle, pale pinkish. Outside with broad, pronounced, broken central red muscle band; convex EL, weakly converging HL; HS along middle of fillet; integument often extensive, pale grey above, silvery white below. Inside flesh medium; belly flap sometimes removed; peritoneum silvery white to translucent. Usually skinned, skin thin, scale pockets small and defined.

Size: Possibly exceeding 140 cm and 18.5 kg (commonly less than 90 cm and about 6 kg).

Habitat: Coastal marine; benthopelagic in tropical habitats, including estuaries and muddy rivers and occasionally venturing into freshwater.

Fishery: Caught mainly by multi-species gillnet fishery off central Queensland.

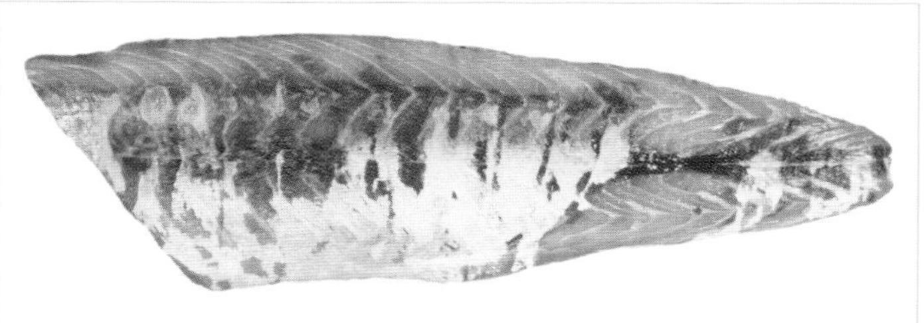

Untrimmed. Protein fingerprint p. 370; oil composition p. 405.

Remarks: An aggressive feeder and active fighter when hooked, it is highly respected by recreational anglers. Sold mostly as gilled or gutted whole fish or as fillets. Blue threadfin is considered to be excellent eating, and has been substituted for barramundi in the past.

King threadfin

Polydactylus sheridani

Previous names: Burnett's salmon, king salmon, threadfin, threadfin salmon

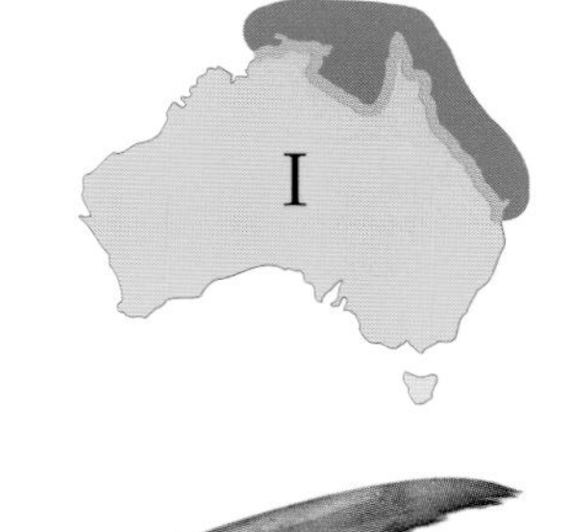

Identifying features: 1 4–5 (usually 5) filaments at bottom of each pectoral fin, longest extending to or slightly past anal-fin origin; 2 anal fin with 3 spines, 9–11 soft rays; 3 mouth behind tip of snout; 4 2 widely separated dorsal fins; 5 pectoral fin inserted very low on body; 6 caudal fin deeply forked.

Comparisons: Largest threadfin, differing from the blue threadfin (*Eleutheronema tetradactylum*, p. 250) in having more and shorter pectoral-fin filaments (usually 5 on each side versus usually 4), a proportionally smaller eye, and fewer anal-fin soft rays (9–11 versus 13–17).

Fillet: Moderately deep, elongate, tapering slightly to deep caudal peduncle, pale pinkish. Outside with broad, intermediate, diffuse central red muscle band; convex EL, converging HL; HS along middle of fillet; integument often extensive, white. Inside flesh coarse; belly flap sometimes removed; peritoneum translucent. Usually skinned, skin medium, scale pockets small and defined.

Size: To 185 cm and 30 kg (commonly 50–90 cm and 1.5–6 kg).

Habitat: Coastal marine; benthopelagic in muddy bays, estuaries and rivers of tropical Australia, mostly in very shallow water. Occasionally venturing into freshwater.

Fishery: Caught mainly by multi-species gillnet fishery in the Gulf of Carpentaria, but is widely distributed across northern Australia in shallow areas where water clarity is poor. Important bycatch of barramundi (*Lates calcarifer*, p. 121) gillnet fishery in the Northern Territory.

Untrimmed. Protein fingerprint p. 370.

Remarks: Bottom feeders; the thread-like pectoral filaments are used to detect small prawns, crabs and worms in the mud. Sold mostly as gilled or gutted whole fish or as fillets. The king threadfin fetches a higher price than blue threadfin but both are excellent foodfishes.

Blue-eye trevalla

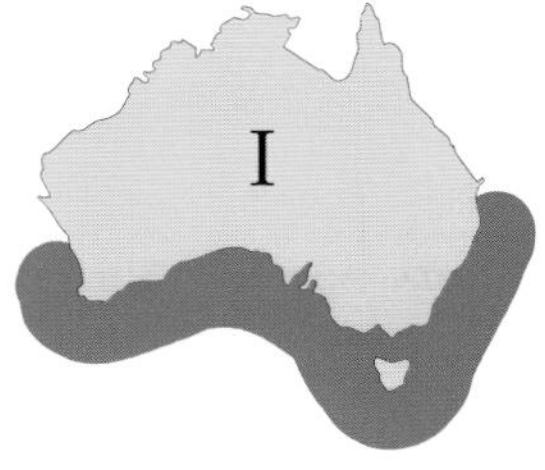

Hyperoglyphe antarctica & *Schedophilus labyrinthica*

Previous names: bigeye, blue-eye, blue-eye cod, deepsea trevalla, deepsea trevally, ocean blue-eye, sea trevally, trevalla

Hyperoglyphe antarctica

Identifying features: ① silvery to bronze or bluish-grey; ② snout blunt or broadly rounded; ③ no dark blotch above pectoral-fin base; ④ body firm, deep, compressed slightly; ⑤ head with numerous small pores; ⑥ dorsal fin with 7–9 spines, 18–21 or 26–29 soft rays; ⑦ anal fin with 3 spines, 13–16 or 18–19 soft rays.

Comparisons: Two species included. The main commercial species (*H. antarctica*) differs from the ocean blue-eye (*S. labyrinthica*) in having a distinct spinous part of the dorsal fin (rather than 1 continuous fin), fewer dorsal-fin and anal-fin rays, and a longer upper jaw (reaching to mid-eye rather than to the front edge of the eye). They differ from the two similarly coloured warehous (*Seriolella* species) in lacking a dark blotch behind the head.

Fillet: *H. antarctica* moderately deep, rather elongate, upper profile slightly convex, taper pronounced, off-white to yellowish. Outside with intermediate, continuous red muscle band; convex EL, converging HL; HS along middle of fillet; integument greyish above, white below. Inside flesh medium; belly flap sometimes present; peritoneum white with melanophores. Usually skinned, skin medium, scale pockets small and defined.

Size: To 140 cm and 50 kg (commonly about 55–90 cm and 2–10 kg).

Habitat: Marine; adults benthopelagic above rocky bottoms and seamounts in 200–600 m, young more midwater often to the surface.

Fishery: Caught mainly using droplines and longlines, but also taken as bycatch of demersal and midwater trawls. Once taken in large quantities from seamounts off the south-east, main fishery now occurs on the continental slope between Sydney (NSW) and Adelaide (SA), including Tasmania.

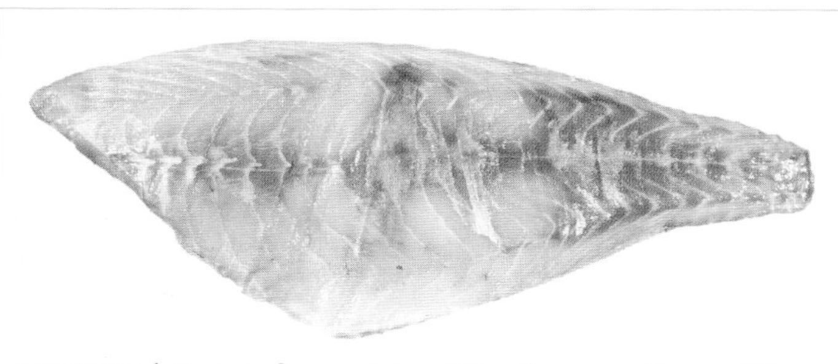

Untrimmed. Protein fingerprint p. 371; oil composition p. 405.

Remarks: Among the most highly regarded temperate foodfishes. The off-white flesh is firm, moist and delicately flavoured.

Blue warehou

Seriolella brama

Previous names: black trevally, sea bream, snotgall, snotgall trevally, snotty trevalla, snottynose trevalla, Tasmanian trevally, trevally

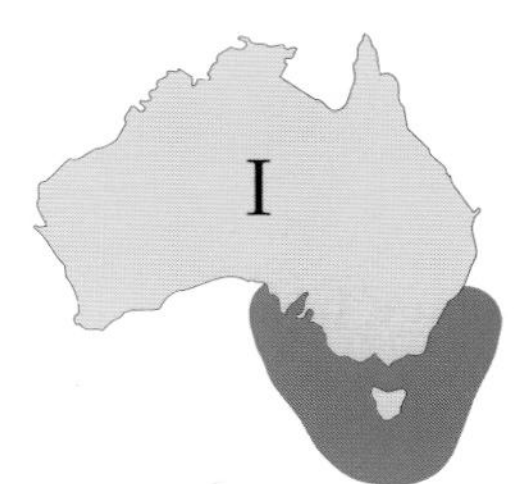

Identifying features: 1 silvery blue without spots; 2 scaled area on midline before dorsal fin shorter than lower jaw; 3 dark blotch above pectoral-fin base; 4 body firm, deep and quite compressed; 5 head with numerous small pores; 6 dorsal fin with 7–9 spines, 25–29 soft rays; 7 anal fin with 3 spines, 19–23 soft rays.

Comparisons: Resembles the related white warehou (*S. caerulea*, p. 256) in body shape but is more bluish (rather than whitish to grey) and has a dark blotch above the pectoral-fin base (otherwise lacking). The more slender silver warehou (*S. punctata*, p. 255) has this blotch but has more dorsal-fin soft rays (35–39 versus 25–29) and small, dark spots on the body.

Fillet: Moderately deep, rather elongate, upper profile slightly convex, taper pronounced, off-white to yellowish. Outside with pronounced, continuous red muscle band; convex EL, converging HL; HS along middle of fillet; integument lumpy, silvery or bluish-white. Inside flesh coarse; belly flap sometimes present; peritoneum transparent. Sometimes skinned, skin thin, scale pockets small and irregular.

Size: To at least 80 cm and possibly 7 kg (commonly about 35–45 cm and 1–2 kg).

Habitat: Marine; adults mostly live near the bottom on the continental shelf and upper slope. Young are pelagic, seeking refuge among the stinging tentacles of jellyfish.

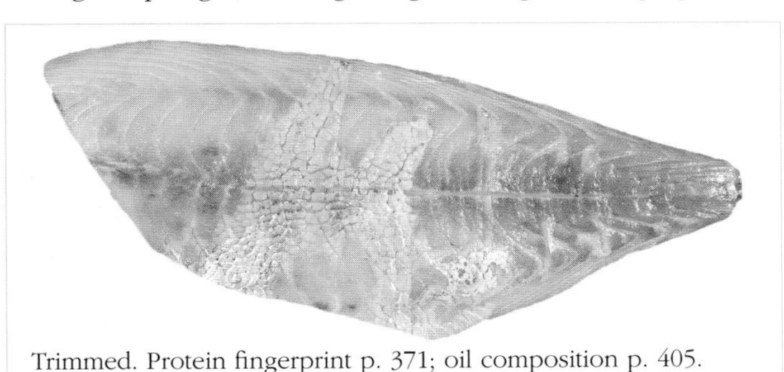

Trimmed. Protein fingerprint p. 371; oil composition p. 405.

Fishery: Caught mainly off southeastern Australia using gillnets, demersal trawls and Danish seines.

Remarks: Blue warehou is an important and popular temperate foodfish. Its off-white flesh has good eating qualities when very fresh but quickly becomes soft if handled poorly.

Rudderfish

Centrolophus, Schedophilus & Tubbia species

Previous names: New Zealand ruffe, ruffe, Tasmanian rudderfish

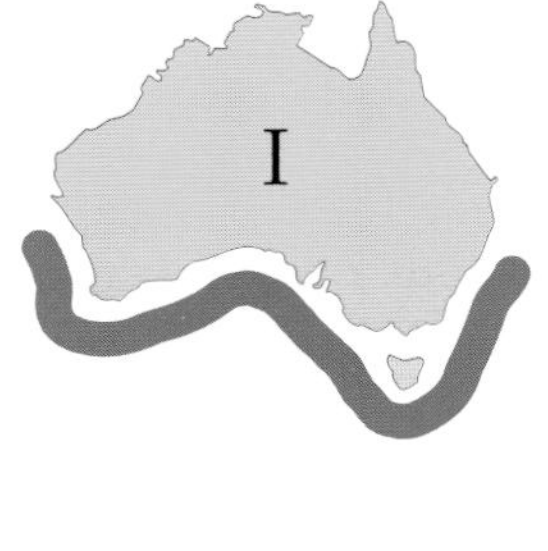

Centrolophus niger

Identifying features: ① dark brown to black; ② pectoral fins relatively short; ③ single dorsal fin without hard spines; ④ no dark blotch above pectoral-fin base; ⑤ body usually soft or flabby, often very compressed; ⑥ head with numerous small pores; ⑦ dorsal fin with 35–63 total rays; ⑧ anal fin with 24–27 or 33–41 total rays.

Comparisons: Distinct from other trevallas in their darker, softer bodies, higher fin-element counts and in having a continuous dorsal fin in which the anterior spiny rays are soft and flexible (rather than hard). The main species (*C. niger*) has a more solid body and fewer anal-fin rays than the other species marketed.

Fillet: *C. niger* moderately deep, elongate, upper profile slightly convex, taper slight, milky white. Outside with intermediate, discontinuous red muscle band; convex EL, weakly converging HL; HS along middle of fillet; integument lumpy, whitish. Inside flesh fine; belly flap sometimes present; peritoneum transparent. Sometimes skinned, skin very thick, scale pockets small and defined.

Size: To 120 cm and 20 kg (commonly about 35–70 cm and 1–4 kg).

Habitat: Marine; adults mostly live in midwater of the open ocean and along the continental slope either solitary or in schools. Young are pelagic, residing with salps and jellyfish near the surface.

Fishery: Probably under-utilised, presently caught sporadically off southern Australia using pelagic and demersal trawls. Resource size unknown.

Untrimmed. Protein fingerprint p. 371.

Remarks: Occasionally marketed foodfishes, with soft, milky, somewhat oily flesh that can have purgative properties if consumed in large quantities. Small wart-like lumps on the outside of skinned fillets are typical of some trevalla species.

Silver warehou

Seriolella punctata

Previous names: snottynose trevally, spotted trevalla, spotted trevally, spotted warehou, trevally

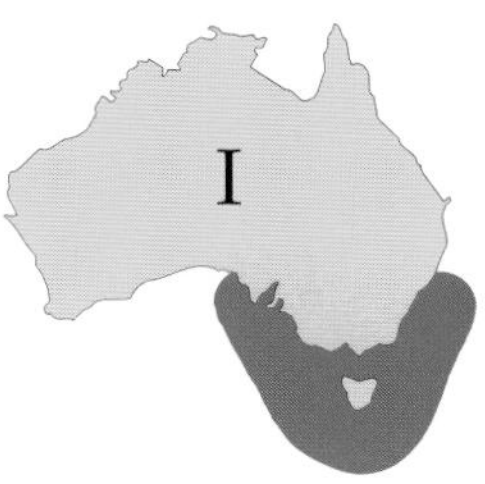

Identifying features: 1 silvery blue with small dark spots (occasionally faint); 2 scaled area on midline before dorsal fin shorter than lower jaw; 3 dark blotch above pectoral-fin base; 4 body firm, relatively slender and quite compressed; 5 head with numerous small pores; 6 dorsal fin with 7–9 spines, 35–39 soft rays; 7 anal fin with 3 spines, 21–24 soft rays.

Comparisons: Resembles the blue warehou (*S. brama*, p. 253) in colouration but is more slender, has small dark spots on the body, and more dorsal-fin soft rays (35–39 versus 25–29). The end of a smooth, scaleless skin patch on top of the head is pointed towards the dorsal fin (otherwise almost square). The paler white warehou (*S. caerulea*, p. 256) lacks a dark blotch above the pectoral-fin base.

Fillet: Moderately deep, rather elongate, upper profile slightly convex, taper slight, off-white to yellowish. Outside with pronounced, continuous red muscle band; parallel EL, converging HL; HS along middle of fillet; integument lumpy, silvery or bluish-white. Inside flesh coarse; belly flap sometimes present; peritoneum pink transparent. Sometimes skinned, skin thick, scale pockets small and irregular.

Size: Exceeding 70 cm and 5.5 kg (commonly about 35–55 cm and 0.4–2.2 kg).

Habitat: Marine; adults demersal on the continental slope in depths of 450–650 m.

Fishery: Caught mainly off southeastern Australia by bottom trawlers, sometimes as bycatch of the blue grenadier (*Macruronus novaezelandiae*, p. 84) fishery. Juveniles occasionally taken in embayments and estuaries in haul nets and by recreational anglers.

Trimmed. Protein fingerprint p. 371; oil composition p. 405.

Remarks: Often sold as 'spotted warehou'. Previously discarded in large quantities but now better received by the markets. The off-white flesh has good eating qualities when very fresh but spoils quickly.

White warehou

Seriolella caerulea

Previous names: ranger trevally, white trevalla

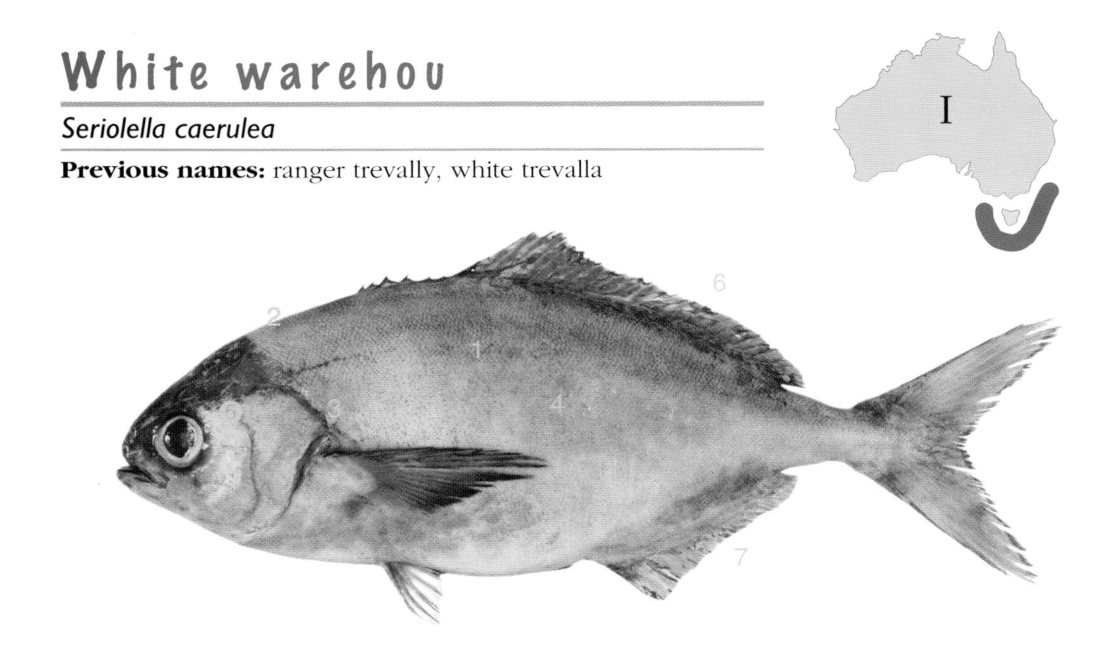

Identifying features: 1 greyish to white without spots; 2 scaled area on midline before dorsal fin longer than lower jaw; 3 no dark blotch above pectoral-fin base; 4 body firm, deep and quite compressed; 5 head with numerous small pores; 6 dorsal fin with 7–8 spines, 30–33 soft rays; 7 anal fin with 2–3 spines, 19–24 soft rays.

Comparisons: Resembles the blue warehou (*S. brama*, p. 253) in body form but is paler and lacks a dark blotch above the pectoral-fin base. It is deeper bodied and lacks the dark spots along the side that typify the silver warehou (*S. punctata*, p. 255). It is the only warehou in which the predorsal scale patch is longer than the lower jaw.

Fillet: Deep, rather elongate, upper profile slightly convex, bottle-shaped, off-white to yellowish. Outside with broad, pronounced, continuous red muscle band; convex EL, converging HL; HS above middle of fillet; integument lumpy, silvery. Inside flesh coarse; belly flap sometimes present; peritoneum transparent. Sometimes skinned, skin thick, scale pockets small and irregular.

Size: To about 80 cm and 6 kg (commonly to about 50 cm and 2 kg).

Habitat: Marine; adults demersal on the continental slope in depths of 500–800 m.

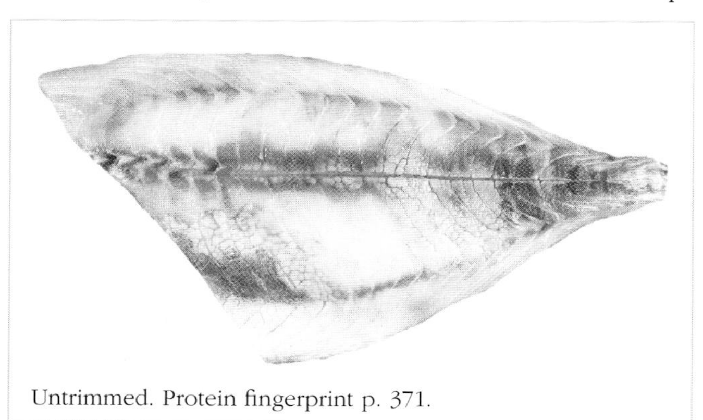

Untrimmed. Protein fingerprint p. 371.

Fishery: Caught in small quantities off southeastern Australia by bottom trawlers as bycatch of the blue grenadier (*Macruronus novaezelandiae*, p. 84) fishery.

Remarks: The use of the extremely confusing marketing name 'trevally', correctly applied to fishes of the family Carangidae, should be avoided for all warehous. The flesh is of good quality when fresh but spoils quickly.

Bigeye trevally

Caranx sexfasciatus

Previous names: great trevally, trevally

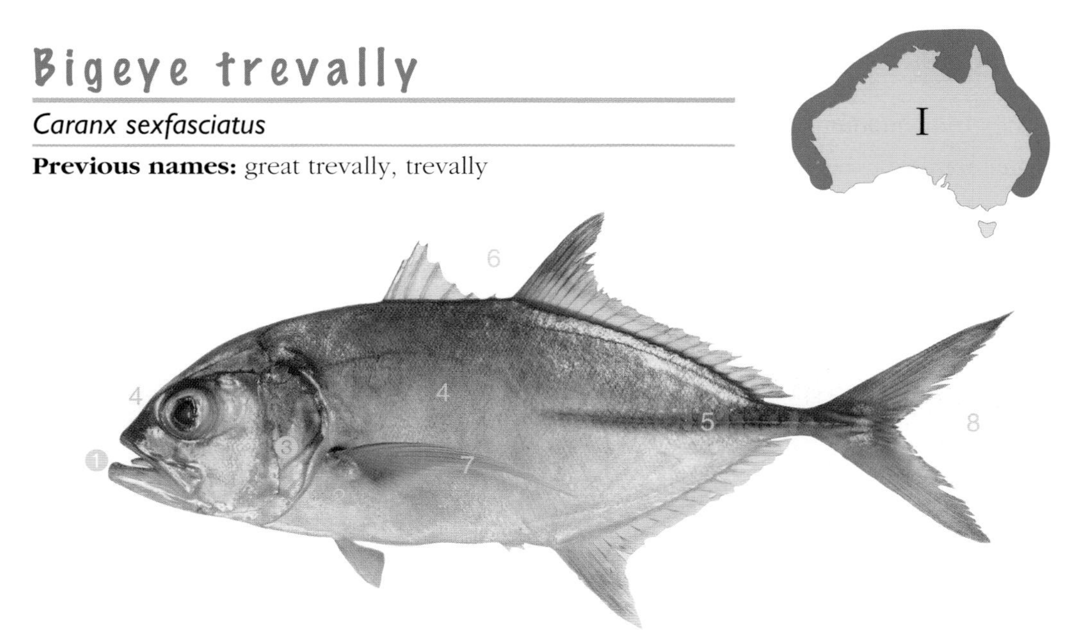

Identifying features: ① teeth in lower jaw enlarged, conical, in single row; ② scales present on breast; ③ gill rakers on outer arch 22–25; ④ body not particularly deep, profile convex in front of eyes; ⑤ 29–34 (mostly black) scute-like scales at end of lateral line; ⑥ dorsal fins separate, first with 8 spines, second usually with white tip; ⑦ pectoral fins very elongated and scythe-like; ⑧ deeply forked caudal fin on a short, slender caudal peduncle.

Comparisons: Medium-sized trevally that has been incorrectly referred to as 'turrum' and 'great trevally' through confusion with the monster of the group, the 'giant trevally' (*Caranx ignobilis*, p. 265). It differs from the turrum (*Carangoides fulvoguttatus*, p. 274) in having a bigger eye and large scutes extending along the entire length of the straight part of the lateral line (rather than smaller scutes almost confined to the caudal peduncle). The giant trevally lacks dark spots on the edge of the operculum and black scutes along the lateral line.

Fillet: Moderately deep, rather elongate, tapering to narrow peduncle, convex above, reddish-brown. Outside with broad, continuous, pronounced central red muscle band; convex EL, converging HL; HS along middle of fillet; integument silvery grey above, silvery white below. Inside flesh medium; belly flap sometimes present; peritoneum translucent. Usually skinned, skin thin, scale pockets small and irregular.

Size: To about 84 cm and at least 5.0 kg (commonly up to 65 cm and about 3.0 kg).

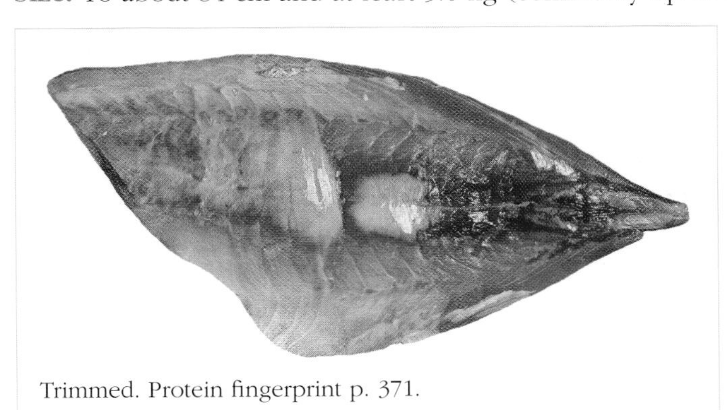

Trimmed. Protein fingerprint p. 371.

Habitat: Marine; pelagic, young common inshore over tidal flats, with adults aggregating near deep drop-offs.

Fishery: Young caught inshore in small quantities using gillnets and tunnel nets. Adults taken by lining or trolling often in large quantities.

Remarks: The firm flesh varies from bland in large fish to delicately flavoured in young.

Black kingfish

Rachycentron canadum

Previous names: cobia, crabeater, sergeant fish

Identifying features: 1 black with white or yellow stripes; 2 head broad and depressed; 3 dorsal-fin spines very short, not connected to each other by a membrane; 4 body very elongate, robust, shark-like; 5 no scute-like scales on lateral line; 6 soft dorsal-fin and anal-fin bases long; 7 pectoral fins not scythe-like; 8 caudal-fin margin rounded to lunar-shaped.

Comparisons: Distinctive relative of the trevallies with a strong, very elongate shark-like body and which lacks scutes along the lateral line. The sucker fish (*Remora remora*) is similar in external appearance but has a sucking disc on top of the head. The black kingfish is quite different in appearance to its namesake the yellowtail kingfish (*Seriola lalandi*, p. 275).

Fillet: Slender, elongate, tapering gently, almost straight above, yellowish-white. Outside with continuous, pronounced central red muscle band; parallel EL, weakly converging HL; HS along middle of fillet; integument translucent above, white below. Inside flesh medium; belly flap sometimes present; peritoneum silvery translucent. Usually skinned, skin very thick, scale pockets minute and embedded.

Size: To almost 200 cm and 70 kg (commonly 65–125 cm and 3–20 kg).

Habitat: Marine; wide-ranging pelagic of tropical seas, extending from inshore in lower estuaries and bays to well offshore around coral reefs. Occurs singly or often aggregates near bottom structures such as wharves, pinnacles and wrecks.

Fishery: Caught as bycatch of coastal net and line fisheries. Significant quantities taken by recreational anglers.

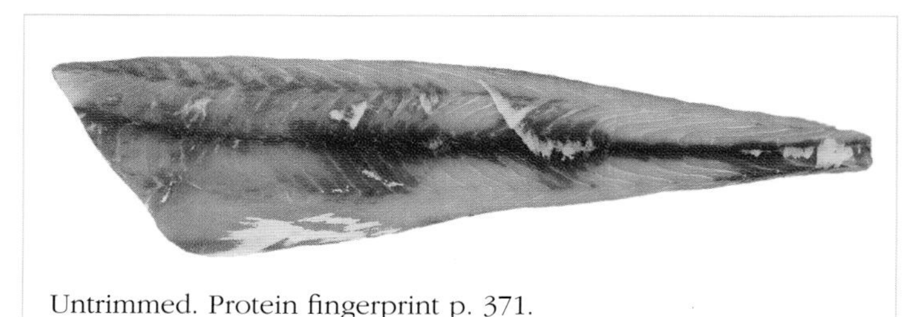

Untrimmed. Protein fingerprint p. 371.

Remarks: Excellent fighter and highly regarded by anglers. A voracious feeder, eating a variety of prey including crabs, squids, reef fishes and even small mackerels and tunas. Superb table-fish with firm, sweet, whitish flesh.

Black pomfret

Parastromateus niger

Previous names: none

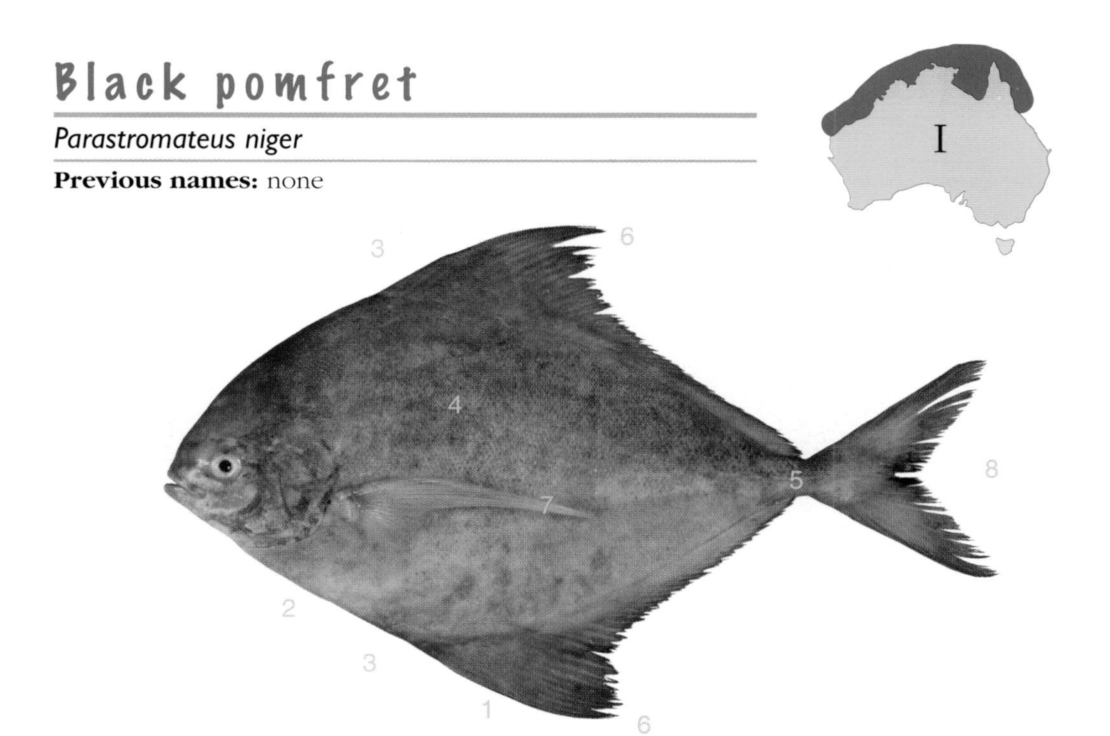

Identifying features: ① no detached anal-fin spines; ② pelvic fins absent in adults; ③ symmetry of dorsal and ventral profiles almost identical; ④ body very deep, compressed; ⑤ weak scute-like scales confined to caudal peduncle; ⑥ dorsal and anal fins tallest anteriorly; ⑦ pectoral fins very elongated and scythe-like; ⑧ deeply forked caudal fin with short, slender caudal peduncle.

Comparisons: Unique member of the trevally family. It closely resembles some relatives of the trevallas (Centrolophidae), the commercially important Asian pomfrets (*Pampus* species). However, unlike pomfrets, it has flat spikey scales on the caudal peduncle typical of trevallies and a dark body (rather than being silvery white).

Fillet: Deep, short, profile convex above and below, reddish-brown. Outside with broad, continuous, very pronounced central red muscle band; converging EL, converging HL; HS along middle of fillet; integument whitish. Inside flesh medium; belly flap sometimes present; peritoneum silvery translucent. Usually skinned, skin very thick, scale pockets small and irregular.

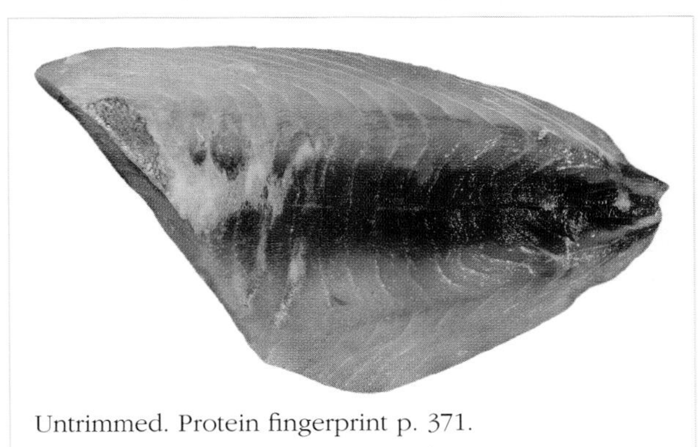

Untrimmed. Protein fingerprint p. 371.

Size: To 55 cm and 2.6 kg (commonly 25–35 cm and 0.3–0.8 kg).

Habitat: Marine; surface dwelling pelagic of tropical seas, commonly occurring inshore and often in very shallow water.

Fishery: Caught mainly as bycatch of coastal net fisheries.

Remarks: Considered excellent eating. Often sold along with pomfrets in Asia and considered to be one of the premium foodfish of the region.

Black trevally

Caranx lugubris

Previous names: none

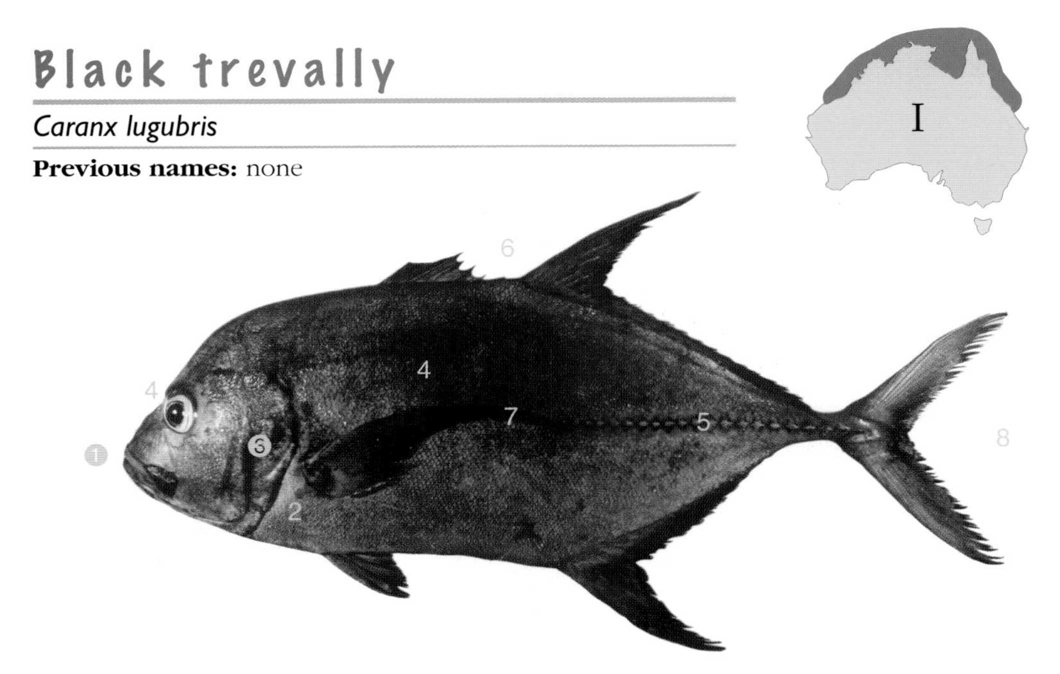

Identifying features: 1 teeth in lower jaw enlarged, conical and in single row; 2 scales present on breast; 3 gill rakers on outer arch 25–28; 4 body dusky brown to black, deep with steep forehead indented in front of eyes; 5 27–30 strong scute-like scales at end of lateral line; 6 dorsal fins well separated, first with 8 spines, second without white tip; 7 pectoral fins very elongated and scythe-like; 8 deeply forked caudal fin with a short, slender caudal peduncle.

Comparisons: Similar to other large tropical trevallies but has a strikingly darker body and fins and very steep forehead with a distinctive indentation forward of the eyes. The giant trevally (*C. ignobilis*, p. 265) is somewhat similar in body shape but is much paler in colour. The bluefin trevally (*C. melampygus*, p. 261) is also dark-bodied but is often more brass-coloured with bluish fins, is usually peppered with small dark spots, and has more numerous, smaller scutes along the lateral line (30–40 versus 27–30). The lateral line of the black trevally is also unusual in having whitish tips on the posterior scutes and in falling very steeply below the origin of the second dorsal fin.

Fillet: Not available.

Size: To about 80 cm and about 7 kg (commonly up to 65 cm and 3.5 kg).

Habitat: Marine; pelagic over inner continental shelf and offshore coral reefs. Large adults often solitary in the company of sharks and rays, young aggregate in huge schools near reef drop-offs.

Fishery: Caught in small quantities, mainly off northern Australia by gillnetting, lining and demersal trawling. Taken occasionally by anglers and spear fishers.

Remarks: Less commercially important than most of the other large tropical trevallies. Landings seem to be spasmodic or more seasonally-based. Other information is provided for protein fingerprints (p. 371).

Bluefin trevally

Caranx melampygus

Previous names: none

Identifying features: 1 teeth in lower jaw enlarged, conical and in single row; 2 scales present on breast; 3 gill rakers on outer arch 25–28; 4 body dark grey or brassy often peppered with small black spots, deep with steep straight forehead in front of eyes; 5 30–40 strong scute-like scales at end of lateral line; 6 dorsal fins bluish well separated, first with 8 spines, second without white tip; 7 pectoral fins very elongated and scythe-like; 8 deeply forked caudal fin with a short, slender peduncle.

Comparisons: Similar in appearance to other large tropical trevallies but has a dark, brass-coloured or greyish body and dark bluish-grey fins. The black trevally (*C. lugubris*, p. 260) is also dark bodied but has less numerous larger scutes along the lateral line (27–30 versus 30–40), and lacks peppery spots and bluish fins (usually dark grey or black).

Fillet: Deep, short, tapering, often to narrow peduncle, convex above, yellowish-white. Outside with broad, continuous, pronounced central red muscle band; convex EL, converging HL; HS along middle of fillet; integument silvery. Inside flesh medium; belly flap sometimes present; peritoneum white. Usually skinned, skin thin, scale pockets small.

Size: To about 70 cm and about 4 kg (commonly up to 55 cm and 1.7 kg).

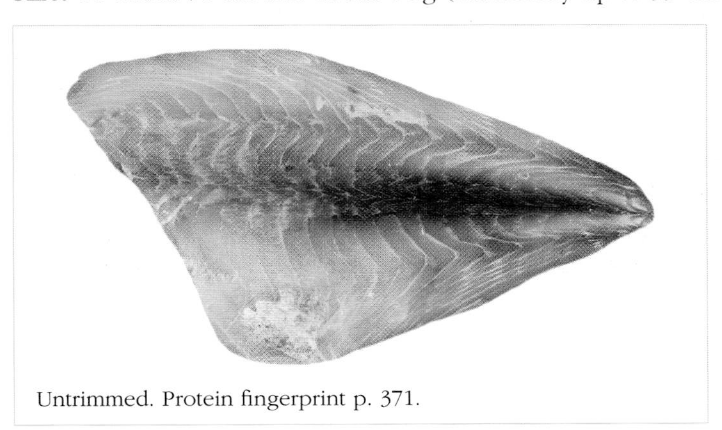

Untrimmed. Protein fingerprint p. 371.

Habitat: Marine; pelagic, young commonly in small schools along the coast over shallow banks, adults around deeper offshore reefs.

Fishery: Small quantities taken mainly off Queensland using gillnets and beach seines; adults taken by lining.

Remarks: Good, firm, delicately flavoured foodfish and among the most important of the tropical trevallies.

Blue-spotted trevally

Caranx bucculentus

Previous names: none

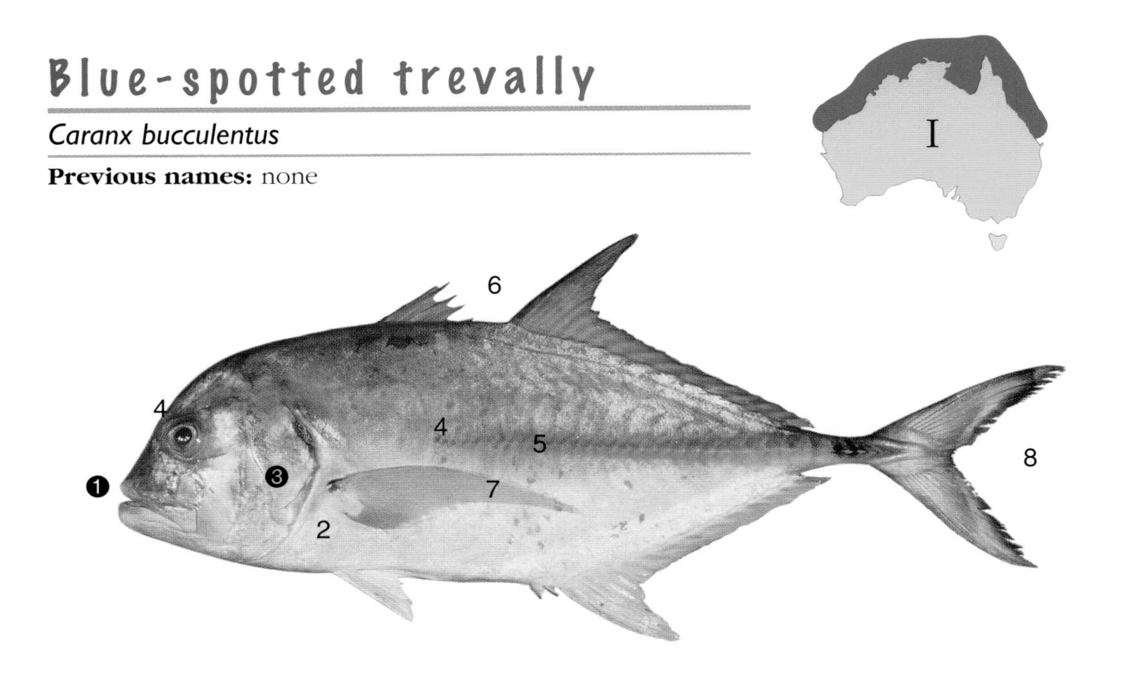

Identifying features: 1 teeth in lower jaw enlarged, conical and in single row; 2 no scales on breast below pectoral fin; 3 about 26 gill rakers on outer arch; 4 body deep with very steep forehead; 5 35–38 strong scute-like scales on lateral line extend forward almost to pectoral-fin base; 6 dorsal fins well separated, first with 8 spines; 7 pectoral fins very elongated and scythe-like; 8 deeply forked caudal fin with short, slender caudal peduncle.

Comparisons: Variable in body shape with deep-bodied young and more elongate adults. Similar to the brassy trevally (*C. papuensis*) but has a relatively short curved anterior portion of the lateral line with a very long, straight, scute-covered posterior portion. A prominent black spot at the top of the pectoral-fin base is also distinctive within the genus.

Fillet: Moderately deep, rather elongate, bottle-shaped, tapering to narrow peduncle, pale pinkish. Outside with broad, continuous, pronounced central red muscle band; convex EL, converging HL; HS along middle of fillet; integument silvery grey above, more whitish below. Inside flesh medium; belly flap rarely present; peritoneum silvery white to translucent. Usually skinned, skin medium, scale pockets small and indistinct.

Size: To about 66 cm and about 3.5 kg (commonly marketed at 30–40 cm and 0.3–0.6 kg).

Trimmed. Protein fingerprint p. 371.

Habitat: Marine; pelagic to demersal over both hard and soft substrates.

Fishery: Juveniles caught inshore but adults sometimes taken in quantity as bycatch of demersal trawling.

Remarks: Considered good eating, the flesh is firm, varying from bland to delicately flavoured.

Dart

Trachinotus species

Previous names: common dart, oyster cracker, pumpkin fish, swallowtail dart

Trachinotus botla

Identifying features: ① fine granular teeth in both jaws; ② scales small, embedded in skin; ③ anterior rays of anal and dorsal fin greatly extended; ④ body deep and angular to oval, forehead not steep; ⑤ no scute-like scales at end of lateral line; ⑥ 5–6 short dorsal spines, not connected to each other by membrane; ⑦ pectoral fins short; ⑧ large, swallow-tailed caudal fin with short caudal peduncle.

Comparisons: Group consisting of six species that are distinctive among the trevallies in body shape and in lacking scutes in the lateral line. Snubnose dart (*T. blochii*) and juvenile giant oystercracker (*T. anak*) each have a deep, angular body similar to diamond trevallies (*Alectis* species, p. 264) but have longer fins and the eye is much closer to the mouth. Most of the other species, such as the largespot dart (*T. botla*), are more oval and compressed in body shape, with a row of spots or blotches along the side.

Fillet: *T. botla* deep, short, tapering, often to narrow peduncle, convex above, yellowish-white. Outside with feeble, continuous central red muscle band; convex EL, converging HL; HS along middle of fillet; integument pale greyish above, white below. Inside flesh medium; belly flap sometimes present; peritoneum silvery white to translucent. Usually skinned, skin thick, scale pockets small and defined.

Size: To about 120 cm and 18 kg (commonly marketed at 35–75 cm and 0.6–3.5 kg).

Habitat: Marine; wide-ranging in pelagic habitats from sheltered bays and estuaries to surf beaches. Also schools offshore around reefs and cays.

Untrimmed. Protein fingerprint p. 371.

Fishery: Caught mainly using seines, gillnets and tunnel nets. Important recreational fishes.

Remarks: Some species have strong bones inside their mouth capable of cracking open oysters and other shellfishes. Very good eating with firm, slightly dry flesh and a characteristic flavour.

Diamond trevally

Alectis species

Previous names: mirrorfish, pennantfish

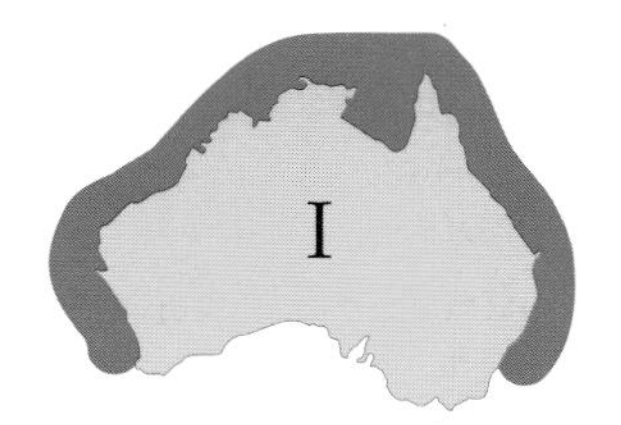

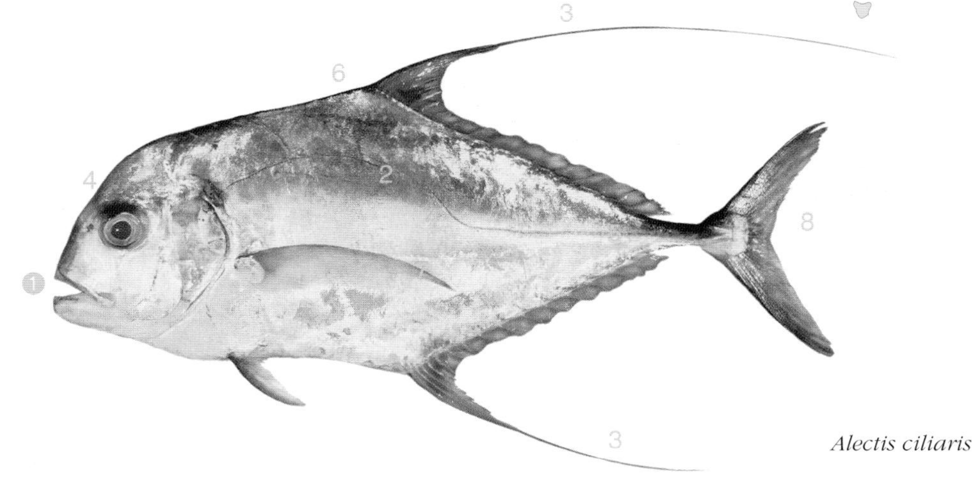

Alectis ciliaris

Identifying features: 1 fine granular teeth in both jaws; 2 scales minute, embedded, skin smooth; 3 anterior rays of anal and dorsal fin long and filamentous; 4 body very deep and angular with very steep forehead; 5 feeble scute-like scales at end of lateral line; 6 dorsal fins well separated, first with 5–6 spines (often embedded and inconspicuous); 7 pectoral fins very elongated and scythe-like; 8 small, forked caudal fin on a short, slender caudal peduncle.

Comparisons: Group consisting of two easily identifiable trevallies. They have a deep angular and compressed body shape with a steep snout, the eye well away from the mouth, and small scutes on a slender caudal peduncle. Some darts (*Trachinotus* species, p. 263) have a similar body shape but have a relatively longer caudal fin and the eye is much closer to the mouth.

Fillet: *A. ciliaris* deep, short, strongly tapering, often to narrow peduncle, convex above, yellowish-white. Outside with broad, continuous, very pronounced central red muscle band; convex EL, weakly converging HL; HS along middle of fillet; integument white. Inside flesh medium; belly flap sometimes present; peritoneum silvery white to translucent. Usually skinned, skin medium, scale pockets minute.

Size: Possibly to about 150 cm and in excess of 40 kg (usually marketed to 100 cm and 13 kg).

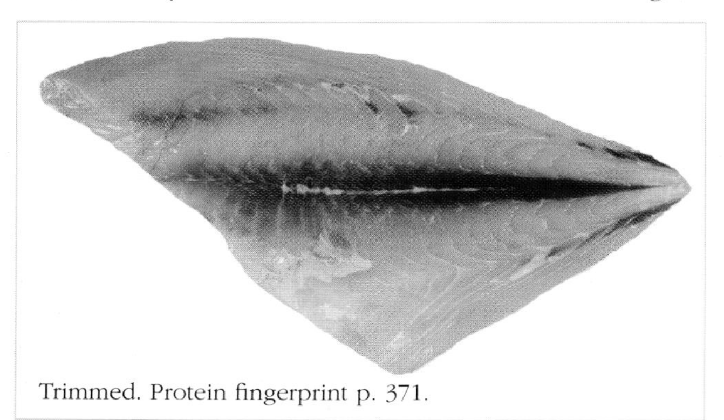

Trimmed. Protein fingerprint p. 371.

Habitat: Marine; pelagic, wide-ranging in habitats from sheltered bays and estuaries to surf beaches. Also schools offshore around coral reefs and cays.

Fishery: Caught mainly using tunnel nets and seines inshore and lines offshore.

Remarks: Well regarded angling species but less well accepted as a foodfish. The flesh is of moderate flavour.

Giant trevally

Caranx ignobilis

Previous name: lowly trevally

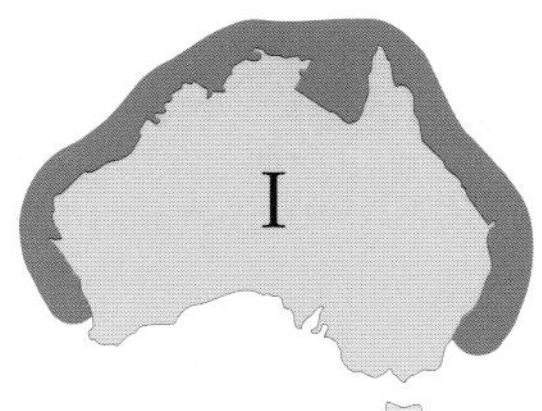

Identifying features: 1 teeth in lower jaw enlarged, conical and in single row; 2 no scales on breast (except immediately before pelvic fins); 3 gill rakers on outer arch 21–24; 4 body very deep with very steep forehead; 5 29–35 strong scute-like scales at end of lateral line; 6 dorsal fins well separated, first with 8 spines; 7 pectoral fins very elongated and scythe-like; 8 deeply forked caudal fin on a short, slender caudal peduncle.

Comparisons: Largest of the tropical trevallies, it has been incorrectly referred to as 'turrum', a name usually reserved for *Carangoides fulvoguttatus* (p. 274). It differs from the turrum in having fewer dorsal-fin soft rays (18–21 versus 25–30) and large scutes extending along the entire length of the straight part of the lateral line (rather than smaller scutes almost confined to the caudal peduncle). It has a distinctive oval patch of scales amid an otherwise scaleless breast, just forward of the pelvic fins.

Fillet: Deep, rather elongate, tapering abruptly, upper profile strongly convex, yellowish-white to brownish. Outside with very pronounced, continuous, central red muscle band; convex EL, converging HL; HS along middle of fillet; integument greyish above, whitish below. Inside flesh medium; belly flap sometimes present, peritoneum silvery white to translucent. Usually skinned, skin thick, scale pockets small and well defined.

Size: To about 170 cm and 62 kg (commonly up to 120 cm and about 30 kg).

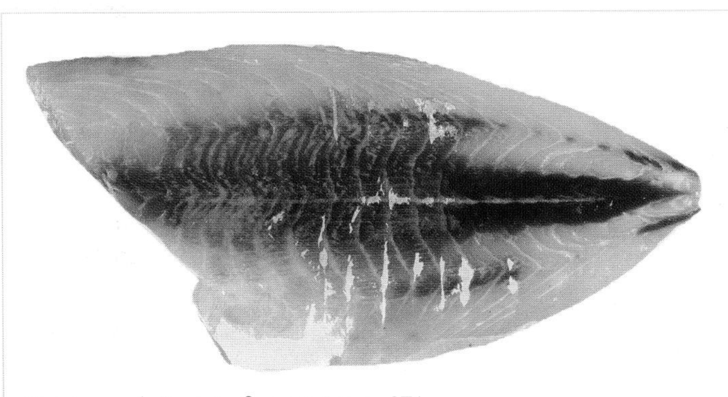

Untrimmed. Protein fingerprint p. 371.

Habitat: Marine; pelagic, adults often solitary or in small groups, over continental shelf and oceanic coral reefs.

Fishery: Young caught inshore in small quantities by gillnets and tunnel nets. Adults taken by lining.

Remarks: Considered an excellent gamefish. Smaller fish below about 15 kg are considered good eating. Flavour strong.

Golden trevally

Gnathanodon speciosus

Previous name: trevally

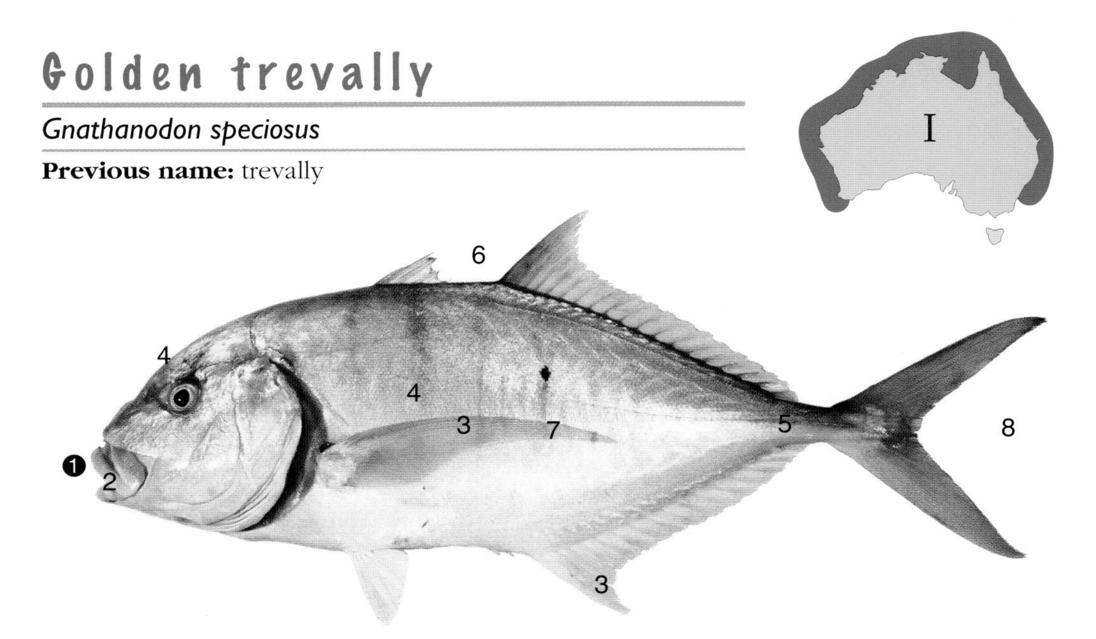

Identifying features: 1 no teeth in jaws of specimens exceeding 80 cm; 2 lips fleshy; 3 fins distinctly yellowish; 4 body oval, forehead not steep, young with dark narrow bands; 5 about 14 feeble, scute-like scales at end of lateral line; 6 dorsal fins well separated, first with 8 spines; 7 pectoral fins very elongated and scythe-like; 8 caudal fin deeply forked with a short, slender caudal peduncle.

Comparisons: Among the most distinctive of the large trevallies with a striking golden pectoral fin (usually also golden over its breast), and unusual mouth with fleshy lips and poorly developed teeth. Young appear more intensely golden with dark bars through eye and down the sides.

Fillet: Deep, short, tapering to narrow peduncle, convex above, yellowish-white. Outside with broad, continuous, pronounced central red muscle band; red muscle at top of shoulder; convex EL, converging HL; HS along middle of fillet; integument silvery. Inside flesh medium; belly flap sometimes present; peritoneum silvery white to translucent. Usually skinned, skin medium, scale pockets small and defined.

Size: To about 120 cm and 37 kg (commonly up to 90 cm and 8 kg).

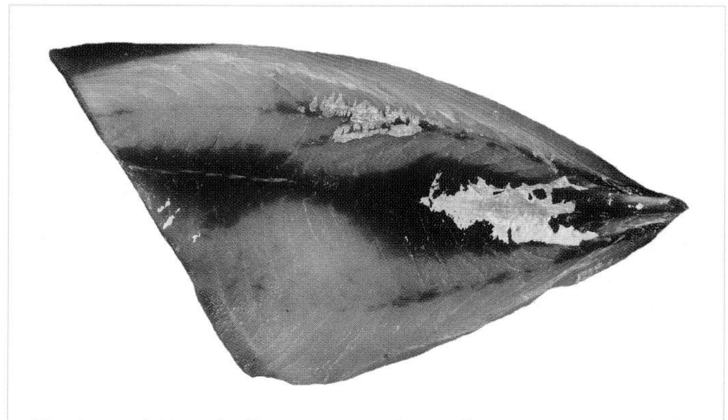

Untrimmed. Protein fingerprint p. 371; oil composition p. 405.

Habitat: Marine; pelagic, in a variety of habitats from inshore to outer reefs. Large adults often solitary over seagrasses in very shallow water.

Fishery: Caught inshore using gillnets, seine nets, and tunnel nets, and also offshore by trolling and trawling.

Remarks: Marketed widely in the tropics as whole fresh carcasses or as fillets. An excellent foodfish with fine textured, sweet flavoured flesh.

Jack mackerel

Trachurus declivis & *T. murphyi*

Previous names: cowanyoung, horse mackerel, scad

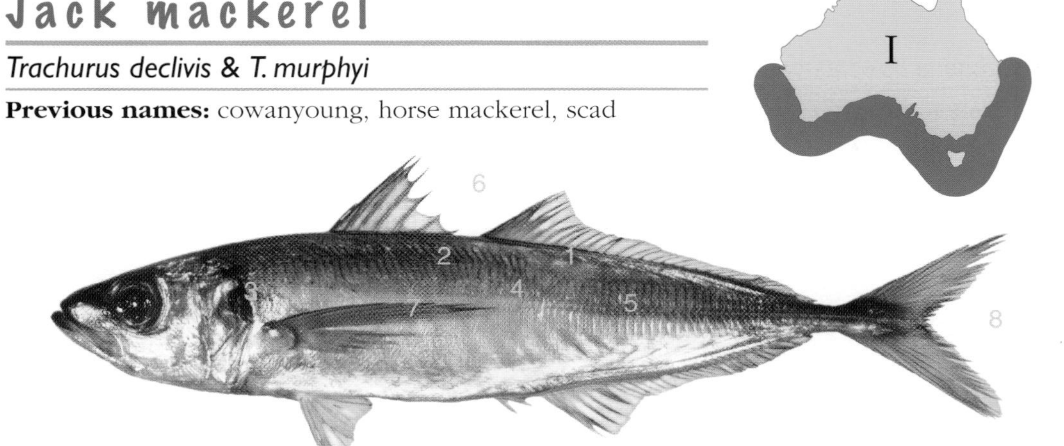

Trachurus declivis

Identifying features: ① upper lateral line ending below dorsal-fin rays 7–9 (in *T. declivis*); ② rarely fewer than 76 scutes in lateral line; ③ black spot on operculum edge above pectoral-fin base; ④ body elongate, with relatively pointed head; ⑤ large scute-like scales along all of lateral line; ⑥ dorsal fins separated slightly, first with 8 spines; ⑦ pectoral fins very elongate; ⑧ deeply forked caudal fin on a short, slender caudal peduncle.

Comparisons: Includes two species that are more slender than most of their relatives and have tall scutes extending along the entire length of their lateral line. The oceanic Peruvian jack mackerel (*T. murphyi*) has more scutes (always more than 90 versus about 80) and more gill rakers (58–63 versus 53–55 on the outer arch) than the common jack mackerel (*T. declivis*). The smaller yellowtail scad (*T. novaezelandiae*, p. 276) has a much shorter upper lateral line that ends under the first few rays of the second dorsal fin and fewer scutes in the lateral line (usually fewer than 76).

Fillet: *T. declivis* moderately deep, rather elongate, tapering gently, weakly convex above, reddish-brown. Outside with broad continuous, very pronounced central red muscle band; weakly convex EL, converging HL; HS along middle of fillet; integument greyish-blue above, white below. Inside flesh medium; belly flap sometimes present; peritoneum pale translucent with melanophores. Sometimes skinned, skin thin, scale pockets small and variable.

Size: To at least 64 cm and about 1.6 kg (commonly 25–40 cm and 0.2–0.6 kg).

Habitat: Marine; pelagic throughout the water column, schooling over the continental shelf and margin. Thought to occasionally descend to about 500 m depth.

Fishery: Caught mainly by purse seine in Tasmanian coastal waters and as bycatch of demersal and midwater trawls. Mostly used for fishmeal, bait or petfood, with small quantities reaching seafood markets.

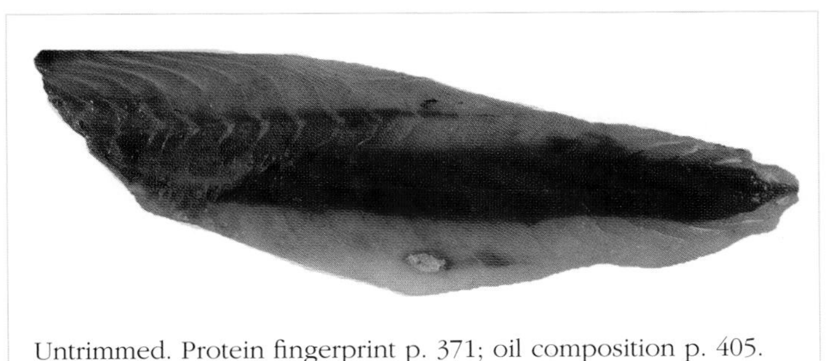

Untrimmed. Protein fingerprint p. 371; oil composition p. 405.

Remarks: Forms huge shoals off the southern coast in summer where the relatively larger Peruvian jack mackerel is taken in small quantities as bycatch. The flesh is somewhat dry and oily.

Mahi mahi

Coryphaena species

Previous name: dolphinfish, dorado

Coryphaena hippurus

Identifying features: 1 dorsal fin originating just behind eye; 2 sides usually yellowish with peppering of black spots; 3 body elongate, very compressed, deepest anteriorly; 4 no scute-like scales on lateral line; 5 distinctly hump-headed; 6 pectoral fins short; 7 large, deeply forked caudal fin with a short, slender caudal peduncle; 8 no true spines in anal fin.

Comparisons: Exquisite fishes unlikely to be confused with any other. Despite some misconceptions they are totally unrelated to dolphins (Cetacea). The raised head, brilliant turquoise, yellow and silver tapering body and fins are characteristic. The dolphinfish, *C. hippurus*, is most common. A second smaller species, the pompano dolphinfish (*C. equiselis*), has a patch of scales covering the full width of the tongue (rather than about half its width).

Fillet: *C. hippurus* narrow, elongate, tapering gradually, pale pinkish. Outside with sharp-edged, pronounced, continuous central red muscle band; weakly convex EL, weakly converging HL; HS along middle of fillet; integument translucent. Inside flesh coarse; belly flap rarely present; peritoneum silvery white to translucent; red muscle evident as red spots and along midline. Usually skinned, skin thin, scale pockets small and indistinct.

Size: To about 200 cm and almost 45 kg (commonly to 160 cm and seldom more than 25 kg).

Habitat: Marine; wide-ranging warm water pelagics occurring mainly in small schools, often in the vicinity of flotsam and jetsam, from well off the coast to the surface of the open ocean.

Fishery: Caught trolling using bait or lures, especially around floating debris or other 'attractants' including trap buoy lines. Increasingly targeted by recreational anglers.

Remarks: The dolphinfish has a rapid growth-rate and is considered an ideal candidate for aquaculture. Considered to be a brilliant gamefish, it is a rapid swimmer capable of leaping clear of the surface. Excellent foodfishes revered overseas and whose popularity locally is increasing. The flesh is tasty, with a firm texture.

Trimmed. Protein fingerprint p. 371; oil composition p. 405.

Queenfish

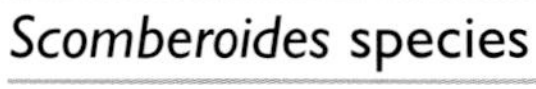

Scomberoides species

Previous names: giant dart, leatherskin, skinnyfish, talang queenfish

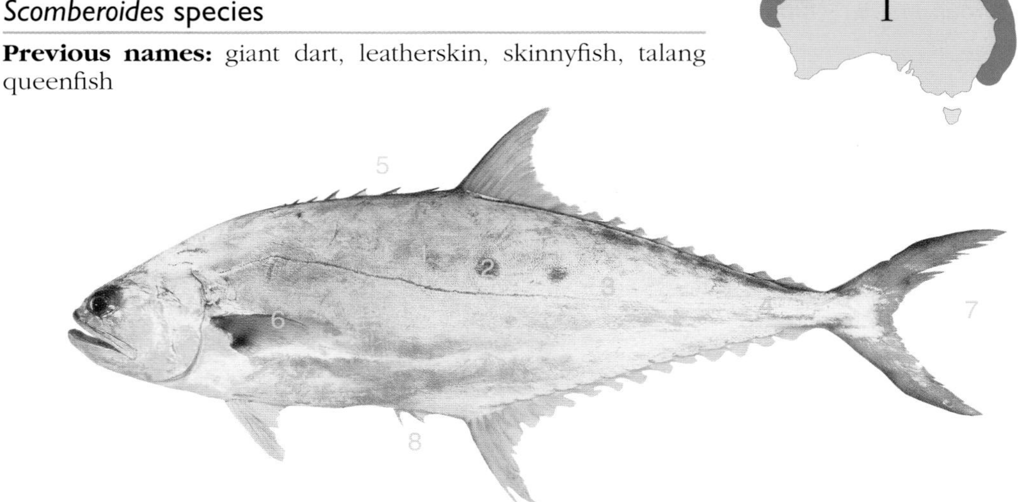

Scomberoides commersonnianus

Identifying features: ① skin leathery with needle-like or elongate scales; ② sides with dark spots or blotches (often faint); ③ body elongate, compressed; ④ no scute-like scales on lateral line; ⑤ dorsal-fin spines short and isolated; ⑥ pectoral fins short; ⑦ large, deeply forked caudal fin with a short, slender caudal peduncle; ⑧ 2 anal-fin spines well forward of main fin.

Comparisons: Group contains four similar species that more closely resemble mackerels than other trevallies. However, they differ from the mackerels (Scombridae) in lacking true finlets behind the dorsal and anal fins (hind rays of soft dorsal fin may resemble finlets) and do not have keels at the base of the tail.

Fillet: *S. commersonnianus* moderately deep, rather elongate, tapering gently, slightly convex above, reddish-brown. Outside with broad continuous, very pronounced central red muscle band; weakly convex EL, converging HL; HS along middle of fillet; integument patchy, white. Inside flesh medium; belly flap sometimes present; peritoneum silvery white to translucent; red muscle evident as red spots and along midline. Sometimes skinned, skin medium, scale pockets small and undetectable.

Size: To about 120 cm and at least 14.3 kg (commonly 50–100 cm and 1–7 kg).

Habitat: Marine; tropical pelagic occurring mainly alone or in small schools over the continental shelf, often in very shallow water.

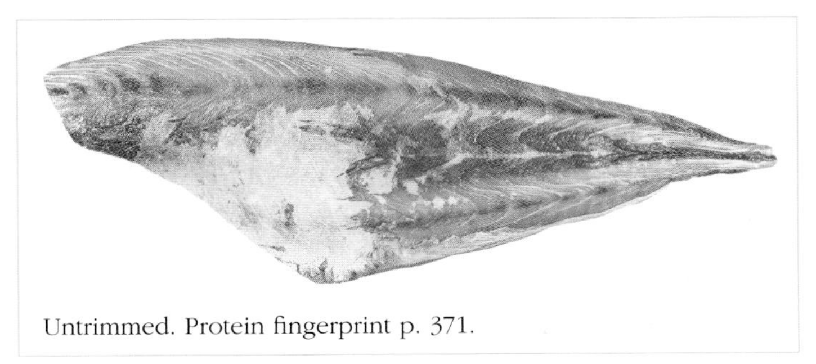

Untrimmed. Protein fingerprint p. 371.

Fishery: Caught using gill-nets, and by lining or trolling.

Remarks: Voracious feeders and strong fighters, large fish are revered by anglers. Fillets tend to be so thin that small fish are not preferred. Otherwise have firm flesh with an excellent flavour but can sometimes be dry.

Samson fish

Seriola dumerili & S. hippos

Previous names: amberjack, samson

Seriola hippos

Identifying features: 1 back mainly greyish-purple with a yellowish stripe through midline; 2 dorsal fin with 6–8 spines, 23–33 soft rays; 3 anal fin with 16–22 soft rays; 4 caudal fin olive to grey; 5 body elongate and robust; 6 no scute-like scales on lateral line; 7 pectoral fins short; 8 forked caudal fin with short peduncle.

Comparisons: Two extremely similar commercial species of the genus *Seriola* can be sold as 'samson fish'. The true samson fish (*S. hippos*) has fewer dorsal-fin soft rays (22–25 versus 32–33) and anal-fin rays (16–17 versus 19–22 rays) than the amberjack (*S. dumerili*). The amberjack, in particular, closely resembles the yellowtail kingfish (*S. lalandi*, p. 275) but has a relatively deeper head and body, and an olive or reddish-grey (rather than yellowish) tail.

Fillet: *S. hippos* moderately deep, rather elongate, tapering to narrow peduncle, weakly convex above, pale pinkish. Outside with continuous, intermediate central red muscle band; parallel EL, weakly converging HL; HS along middle of fillet; integument silvery grey above, white below. Inside flesh medium; belly flap sometimes present; peritoneum translucent. Usually skinned, skin thick, scale pockets small and defined.

Size: To 170 cm and 53 kg (commonly marketed at 50–100 cm and 1–15 kg).

Habitat: Marine; pelagic in warm temperate and tropical waters, mainly aggregating in small groups over the inner continental shelf.

Fishery: Samson fish are mostly caught off temperate Western Australia using handlines and droplines with smaller landings from bycatch of local shark fisheries. Also a common bycatch species of line fishing off southeastern Queensland and northern New South Wales. Smaller catches of amberjack are made offshore in Queensland.

Trimmed. Protein fingerprint p. 371.

Remarks: Important angling species which are best eaten when small.

Silver trevally

Pseudocaranx dentex & P. wrighti

Previous names: sand trevally, silver bream, skipjack trevally, skippy, white trevally

Pseudocaranx dentex

Identifying features: 1 teeth in both jaws in single row (occasionally additional row in front of upper jaw); 2 black spot on operculum edge above pectoral-fin base; 3 sides often with yellowish stripe, but lacking spots or bands; 4 body moderately elongate with relatively pointed head; 5 24–46 scute-like scales at end of lateral line; 6 dorsal fins well separated, first with 8 spines; 7 pectoral fins very elongated and scythe-like; 8 deeply forked caudal fin on a short, slender caudal peduncle.

Comparisons: Similar in appearance to other main commercial trevallies (*Caranx* and *Carangoides* species) but usually have blunt conical teeth in upper jaw confined to a single row (rather than being in multiple rows). The two species included, *P. dentex* and *P. wrighti*, can be distinguished by the number of scutes in the lateral line (34–46 versus 24–35 respectively).

Fillet: *P. dentex* deep, short, tapering to narrow peduncle, strongly convex above, pale reddish-brown. Outside with broad, continuous, very pronounced central red muscle band; convex EL, converging HL; HS along middle of fillet; integument silvery grey above, white below. Inside flesh medium; sometimes with belly flap; peritoneum pale translucent with melanophores; holes above midline. Usually skinned, skin medium, scale pockets small, defined.

Size: To about 94 cm and at least 10 kg (commonly 35–60 cm and 0.4–2.5 kg).

Habitat: Marine; mostly pelagic, young school inshore, with *P. dentex* adults occurring deeper on the shelf.

Untrimmed. Protein fingerprint p. 371.

Fishery: Mainly taken by demersal trawls and traps off New South Wales. Also taken in smaller quantities in estuaries using seines and lines. Important angling species.

Remarks: Sometimes marketed under the confusing name 'silver bream'. Regarded as good foodfishes, usually marketed whole or as fillets.

Tailor

Pomatomus saltatrix

Previous names: bluefish, skipjack

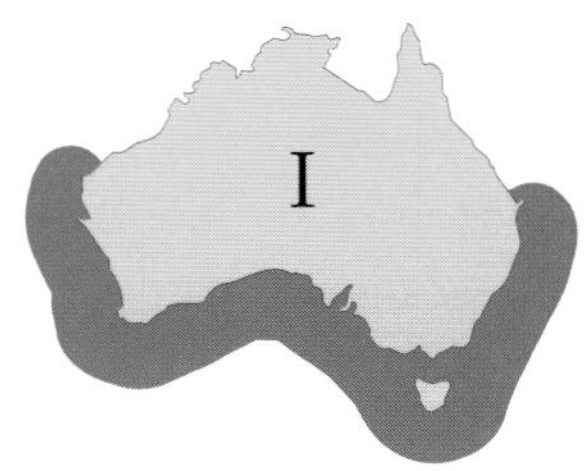

Identifying features: ① 1–2 pairs of canine teeth at the front of each jaw; ② mouth oblique with prominent lower jaw; ③ 7–8 short dorsal-fin spines, depressible into a groove and connected to each other by a membrane; ④ body silvery, elongate and compressed; ⑤ no scute-like scales along lateral line; ⑥ second dorsal fin with 1 spine, 24–26 soft rays; ⑦ pectoral fins short; ⑧ caudal fin margin forked.

Comparisons: The tailor is not a member of the trevally family (Carangidae) although it is often grouped with these fishes. Its body shape, which is unlike the trevallies, resembles that of the Australian salmons (*Arripis* species, pp 56–57) but it has a smaller spinous dorsal fin that is not connected to the rayed portion and is plain-coloured (rather than spotted or with yellow pectoral fins). Its head is similar to that of the teraglin (*Atractoscion aequidens*, p. 154) but lacks the characteristic yellow mouth of the latter.

Fillet: Moderately deep, rather elongate, tapering gently, slightly convex above, yellowish to brownish. Outside with continuous, intermediate central red muscle band; convex EL, converging HL; HS along middle of fillet; integument silvery grey above, white below. Inside flesh coarse; belly flap sometimes present; peritoneum silvery white to translucent. Usually skinned, skin thin, scale pockets small and defined.

Size: To about 120 cm and at least 14 kg (commonly 30–60 cm and 0.4–2.2 kg).

Habitat: Marine; pelagic near the surface off beaches and in bays extending on the shelf to about 50 m depth.

Untrimmed. Protein fingerprint p. 371; oil composition p. 405.

Fishery: The main commercial catches are off southern Queensland using beach seines and gillnets. Very popular target of recreational anglers.

Remarks: Good foodfish but not considered to have a long shelf life. Sold mainly whole fresh or as fillets.

Trevally

Caranginae

Previous names: none

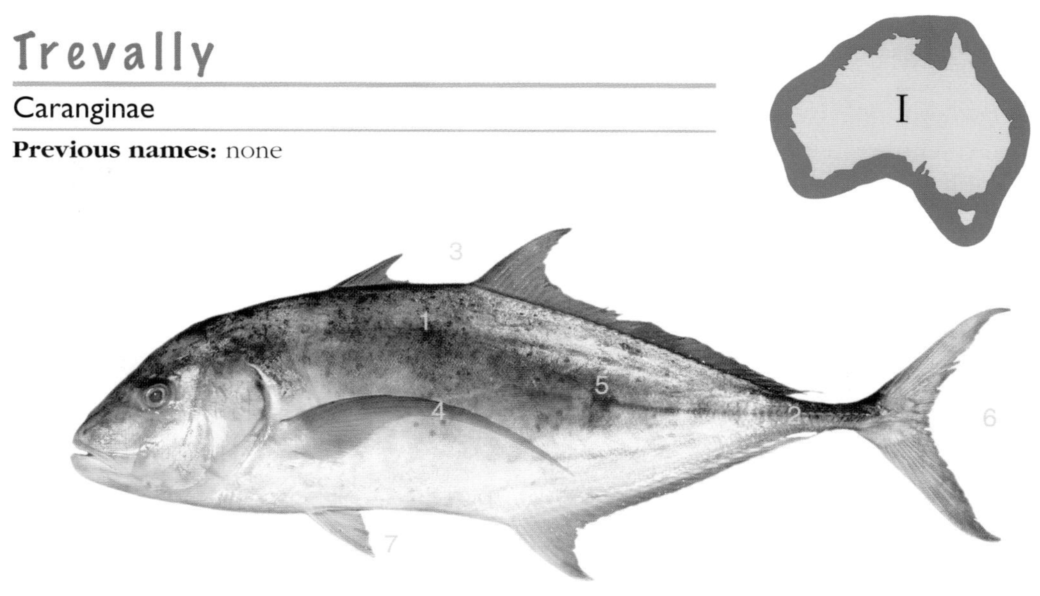

Carangoides gymnostethus

Identifying features: ① body usually silvery, and deep and compressed; ② enlarged, scute-like scales at end of lateral line; ③ usually with 2 dorsal fins, the first dorsal fin with 5–8 spines; ④ pectoral fins mostly elongated, scythe-like; ⑤ body covered in small cycloid scales; ⑥ deeply forked caudal fin on a short, slender peduncle; ⑦ pelvic fins small, base beneath pectoral-fin base.

Comparisons: Members of the trevally family (Carangidae) are also known as 'jacks'. Of these, only members of the diverse subfamily Caranginae may be marketed as 'trevally'. However, some 13 species are considered important enough to be marketed under separate names. Other carangid subfamilies, the amberjacks (Seriolinae), queenfishes (Scomberoidinae), and darts (Trachinotinae), differ in appearance and do not have a row of enlarged scutes along the lateral line.

Fillet: Tend to be rather deep, with pinkish or whitish flesh and a pronounced central red muscle band on the outer fillet. See descriptions of other trevallies.

Size: To at least 170 cm and exceeding 60 kg (adults usually 30–100 cm and 0.3–15 kg).

Habitat: Marine; pelagic, mainly over the tropical continental shelf but some venture further offshore and into southern waters. Most occur in large schools.

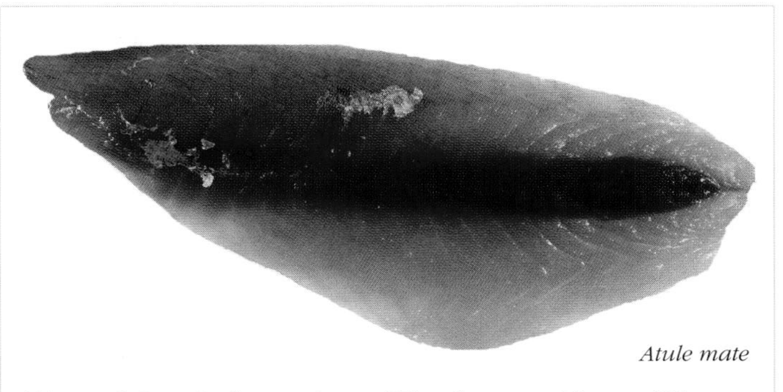

Atule mate

Trimmed. Protein fingerprint p. 371; oil composition p. 405.

Fishery: Most species are edible and are landed in varying quantities as bycatch of other fisheries. Usually caught using nets but some are important angling species.

Remarks: The flesh quality varies between species from being pale and dry to dark and oily.

Turrum

Carangoides fulvoguttatus

Previous names: gold-spotted trevally, yellow-spotted trevally

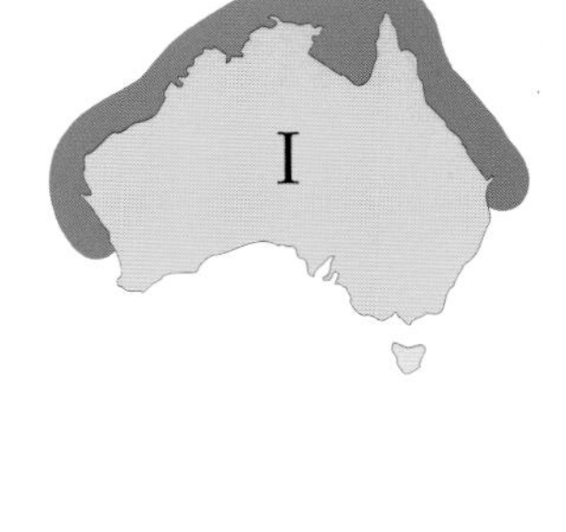

Identifying features: ① teeth in lower jaw small and in more than 1 row; ② scaleless patch on breast extending halfway to pectoral fin; ③ gill rakers on outer arch 24–25; ④ body rather elongate; ⑤ 14–20 scute-like scales at end of lateral line; ⑥ dorsal fins well separated, first with 8 spines; ⑦ pectoral fins very elongated and scythe-like; ⑧ deeply forked caudal fin on a short, slender caudal peduncle.

Comparisons: Can be confused with the bludger trevally (*C. gymnostethus*), which is sold as 'trevally' but has a steeper head with the eye relatively higher above the mouth, a smaller scaleless patch on the breast (about halfway to pectoral-fin base versus reaching the pectoral-fin base), and fewer gill rakers (24–25 on outer arch). The larger giant trevally (*Caranx ignobilis*, p. 265), which is occasionally misnamed as 'turrum', has fewer dorsal-fin soft rays (18–21 versus 25–30).

Fillet: Moderately deep, rather elongate, bottle-shaped, tapering to narrow peduncle, pale reddish-brown. Outside with broad, continuous, pronounced central red muscle band; parallel EL, converging HL; HS along middle of fillet; integument silvery white. Inside flesh medium; belly flap often present; peritoneum silvery white to translucent. Usually skinned, skin thin, scale pockets small and well defined.

Size: To 126 cm and about 12 kg (commonly up to 80 cm and about 5 kg).

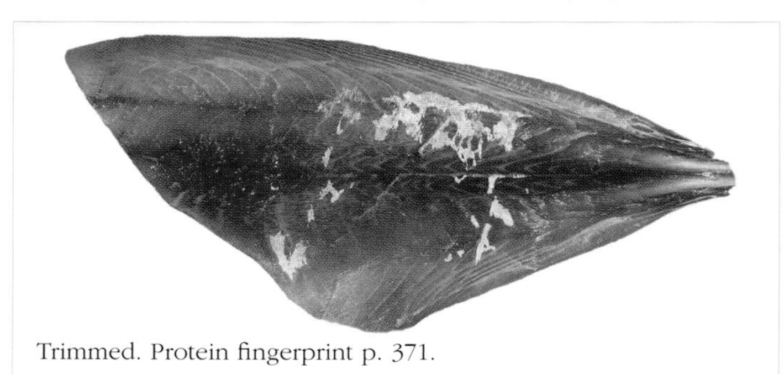

Trimmed. Protein fingerprint p. 371.

Habitat: Marine; pelagic over inner continental shelf and offshore coral reefs.

Fishery: Caught mainly off Queensland by gillnetting, lining and demersal trawling.

Remarks: A sought-after angling fish considered to be among the best eating of the tropical trevallies.

Yellowtail kingfish

Seriola lalandi

Previous names: kingfish, kingie, Tasmanian yellowtail, yellowtail

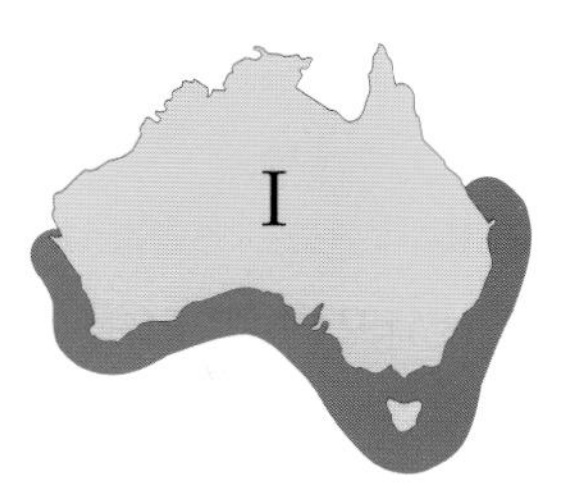

Identifying features: ① back bluish-green with a yellowish stripe through midline; ② dorsal fin with 6–7 spines, 30–37 soft rays; ③ anal fin with 19–21 soft rays; ④ caudal fin yellowish; ⑤ body elongate and robust; ⑥ no scute-like scales on lateral line; ⑦ pectoral fins rather short; ⑧ forked caudal fin with short caudal peduncle.

Comparisons: Members of the genus *Seriola* are distinctive within the trevallies. Resembles related fishes sold as 'samson fish' (*S. dumerili* and *S. hippos*, p. 270) but has a more slender head and body, and a yellowish (rather than olive or greyish) tail. Also has more dorsal-fin soft rays (30–37 versus 22–25) and anal-fin soft rays (19–21 versus 16–17 rays) than *S. hippos*.

Fillet: Moderately deep, rather elongate, tapering gently, weakly convex above, reddish-brown. Outside with continuous, pronounced central red muscle band; parallel EL, weakly converging HL; HS along middle of fillet; integument silvery grey above, white below. Inside flesh coarse; belly flap sometimes present; peritoneum translucent. Usually skinned, skin medium, scale pockets small and barely detectable.

Size: To about 190 cm and 70 kg (commonly up to 100 cm and 10–15 kg).

Habitat: Marine; pelagic, solitary or in small schools mainly near the coast and around offshore islands and reefs in warm temperate water.

Fishery: Caught mainly off New South Wales using handlines with lures or live baits, occasionally taken using droplines and demersal trawls. Also targeted elsewhere in its range. Very significant recreational catches.

Untrimmed. Protein fingerprint p. 371; oil composition p. 405.

Remarks: Important gamefish. Small individuals considered to be excellent eating. Marketed whole, gilled and gutted or as cutlets or fillets. Premium grade fish used for sashimi. Good smoking qualities.

Yellowtail scad

Trachurus novaezelandiae

Previous names: scad, yellowtail, yellowtail horse mackerel

Identifying features: 1 upper lateral line ending below dorsal-fin soft rays 1–2; 2 rarely more than 76 scutes in lateral line; 3 black spot on operculum edge above pectoral-fin base; 4 body elongate, compressed with relatively pointed head; 5 large scute-like scales along all of lateral line; 6 dorsal fins separated slightly, first with 8 spines; 7 pectoral fins very elongate; 8 deeply forked caudal fin with short, slender caudal peduncle.

Comparisons: Similar to a larger species marketed as 'jack mackerel' (*T. declivis*, p. 267) but has a much shorter upper lateral line that ends under the first few soft rays of the second dorsal fin (rather than ending under the 7–9th soft ray) and fewer scutes in the lateral line (usually fewer than 76 rather than usually more than 76). Other jack mackerels have a short lateral line but also have more lateral-line scutes. Tropical relatives known as 'mackerel scads' (*Decapterus* species) have a similar body form but the enlarged scutes are lacking from the curved part of the lateral line.

Fillet: Moderately deep, rather elongate, tapering gently, weakly convex above, reddish-brown. Outside with continuous, very pronounced central red muscle band; convex EL, converging HL; HS along middle of fillet; integument greyish-blue above, white below. Inside flesh medium; belly flap sometimes present; peritoneum white translucent with melanophores. Sometimes skinned, skin thin, scale pockets small and variable.

Size: To about 50 cm and about 1.0 kg (commonly to 30 cm and 0.3 kg).

Habitat: Marine; pelagic, schools mainly inshore over the inner shelf and in estuaries.

Untrimmed. Protein fingerprint p. 371.

Fishery: Main commercial catches off New South Wales using purse seines but also taken using haul nets and demersal trawls. Caught by line in South Australia. Very popular with anglers for use as bait.

Remarks: Used as bait for game-fishing. Marketed whole fresh in Sydney. May have export potential.

Striped trumpeter

Latris lineata

Previous names: common trumpeter, real trumpeter, stripey, Tasmanian trumpeter, trumpeter

Identifying features: ① 3 broad, dark stripes on upper sides; ② pectoral fins rounded, middle rays longest; ③ dorsal fin with 17–18 spines, 34–36 soft rays; ④ anal fin with 3 spines, 31–32 soft rays; ⑤ distinct notch between spinous and soft-ray sections of dorsal fin; ⑥ body elongate and compressed; ⑦ body scales small and numerous.

Comparisons: Distinguished from other trumpeters (*Latridopsis* species, p. 278) in having 3 broad, dark stripes on the upper sides and rounded pectoral fins. Morwongs (Cheilodactylidae) are similar but have much larger pectoral fins (usually with the central rays greatly elongated) and fewer anal-fin soft rays (8–19 versus 31–32).

Fillet: Moderately deep, rather elongate, almost bottle-shaped, off-white to yellowish. Outside with feeble continuous central red muscle band; top of shoulder without red muscle patch; convex EL, converging HL; HS along middle of fillet; integument greyish above, white below. Inside flesh fine; belly flap usually removed; peritoneum black, often with fat layers. Usually skinned, skin medium, scale pockets small and defined.

Size: To 120 cm and 25 kg (commonly to 88 cm and 7 kg).

Habitat: Marine; live around rocky reefs from shallow water to depths of 300 m (most common deeper than 120 m).

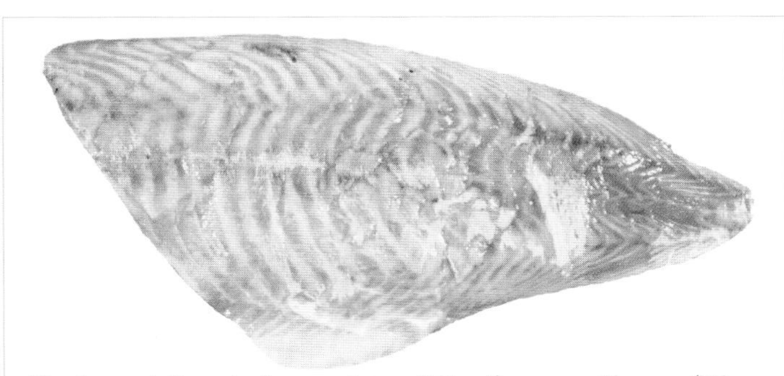

Untrimmed. Protein fingerprint p. 371; oil composition p. 405.

Fishery: Adults taken by line and sometimes in craypots off Tasmania, South Australia and Victoria. Juveniles taken inshore using gillnets. Aquaculture trials are progressing in Tasmania.

Remarks: Highly esteemed as one of the best eating fishes in Australia. The flesh is firm, tasty and fatty.

Trumpeter

Latridopsis species

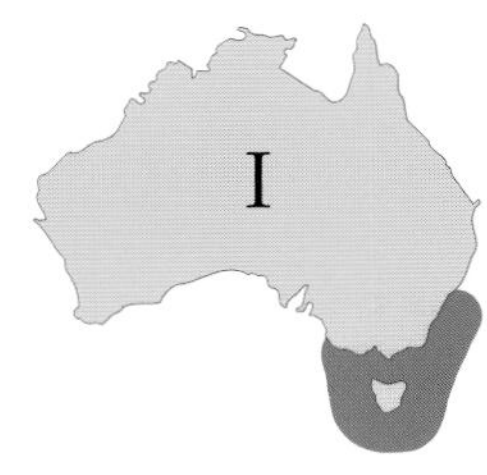

Previous names: bastard trumpeter, blue moki, moki, silver trumpeter, Tassie trumpeter, Tazzie

Latridopsis forsteri

Identifying features: 1 no broad, dark stripes on upper sides but may have several thin golden brown stripes; 2 pectoral fins bluntly pointed, upper rays longest; 3 dorsal fin with 17 spines, 38–40 soft rays; 4 anal fin with 3 spines, 32–35 soft rays; 5 distinct notch between spinous and soft-ray sections of dorsal fin; 6 body moderately elongate and compressed; 7 body scales small and numerous.

Comparisons: Two species. The bastard trumpeter (*L. forsteri*) has numerous thin, golden brown stripes on the upper sides and dark edges to the dorsal, pectoral and caudal fins while a second species, the blue moki (*L. ciliaris*), is uniform grey blue. They both lack the broad, dark stripes of the striped trumpeter (*Latris lineata*, p. 277). Morwongs (Cheilodactylidae) have larger pectoral fins and fewer anal-fin soft rays (8–19 versus 32–35).

Fillet: *L. forsteri* moderately deep, rather elongate, slightly convex above, off-white to pinkish. Outside with broad, very pronounced continuous central red muscle band; convex EL, converging HL; HS along middle of fillet; integument white with melanophores. Inside flesh medium; top of shoulder with inner red muscle patch; belly flap usually removed; peritoneum black. Usually skinned, skin thick, scale pockets small and defined.

Size: Bastard trumpeter reach at least 70 cm and 4.3 kg (commonly to 50 cm and 1.5 kg). Blue moki reach 100 cm and 10 kg.

Habitat: Marine; shallow coastal waters to depths of over 100 m. Juveniles form schools around coastal reefs, whereas adults are more solitary.

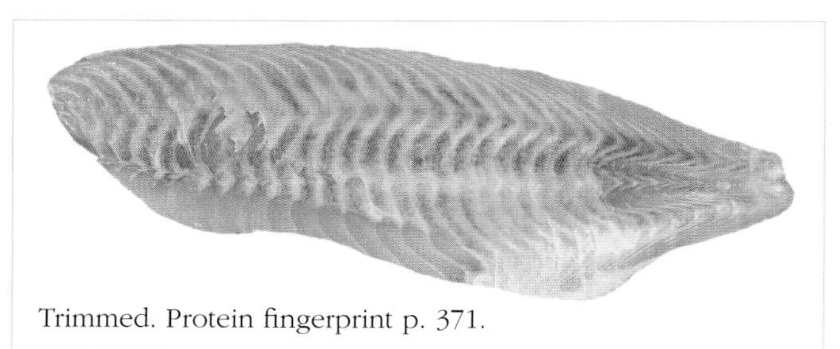

Trimmed. Protein fingerprint p. 371.

Fishery: Caught inshore in gillnets and offshore as trawl and net bycatch, mostly in Tasmanian waters. Speared by sport divers but rarely caught by anglers.

Remarks: Blue moki, predominantly a New Zealand species, is periodically taken in the southeast. Very good eating.

King George whiting

Sillaginodes punctata

Previous names: black whiting, South Australian whiting, spotted whiting

Identifying features: 1 scales very small (129–147 in lateral line); 2 numerous small, dark spots on sides; 3 2 dorsal fins, often slightly separated, the first with 11–13 spines and the second with 1 spine and 25–27 soft rays; 4 caudal fin forked; 5 body elongate with pointed snout.

Comparisons: Only member of the genus *Sillaginodes* and easily distinguished from other whitings (*Sillago* species) by the tiny scales and dark spots on the sides. Differs from the similarly shaped grass whiting (*Haletta semifasciata*, p. 287)—which is not a true whiting—in having 2 usually separate dorsal fins and a forked caudal fin.

Fillet: Moderately deep, elongate, tapering gently, almost straight above, yellowish to greyish with dark veins. Outside with narrow, sharp-edged continuous, central red muscle band; weakly converging EL, weakly converging HL; HS along middle of fillet; integument silvery grey above, silvery white below. Inside flesh fine; belly flap sometimes present; peritoneum greyish. Usually skinned, skin thin, scale pockets small and irregular.

Size: To 72 cm and possibly 4.8 kg (commonly 35–60 cm and 0.3–1.4 kg).

Habitat: Coastal marine; usually in depths less than 50 m but reported to 200 m. Juveniles prefer seagrass beds in shallow, sheltered bays and estuaries, whereas adults are usually found deeper over sand or weed.

Fishery: Seine nets, gillnets or handlines, mainly off South Australia where it is the most valuable scalefish caught. Smaller fisheries exist off Victoria and southwestern Western Australia. Mostly taken in autumn and winter. Popular target of recreational anglers throughout its range. In South Australia the recreational catch is equivalent to about half the commercial catch. Aquaculture experiments are proving very successful.

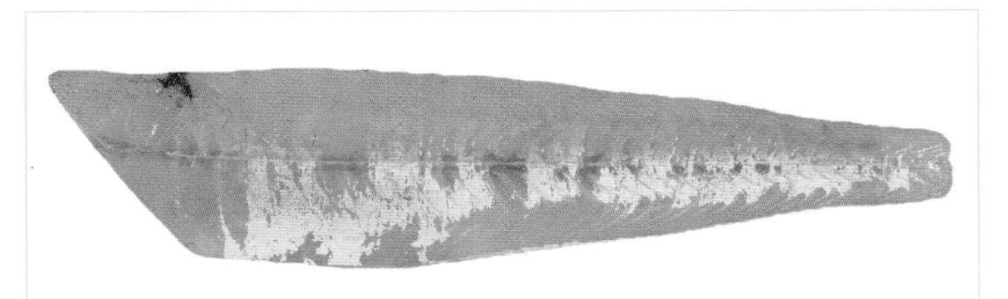

Untrimmed. Protein fingerprint p. 371; oil composition p. 406.

Remarks: Largest whiting and makes excellent eating. The flesh is white, with fine texture and mild flavour.

Sand whiting

Sillago ciliata

Previous names: silver whiting, summer whiting, whiting

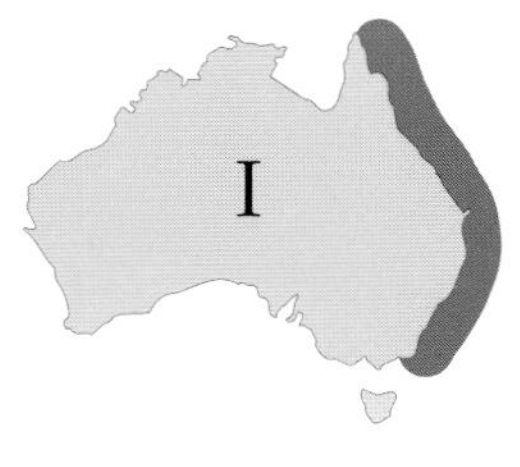

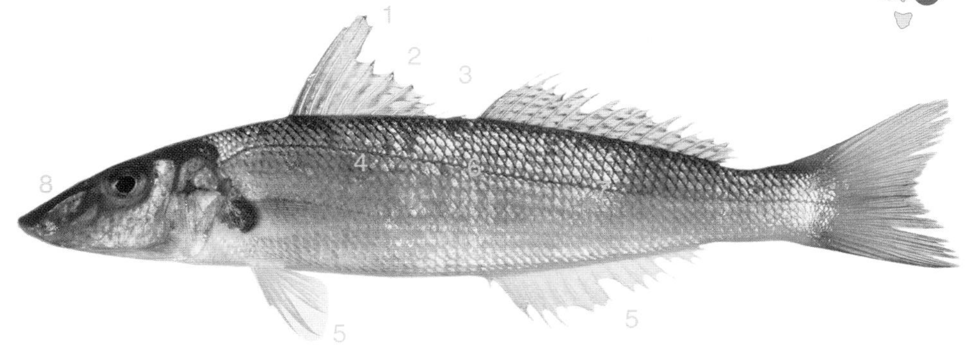

Identifying features: ① first dorsal-fin height much longer than snout length; ② first dorsal-fin height decreasing rapidly after fourth spine; ③ 2 slightly separated dorsal fins, the first with 11 spines, the second with 1 spine and 16–18 soft rays; ④ scales slightly deciduous; ⑤ pelvic and anal fins bright yellow; ⑥ sides plain, silvery; ⑦ lateral-line scales 60–69; ⑧ body elongate with pointed snout.

Comparisons: Similar to yellowfin whiting (*S. schomburgkii*, p. 285), which also has bright yellow pelvic and anal fins, but has a higher first dorsal fin and fewer second dorsal-fin soft rays (16–18 versus 19–22). Lacks colour markings such as spots, blotches and stripes on the sides of the body. Another similar species, the goldenline whiting (*S. analis*), lacks a dark spot on the pectoral-fin base and has a golden or yellow stripe below the lateral line.

Fillet: Moderately deep, elongate, tapering gently, slightly convex above, yellowish-white with some dark veins. Outside with narrow, sharp-edged continuous, central red muscle band; parallel EL, weakly converging HL; HS along middle of fillet; integument silvery white. Inside flesh fine; belly flap sometimes present; peritoneum silvery white with melanophores. Usually skinned, skin thin, scale pockets moderate and defined.

Size: To about 55 cm and 1.4 kg (commonly to 35 cm and 0.3 kg).

Habitat: Marine and estuarine; schools over sandy bottoms of open bays and estuaries and along ocean beaches in depths of 0–45 m (adults typically in 10–30 m). Penetrates to tidal limits of estuaries.

Fishery: Mainly caught off southern Queensland and New South Wales using beach seines and, in estuaries, haul nets and gillnets. Incidental catches are made by inshore prawn and fish trawlers. Esteemed angling fish taken in large quantities. Research into aquaculture is continuing.

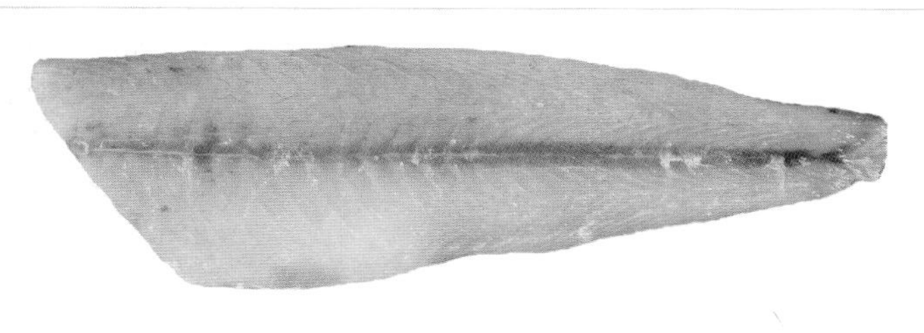

Untrimmed. Protein fingerprint p. 371; oil composition p. 406.

Remarks: Often marketed whole. The flesh is firm, flaky and tasty.

School whiting (page 1 of 2)

Sillago bassensis, S. flindersi & S. robusta

Previous names: redspot whiting, silver whiting, trawl whiting, whiting

Sillago flindersi

Identifying features: ① silvery white stripe along middle of sides; ② lateral-line scales 64–73; ③ 2 slightly separated dorsal fins, the first with 10–12 spines and the second with 1 spine and 16–19 soft rays; ④ body elongate with pointed snout; ⑤ mouth small.

Comparisons: Includes three regional species, each with a silvery white stripe along the middle of the sides and pale pelvic and anal fins. May have narrow, rusty brown bars on the upper sides but lack the dark brown oval or round blotches of the trumpeter whiting (*S. maculata*, p. 283).

Fillet: *S. flindersi* moderately deep, elongate, tapering gently, almost straight above, white with a few dark veins. Outside with narrow, sharp-edged continuous, central red muscle band (overlain with silver stripe); parallel EL, weakly converging HL; HS along middle of fillet; integument silvery. Inside flesh fine; belly flap sometimes present; peritoneum silvery white. Usually skinned, skin thin, scale pockets moderate and indistinct.

Size: To 40 cm and 0.6 kg (commonly to 28 cm and 0.2 kg).

Habitat: Marine; forms schools close to sandy bottoms, usually inshore but also to depths of 170 m. Juveniles prefer coastal waters and estuaries, often near flotsam and beaches.

Fishery: Mainly caught by Danish seine vessels in eastern Bass Strait in depths less than 50 m. Also taken in large quantities with otter trawls off Queensland, New South Wales, and Western Australia, and in smaller quantities off Victoria and Tasmania. Most product is exported to Asia. Sometimes caught by recreational anglers.

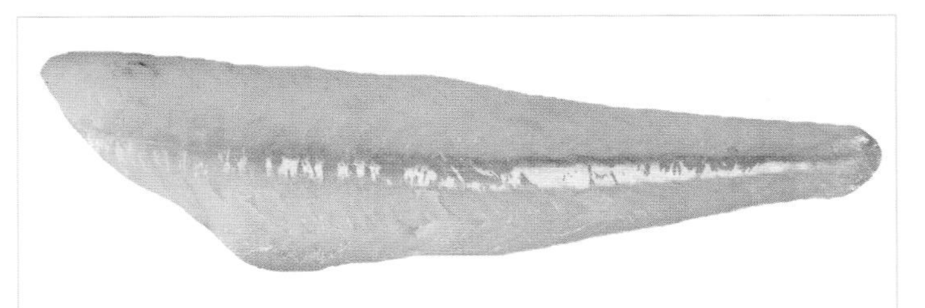

Untrimmed. Protein fingerprint p. 371; oil composition p. 406.

Remarks: The remnant silver integument wholly or partly overlaying the red muscle band on the outer fillet is typical of garfishes and most species of whiting. These species are good eating and keep well when chilled. Their flesh is moderately firm with a subtle flavour.

School whiting (page 2 of 2)

Sillago bassensis, S. flindersi & S. robusta

Sillago flindersi

Remarks: Commonly called 'eastern school whiting', this species is distributed off the east and south coasts between Noosa (Qld) and Port Lincoln (SA) in depths of 0–170 m (typically less than 50 m). Best distinguished in having a row of rusty brown blotches on the sides just above a silver white stripe and oblique, narrow, brownish bars on the upper sides. Reaches 33 cm (rarely over 25 cm) and 0.4 kg.

Sillago bassensis

Remarks: Commonly called 'western school whiting', this species is distributed along the south and west coasts between Western Port (Vic.) and Geraldton (WA) in 0–55 m depth. Best distinguished in having dark edges on the caudal fin and no row of rusty brown blotches on the sides just above the silver white stripe. To about 40 cm and 0.6 kg.

Sillago robusta

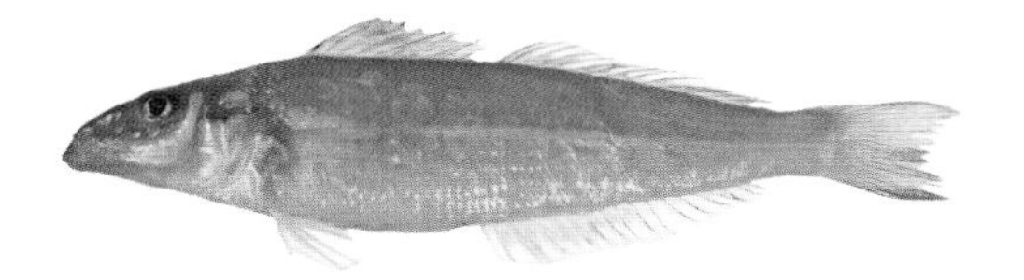

Remarks: Commonly called 'stout whiting', this species is distributed between Fremantle and at least Shark Bay (WA) and between Bowen (Qld) and Newcastle (NSW) in depths of 10–70 m. It may also occur in Northern Territory waters. Best distinguished in having a mostly black first dorsal-fin spine which has a keel on the leading edge. Also has a yellow blotch on the operculum (between the eye and the pectoral fin). To 30 cm and 0.2 kg.

Trumpeter whiting

Sillago maculata

Previous names: diver whiting, spotted whiting, winter whiting

Identifying features: 1 numerous brown oval or roundish blotches on sides of the body; 2 black spot on pectoral-fin base; 3 lateral-line scales 70–75; 4 2 slightly separated dorsal fins, the first with 11–12 spines and the second with 1 spine and 19–21 soft rays; 5 body elongate with pointed snout.

Comparisons: Differs from all other whitings that have a unique marketing name by the presence of numerous brown oval or roundish blotches on the sides of the body. The upper blotches are often joined with the lower blotches, particularly posteriorly. In the similar western trumpeter whiting (*S. burrus*) the upper and lower blotches are usually separate.

Fillet: Moderately deep, elongate, tapering gently, almost straight above, white with a few dark veins. Outside with narrow, sharp-edged continuous, central red muscle band (overlain with silver stripe); weakly converging EL, converging HL; HS along middle of fillet; integument silvery. Inside flesh fine; belly flap sometimes present; peritoneum silvery white. Usually skinned, skin thin, scale pockets moderate and indistinct.

Size: To 30 cm and 0.2 kg (commonly to 25 cm and 0.1 kg).

Habitat: Marine; sheltered bays and estuaries on mud and silt bottoms in depths to 30 m. Especially common in river mouths and mangrove creeks.

Fishery: Mostly in southern Queensland (Moreton Bay) and central and northern New South Wales. Taken as bycatch of prawn trawling, by haul seine in shallow water and with hook-and-line. Popular recreational species, especially in Hervey and Moreton Bays (Qld).

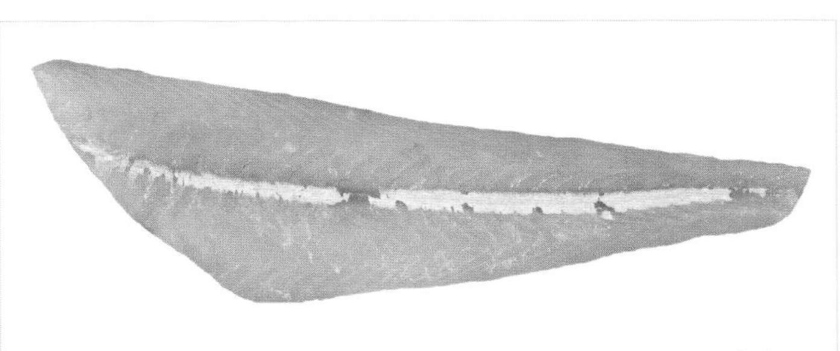

Untrimmed. Protein fingerprint p. 371; oil composition p. 406.

Remarks: Often marketed in winter and hence the common name 'winter whiting'. Trawled fish are easily damaged but, if treated carefully, the flesh is firm and flaky when cooked, with a delicate flavour.

Whiting

Sillaginidae

Previous names: none

Sillago sihama

Identifying features: 1 body elongate with pointed snout; 2 2 slightly separated dorsal fins, the first with 10–13 spines and the second with 1 spine and 16–27 soft rays; 3 lateral-line scales 54–147.

Comparisons: Thirteen species occur in Australian waters and accurate identification often requires expert knowledge and dissection of internal organs such as the swim bladder. Differ from the similarly shaped grass whiting (*Haletta semifasciata*, p. 287)—which is not a true whiting—and tilefishes (Branchiostegidae) in having 2 separate dorsal fins. Unrelated to the imported North Sea whiting (*Merlangius merlangus*) which has 3 dorsal fins and 2 anal fins.

Fillet: *Sillago sihama* moderately deep, elongate, tapering gently, slightly convex above, white without dark veins. Outside with narrow, sharp-edged continuous, central red muscle band; parallel EL, weakly converging HL; HS along middle of fillet; integument silvery. Inside flesh fine; belly flap sometimes present; peritoneum white. Usually skinned, skin thin, scale pockets moderate and indistinct.

Size: Possibly to 72 cm and 4.8 kg (commonly to 35 cm and 0.3 kg).

Habitat: Marine; demersal on the continental shelf. Form schools in coastal bays and estuaries (often over sand or mud) with some species also occurring to depths of 200 m.

Fishery: Most species are caught commercially, often as bycatch of other trawl fisheries but some species are targeted using Danish seine vessels, seine nets, gillnets and handlines. Also taken in large quantities by recreational anglers.

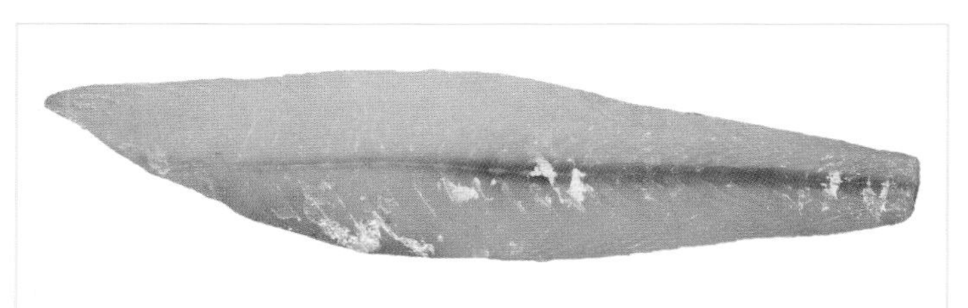

Trimmed. Protein fingerprint p. 371; oil composition p. 406.

Remarks: Many tropical species are hard to distinguish from each other and do not have unique marketing names. They are simply marketed as 'whiting'. Popular foodfishes throughout their range.

Yellowfin whiting

Sillago schomburgkii

Previous names: fine-scale whiting, silver whiting, western sand whiting

Identifying features: 1 first dorsal-fin height about equal to snout length; 2 first dorsal-fin height grading evenly posteriorly; 3 2 slightly separated dorsal fins, the first with 10–12 spines and the second with 1 spine and 19–22 soft rays; 4 scales very deciduous; 5 pelvic and anal fins bright yellow; 6 sides plain, silvery; 7 lateral-line scales 66–76; 8 body elongate with pointed snout.

Comparisons: Distinguished from eastern Australian sand whiting (*S. ciliata*, p. 280), which also has bright yellow pelvic and anal fins, by having a lower first dorsal fin and more second dorsal-fin soft rays (19–22 versus 16–18). No spots, blotches or stripes on the sides of the body.

Fillet: Moderately deep, elongate, tapering gently, slightly convex above, yellowish-white with dense veins. Outside with narrow, sharp-edged continuous, central red muscle band; parallel EL, weakly converging HL; HS along middle of fillet; integument silvery white. Inside flesh fine; belly flap sometimes present; peritoneum silvery white with melanophores. Usually skinned, skin thin, scale pockets moderate and indistinct.

Size: To 42 cm and 0.7 kg (commonly 23–32 cm and 0.1–0.2 kg).

Habitat: Marine; occurs in sandy areas around estuary mouths and other inshore waters. Occasionally enters brackish water.

Fishery: Caught mainly from Spencer Gulf and Gulf St Vincent (SA) and Shark Bay (WA) but also from southwestern Western Australia. Main methods include haul seining and gillnetting. The South Australian catch is consumed locally but much of the Western Australian catch is exported to South Australia and the eastern states. Recreational fishers mostly use handlines from the shore. Has aquaculture potential.

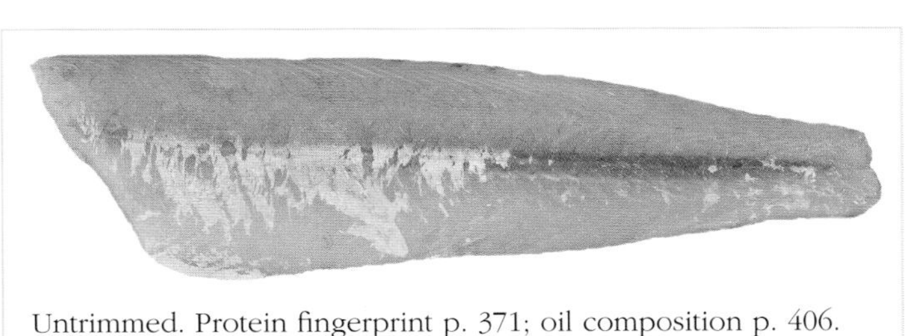

Untrimmed. Protein fingerprint p. 371; oil composition p. 406.

Remarks: Move onto sand flats with the rising tide. Highly esteemed foodfish with a sweet, delicate flavour.

Baldchin groper

Choerodon rubescens

Previous names: parrotfish, tuskfish

Identifying features: 1 chin white, body not brightly coloured; 2 snout blunt with short head and eyes well above mouth; 3 2 pairs of thick, forward pointing, tusk-like teeth at the front of each jaw; 4 lateral line curved evenly (with 27 pored scales); 5 no scales immediately below eye; 6 very large body scales; 7 dorsal fin with 13 spines, 6–8 soft rays; 8 anal fin with 3 spines, 8–10 soft rays.

Comparisons: Distinguished from other tuskfishes by the colouration. Several other members of the genus are marketed but these have brilliant fin and body colours or distinctive markings of lines, spots and blotches. In comparison, the baldchin groper is more plain-coloured and has a large white area on the chin.

Fillet: Deep, rather elongate, upper profile straight, taper pronounced, off-white to yellowish. Outside with feeble, diffuse central red muscle band; convex EL, parallel HL; HS along middle of fillet; integument translucent above, whitish below. Inside flesh fine; belly flap sometimes present; peritoneum silvery white to translucent. Usually skinned, skin medium, scale pockets large and well defined.

Size: To 90 cm and 14 kg (commonly to 80 cm and about 10 kg).

Habitat: Marine; demersal, solitary or in small groups, over seagrass beds and coral reefs on the inner continental shelf to at least 40 m depth.

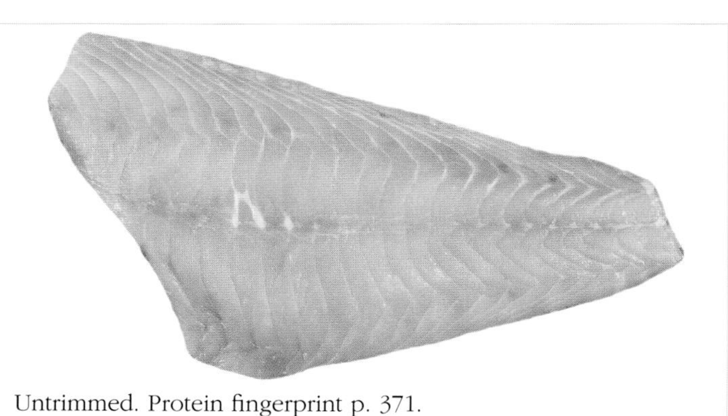

Untrimmed. Protein fingerprint p. 371.

Fishery: Caught mainly using handlines and in smaller quantities by gillnets, spears and traps.

Remarks: Largest tuskfish, it is highly sought-after by recreational spear and line fishers in Western Australia. Baldchin groper is highly regarded as a table-fish, with firm, white, flaky flesh.

Grass whiting

Haletta semifasciata

Previous names: blue rock whiting, rock whiting, stranger, weed whiting

Identifying features: 1 teeth in front of jaws joined to form bleak-like plates; 2 mouth very small; 3 body long and thin; 4 caudal fin rounded slightly; 5 scales cycloid, usually large; 6 dorsal fin usually with 17–18 spines, 12–14 soft rays; 7 anal fin usually with 3 spines, 10–12 soft rays.

Comparisons: Named because of its resemblance to the unrelated whitings (Sillaginidae), particularly in head shape. However, the grass whiting's dorsal fin is not notched, its caudal fin is never forked, and it has a more brightly coloured body (rather than silvery). It is a close relative of wrasses (Labridae) but has beak-like teeth, more dorsal-fin spines, and only 4 soft rays in the anal-fin (rather than 5). Ten similar species occur locally but most are small and only one, the grass whiting, is marketed regularly.

Fillet: Moderately deep, elongate, upper profile almost straight, taper slight, bluish-white. Outside without red muscle band, pinkish on shoulder; weakly converging EL, weakly converging HL; HS along middle of fillet; integument translucent. Inside flesh fine; belly flap usually present; peritoneum translucent. Often skinned, skin medium, scale pockets small and defined.

Size: To 40 cm and 0.8 kg (commonly marketed at 25–35 cm and 0.2–0.5 kg).

Habitat: Marine; demersal among seagrasses in shallow sheltered bays of southern Australia.

Fishery: Caught as bycatch of inshore fisheries for garfishes and true whitings using haul nets.

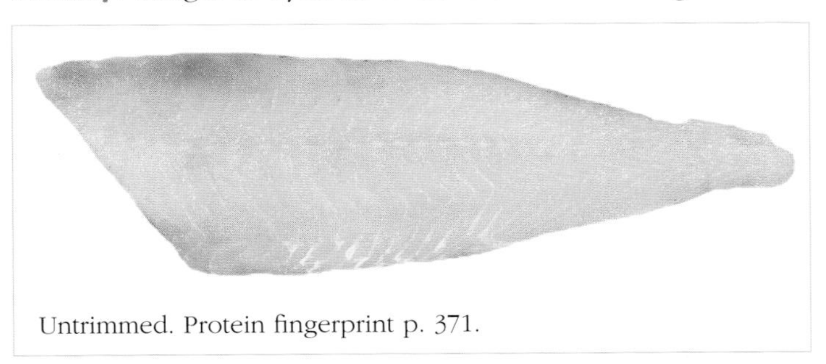

Untrimmed. Protein fingerprint p. 371.

Remarks: An opportunistic feeder, using its small mouth to selectively ingest a variety of small animals, of which sand worms are the major food. Flesh is translucent white, the inner fillet appearing bluish when contrasted against the skin. Considered to be of intermediate quality.

Maori wrasse (page 1 of 2)

Cheilinus & *Oxycheilinus* species

Previous names: giant wrasse, humphead wrasse, napolean, parrotfish

Cheilinus undulatus

Identifying features: 1 snout usually bluntly pointed with a long head; 2 2 pairs of thick, forward pointing, canine teeth at the front of each jaw; 3 lateral line interrupted below end of dorsal fin; 4 2 rows of scales on cheek and operculum; 5 extremely colourful; 6 body scales cycloid, very large; 7 dorsal fin with 9–10 spines, 8–10 soft rays; 8 anal fin with 3 spines, 8 soft rays.

Comparisons: Richly coloured, slender to deep-bodied fishes that can be identified from other wrasses by the combination of characters above. All five Australian species, distinguishable from each other by their body shape and colour, are marketed in small quantities.

Fillet: *O. digrammus* deep, rather elongate, upper profile slightly convex, tapering gradually, brilliant white. Outside without red muscle bands; convex EL, converging HL; HS slightly below middle of fillet; integument transparent. Inside flesh fine; belly flap sometimes present; peritoneum silvery white to translucent. Usually skinned, skin medium, scale pockets large and well defined.

Size: To 230 cm and 190 kg (commonly marketed at 30–60 cm and 0.4–3 kg).

Habitat: Marine; demersal in tropical regions on coral reefs and inner continental shelf. Often occur in shallow water over reef flats but also down to depths of at least 160 m.

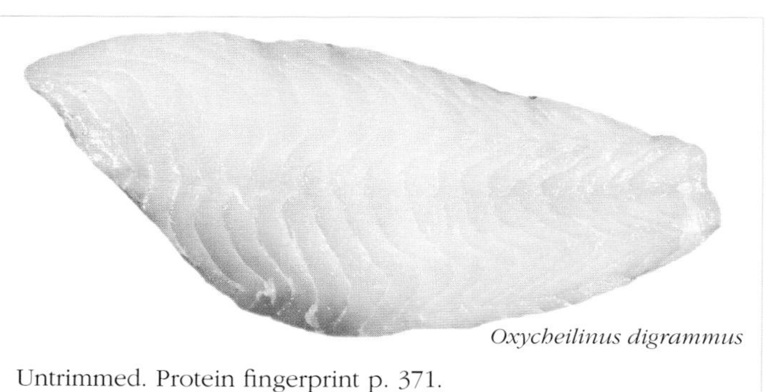

Oxycheilinus digrammus

Untrimmed. Protein fingerprint p. 371.

Fishery: Caught mainly using spears, lines and traps.

Remarks: Maori wrasses are high priced fishes with strong demand both as aquarium fishes and for the Asian live fish trade. The flesh of smaller individuals is firm, white and flaky.

Maori wrasse (page 2 of 2)

Cheilinus & *Oxycheilinus* species

Cheilinus undulatus

Remarks: Commonly called 'humphead Maori wrasse', this giant fish is distributed in coral habitats around northern Australia between southern Queensland and the offshore reefs of northern Western Australia in depths to 100 m. Best distinguished in having relatively deep, olive green body with wavy lines, long snout, and rounded caudal fin. Adults develop a characteristic bump immediately above their eyes. Reaches about 230 cm and 190 kg.

Oxycheilinus digrammus

Remarks: Commonly called 'violetline Maori wrasse', this species is distributed in coral habitats around northern Australia between southern Queensland and the Kimberley region (WA) in depths to 30 m. Best distinguished in having relatively slender body, deep jaws, and pronounced bluish-pink pencil thin markings on the head. Reaches about 40 cm and about 1.0 kg.

Cheilinus chlorourus

Remarks: Commonly called 'floral Maori wrasse', this attractive fish is distributed in coral habitats around northern Australia between southern Queensland and Point Quobba (WA) in depths to at least 20 m. Best distinguished in having a relatively deep body, mottled colour pattern overlain with pale spots, and caudal fin with the middle and outer rays longer than those adjacent. Reaches about 36 cm and 0.9 kg.

Parrotfish

Scaridae

Previous names: blue parrot, tuskfish

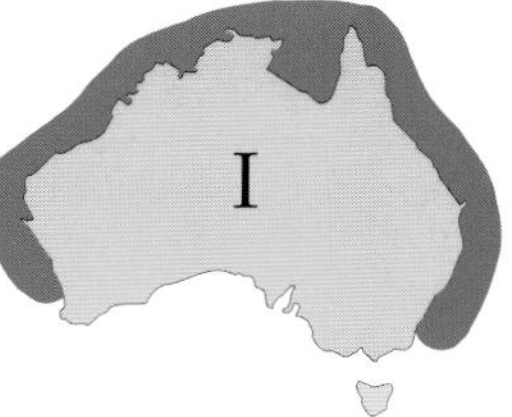

Scarus ghobban

Identifying features: 1 plate-like teeth in front of jaws joined to form beak-like cutting edges; 2 mouth very small; 3 usually extremely colourful; 4 caudal fin truncated or lunate (rarely rounded); 5 scales large and cycloid; 6 dorsal fin usually with 9 spines, 10 soft rays; 7 anal fin usually with 3 spines, 9 soft rays.

Comparisons: Variable in colour with adult males and females usually quite different from each other but the fin-ray counts are totally uniform within the group. They have a distinctive appearance with a rounded or bump-headed snout and small mouth. The jaws cannot be protruded forward and the plate-like teeth are joined to form a structure resembling a parrot's beak. Wrasses (Labridae), which have been incorrectly called 'parrotfish', have conical or peg-like teeth which are separate and which do not form a cutting edge.

Fillet: *Scarus ghobban* deep, rather elongate, upper profile slightly convex, taper pronounced, pale pinkish. Outside with feeble, diffuse central red muscle band; convex EL, converging HL; HS along middle of fillet; integument translucent above, whitish below. Inside flesh fine; belly flap sometimes present; peritoneum white with melanophores. Usually skinned, skin medium, scale pockets large and well defined.

Size: To 130 cm and 68 kg (commonly marketed at 30–50 cm and 0.4–2.5 kg).

Habitat: Marine; demersal in shallow tropical seas, mainly in coral reef habitats.

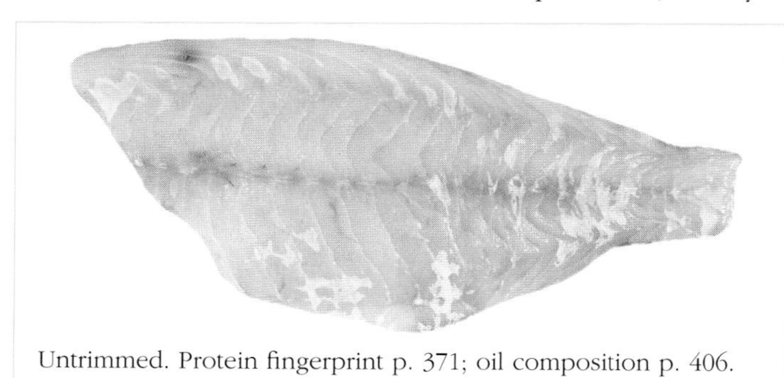

Untrimmed. Protein fingerprint p. 371; oil composition p. 406.

Fishery: Caught mainly using spears and nets as they rarely take bait. Occasionally sold for live fish market or the aquarium trade.

Remarks: The beak-like teeth are used to scrape and bite coralline algae and hard coral. Extremely highly regarded as foodfishes with firm, white and delicate-tasting flesh.

Pigfish

Bodianus species

Previous names: hogfish, wrasse

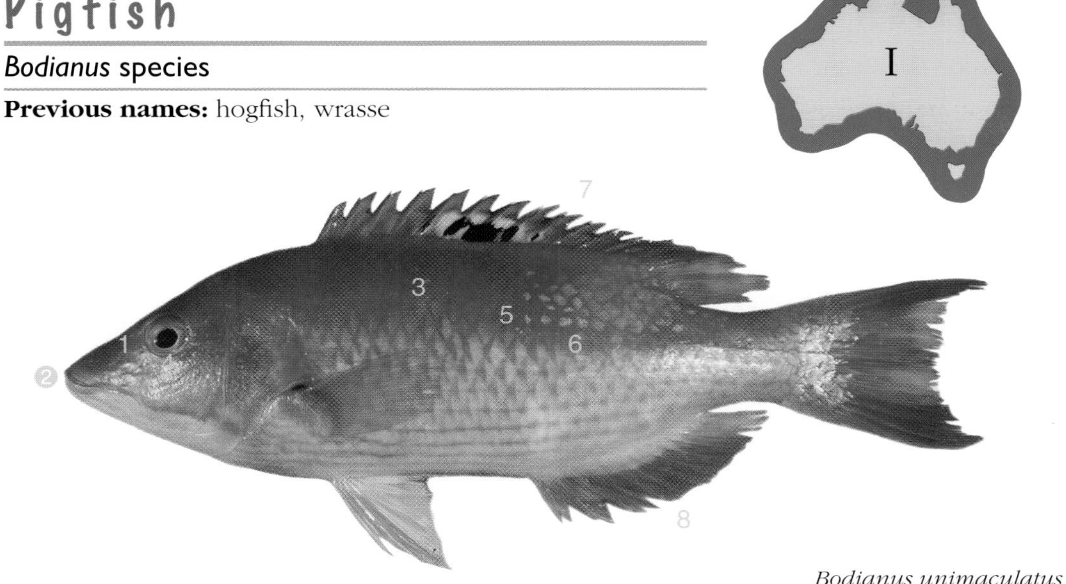

Bodianus unimaculatus

Identifying features: ① snout usually pointed, eyes not well away from mouth; ② 2 pairs of thick, forward pointing, canine teeth at the front of each jaw; ③ lateral line curved evenly (mostly with 30–32 pored scales); ④ scales present immediately below eye; ⑤ usually very colourful; ⑥ body scales cycloid, moderately large; ⑦ dorsal fin usually with 12 spines, 9–11 soft rays; ⑧ anal fin usually with 3 spines, 10–12 soft rays.

Comparisons: Brightly coloured fishes that are distinguished from other wrasses by the combination of characters above. Several species occur but few are marketed regularly. The reddish blackspot pigfish (*B. unimaculatus*) is probably the most important member of the group. The sexes differ in colour.

Fillet: *B. perditio* moderately deep, rather elongate, upper profile almost straight, barely tapering, brilliant white. Outside without red muscle bands; parallel EL, parallel HL; HS slightly below middle of fillet; integument silvery white. Inside flesh medium; belly flap sometimes present; peritoneum silvery white to translucent. Usually skinned, skin thick, scale pockets large and well defined.

Size: To 60 cm and 3 kg (commonly marketed to 50 cm and 1.5 kg).

Habitat: Marine; demersal over coral and rocky reefs on the inner continental shelf to depths of at least 200 m.

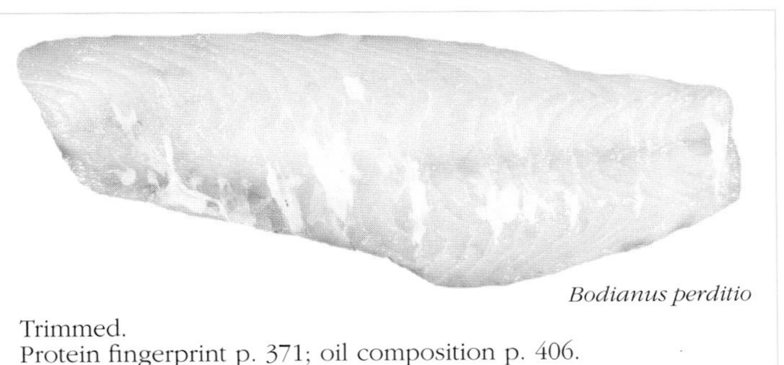

Bodianus perditio

Trimmed.
Protein fingerprint p. 371; oil composition p. 406.

Fishery: Caught mainly using traps and as bycatch of prawn and fish trawls.

Remarks: The firm, white, flaky flesh is in extremely high demand by the Asian community. Sydney prices for pigfishes are among the highest locally for any Australian fish.

Tuskfish

Choerodon species

Previous names: blackspot tuskfish, blue parrot, bluebone, purple tuskfish, venus tuskfish

Choerodon venustus

Identifying features: 1 snout blunt with short head and eyes well above mouth; 2 2 pairs of thick, forward pointing, tusk-like teeth at the front of each jaw; 3 lateral line curved evenly (with 27 pored scales); 4 no scales immediately below eye; 5 usually very colourful; 6 body scales cycloid, very large; 7 dorsal fin usually with 12–13 spines, 7–8 soft rays; 8 anal fin usually with 3 spines, 9–10 soft rays.

Comparisons: Distinguished from the baldchin groper (*Choerodon rubescens*, p. 286), which is really another tuskfish, by their colouration. They have more brilliant fin and body colours or distinctive markings of lines, spots and blotches. Some species have a whitish patch on the chin but it is much less extensive than in the baldchin groper.

Fillet: *C. venustus* deep, rather elongate, upper profile straight, taper pronounced, off-white to yellowish. Outside with feeble, diffuse central red muscle band; convex EL, converging HL; HS just below middle of fillet; integument translucent above, whitish below. Inside flesh fine; belly flap sometimes present; peritoneum silvery white to translucent with melanophores. Usually skinned, skin thin, scale pockets large and indistinct.

Size: To about 90 cm and 16 kg (commonly 35–55 cm and 0.8–3 kg).

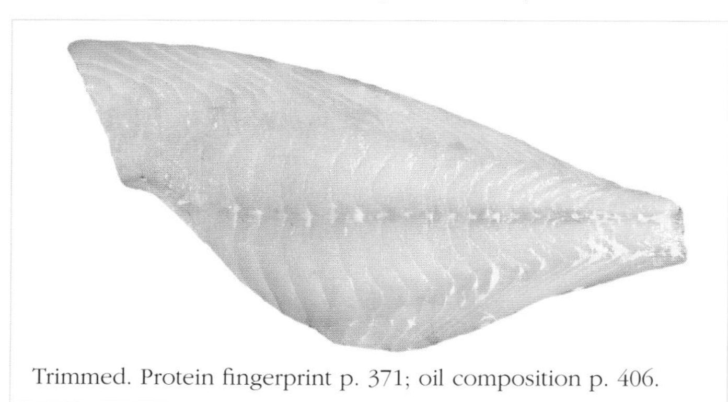

Trimmed. Protein fingerprint p. 371; oil composition p. 406.

Habitat: Marine; demersal over seagrass beds and along coral reef fringes, mainly on the inner continental shelf but to at least 200 m depth.

Fishery: Caught using handlines, gillnets, spears and traps.

Remarks: Sought-after recreationally for their fighting and eating qualities. Flesh varies within each species but is firm, white and flaky in most.

Wrasse (page 1 of 2)

Labridae

Previous names: bluethroat parrot fish, brown-spotted wrasse, parrotfish

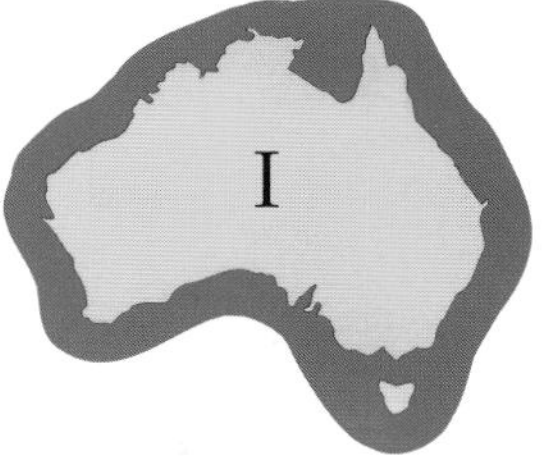

Notolabrus tetricus

Identifying features: ① large teeth in front of jaws (usually visible when mouth open slightly); ② mouth usually small; ③ usually very colourful; ④ caudal fin truncated or rounded (rarely forked); ⑤ scales cycloid, usually large; ⑥ dorsal fin usually with 8–14 spines, 6–21 soft rays; ⑦ anal fin usually with 3 spines, 7–18 soft rays.

Comparisons: Possibly the most variable fish group in terms of appearance, size and colour with growth stages and adult sexes mostly quite different from each other. Hence the group as a whole is difficult to typify but each of the species can usually be identified through unique features, mostly relating to their colour pattern. Parrotfishes (Scaridae, p. 290), often confused with wrasses, also share these features but their teeth, rather than being separate, are joined to form a 'beak'.

Fillet: *Notolabrus gymnogenis* moderately deep, rather elongate, upper profile straight, taper pronounced, off-white to reddish. Outside with feeble, diffuse central red muscle band; parallel EL, converging HL; HS just below middle of fillet; integument translucent. Inside flesh coarse; belly flap sometimes present; peritoneum silvery white to translucent. Usually skinned, skin medium, scale pockets large and well defined.

Size: To 160 cm and about 40 kg (commonly marketed at 25–50 cm and 0.3–2.0 kg).

Habitat: Marine; demersal in kelps, seagrasses and over coral and rocky reefs on the continental shelf to at least 200 m.

Notolabrus gymnogenis

Trimmed. Protein fingerprint p. 371; oil composition p. 406.

Fishery: Caught mainly using lines, traps and gillnets. Several species targeted in recent years for Asian live fish market.

Remarks: Mostly white flesh, varies considerably in quality between species from soft to firm and bland to delicately flavoured.

Wrasse (page 2 of 2)

Labridae

Notolabrus gymnogenis

Remarks: Commonly called 'crimsonband wrasse', this species is distributed near shallow inshore reefs between southern Queensland and Mallacouta (Vic.) in depths to 40 m. Best distinguished in having adult males with red dorsal and anal fins and striking reddish band on the mid-body. Females are reddish-brown and covered in whitish spots. Reaches 48 cm and 1.9 kg.

Achoerodus gouldii

Remarks: Commonly called 'blue groper', this species is distributed around southern Australia between Port Phillip Bay (Vic.) and Houtman Abrolhos (WA) in depths to about 40 m. Among the largest of wrasses, a similar giant species (*Achoerodus viridis*) occurs off eastern Australia but is protected. Males best distinguished by their bluish or greenish colour, blue eye, large fleshy lips and peg-like teeth. Reaches about 160 cm and 39 kg.

Anampses lennardi

Remarks: Commonly called 'blue and yellow wrasse', this brightly coloured species is distributed from north-west Australia to the Gulf of Carpentaria on coral reefs in depths to about 30 m. It is typical of more than 50 small tropical wrasses that infrequently reach domestic markets. Most are good table-fishes but are generally more sought-after by the aquarium trade. *A. lennardi* reaches at least 30 cm and 0.4 kg.

Crustaceans

G. K. Yearsley and P. R. Last

Balmain bug

Ibacus species

Previous names: bug, flapjack, flying saucer, shovelnose lobster, slipper lobster, southern baylobster, squagga

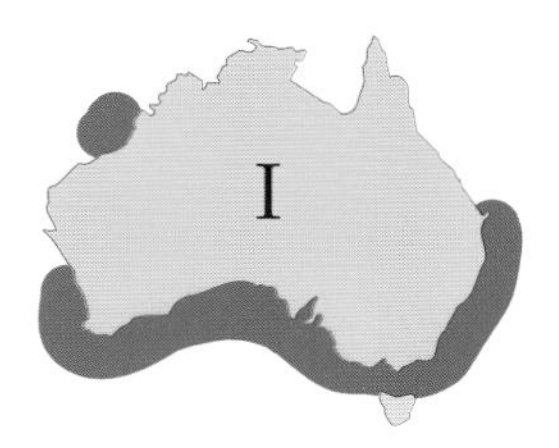

Ibacus peronii

Identifying features: 1 eyes closer to body midline than to carapace margin; 2 carapace with prominent marginal spines; 3 carapace broader than long; 4 legs all of similar size and shape; 5 short, broad and flattened antennae; 6 body strongly depressed; 7 telson broadly convex.

Comparisons: Australia's commercial crustaceans belong to the order Decapoda (meaning 'ten feet') and all have 5 pairs of legs and a large carapace. Balmain bugs are easily distinguished from similar species in having their eyes close to the body midline—eyes at carapace margin in Moreton Bay bugs (*Thenus* species, p. 299) and eyes closer to carapace margin than to the body midline in slipper lobsters (*Scyllarides* species, p. 300).

Size: To a carapace length of nearly 9 cm (total length over 25 cm) and nearly 0.4 kg.

Habitat: Marine; often buried in sand or mud during the day and becoming active at night. Occur in water depths of 15 m to greater than 650 m but most common on the continental shelf in depths less than 150 m.

Fishery: Mostly off the eastern seaboard as a bycatch of fish and prawn trawling but also targeted in some areas. Common at wholesale markets and popular on restaurant menus. About 5 per cent of bugs landed in Australia are exported to the USA.

Remarks: Seven species have been recorded from Australia. The largest and most well known is the eastern Balmain bug (*I. peronii*). Bugs travel backwards when they're in a hurry. Reverse gear is engaged by a 'flip' of the tail and pitch is controlled or induced by raising or lowering the short, broad antennae. Considered by some seafood enthusiasts to be inferior to Moreton Bay bugs, but tasty nonetheless, with a strong flavour. Other information is provided for protein fingerprints (p. 372) and oil composition (p. 407); the correct name of the species referred to as *I. ciliatus* is probably *I. pubescens*.

Champagne lobster

Linuparus species

Previous name: spearlobster

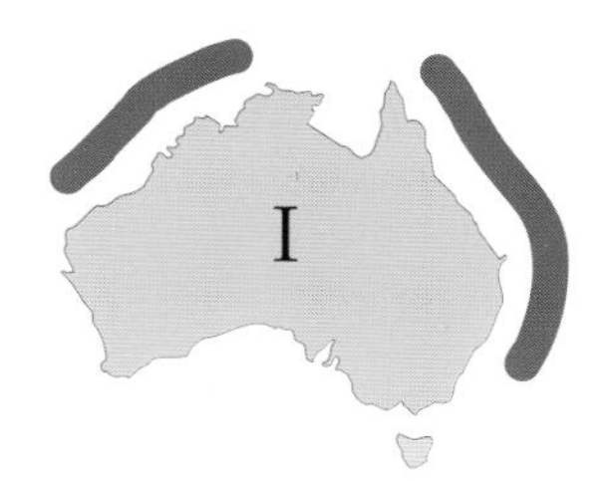

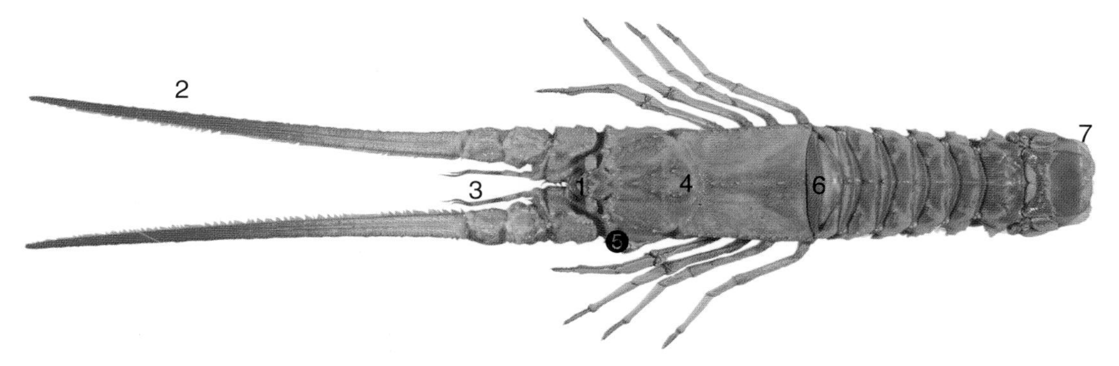

Linuparus trigonus

Identifying features: 1 2 or more very short spines between the eyes (outer spines largest); 2 antennae inflexible and very long; 3 antennules with short flagella; 4 carapace much longer than broad; 5 first pair of legs slightly more robust than those following; 6 body more-or-less circular in cross-section; 7 telson broadly convex.

Comparisons: Most similar to other rocklobsters but the antennule flagella and the spines between the eyes are short, and the antennae inflexible. Two species occur in Australian waters. The red champagne lobster (*L. trigonus*) has a pair of prominent ventral-surface 'teeth' near the origin of the antennules that are absent in the white champagne lobster (*L. sordidus*).

Size: To 47 cm in body length (commonly 20–35 cm).

Habitat: Marine; prefer sand or mud bottoms on the upper continental slope in depths of 200–500 m; also found on the continental shelf as shallow as 30 m. The red champagne lobster occurs off both the east and west coasts whereas the white champagne lobster is restricted to Western Australia.

Fishery: Although once targeted by trawlers off Queensland during prawn fishery closed seasons, today's catches are incidental. Product is marketed locally or exported interstate. Similar incidental fisheries occur elsewhere in Asia.

Remarks: The inflexible antennae have given rise to the common name of 'spearlobster'. The terms lobster and crayfish have been used interchangeably in Australia but the international convention is to call marine species 'lobsters' and freshwater species 'crayfishes'. Flesh coarse compared with Australia's other rocklobsters. Other information is provided for protein fingerprints (p. 372) and oil composition (p. 407).

Eastern rocklobster

Jasus verreauxi

Previous names: crayfish, green rocklobster, packhorse crayfish, rocklobster, Sydney crayfish

Identifying features: ① abdominal segments smooth; ② distinct rostrum, flanked both sides by a similar-sized forward-projecting spine; ③ antennae flexible and very long; ④ antennules with short flagella; ⑤ fresh body greenish; ⑥ carapace much longer than broad; ⑦ first pair of legs slightly more robust than those following (much larger in large males); ⑧ body slightly depressed; ⑨ telson broadly convex.

Comparisons: Very similar to southern rocklobster (*J. edwardsii*, p. 301) but is green before cooking (rather than red or orange) and has smooth (rather than sculptured) abdominal segments. The short flagella on the antennules distinguishes it from the other rocklobsters included here. Differs from champagne lobsters (*Linuparus* species, p. 297) in having long spines between the eyes and flexible antennae.

Size: Largest rocklobster in the world, reaching a massive 8 kg, over 40 cm in carapace length and almost a metre in body length. However, the largest Australian specimen recorded was about 26 cm carapace length (commonly less than 15 cm carapace length).

Habitat: Marine; occurs from close inshore to depths of over 200 m, preferring exposed reef areas.

Fishery: Small compared with other southern Australian rocklobster fisheries but relatively valuable as they are keenly sought in the Sydney market. Caught using baited traps, predominantly off New South Wales, where the fishery is managed by a Total Allowable Catch (TAC).

Remarks: Members of the genera *Jasus* and *Panulirus* are known as 'rocklobsters' in Australia. The term 'crayfish' should be reserved for freshwater species. Flesh is firm, moist and tasty. Other information is provided for protein fingerprints (p. 372).

Moreton Bay bug

Thenus species

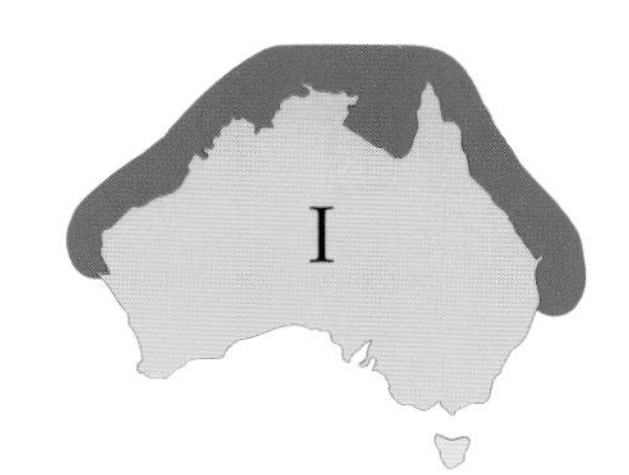

Previous names: baylobster, bug, gulf-lobster, mudbug, northern baylobster, sandbug, shovelnose lobster

Thenus orientalis

Identifying features: 1 eyes at carapace margin; 2 most of carapace margin without prominent spines; 3 carapace broader than long; 4 legs all of similar size and shape; 5 short, broad and flattened antennae; 6 body strongly depressed; 7 telson broadly convex.

Comparisons: Two species recorded from Australian waters. The sandbug (*T. orientalis*) differs from the mudbug (*T. indicus*) in having conspicuous spots on the legs and a brown tail fan (no spots and a yellow tail fan in the mud bug). Differ from similar species in having eyes at the carapace margin—eyes near the body midline in Balmain bugs (*Ibacus* species, p. 296) and closer to the carapace margin than the body midline in slipper lobsters (*Scyllarides* species, p. 300).

Size: Reach 10 cm in carapace length and over 0.5 kg (commonly 4–6 cm carapace length and 0.1–0.15 kg).

Habitat: Marine; on mud or sand bottoms, in depths of 10–60 m. Usually buried during the day and active at night.

Fishery: Mostly as a bycatch of prawn and scallop fisheries off Queensland, predominantly between Cairns and Bundaberg. However, due to their popularity as food, they are a sought-after bycatch species throughout the north. Presence in the markets reflects the seasonal nature of local prawn fisheries.

Remarks: Recent research suggests that similar bugs in Asia are different species, and the scientific names of Australian forms may change. Females carry broods of up to 60 000 eggs under their tail; berried (egg bearing) females should be released. Bugs were once unpopular as food species due to their bug-like appearance. However, they are now highly prized and very popular in restaurants. The tail flesh is dryer and with a stronger taste than that of rock-lobsters. Other information is provided for protein fingerprints (p. 372) and oil composition (p. 407).

Slipper lobster

Scyllarides species

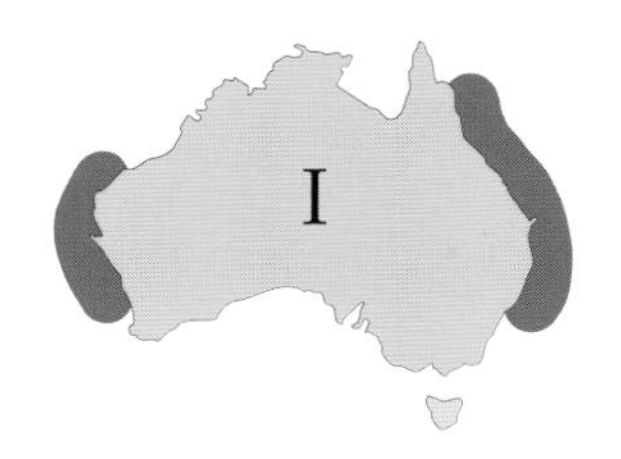

Previous names: Aesop slipper lobster, blunt slipper lobster, flat lobster, slipper bug

Scyllarides squammosus

Identifying features: 1 eyes closer to carapace margin than to body midline; 2 carapace length about equal to or slightly more than breadth; 3 first pair of legs slightly more robust than those following; 4 short, broad and flattened antennae; 5 body depressed; 6 telson broad, convex or slightly concave.

Comparisons: Two Australian species. The Aesop slipper lobster (*S. haanii*) has more pronounced humps on the abdomen than the blunt slipper lobster (*S. squammosus*). Slipper lobsters differ significantly in body shape from their relatives the rocklobsters (Palinuridae) and also have short, broad plate-like antennae rather than very long, almost cylindrical antennae with a spiny base. Differ from bugs in eye position—eyes closer to the carapace margin than the body midline in slipper lobsters, close to the body midline in Balmain bugs (*Ibacus* species, p. 296) and at the carapace margin in Moreton Bay bugs (*Thenus* species, p. 299). Unlike bugs, the first pair of legs is distinctly more robust than those following.

Size: Carapace length to over 21 cm (body length to 50.5 cm).

Habitat: Marine; rocky reef areas in depths of 10–135 m.

Fishery: Not fished regularly in Australia; however, they do appear at provincial markets, particularly in Queensland where they are taken in a developmental pot fishery and as prawn trawl bycatch.

Remarks: These elusive nocturnal creatures have a patchy distribution and are never caught in great quantities. The flesh, which is highly regarded locally and overseas, is white when cooked and very tasty. Other information is provided for protein fingerprints (p. 372).

Southern rocklobster

Jasus edwardsii

Previous names: cray, crayfish, Melbourne crayfish, rocklobster, southern spinylobster, spinylobster, Tasmanian crayfish

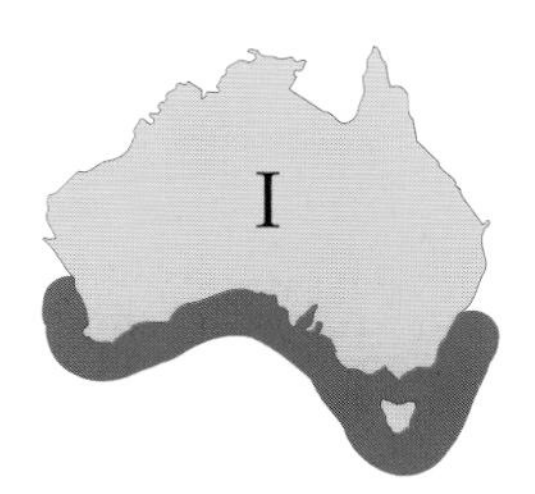

Identifying features: ① abdominal segments sculptured; ② minute rostrum (central spine) flanked by 2 prominent spines; ③ antennae flexible, very long; ④ antennules with short flagella; ⑤ fresh body colour variable, usually reddish or orange; ⑥ carapace much longer than broad; ⑦ first pair of legs slightly more robust than those following (much larger in large males); ⑧ body slightly depressed; ⑨ telson broadly convex.

Comparisons: Distinguished from the similar looking eastern rocklobster (*J. verreauxi*, p. 298) in having sculptured (rather than smooth) abdominal segments, and is reddish to orange rather than green before cooking. Champagne lobsters (*Linuparus* species, p. 297) are also similar in appearance but have characteristic inflexible antennae. These three species all have short flagella on the antennules whereas other harvested rocklobsters have long, forked flagella on each antennule.

Size: Carapace length reaches at least 23 cm and has been reported to 30 cm. Large males can weigh in excess of 6 kg.

Habitat: Marine; shelters in caves, under rocks and in crevices in both sheltered and exposed areas from close inshore to depths of 200 m. Nocturnal and carnivorous, feeding on bottom-dwelling invertebrates.

Fishery: Supports a valuable pot fishery, mainly off southeastern Australia; the majority of the catch comes from South Australia. In addition to the pot fishery, small numbers are taken from shallow water with ring nets. Recreational catch is high, equivalent to over 10 per cent of the commercial catch in Tasmania. Suitability for aquaculture is being assessed.

Remarks: *J. lalandii* and *J. novaehollandiae* are old names for this species locally. It is very popular and well known throughout its range, which extends to New Zealand. The cooked flesh is firm, white and very tasty; flesh from the legs is a little sweeter than tail flesh. Other information is provided for protein fingerprints (p. 372) and oil composition (p. 402).

Tropical rocklobster

Panulirus species except *P. cygnus*

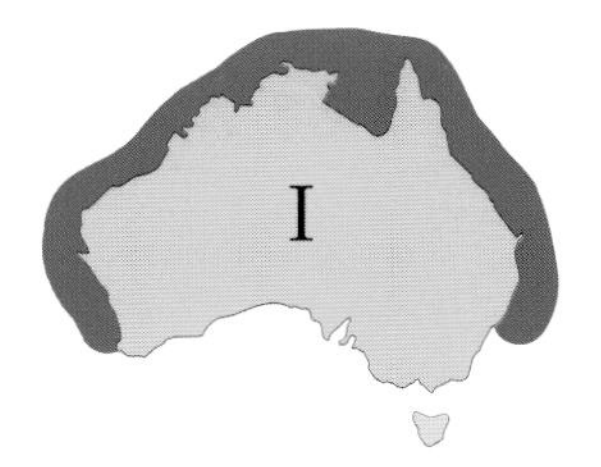

Previous names: coral crayfish, doublespine rocklobster, green crayfish, ornate rocklobster, painted crayfish, rock crayfish, scalloped lobster, tropical spinylobster

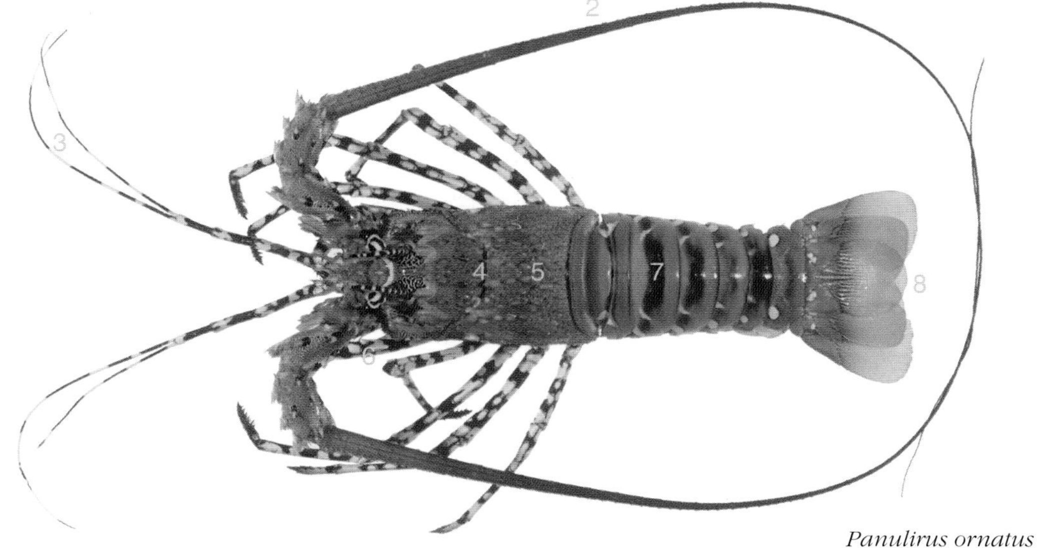

Panulirus ornatus

Identifying features: ① 2 prominent spines between eyes; ② antennae flexible and extremely long; ③ antennules with very long, forked flagella; ④ fresh body colour variable, often bright and ornately patterned; ⑤ carapace much longer than broad; ⑥ first pair of legs slightly more robust than those following; ⑦ body slightly depressed; ⑧ telson broad, usually convex.

Comparisons: Includes five species that are generally more vividly coloured than the closely related western rocklobster (*P. cygnus*, p. 303). Differ from the remaining rocklobsters included here by the presence of long flagella on the antennules (flagella short in other species).

Size: To 20 cm in carapace length and 50 cm in body length. Usually marketed at 30–35 cm in body length.

Habitat: Marine; although most common in shallow water (less than 20 m depth) some species reach depths exceeding 100 m. Most species live on rock and coral reefs but the ornate rocklobster (*P. ornatus*) is often found on sand or mud.

Fishery: Centered in Torres Strait where it is jointly managed by Australia and Papua New Guinea. These rocklobsters rarely enter pots so are collected mainly by spear or hand; also taken incidentally by trawlers.

Remarks: The ornate rocklobster is the main commercial tropical rocklobster and is also the largest. Starting in August each year, thousands of individuals migrate from inshore to deeper water east of Torres Strait, some travelling over 500 km across the Gulf of Papua to near Yule Island (Papua New Guinea), where they eventually breed and then die. Popular food species with a firm, meaty texture and fished virtually wherever they occur. Other information is provided for protein fingerprints (p. 372) and oil composition (p. 407).

Western rocklobster

Panulirus cygnus

Previous names: West Australian crayfish, western cray

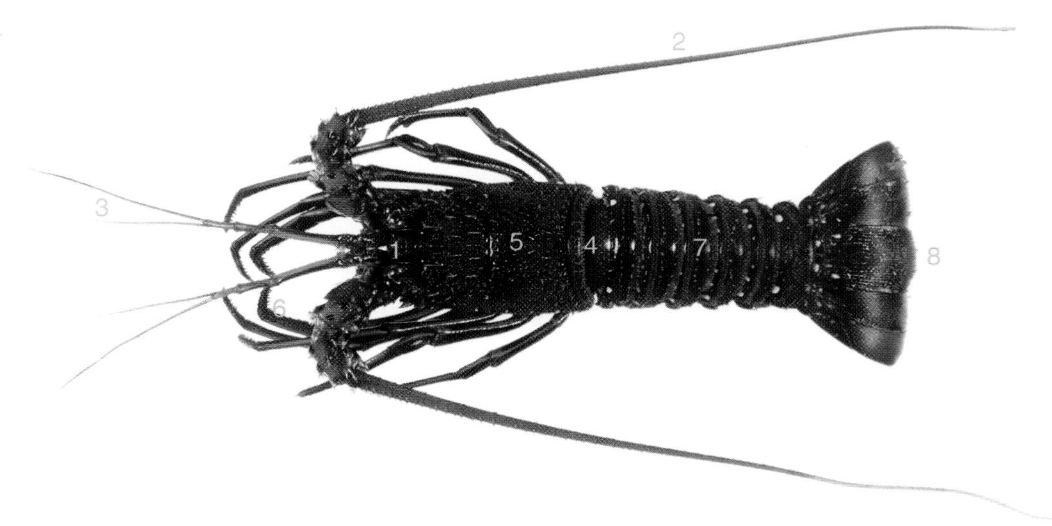

Identifying features: ① 2 prominent spines between eyes; ② antennae flexible and very long; ③ antennules with long, forked flagella; ④ fresh body colour reddish-purple or, in the case of recently moulted pre-adults, pale pink; ⑤ carapace much longer than broad; ⑥ first pair of legs slightly more robust than those following; ⑦ body slightly depressed; ⑧ telson broadly convex.

Comparisons: Closely related to tropical rocklobsters (other *Panulirus* species, p. 302) but are comparatively drab in colour, plain or with light streaks (rather than pronounced leg markings), and medium-length antennules (rather than long). Other commercial lobsters have antennules that are shorter and not prominently forked.

Size: Reported to an impressive 5.5 kg but animals over 3 kg are rare. Commonly taken at about 8–10 cm carapace length (just over the minimum size limit), and about 0.5 kg.

Habitat: Marine; shelters under rocks, on ledges and among coral in reef areas to depths of 200 m; adults most common between 35–60 m depth. Like other lobsters and bugs, most active after dark.

Fishery: Confined to the Western Australian coast. By far the most valuable single-species commercial fishery in Australia with its dollar value over half that of all finfishes combined. Also targeted by recreational fishers who use pots or dive to catch their prey. A licence is required for both commercial and recreational fishing.

Remarks: Most of the catch is exported live or frozen to Taiwan, Japan or China. Females can produce close to one million eggs, often twice per season. The larvae are planktonic and have been found as far as 1500 km off the coast. Rocklobsters from cooler more temperate waters are considered to provide superior flesh to those from the tropics so the western rocklobster is very popular both locally and overseas. Other information is provided for protein fingerprints (p. 372) and oil composition (p. 407).

Blue swimmer crab

Portunus pelagicus

Previous names: blue crab, blue manna crab, bluey, sand crab, sandy

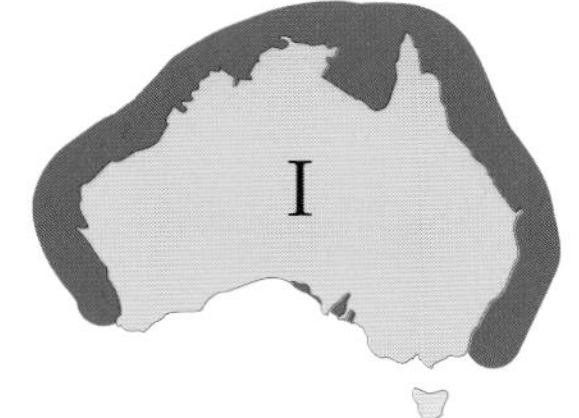

Identifying features: 1 each side of carapace with 9 sharp spines, the last long and projecting laterally; 2 dorsal surface colour variable—generally mottled blue in males and mottled brown in females; 3 first pair of legs much longer than second and with prominent, slender claws; 4 tips of posterior pair of legs broadly flattened (disc-shaped); 5 carapace broader than long; 6 abdomen short and tucked under carapace.

Comparisons: Most of Australia's edible crabs, including the blue swimmer crab, belong to the family Portunidae and the tips of the last pair of legs are disc-shaped (for swimming). This species differs from many other portunids in having 9 sharp spines along each side of the carapace, the last very prominent. The three-spotted crab (*P. sanguinolentus*, p. 307) is similar but has 3 white-edged, red spots towards the rear of the carapace. In mud crabs (*Scylla* species, p. 309), the 9 carapace spines on each side are all of similar size.

Size: Recorded to nearly 22 cm in carapace width and over 1 kg (common size varies between populations and commercial size depends on local state regulations).

Habitat: Coastal marine; occurs in bays, estuaries, and intertidal areas to depths of about 60 m but shallower in South Australia. Prefers muddy or sandy bottoms but also found on rubble, seagrass and seaweed.

Fishery: Both commercial and recreational fisheries have management controls that vary throughout Australia. About half the commercial catch is taken from southern Queensland with most of the remainder from central New South Wales, South Australia and Western Australia north to Shark Bay. Crabs are caught using traps (mostly cylindrical), dillies and entangling devices, and as trawl bycatch. A very popular target of recreational fishers, they are caught using a variety of traps, rakes, dip nets and drop nets.

Remarks: Females (known as 'jennies') may spawn several times per season and can produce up to two million eggs per batch. Berried females are protected. Usually marketed cooked whole or as crab meat. The soft, sweet flesh is very highly regarded. Other information is provided for protein fingerprints (p. 372) and oil composition (p. 407).

Coral crab

Charybdis feriata

Previous name: crucifix crab

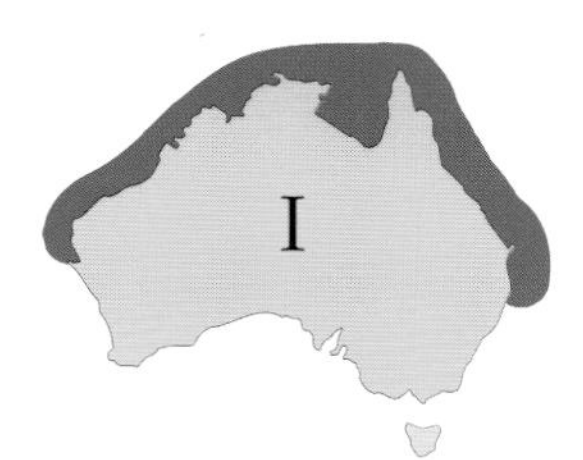

Identifying features: ① prominent stripes on dorsal surface of carapace; ② each side of carapace with 6 similar-sized spines; ③ first pair of legs much longer than second, with prominent, slender claws; ④ tips of posterior pair of legs broadly flattened (disc-shaped); ⑤ carapace broader than long; ⑥ abdomen short and tucked under carapace.

Comparisons: A distinctive species with 6 similar-sized spines on each side of a prominently striped carapace. Most closely related to other portunid crabs that all have broadly flattened tips on the last pair of legs. The blue swimmer crab (*Portunus pelagicus*, p. 304), with which it is caught, has more spines along each side of the carapace (9 rather the 6). Sand crabs (*Ovalipes* species, p. 310) have only 5 spines along each side of the carapace.

Size: To 20 cm in carapace width and possibly 1 kg.

Habitat: Marine; benthic in coastal waters on a range of bottom types including mud, sand, rocks and seagrasses, in depths to 60 m.

Fishery: Usually taken as a bycatch of the blue swimmer crab fishery in Queensland. It is marketed frequently but only in small quantities. Also harvested overseas; for example, it supports an important fishery off India.

Remarks: Has been referred to incorrectly as '*Charybdis cruciata*' in much recent literature. Like many crab species, females have smaller claws and a wider abdomen than males; they can carry over three million eggs per batch. The flesh has a delicate flavour. Other information is provided for protein fingerprints (p. 372).

Crab (page 1 of 2)

Infraorder Brachyura

Previous names: none

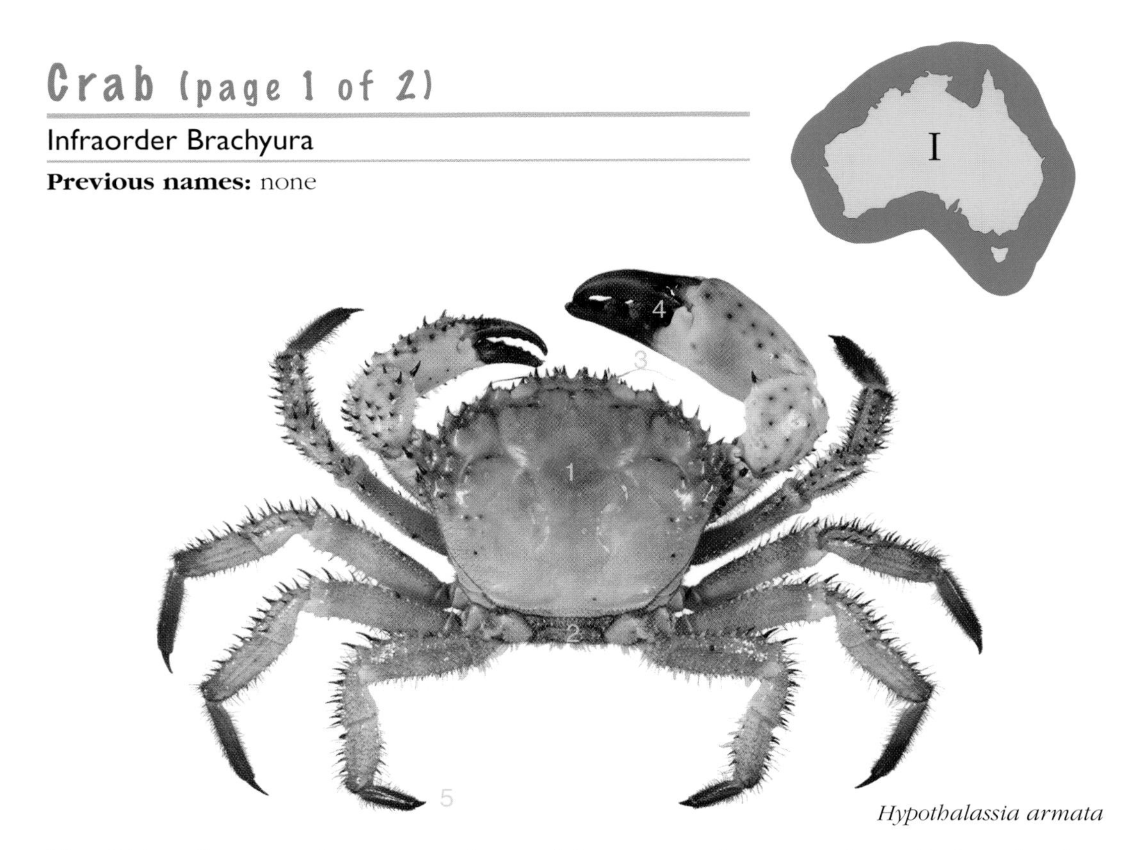

Hypothalassia armata

Identifying features: ① carapace flattened from top to bottom and usually broader than long; ② abdomen short and usually tucked completely under carapace; ③ antennae short and thin; ④ first pair of legs robust and with prominent claws; ⑤ tips of posterior pair of legs often broadly flattened (disc-shaped).

Comparisons: The infraorder Brachyura contains the 'true crabs', a few of which are sold commercially in Australia. For the purposes of seafood identification they are best described as decapod crustaceans with a depressed carapace (flattened from top to bottom) and a very short abdomen. In most cases the abdomen is tucked completely under the carapace.

Size: The largest crab in Australia is the giant crab (*Pseudocarcinus gigas*, p. 308), which reaches 45 cm in carapace width and over 17.5 kg. However, most crabs marketed are much smaller, usually less than 15 cm in carapace width and less in 0.5 kg.

Habitat: Marine; benthic in a wide range of habitats from intertidal flats and rockpools to depths of more than 500 m. They occur on many bottom types including sand, mud, rock, seaweed, seagrass and coral.

Fishery: Most species are caught incidentally around Australia and few are targeted. Others, such as the velvet crab (*Nectocarcinus tuberculosus*) in southeastern Australia, have recently become the focus of developmental fisheries. Methods used include traps, entangling devices and dillies. Also taken incidentally by demersal trawls.

Remarks: All the crabs included in this guide can be sold under the marketing name of 'crab'. However, a higher price is usually gained by using the correct, specific marketing name, such as 'blue swimmer crab' or 'spanner crab', where one exists. Crab flesh is usually soft, moist and tasty. Other information is provided for protein fingerprints (p. 372) and oil composition (p. 407).

Crab (page 2 of 2)

Infraorder Brachyura

Hypothalassia armata

Remarks: Commonly called 'champagne crab', this species is distributed around southern Australia between Mackay (Qld) and Port Hedland (WA) in 30–540 m depth. Best distinguished in having numerous spines on the edges of the carapace and on the legs. Reaches at least 15 cm in carapace width.

Portunus sanguinolentus

Remarks: Commonly called 'three-spotted crab', this mainly tropical species of swimmer crab is distributed between northern New South Wales and southern Western Australia in depths of 0–30 m. It is easily identified by the 3 white-edged, red spots towards the rear of the carapace. Reaches 20 cm in carapace width.

Charybdis natator

Remarks: Commonly called 'hairyback crab', this species of swimmer crab is distributed off northern Australia between northern New South Wales and Shark Bay (WA) in depths of 5–40 m. Best distinguished, as its name suggests, in having a covering of fine hairs on the carapace. Reaches 17 cm in carapace width.

Giant crab

Pseudocarcinus gigas

Previous names: giant deepwater crab, giant Tasmanian crab, king crab

Identifying features: 1 carapace broader than long, tapering strongly posteriorly; 2 broadest point of carapace above first pair of legs; 3 first pair of legs exceptionally robust, much longer than second, and with huge claws; 4 dorsal surface red or orange with cream-coloured flecks and edges; 5 'fingers' of claws black; 6 tips of posterior pair of legs not broadly flattened like a disc; 7 abdomen short and tucked under carapace.

Comparisons: Has numerous characteristics that distinguish it from Australia's other commercial crabs; for example, a strongly tapering carapace that is widest towards the front, huge claws, and no disc-shaped tip on the last pair of legs.

Size: Reaches a massive 45 cm in carapace width and over 17.5 kg (commonly marketed at less than 20 cm in carapace width and less than 4 kg as smaller crabs are favoured).

Habitat: Marine; usually in depths of 150–275 m although occasionally caught shallower. Females apparently move onto the continental shelf (less than 200 m depth) to incubate their eggs over winter.

Fishery: A pot fishery extends across the southern coastline. It expanded rapidly during the early 1990s but catches have declined in recent years. These long-lived, slow-growing crabs may require strict management controls to ensure sustainable fishing. Research into aquaculture has been conducted in Tasmania.

Remarks: The world's heaviest crab. It was once referred to locally as 'king crab', causing confusion with members of the family Lithodidae which are traditionally marketed internationally as 'king crab'. The flesh is very succulent and is highly sought-after, particularly among Asians. Other information is provided for protein fingerprints (p. 372) and oil composition (p. 407).

Mud crab

Scylla species

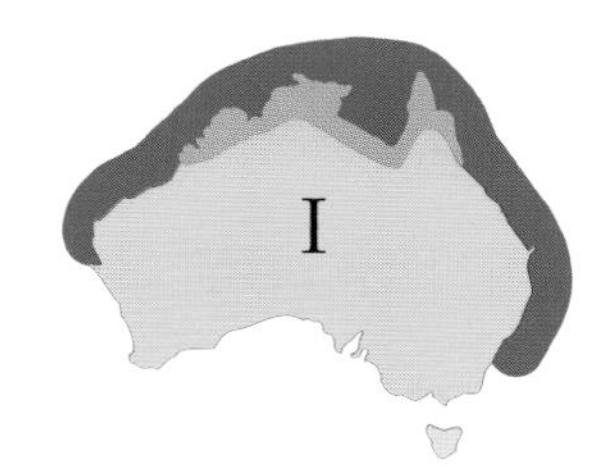

Previous names: black crab, brown crab, giant mud crab, green crab, mangrove crab, orange mud crab

Scylla serrata

Identifying features: 1 each side of carapace with 9 similar-sized spines; 2 carapace dark green or brown, sometimes mottled but without prominent spots or stripes; 3 first pair of legs very robust, much longer than second, and with prominent claws; 4 tips of last pair of legs broadly flattened (disc-shaped); 5 carapace broader than long; 6 abdomen short and tucked under carapace.

Comparisons: As with other members of the family Portunidae, or swimmer crabs, the last pair of legs are flattened at the tip. Mud crabs differ from other portunids in having very robust claws and 9 similar-sized spines on each side of the carapace. Two species occur in Australia. The giant mud crab (*S. serrata*) has sharper spines along the front margin of the carapace than the orange mud crab (*S. olivacea*).

Size: Reaches 28 cm in carapace width and 3 kg (commonly about 17 cm in width and 0.8 kg).

Habitat: Marine and estuarine; prefer muddy bottoms among mangroves, in sheltered estuaries and on tidal flats; can tolerate reduced salinity for extended periods. Usually found in shallow water but berried females occur well offshore.

Fishery: Pots, drop nets and dillies are among the main methods used by commercial and recreational fishers. Commonly taken in Queensland (male crabs only) and the Northern Territory but also in Western Australia and New South Wales. Research into aquaculture is ongoing.

Remarks: Until recently, both species were lumped under the name '*Scylla serrata*' in Australia. Marketed live and usually tied up to prevent injury to their captors (and each other). They are considered by some to be not only the best eating local crabs but among the best crustaceans. The cooked flesh is moist, flaky and sweet. Other information is provided for protein fingerprints (p. 372) and oil composition (p. 407).

Sand crab

Ovalipes species

Previous names: coral crab, rock crab, red-spotted crab, sandy crab, South Australian sand crab, surf crab, two-spotted crab

Ovalipes australiensis

Identifying features: ① each side of carapace with 5 similar-sized sharp spines; ② first pair of legs only about as long as second pair and with prominent claws; ③ carapace only slightly broader than long; ④ tips of posterior pair of legs broadly flattened (disc-shaped); ⑤ abdomen short and tucked under carapace.

Comparisons: Most similar to other swimmer crabs (Portunidae) and have broadly flattened, disc-shaped tips on the last pair of legs. The number and relative size of carapace spines and the carapace colour pattern distinguish them from the other portunids included here such as the blue swimmer crab (*Portunus pelagicus*, p. 304). Although a number of *Ovalipes* species occur in Australia, the fishery is dominated by the common sand crab (*O. australiensis*), which has 2 prominent red spots at the base of the carapace.

Size: To a carapace width of 15 cm and about 0.7 kg (commonly 11–12 cm in carapace width and 0.3–0.5 kg).

Habitat: Marine; bottom-dwelling on soft substrates in intertidal areas and off sandy beaches. Reported to depths of 100 m.

Fishery: A trap, hoop net and drop net fishery based mainly in South Australia, and now developing in Victoria. Also taken opportunistically by trawlers and coastal net fishers off South Australia and Western Australia. Recreational fishers capture sand crabs from boats or off jetties, with greatest effort in South Australia.

Remarks: Distribution shown above for *O. australiensis* only. Sometimes confused in the marketplace with the larger blue swimmer crab and both have been sold as 'sand crab'. Sweet flesh, soft and flaky, usually a little dryer than the blue swimmer crab. Other information is provided for protein fingerprints (p. 372).

Spanner crab

Ranina ranina

Previous names: frog crab, red frog crab

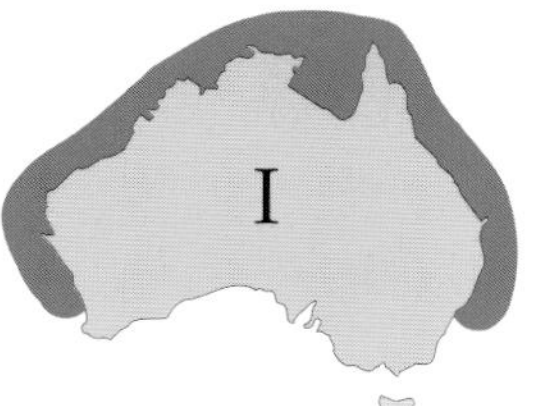

Identifying features: ① carapace longer than broad; ② carapace squared off anteriorly, with numerous spines along anterior margin; ③ abdomen short but not tucked under carapace; ④ first pair of legs very robust and with prominent, inwardly-directed (spanner-shaped) claws; ⑤ bright orange or brick-red, usually with a few white or pale blue spots on the carapace; ⑥ tips of posterior 4 pairs of legs somewhat flattened.

Comparisons: Easily distinguished from other crabs by the curved, spanner-shaped claws, short hind legs, dorsally visible abdomen, long carapace and bright orange or red colour. Younger animals are more rounded anteriorly and paler than adults.

Size: To 15 cm in carapace width and 0.9 kg (commonly about 8.5 cm in width and 0.4 kg).

Habitat: Marine; occurs from close inshore to depths of at least 100 m. Usually buried in sand from where they launch lightning attacks on passing prey items such as bottom-dwelling, small fishes.

Fishery: Caught commercially off New South Wales and Queensland where the fishery has grown rapidly since the early 1980s. Most crabs are caught using dillies and occasionally as a bycatch of prawn trawling. The recreational fishery is small.

Remarks: Males grow larger than females and are more common in the marketplace; females are mostly excluded by local size restrictions. Usually sold whole and cooked. Highly regarded food species in many countries. The white, soft flesh differs noticeably in taste from that of other crabs. Other information is provided for protein fingerprints (p. 372) and oil composition (p. 407).

Banana prawn

Fenneropenaeus indicus & *F. merguiensis*

Previous names: Indian banana prawn, redleg banana prawn, white banana prawn, white prawn

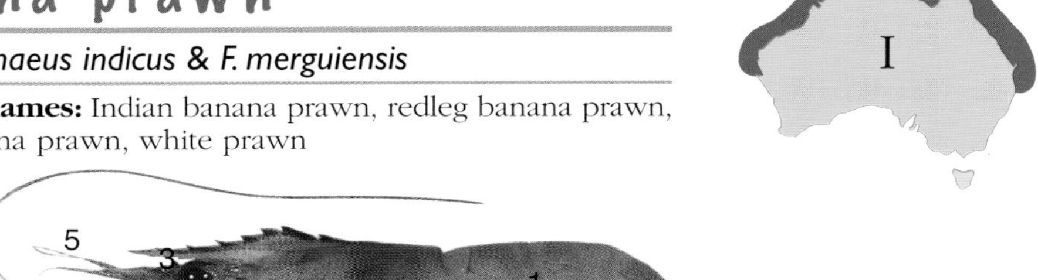

Fenneropenaeus merguiensis

Identifying features: 1 body translucent grey or yellow, speckled with small, dark spots; 2 hepatic ridge absent; 3 lower margin of rostrum with 2–5 spines; 4 telson pointed, without spines; 5 antennule flagella shorter than carapace; 6 second pair of legs not differing dramatically in size from first pair; 7 body compressed.

Comparisons: The two species of banana prawn differ in leg colour. The white banana prawn (*F. merguiensis*) has cream or yellow legs whereas the redleg banana prawn (*F. indicus*) has pink or red legs. Many other prawns have a similar body colour—generally pale and without the bands characteristic of tiger prawns. Banana prawns can be distinguished by the presence of 2–5 spines on the lower margin of the rostrum and the absence of spines on the telson. In addition to colour differences, they differ from tiger prawns (p. 323) of the genus *Penaeus* by the lack of a hepatic ridge.

Size: White banana prawns reach 25 cm in body length (about 6.3 cm in carapace length) and 75 g (commonly about 30 g). Redleg banana prawns are slightly smaller.

Habitat: Mostly marine. Juvenile white banana prawns inhabit sheltered and often turbid estuaries and rivers, usually among mangroves. Adults of both species are found over muddy or sandy bottoms in coastal waters, with redleg banana prawns preferring deeper water to depths of 90 m. Both species form large aggregations.

Fishery: Support extensive fisheries throughout their range. In northern Australia, trawlers target aggregations between Exmouth Gulf (WA) and Brisbane (Qld) with the bulk of the catch coming from the Gulf of Carpentaria. Most fishing is done during the day and good catches are often linked with heavy rainfall.

Remarks: Most of Australia's commercial prawns belong to the family Penaeidae and, until recently, many were classified in the genus *Penaeus*. However, a recent major revision has reclassified many species into new genera. For example, the two banana prawns were transferred from *Penaeus* to *Fenneropenaeus*. The flesh is sweet and moist, with a medium texture. Other information is provided for protein fingerprints (p. 372) and oil composition (p. 408).

Bay prawn

Metapenaeus bennettae & *M. insolitus*

Previous names: emerald shrimp, greasyback prawn, greentail prawn, inshore greasyback prawn, river prawn

Metapenaeus bennettae

Identifying features: 1 telson pointed, without obvious spines; 2 ischial spine on inside edge of first pair of legs reduced to a small blunt angle; 3 fine hairs on abdomen often giving a 'greasy' feel; 4 lower margin of rostrum without spines; 5 body translucent brown or green with dark brown speckling, tips of tail fan green; 6 antennule flagella shorter than carapace; 7 second pair of legs not differing dramatically in size from first pair; 8 body compressed.

Comparisons: Most similar to other *Metapenaeus* species—Endeavour prawns (p. 315) and the school prawn (p. 322)—but the telson lacks defined spines and the ischial spine is relatively minute. Distinguished from most other commercial prawns by the lack of lower spines on the rostrum. The greentail prawn (*M. bennettae*) is distinguished from the greasyback prawn (*M. insolitus*) in having more distinct patches of hairs on the abdomen.

Size: To 13 cm in body length. The average commercial size is small, about 8 g.

Habitat: Marine, estuarine and freshwater. Small juveniles have been found in rivers as far as 35 km from the sea. Larger juveniles are usually found nearer the coast in shallow mangrove areas and adults are most common in coastal marine waters to depths of 35 m. Adults prefer mud bottoms but are also taken over sand.

Fishery: Greentail prawns are found along Australia's most densely populated coastline—the southeastern seaboard—and have been the target of both commercial and recreational fishers for decades. Most of the commercial catch is trawled at night from the Brisbane River and Moreton Bay (Qld). Greasyback prawns are taken mostly near Darwin (NT).

Remarks: Marketed mainly fresh and often lower priced than other prawns. The flesh is very sweet, firm and moist. Other information is provided for protein fingerprints (p. 372) and oil composition (p. 408).

Black tiger prawn

Penaeus monodon

Previous names: blue tiger prawn, giant tiger prawn, jumbo tiger prawn, leader prawn, panda prawn, tiger prawn

Identifying features: 1 base colour brown with bands of dark brown, black or blue and cream; 2 antennae uniform brown or red; 3 upper margin of rostrum with 7–8 spines and lower margin with 2–3 spines; 4 adrostral ridge extending back to about the level of first rostral spine; 5 telson pointed, without spines; 6 antennule flagella shorter than carapace; 7 second pair of legs not differing dramatically in size from first pair; 8 body compressed.

Comparisons: Three other species marketed collectively as 'tiger prawn' (p. 323), are banded. The black tiger prawn is distinguished from the brown (*P. esculentus*) and grooved (*P. semisulcatus*) tiger prawns, by the presence of 7–8 upper and 2–3 lower rostral spines (5–7 upper and 3–4 lower in the brown tiger prawn) and in having the adrostral ridge extending only to the level of the first rostral spine (extending beyond the first rostral spine in the grooved tiger prawn). In the third tiger prawn, the kuruma prawn (*Marsupenaeus japonicus*), the adrostral ridge extends back to near the carapace margin.

Size: One of the largest prawns in Australian waters, reaching a body length of 35 cm and exceeding 150 g. Usually harvested between 10 and 13 cm total length (20–30 g).

Habitat: Marine and estuarine; adults marine, preferring sandy and muddy bottoms in coastal waters to depths of 150 m but usually in less than 30 m. Juveniles live in estuaries, among seagrass beds and mangroves.

Fishery: Occasionally trawled off the coasts of Queensland, the Northern Territory and Western Australian but most production is from aquaculture. Black tiger prawns and kuruma prawns account for most of the prawn aquaculture production in Australia. A tolerant, fast-growing species farmed mostly between Cooktown and Brisbane (Qld) but also in New South Wales and the Northern Territory.

Remarks: Farmed animals sometimes retain their juvenile colouration—grey and with less distinct bands. Almost all local product is sold in Australia. The flesh is firm with a sweet flavour. Other information is provided for protein fingerprints (p. 372) and oil composition (p. 408).

Endeavour prawn

Metapenaeus endeavouri & *M. ensis*

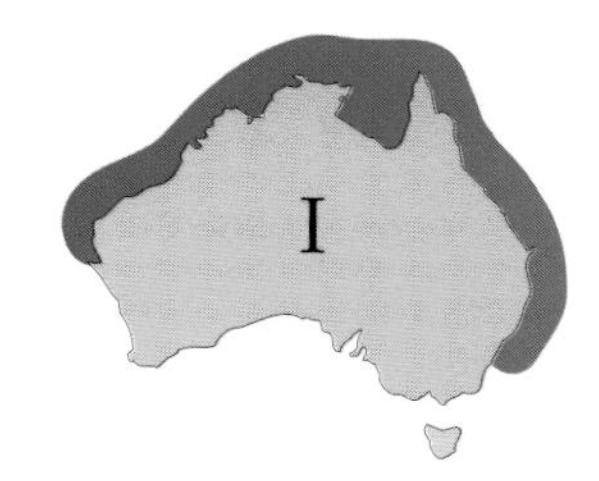

Previous names: blue Endeavour prawn, bluetail Endeavour prawn, Endeavour shrimp, greasyback prawn, red Endeavour prawn, redtail Endeavour prawn

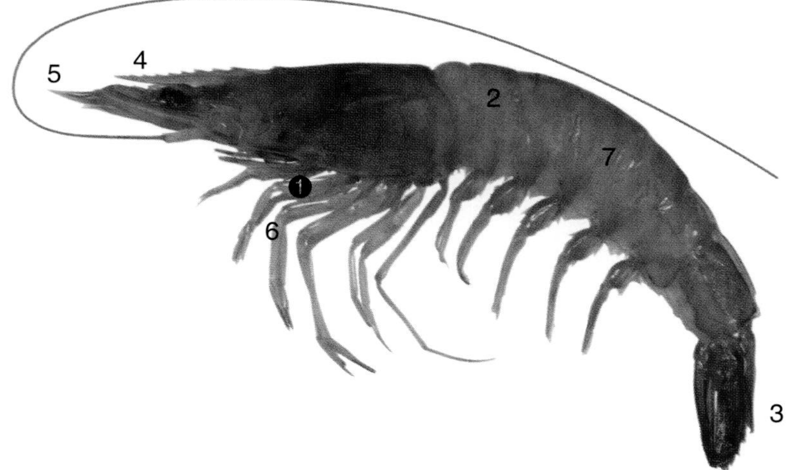

Metapenaeus ensis

Identifying features: 1 distinct ischial spine on inside of first pair of legs; 2 body pale brown or pink, with bright blue or red edge to the tail fan; 3 telson pointed, with or without 3 pairs of movable spines; 4 lower margin of rostrum without spines; 5 antennule flagella shorter than carapace; 6 second pair of legs not differing dramatically in size from first pair; 7 body compressed.

Comparisons: Easily distinguished from most other commercial prawns in lacking spines on the lower margin of the rostrum. However, multiple features are required to separate them from other *Metapenaeus* species, such as bay prawns (*M. bennettae* and *M. insolitus*, p. 313). These include colour, the number of pairs of spines on the telson and the prominence of the ischial spine. Common names are taken from the colour of the tail fan. The blue Endeavour prawn (*M. endeavouri*) has a blue edge to the tail fan while in the red Endeavour prawn (*M. ensis*) the tail fan is red. Blue Endeavour prawns have 3 pairs of spines on the telson which are absent in red Endeavour prawns.

Size: Blue Endeavour prawns reach at least 19 cm in body length (commonly 7–14 cm). Red Endeavour prawns are slightly smaller. Average commercial weight is about 30 g.

Habitat: Adults marine; benthic in turbid coastal waters to depths of 95 m on sandy or muddy bottoms. Juveniles usually associated with seagrass beds in estuaries, but juvenile red Endeavour prawns are also found among mangroves and on mud flats.

Fishery: Taken with otter and beam trawls at night in rivers, inshore areas or offshore continental shelf waters of Western Australia, the Northern Territory and northern Queensland. The largest catches are taken in the Gulf of Carpentaria. In Torres Strait, blue Endeavour prawns comprise nearly half the total prawn catch.

Remarks: The flesh has a stronger flavour than most other prawns. Other information is provided for protein fingerprints (p. 372) and oil composition (p. 408).

Freshwater prawn

Macrobrachium species

Previous names: cherabim, cherabin, cherabun

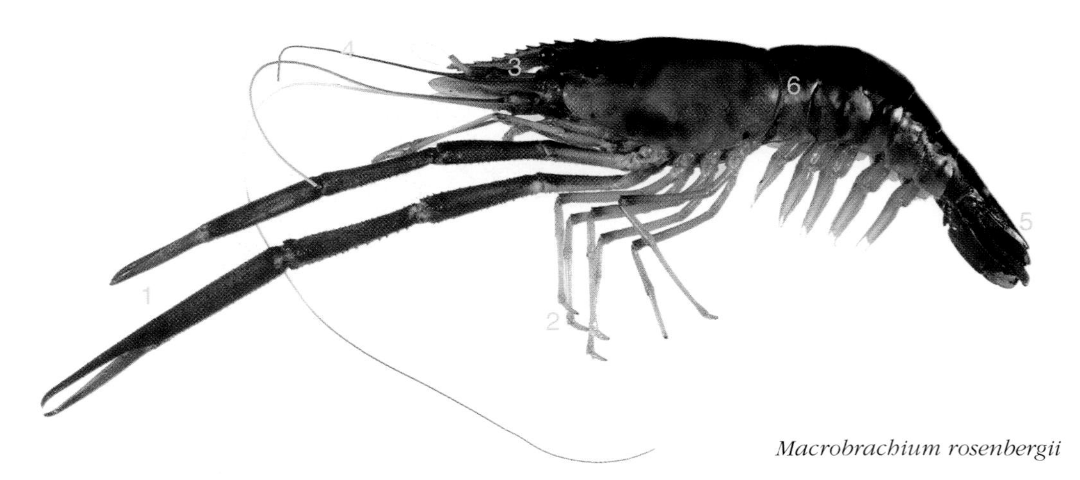

Macrobrachium rosenbergii

Identifying features: 1 second pair of legs much larger than the first and with prominent claws; 2 third pair of legs without claws; 3 lower margin of rostrum with spines; 4 antennule flagella just shorter than carapace length; 5 telson pointed, with 2 pairs of spines; 6 body compressed.

Comparisons: Belong to the family Caridae and, unlike all the other prawns included here, have a pair of extremely long claws and lack small claws on the third pair of legs. Sometimes confused with freshwater crayfishes (*Cherax* species, pp 324–326) and scampi (Nephropidae, p. 321) but have the second pair of legs enlarged (rather than the first) and a pointed telson (rather than broadly convex).

Size: Reaches over 11 cm in carapace length in Australia (commonly 10–20 cm in body length). Elsewhere, body length reported to 34 cm.

Habitat: Freshwater, estuarine and marine. A large freshwater species, the cherabin (*M. rosenbergii*), occurs in permanent streams, lakes and ponds; other smaller species are found in freshwater, estuaries and inshore marine waters to depths of 20 m and often among seagrass beds. Most freshwater species migrate to estuaries to breed.

Fishery: Several species occur in Australia but only the cherabin is marketed. It is regarded as an extremely important commercial crustacean in south-east Asia. An aquaculture industry has developed rapidly in many countries since the 1960s. In Australia, the cherabin is farmed on a small scale in Western Australia and has great potential for expansion.

Remarks: Distribution shown above for *M. rosenbergii* only. The smaller coastal marine species, which are sometimes taken by recreational fishers netting for school and bay prawns, are extremely abundant. The flesh tastes slightly sweet and delicate. Other information is provided for protein fingerprints (p. 372).

King prawn

Melicertus latisulcatus & *M. plebejus*

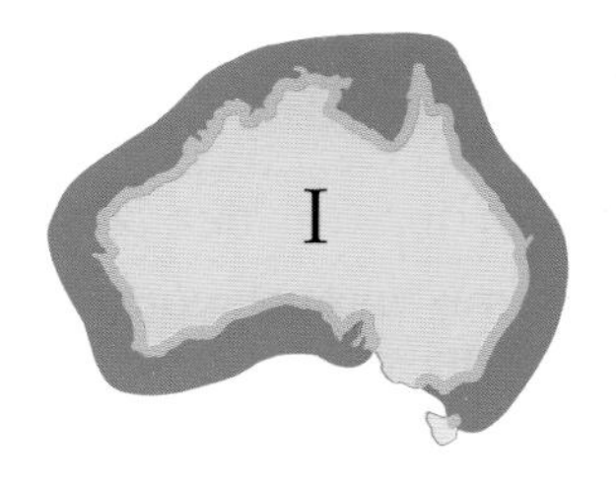

Previous names: blueleg prawn, eastern king prawn, eastern prawn, ocean king prawn, sand prawn, western king prawn, western prawn

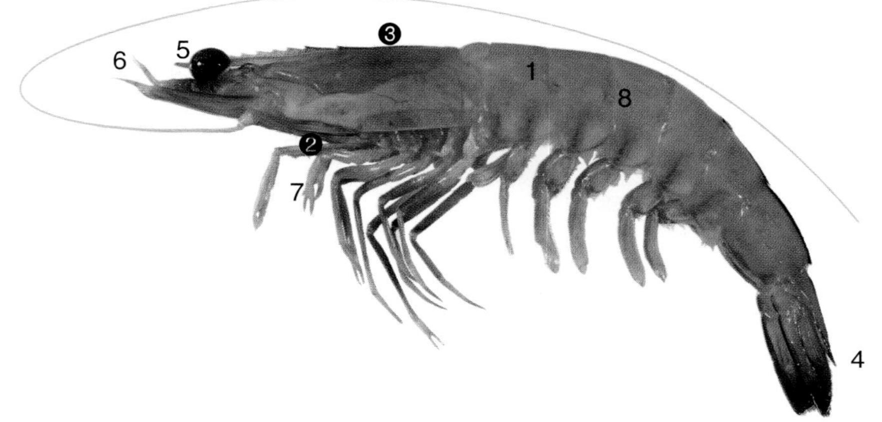

Melicertus latisulcatus

Identifying features: 1 body yellow, cream or light brown, without prominent bands or spots; 2 no ischial spine on first pair of legs; 3 top of carapace with a pair of longitudinal parallel adrostral grooves (and associated ridges) extending back almost to rear margin of carapace; 4 telson pointed, with 3 pairs of movable lateral spines; 5 lower margin of rostrum with 1 small spine; 6 antennule flagella much shorter than carapace; 7 second pair of legs not differing dramatically in size from first pair; 8 body compressed.

Comparisons: Together with the kuruma prawn (*Marsupenaeus japonicus*, p. 323) and the redspot king prawn (*Melicertus longistylus*, p. 319), they differ from other commercial species in having 3 pairs of movable spines on the telson and parallel grooves running right along the top of the carapace. Similarly coloured to the blue Endeavour prawn (*Metapenaeus endeavouri*, p. 315) but have a lower rostral spine (otherwise absent). Colour patterns distinguish western (*Melicertus latisulcatus*) and eastern (*M. plebejus*) king prawns from each other and from the redspot king prawn and kuruma prawn. The legs are blue in the western king prawn but usually cream in the eastern king prawn.

Size: Eastern king prawn to 30 cm body length (commonly 14–21 cm); western king prawn to 20 cm (commonly 10–16 cm). Average commercial weight is about 50 g.

Habitat: Marine; adults are found on a range of bottom types, including rock, sand, mud or gravel, in depths to over 220 m. Juveniles prefer shallow coastal waters and estuaries.

Fishery: Western king prawns are usually trawled at night, mostly off South Australia and Western Australia but also off the Northern Territory and Queensland. Eastern king prawns are of considerable commercial importance off Queensland, New South Wales and, to a lesser degree, Victoria. Juveniles are targeted inshore by trawling and a variety of nets, whereas adults are trawled offshore. Recreational fishers use scoop, drag and scissor nets.

Remarks: Recently transferred from *Penaeus* to *Melicertus*. The flesh is highly regarded but can be tough if poorly handled. Other information is provided for protein fingerprints (p. 372) and oil composition (p. 408).

Prawn

Infraorder Caridea & superfamily Penaeoidea

Previous names: coral prawn, giant red prawn, red prawn, velvet prawn

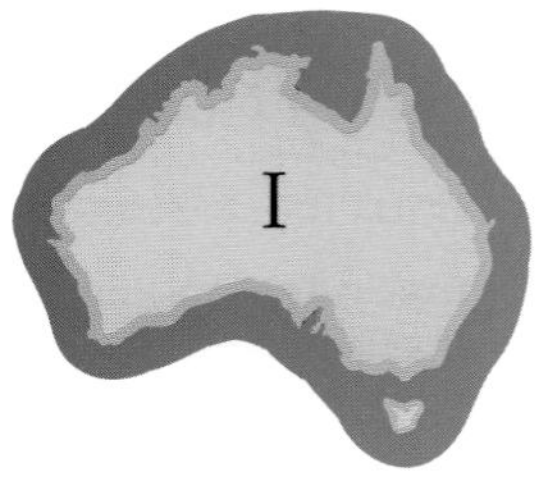

Aristaeopsis edwardsiana

Identifying features: ① rostrum well developed; ② abdomen longer than carapace and compressed; ③ telson pointed; ④ antennae very long, thin and roughly cylindrical in cross-section.

Comparisons: Several characteristics distinguish members of the superfamily Penaeoidea (tropical prawns) from members of the infraorder Caridea (freshwater prawns, p. 316). In penaeoid prawns, the third pair of legs has claws and the first abdominal segment overlaps the second. In carid prawns, which are usually smaller, the third pair of legs is not clawed and the second abdominal segment overlaps the first. Distinguished from scampi (Nephropidae, p. 321) in having a pointed (rather than broadly convex) telson.

Size: Perhaps the largest species marketed in Australia is the scarlet prawn, *Aristaeopsis edwardsiana*, which reaches 35 cm in body length and 180 g. However, many species (such as coral prawns, *Metapenaeopsis* species) are much smaller.

Habitat: Marine, estuarine and freshwater. Occupy a vast range of habitats, including freshwater ponds, rivers, estuaries, coastal inshore areas and deep ocean waters, to depths exceeding 700 m. Bottom types are usually soft (sand or mud).

Fishery: Prawn harvesting (both wild and farmed) contributes enormously to the Australian economy. Most populated sections of the mainland Australian coastline are home to a prawn fishing fleet and the farming sector accounts for an ever increasing portion of total prawn production.

Remarks: Any species of shrimp or prawn can be sold simply as 'prawn' although there may be an economic disadvantage in doing so. In Australia, small prawns are sometimes called 'shrimps'. However, internationally, the term 'shrimp' is often used for all marine species (regardless of size) and the 'prawn' is reserved for freshwater species. Very popular food species. Other information is provided for protein fingerprints (p. 372) and oil composition (p. 408).

Redspot king prawn

Melicertus longistylus

Previous name: redspot prawn

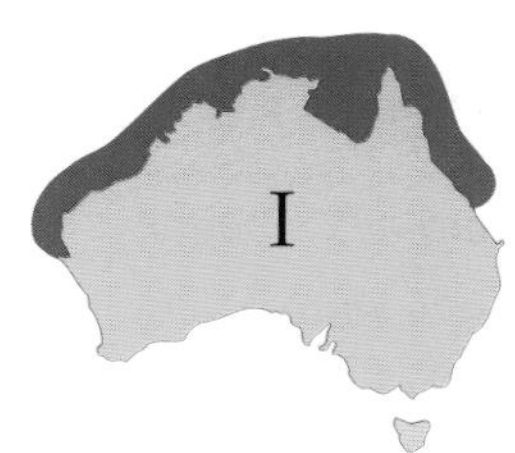

Identifying features: 1 prominent red spot on each side of the third abdominal segment; 2 ischial spine present on the first pair of legs; 3 top of carapace with a pair of longitudinal parallel adrostral grooves extending back almost to rear margin of carapace; 4 telson pointed, and with 3 pairs of movable lateral spines; 5 lower margin of rostrum with 1 spine; 6 antennule flagella much shorter than carapace; 7 second pair of legs not differing dramatically in size from first pair; 8 body compressed.

Comparisons: Easily distinguished from the closely related king prawns (*M. latisulcatus* and *M. plebejus*, p. 317) by the presence of a red spot on each side of the abdomen and a small spine on the third segment of the first pair of legs. Like king prawns and the kuruma prawn (*Marsupenaeus japonicus*, p. 323), has 3 pairs of movable spines on the telson and parallel grooves running right along the top of the carapace. These features separate it from other local commercial prawns.

Size: Reported to 21 cm in body length (commonly 10–15 cm).

Habitat: Marine; usually found in the vicinity of coral reefs. Juveniles frequent shallow lagoons whereas adults migrate to deeper water (to depths of 60 m or so) and settle on hard sand or mud.

Fishery: Caught by trawlers operating at night mainly north of Gladstone along Queensland's east coast. Also targeted off the eastern tip of Cape York (Qld) and taken incidentally elsewhere in its range. Exported and consumed locally.

Remarks: Until recently, classified in the genus *Penaeus*. Its flesh is popular but needs to be treated carefully to prevent it from becoming tough. Other information is provided for protein fingerprints (p. 372) and oil composition (p. 408).

Royal red prawn

Haliporoides sibogae

Previous name: pink prawn

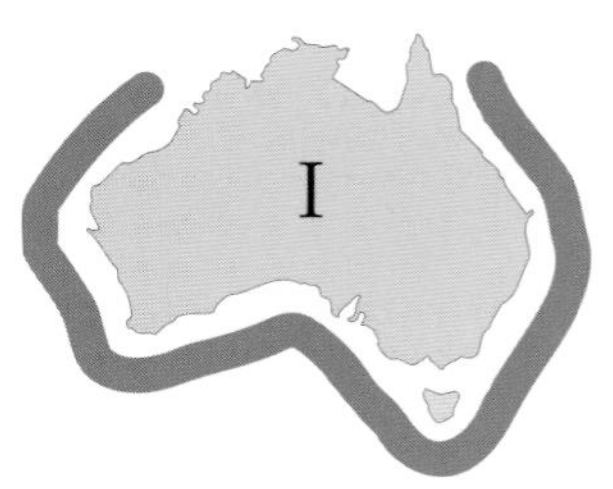

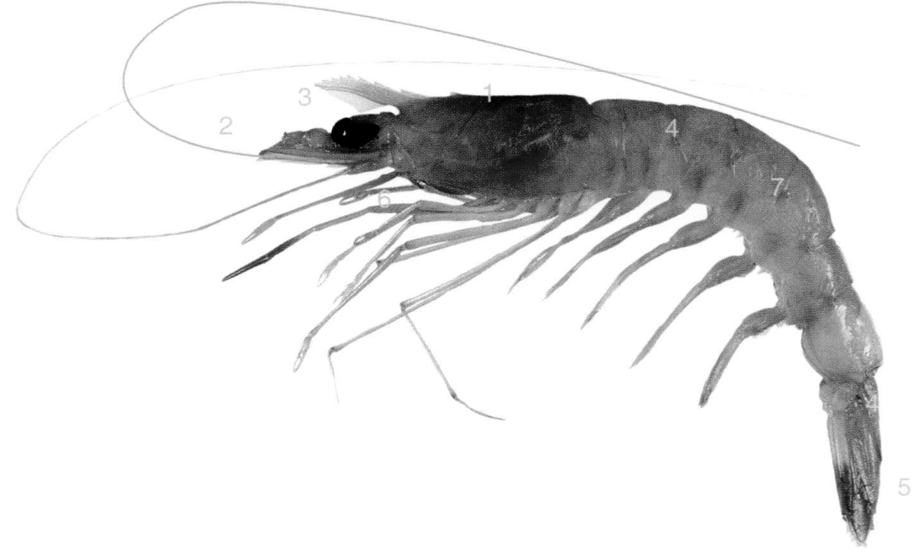

Identifying features: ① cervical groove reaching to top of carapace; ② antennule flagella very long, not less than three times carapace length; ③ lower margin of rostrum with 1–2 spines; ④ body and tail fan uniformly pink or red; ⑤ telson pointed, and with 1 pair of obvious fixed lateral spines; ⑥ second pair of legs not differing dramatically in size from first pair; ⑦ body compressed.

Comparisons: The only species of the family Solenoceridae of commercial importance in Australia. Distinguished from other commercial prawns and shrimps by having cervical grooves reaching the top of the carapace. The cervical grooves of species of the family Penaeidae, which includes most of the other prawns included here, never reach the top of the carapace.

Size: Reaches 20 cm in body length (commonly 7–10 cm), 5 cm in carapace length and about 25 g. As with most prawns, females are larger than males.

Habitat: Marine; found on mud bottoms in depths of 220–820 m (typically 365–550 m). Reported to occur at shallower end of range during winter.

Fishery: Mostly caught using demersal trawl nets off New South Wales between Sydney and Ulladulla in depths of 400–500 m. However, probably widely distributed and abundant across southern Australia. About 300 tonnes are caught each year with small quantities exported to Japan. Also of minor commercial importance off northwestern Western Australia.

Remarks: Considered to be of good quality and makes excellent eating if handled properly. The flesh spoils quickly so the prawns must be refrigerated or frozen quickly after capture. Other information is provided for protein fingerprints (p. 372).

Scampi

Nephropidae

Previous names: none

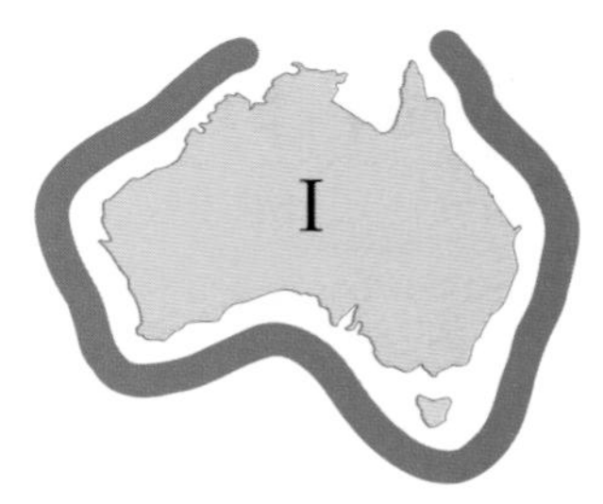

Metanephrops boschmai

Identifying features: 1 first pair of legs greatly enlarged; 2 rostrum with pairs of spines along its upper sides; 3 antennule flagella shorter than carapace; 4 telson broadly convex; 5 body tubular.

Comparisons: Although similar in appearance to prawns, scampi are most closely related to freshwater crayfishes (*Cherax* species) and rocklobsters (Palinuridae). The rostrum is better developed and has more pronounced spines (some immediately posterior to rostrum) than that of freshwater crayfishes. The greatly enlarged first pair of legs, the broadly convex telson, and the paired spines along the sides of the rostrum distinguish scampi from prawns.

Size: Reported to 25 cm in body length (12 cm carapace length) but commonly caught in Australian waters at less than half that size.

Habitat: Marine; usually found on soft bottoms, including both mud and sand, in depths of 170–1000 m. Depth preference varies with species but is often within a narrow range on the continental slope.

Fishery: Targeted by trawlers, mostly off northwestern Australia, since 1985. Recent annual landings vary greatly depending on shifts in effort to other fisheries. About six species are caught; the velvet scampi, *Metanephrops velutinus*, is the most common.

Remarks: Despite small catches, highly regarded both locally and overseas. They are often served whole (as a garnish) and in Japan are eaten raw. Other information is provided for protein fingerprints (p. 372).

School prawn

Metapenaeus species

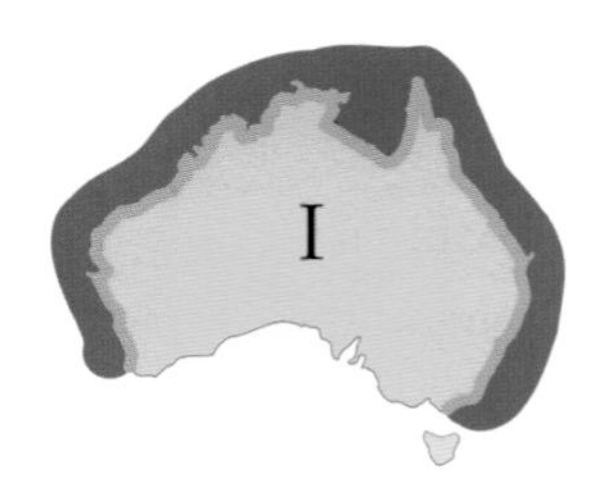

Previous names: bay prawn, Endeavour prawn, greasyback bay prawn, greentail prawn, inshore greasyback prawn, river prawn, schoolie, Stockton Bight prawn, western school prawn, white river prawn, york prawn

Metapenaeus macleayi

Identifying features: ① lower margin of rostrum without spines; ② antennule flagella shorter than carapace; ③ second pair of legs not differing dramatically in size from first pair; ④ telson pointed; ⑤ body compressed.

Comparisons: Although any *Metapenaeus* species, including bay prawns (p. 313) and Endeavour prawns (p. 315), can be sold under the name of 'school prawn', it is mostly used for *M. macleayi*, commonly called the 'school prawn'. *M. macleayi* is translucent with green or brown speckling and has 4 pairs of movable spines on the telson. Also, the tip of the rostrum is curved upward and lacks spines.

Size: Although reaching 17.5 cm in body length, commonly harvested at about half that size.

Habitat: Marine and estuarine; juveniles on estuarine seagrass beds or sandy bottoms in shallow water, larger prawns prefer coarse sandy bottoms near river mouths to depths of about 55 m.

Fishery: Although previously farmed in New South Wales, most of the school prawn (*M. macleayi*) catch is now trawled or netted in estuaries south of Noosa (Qld). These prawns are sold locally. A small recreational fishery exists in Western Australia for the western school prawn, *M. dalli*, using scoop and drag nets.

Remarks: The flesh varies, but usually has a mild taste and a medium texture. Other information is provided for protein fingerprints (p. 372).

Tiger prawn

Marsupenaeus japonicus, Penaeus esculentus & P. semisulcatus

Previous names: brown tiger prawn, common tiger prawn, green tiger prawn, grooved tiger prawn, Japanese king prawn, Japanese tiger prawn, kuruma prawn, northern tiger prawn

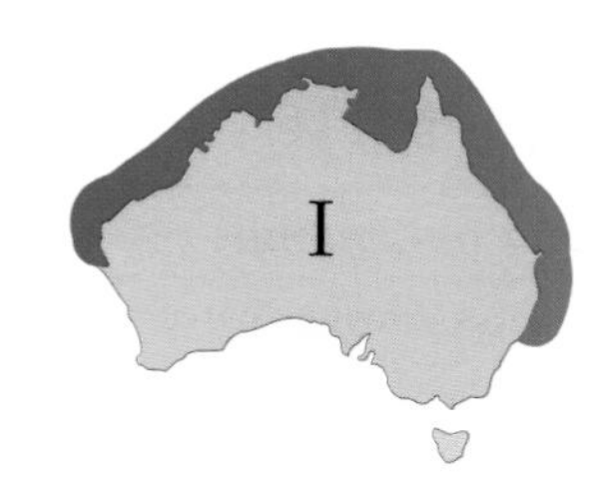

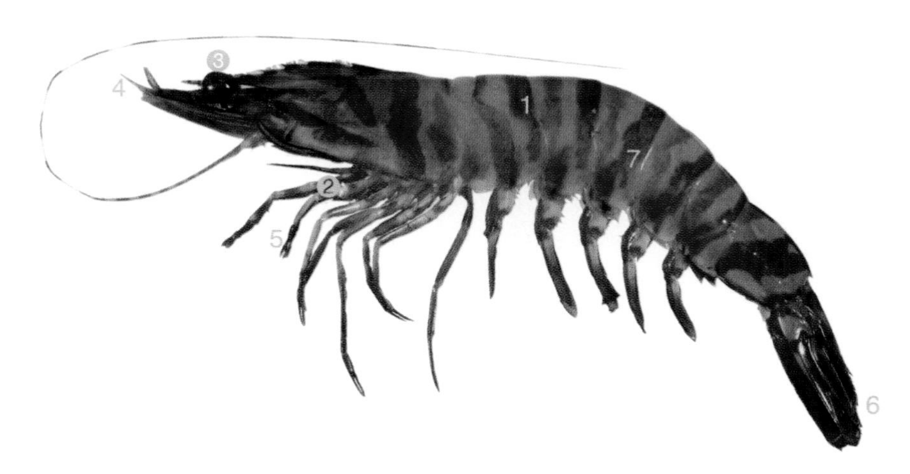

Penaeus esculentus

Identifying features: ➊ light brown or pink (sometimes greenish) with many dark brown or red bands; ➋ ischial spine absent from first pair of legs; ➌ lower margin of rostrum with 1–5 spines; ➍ antennule flagella shorter than carapace; ➎ second pair of legs not differing dramatically in size from first pair; ➏ telson pointed; ➐ body compressed.

Comparisons: Although banded, the kuruma prawn (*M. japonicus*) is most closely related to the king prawns (*Melicertus latisulcatus* and *M. plebejus*, p. 317). It differs from the brown (*P. esculentus*) and grooved (*P. semisulcatus*) tiger prawns in having non-banded antennae and 3 pairs of movable spines on the telson. The black tiger prawn (*P. monodon*, p. 314) also has uniformly coloured antennae. The carapace banding patterns of the two *Penaeus* tiger prawns differ but these species are more reliably distinguished by the posterior extension of the adrostral ridge. In the brown tiger prawn this ridge reaches or just falls short of the first rostral spine while in the grooved tiger prawn it extends beyond the first rostral spine.

Size: To 30 cm in body length (commonly 11–20 cm).

Habitat: Mostly marine; juveniles prefer shallow estuarine waters, often among seagrass beds, adults are usually found in coastal waters over sandy or muddy bottoms in depths less than 20 m but do occur to 200 m.

Fishery: Brown and grooved tiger prawns are mostly trawled at night in northern Australia between Shark Bay (WA) and northern New South Wales. They are often caught with other commercial prawns. Kuruma prawns are not caught in the wild but are farmed in Queensland (between Bundaberg and Brisbane) and northern New South Wales. Most are exported live to Japan where they can fetch prices as high as $200 per kg.

Remarks: Until recently, *Marsupenaeus japonicus* was classified in the genus *Penaeus*. The flesh has reddish bands and its delicate, sweet flavour is highly regarded. Other information is provided for protein fingerprints (p. 372) and oil composition (p. 408).

Marron

Cherax tenuimanus

Previous name: West Australian marron

Identifying features: ① top of head with 5, almost parallel ridges; ② rostrum with 3 pairs of spines; ③ telson broadly convex, with 2 small spines on base; ④ first pair of legs greatly enlarged and bearing prominent claws; ⑤ abdomen long and not tucked under body; ⑥ carapace about twice as long as broad.

Comparisons: Australia's commercially important freshwater crayfishes all belong to the genus *Cherax* (Parastacidae). The marron, a native of Western Australia, is distinguished from other species by having 5 parallel ridges on top of the head, 3 pairs of rostral spines and 2 small spines on the telson.

Size: Attains a maximum carapace length of 18 cm (total length 38.5 cm) and 2.2 kg.

Habitat: Freshwater; prefers sandy bottoms in deeper sections of permanent rivers and streams. Also survives in well-oxygenated ponds and dams.

Fishery: Commercial culture began in the 1970s and is now well established. While based primarily in Western Australia, there are also farms in New South Wales, South Australia and a number of locations overseas. Growers in Queensland now farm their local species, redclaw (*C. quadricarinatus*, p. 325), which is better suited to the local climate. Wild marron populations are targeted by recreational fishers using drop nets, scoop nets and pole snares. Strict management regulations apply.

Remarks: The largest *Cherax* species and one of the world's largest freshwater crayfishes. The term 'crayfish' is usually used for freshwater species while 'lobster' is reserved for marine species. Freshwater crayfishes are considered a delicacy by many Australians. The cooked flesh is moist and firm. Other information is provided for protein fingerprints (p. 373).

Redclaw

Cherax quadricarinatus

Previous names: clear-water crayfish, marron, Queensland marron, tropical blue

Identifying features: ① top of head with 4, almost parallel ridges, the 2 middle ones very long; ② rostrum usually with 3 pairs of small spines; ③ telson broadly convex, without spines; ④ first pair of legs greatly enlarged and bearing prominent claws; ⑤ abdomen long and not tucked under body; ⑥ carapace about twice as long as broad.

Comparisons: Distinguished from the other commercial freshwater crayfishes by having 4 ridges on top of the head (5 in marron, *C. tenuimanus*, p. 324) and 3 pairs of rostral spines (one poorly-developed pair in yabby, *C. destructor*, p. 326). Males have a characteristic red patch on the outer surface of each claw.

Size: To at least 10 cm in carapace length and 0.6 kg (average size about 15 cm in body length and less than 0.1 kg).

Habitat: Freshwater; a very tolerant crayfish, found naturally in a variety of habitats to depths of 5 m in north-flowing rivers and streams of Queensland and the Northern Territory. Also contained in culture ponds along Queensland's east coast, in northern New South Wales and, to a lesser extent, in the Northern Territory and Western Australia.

Fishery: Numerous attributes, such as fast growth rate and tolerance to low oxygen levels, make for excellent aquaculture potential. Commercial operations began in Queensland in the mid-1980s. Usually sold live locally and exported to Asia. The recreational fishery is very small and, in Queensland, fishers are required to consume their prey at the site of capture.

Remarks: Only natural distribution shown above. Disease outbreak is one of the risks involved with aquaculture and strict import regulations apply throughout Australia. Edible qualities compare favourably with those of marine rocklobsters. Other information is provided for protein fingerprints (p. 373) and oil composition (p. 408).

Yabby

Cherax species except *C. quadricarinatus* & *C. tenuimanus*

Previous names: crawbob, freshwater crayfish, gilgie, gilgy, koonac, lobby, yabbie

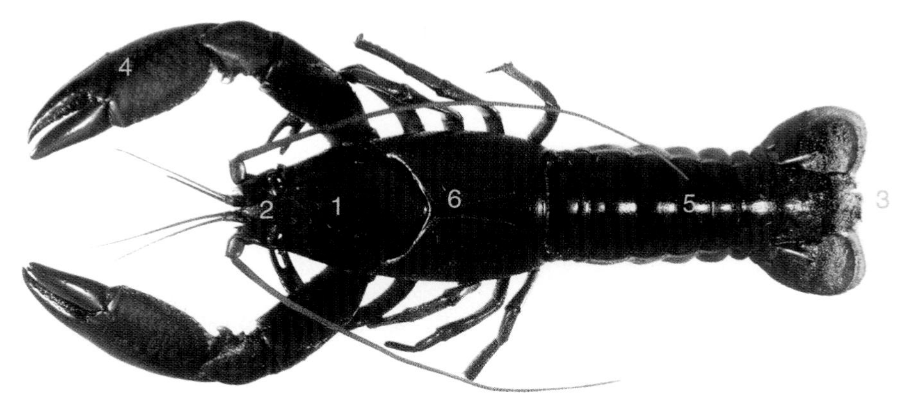

Cherax destructor

Identifying features: ① ridges on top of the head variable in number and often not parallel; ② rostrum smooth or with up to 2 pairs of spines; ③ telson broadly convex, usually without small spines; ④ first pair of legs greatly enlarged and bearing prominent claws; ⑤ abdomen long and not tucked under body; ⑥ carapace about twice as long as broad.

Comparisons: The yabby (*C. destructor*) is distinguished from marron (*C. tenuimanus*, p. 324) by having 4 ridges on the head (rather than 5) and by lacking spines on the telson. It differs from redclaw (*C. quadricarinatus*, p. 325) by having only 1 poorly-developed pair of rostral spines (rather than 3).

Size: The main commercial yabby reaches 7 cm in carapace length and 16 cm in body length (commonly 12 cm in body length and less than 70 g). Some other species, such as the koonac (*C. preissii*), reach 9 cm carapace length and 20 cm body length.

Habitat: Freshwater; wide variety of habitats such as rivers, streams, pools, swamps, billabongs, dams and temporary water bodies. Yabbies prefer slow-flowing water and are found to depths of 5 m. Some species can survive periods of drought by burrowing into river beds where they wait for more favourable conditions to return before emerging.

Fishery: The yabby (*C. destructor*) is farmed in New South Wales, South Australia, Victoria and Western Australia and a number of locations overseas. Western Australia is the largest producer. Additional species are farmed; *C. plebejus* and *C. glaber* in Western Australia and *C. albidus* in Victoria. Yabbies are also fished commercially in New South Wales, Victoria and South Australia using baited pots and drop nets. A number of species are caught recreationally using pots and hoop nets.

Remarks: The marketing name 'yabby' is used collectively for all *Cherax* species except marron and redclaw. The common name of the most well known species, *C. destructor*, is also 'yabby'. The taxonomy and distribution of some of Australia's *Cherax* species remains unclear and more research is needed. Most yabbies are sold cooked; the flesh is sweet and well regarded. Other information is provided for protein fingerprints (p. 373) and oil composition (p. 408).

Molluscs

G. K. Yearsley and P. R. Last

Blacklip abalone

Haliotis rubra

Previous names: abalone, muttonfish

Identifying features: 1 foot with black lip; 2 outer shell surface rough, with deep outward-radiating grooves; 3 exposed outer shell surface reddish-brown or reddish-green; 4 spiral ridges on outer surface very fine; 5 single, ear-shaped shell, with low spire and broad aperture occupying most of the underside.

Comparisons: Molluscs all have soft bodies without an internal skeleton but vary tremendously in body form. Abalones are gastropod (single shell) molluscs and have their foot (the edible part) and internal organs protected by one, spirally-coiled shell. The shell is flat relative to some other gastropods such as periwinkles (p. 346) and garden snails. Blacklip abalone is distinguished from other commercial abalones by having a black lip and deep outward-radiating grooves on the outer surface of the shell.

Size: Shell length (maximum diameter) to about 22 cm (commonly 14–16 cm).

Habitat: Marine; exposed or high wave energy coastal waters to depths of at least 40 m. Concealed in rocky gutters, fissures or overhangs, and similar bottoms covered with kelp. Abalone larvae are briefly pelagic before settling on reefs.

Fishery: Abalone are harvested commercially by divers using independent vessels or mother boats. The Australian abalone fishery expanded rapidly after 1960, with blacklip abalone by far the dominant species. It is mainly taken off Tasmania and Victoria, with smaller catches off New South Wales and South Australia. The majority of Australian product is frozen or canned and exported to Asia. Abalone are a popular target of sport divers. Commercial and recreational fisheries are strictly controlled. Expanding aquaculture industry in southern states.

Remarks: Found only in southern Australian waters where they feed mainly on drift algae. Abalone flesh can be tough if it is improperly prepared or overcooked. Other information is provided for protein fingerprints (p. 373) and oil composition (p. 409).

Brownlip abalone

Haliotis conicopora

Previous names: abalone, muttonfish

Identifying features: ① foot lip with light brown and dark brown patches; ② outer shell surface rough, with shallow outward-radiating grooves; ③ exposed outer shell surface very dark, reddish-brown or reddish-green; ④ spiral ridges on outer surface very fine; ⑤ single, ear-shaped shell, with low spire and broad aperture occupying most of the underside.

Comparisons: Very similar to the blacklip abalone (*H. rubra*, p. 328) but with a darker shell (algae usually needs to be removed before shell colour can be seen) and the outward-radiating ribs are not as deep. Distinguished from other commercial abalone in having a brown lip around the edge of their foot (green in the greenlip abalone, *H. laevigata*, p. 330) and very fine spiral ridges on shell (very pronounced in Roe's abalone, *H. roei*, p. 331).

Size: Shell length reaches about 22 cm (commonly 14–16 cm).

Habitat: Marine; cryptic, lives in coastal waters to depths of 30 m. Prefers areas of reduced wave action and characterised as a 'calm water' species. Favoured water temperatures are 12–22°C.

Fishery: Small quantities are taken off southwestern Western Australia. Strict regulations apply to both commercial and amateur fishers. Overseas, there has been some success in cultivating calm water species and those techniques may be applicable to brownlip abalone.

Remarks: Considered by some to be a subspecies of blacklip abalone rather than a distinct species. Forensic (genetic) tests that are more sensitive than the protein fingerprint tests used here (p. 373) may resolve the issue. The flesh is very similar to blacklip abalone. Several smaller abalone species are found in tropical Australia but are not commercially harvested. At least one, *H. asinina*, has aquaculture potential.

Greenlip abalone

Haliotis laevigata

Previous names: abalone, muttonfish

Identifying features: ① foot with bright green lip; ② outer shell surface smooth, with very fine outward-radiating grooves; ③ exposed shell colour pale (often white) with orange or scarlet radiating stripes; ④ spiral ridges on outer surface indistinct; ⑤ single, ear-shaped shell, with low spire and broad aperture occupying most of the underside.

Comparisons: Distinguished from other abalones by the green lip of the foot muscle and the smooth outer surface of the shell.

Size: Shell length reaches at least 24 cm (commonly about 14–17 cm).

Habitat: Marine; mainly on shallow, rocky reefs, in the vicinity of seagrass beds, in coastal waters less than 40 m deep. In South Australia they are most abundant at depths of 20 m or deeper. Prefers turbulent and exposed areas, often on barren rocks when tidal current flow is high. Water temperatures of 12–22°C are preferred.

Fishery: Targeted by commercial and recreational divers along the coasts of northern Tasmania, Victoria, South Australia and southern Western Australia. The commercial catch is exported, mainly to Japan and other Asian countries. Aquaculture production occurs in a number of locations. Greenlip abalone and blacklip abalone (*H. rubra*, p. 328) naturally hybridise and the hybrid is cultured, often under the name 'tiger abalone'.

Remarks: Preferred abalone species in China, Hong Kong and Singapore, where highest prices are gained. The flesh is stronger flavoured and considered by some to be better eating than that of other abalones. Other information is provided for protein fingerprints (p. 373) and oil composition (p. 409).

Roe's abalone

Haliotis roei

Previous names: abalone, redlip, redlip abalone

Identifying features: ① foot with dark red or black lip; ② outer shell surface moderately rough, outward-radiating grooves indistinct; ③ exposed shell colour white with olive-green and red blotches and stripes; ④ spiral ridges on outer surface very pronounced; ⑤ single, ear-shaped shell, with low spire and broad aperture occupying most of the underside.

Comparisons: Distinguished from other commercial abalones by its small size and obvious spiral ridges on shell.

Size: Up to 12.5 cm in length (usually harvested at 7–8 cm).

Habitat: Marine; lives on shallow, rocky outcrops and reef platforms. Abundant off south-western Australia in the surf zone and where there is intense wave action. Found intertidally to depths of 4 m and prefers water temperatures of 14–26°C.

Fishery: Large recreational fishery, collected by people wading on rock platforms and shallow reefs around Perth, Greenough and Kalbarri (WA). Also harvested commercially further south off Western Australia. Commercial divers operate in waters shallower than 2 m. Currently harvested legally at a diameter of 6 cm but commercial divers have agreed to size limits of 7 cm or 7.5 cm depending on the location; other regulations also apply. The annual commercial catch is about 100 tonnes (more than $1 million) per year, and much of this is exported frozen whole in the shell, to Japan and south-east Asia. An aquaculture industry is being established in Western Australia.

Remarks: Considered to spawn year-round, although Western Australian stocks appear to spawn only in winter (July–August). Their planktonic larvae may be carried up to 50 km before settling inshore, preferably on a substrate of coralline algae. They feed at night and eat mainly red algae. The flesh is darker than greenlip abalone (*H. laevigata*, p. 330), with a stronger flavour. Other information is provided for protein fingerprints (p. 373).

Cockle

Anadara & *Katelysia* species

Previous names: arc cockle, blood cockle, clam, mud arc, mud cockle, sand cockle, Sydney cockle, venus shell

Katelysia scalarina

Identifying features: ① valve surfaces sculptured, with about 26, outward-radiating ribs in the Sydney cockle, or with numerous, concentric ridges in sand cockles; ② valves almost oval, usually oblique and longer than high; ③ external colour white, cream or light brown, sometimes with darker zigzag patterns; ④ internal colour white or yellow and sometimes with purple patches.

Comparisons: Bivalve molluscs have 2 valves joined at a hinge. Cockles represent two families: Arcidae, which includes *Anadara* species (the main one is *A. trapezius*, the Sydney cockle), and Veneridae, which includes *Katelysia* species (sand cockles). The Sydney cockle can be distinguished from the other commercial clams by their prominent outward-radiating ribs. In sand cockles, concentric ridges are the striking feature on the outer surface of each valve. Sand cockles can be distinguished from surf clams (*Dosinia* species, p. 334) in having oval, oblique (rather than circular) valves. Pipis (*Donax* species, p. 333) have smooth valves.

Size: Sydney cockle attains 8 cm in length (maximum diameter), sand cockles reach 4–5 cm.

Habitat: Estuarine and marine. The Sydney cockle is most common in estuaries and on mudflats and seagrass beds. Sand cockles inhabit tidal flats and estuary mouths on sheltered or sandy subtidal sediment underlying seagrass beds to about 5 m depth.

Fishery: Collected for food and bait by wading on sandy shores of bays and estuaries, digging with large forks, or by snorkelling. The South Australian sand cockle fishery, although prone to sudden collapse, is developing rapidly in some areas due to increased demand. The Tasmanian industry is also expanding. Sand cockles are sold live in the shell to local distributors and restaurants. Aquaculture potential—*K. rhytiphora* was the first hatchery produced cockle, in 1992 at Port Stephens (NSW).

Remarks: Only a few of the many Australian cockle species are harvested; their distribution only is shown above. Specific marketing names may be required for some in the future. Meat is rich in flavour, particularly when roasted in the shell. Other information is provided for protein fingerprints (p. 373).

Pipi

Donax species

Previous names: beach pipi, clam, Coorong cockle, eugarie, Goolwa cockle, ugari

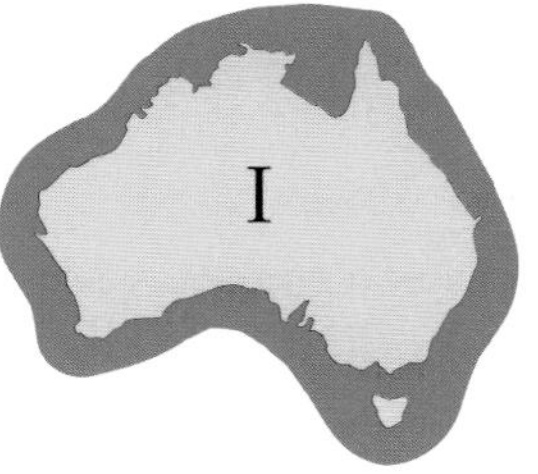

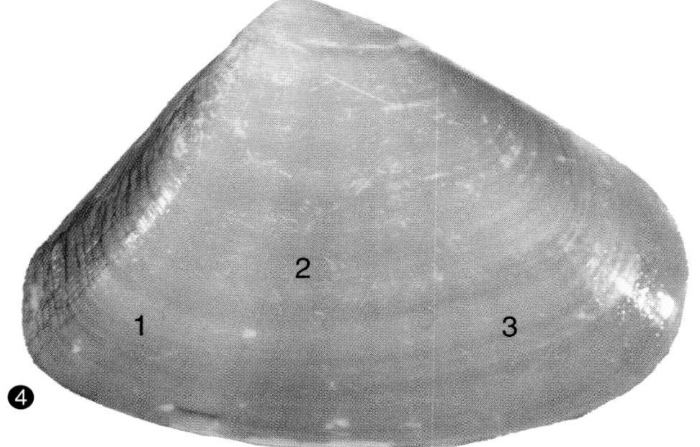

Donax deltoides

Identifying features: ❶ valve surfaces smooth; ❷ valves wedge-shaped, much longer than high; ❸ external colour variable (white, pale brown, green or yellow) and sometimes tinged with bands of lavender or rose; ❹ internal colour variable, often blue or purple.

Comparisons: Belong in the family Donacidae. There are at least seven species, with *D. deltoides* (pipi) the largest growing and the only one commercially harvested. It can be easily distinguished from other clam species by its smooth, wedge-shaped shell. Cockles and surf clams have ribs or distinct concentric ridges on the outer surface of each valve.

Size: To a maximum diameter of about 8 cm (commonly 5–6 cm).

Habitat: Marine; sandy surf beaches in the intertidal zone, often on high energy coastlines. Abundant from about 10 cm below the surface.

Fishery: Have been harvested, along with some other bivalve molluscs such as the Sydney cockle (*Anadara trapezius*, p. 332), by coastal Aborigines for thousands of years. Today, taken by hand in southeastern Australia between southern Queensland and South Australia. The commercial catch favours larger individuals and comes mainly from New South Wales and South Australia. The South Australian commercial fishery topped 1000 tonnes in 1997–98; minimum legal lengths and a closed season (during spawning) apply. Recreational fishers are generally less concerned about size.

Remarks: Widely distributed and locally common—they are the most common, sizeable molluscs living in the sand along ocean beaches of New South Wales. Population sizes can vary greatly year to year. Used as bait as well as for food (soups, chowders, fresh and pickled) and can be stored live, frozen or pickled. Other information is provided for protein fingerprints (p. 373) and oil composition (p. 409).

Surf clam

Dosinia species

Previous name: dosinia

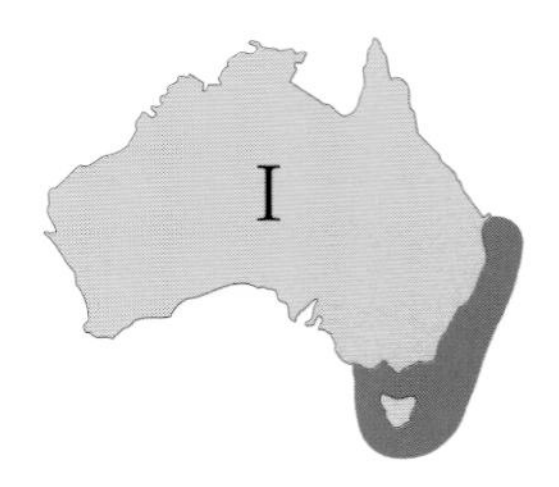

Dosinia caerulea

Identifying features: 1 valve surfaces sculptured, with numerous concentric ridges; 2 valves almost circular and almost as high as long; 3 external colour cream, greyish-white, yellow or light brown, often with darker patterning and tinged with pink, orange or blue near the hinge; 4 internal colour white or cream.

Comparisons: Like sand cockles, surf clams belong in the family Veneridae. There are more than 20 species, distinguished by colour and number and prominence of concentric ridges. The presence of these ridges makes it easy to separate them from pipis (*Donax* species, p. 333), which have a smooth shell, and the Sydney cockle (*Anadara trapezius*, p. 332), which has pronounced outward-radiating ribs. Rounder, relatively higher and more robust than sand cockles (*Katelysia* species, p. 332).

Size: Maximum diameter varies from 3–4.4 cm.

Habitat: Marine and estuarine; live in sand and silt along medium energy coastlines with moderate current flow. Sometimes off ocean beaches and often in association with *Posidonia* seagrass beds.

Fishery: Based on an eastern temperate species, *Dosinia caerulea*. Collected by hand in New South Wales and sold commonly at the Sydney Fish Market. Harvested elsewhere in south-eastern Australia in small quantities, particularly by recreational fishers. Used for food and bait.

Remarks: As with other clams, usually need to be 'staged' or 'flushed' in clean seawater for several hours to reduce the amount of sand in the gut and mantle cavity. Found in marine waters of all Australian states but few species are large enough to be edible. Other information is provided for protein fingerprints (p. 373) and oil composition (p. 409).

Blue mussel

Mytilus edulis

Previous name: mussel

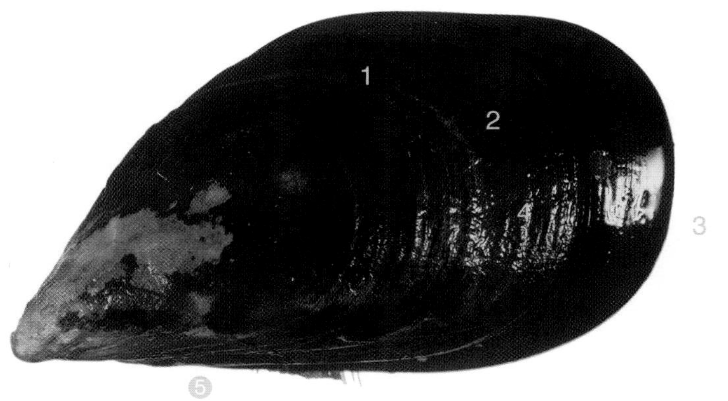

Identifying features: ① valve surfaces smooth or with numerous concentric fine lines, and sometimes hairy; ② valves wedge-shaped, about the same size and much longer than high; ③ posterior ventral margin (opposite hinge) rounded; ④ external colour brown, blue, purple or black; ⑤ internal colour bluish-white, darker towards margin.

Comparisons: Bivalve molluscs of the family Mytilidae are known as 'mussels'. Only one local species is marketed in quantity. It has a distinctive shape and colour and is easily distinguished from other commercial bivalves such as oysters (Ostreidae and Pteriidae, pp 337–341) and cockles (*Anadara* and *Katelysia* species, p. 332).

Size: To 12.7 cm in length (commonly less than 9 cm).

Habitat: Marine; intertidal to depths of at least 20 m. Mainly on exposed reefs, rocks and jetty pylons, often in dense groups or clumps attached by byssal threads.

Fishery: Collected by divers off New South Wales, Victoria and Western Australia. However, mainly aquacultured. Aquaculture industry established in 1976 in New South Wales and has since spread to Victoria, Tasmania, and Western Australia with pilot studies being undertaken in South Australia. Grown in clean and sheltered water 5–20 m deep. The spat are collected on ropes and raised in long 'socks', suspended from horizontal ropes with buoys attached ('subtidal suspended culture'). Some hatchery spat are produced in Tasmania. Harvested at 6–9 cm length (at 12–18 months age) when gonads are ripe and full.

Remarks: Also referred to as '*Mytilus planulatus*', the correct species name is unclear. During spawning, may produce up to 8 million eggs, each 0.07 mm in diameter. The larvae are planktonic for up to three months before settling. Sold live and whole in the shell for local consumption. The edible parts have a distinctive and enjoyable 'sea flavour' but shrivel and can be tough if overcooked. The green mussel (*Perna canaliculus*), often sold locally, is imported from New Zealand. Other information is provided for protein fingerprints (p. 373) and oil composition (p. 409).

Octopus

Octopus species

Previous name: baby octopus

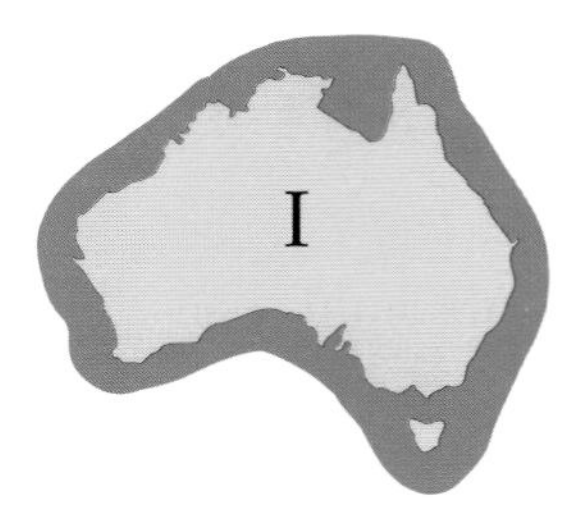

Octopus pallidus

Identifying features: ❶ 8 arms and no tentacles; ❷ no fins on head; ❸ mantle variable in shape, usually ovate or pear-shaped; ❹ suckers without small teeth-like spines and hooks.

Comparisons: Octopuses and squids are cephalopod molluscs, having either no shell or an internal shell, and a circle of arms surrounding a well-developed head. Octopuses usually have 8 arms whereas squids have 8 arms plus 2 longer tentacles. Unlike squids, octopuses lack both an internal shell and teeth or hooks on their suckers. Commercial species, such as the Maori octopus (*O. maorum*), the southern octopus (*O. australis*) and the pale octopus (*O. pallidus*), are distinguished from each other by relative arm length, and mantle shape and size.

Size: To over 250 cm in total length and 9 kg (commonly less than 2 kg).

Habitat: Marine; occur from shallow tide pools to depths over 3000 m. However, commercial species are generally found closer inshore among seagrass and on muddy, sandy or reefy bottoms on the continental shelf (less than 200 m in depth). Most species are solitary, either burying in the substrate or living in holes or under rocks.

Fishery: Mostly caught off eastern and southern Australia, although one species is trapped and taken as bycatch of the rocklobster fishery off Western Australia. Other commercial species are taken (often as bycatch) using trawls, dredges, pots and nets. They are sold as food or bait. Occasionally taken by sport divers.

Remarks: Most of the 'baby octopus' sold in Australia are small tropical species imported from Asia. The firm arm-flesh of some species should be tenderised before cooking. Smaller species are sometimes tenderised in small cement-mixers. Smoked and marinated octopus have become popular with the restaurant trade. When cooked, octopus are firm and dry with a delicate flavour. Other information is provided for protein fingerprints (p. 373) and oil composition (p. 409).

Native oyster

Ostrea angasi

Previous names: flat oyster, mud oyster, native flat oyster

Identifying features: ① roughly egg-shaped (circular to triangular) and sometimes rather pointed dorsally; ② valves of unequal size and shape; ③ outer surface purplish-green, olive-brown or grey; ④ inner surface white or cream with darker margins; ⑤ valve margins slightly irregular.

Comparisons: Oysters are bivalve molluscs, with most commercial species in the family Ostreidae. Their two valves (usually of unequal size) can vary in shape even within a species. The native oyster is roughly egg-shaped with a slightly irregular margin.

Size: Attains about 25 cm in greatest diameter (but commonly less than 10 cm).

Habitat: Marine and estuarine. Usually attached to hard substrates, later breaking free to settle on soft mud or sand. Subtidal from about 2–20 m.

Fishery: A few oyster species have been a common food of Aborigines for thousands of years. Commercial fisheries were established with European settlement when the native oyster was targeted, particularly off Victoria and South Australia. Now, commercial and recreational fisheries are small. Culture attempts have had only limited success in all southern states but are continuing.

Remarks: An oyster's shape is determined largely by its environment. They can be very flat when attached to soft bottoms or irregular on rocky bottoms. Hence, identification is difficult and detailed oyster identification can require specialist knowledge and examination of internal characters. Interested readers should consult the reference list (p. 411) for publications that offer more detail. The European oyster, *O. edulis*, which is highly regarded in Europe, has very similar edible qualities. These differ considerably from the edible qualities of Pacific oyster (*Crassostrea gigas*, p. 339) and Sydney rock oyster (*Saccostrea glomerata*, p. 341). Other information is provided for protein fingerprints (p. 373) and oil composition (p. 410).

Oyster

Ostreidae & Pteriidae

Previous names: bay oyster, blacklip oyster, milky oyster, western rock oyster

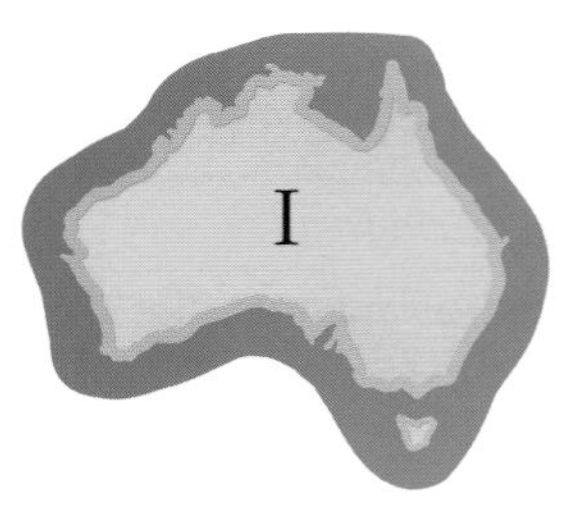

Crassostrea gigas

Identifying features: ① shape very variable from round or oval to triangular or trapezoidal; ② valves of variable (often uneven) relative size and shape; ③ valve margins usually at least partly irregular; ④ usually higher (measured from hinge to opposite margin) than long.

Comparisons: Many Australian oyster species fall into this category but only a few are harvested. Oysters are distinguished from other bivalve molluscs, such as the blue mussel (*Mytilus edulis*, p. 335), by the shape and sculpturing of the valves and the methods by which they attach to the substrate. Four species or groups have separate marketing names.

Size: Varies enormously but those harvested usually have a maximum diameter of 10 cm or more. The Pacific oyster (*Crassostrea gigas*, p. 339) is reported to grow to 45 cm in maximum diameter.

Habitat: Estuarine and coastal marine; usually attached to hard substrates including rocks, shells and dead coral. However, some are more mobile and occur on soft bottoms. Rare below depths of 50 m.

Fishery: Recreational fishers target a variety of oysters found along the coastline. For example, the milky oyster (*Saccostrea cucculata*) is a favourite of locals in Queensland. Other than those with separate marketing names, commercial operations focus on aquaculture development of the milky oyster and the blacklip oyster (*S. echinata*) in Queensland.

Remarks: A versatile seafood immensely popular the world over. The edible qualities vary greatly from species to species, and within species depending on growing location. Other information is provided for protein fingerprints (p. 373) and oil composition (p. 410).

Pacific oyster

Crassostrea gigas

Previous names: Coffin Bay oyster, Japanese oyster, oyster

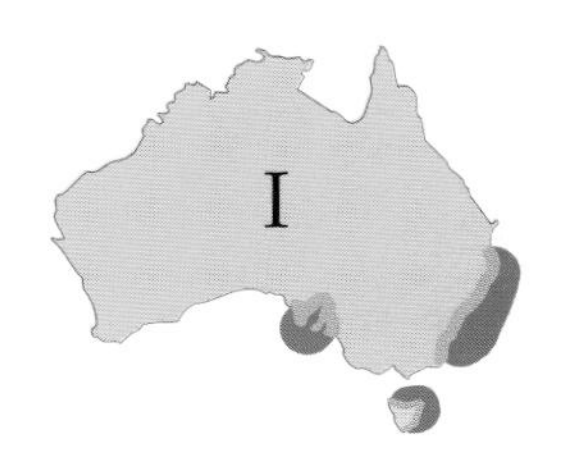

Identifying features: 1 often almost oval, sometimes elongate; 2 valves of roughly similar size but unequal shape; 3 outer surface often whitish with occasional purple or red patches; 4 inner surface opalescent or chalky white with mauve or brown patches; 5 valve margins slightly irregular; 6 outer surface chalky, often with spikey lobes and protrusions; 7 mantle edges black.

Comparisons: Introduced, and the only *Crassostrea* species resident in Australia. Distinguished from the Sydney rock oyster (*Saccostrea glomerata*, p. 341) in having a chalky outer surface often with spikey lobes and protrusions (rather than smooth) and black edges to the mantle (rather than pale).

Size: Reported to a massive 45 cm in maximum length but rarely sold in markets over 10 cm. A standard size shell (7–8 cm) produces about 14 g of edible tissue.

Habitat: Estuarine and marine; sheltered rocky shores and intertidal areas, attaches to rock and debris.

Fishery: First introduced to Tasmania from Japan and now farmed extensively in Tasmania and South Australia. Australian production has risen sharply in recent years and in 1997 the harvest value was $26 million. Grows more quickly than the Sydney rock oyster, reaching marketable size within 1–2 years (rather than 2–3 years).

Remarks: Selective breeding is underway which should lead to the production of Pacific oysters with a faster growth rate, a more symmetrical shell, and higher meat content. Oysters in spawning mode tend to be emaciated and less flavoursome. Stronger tasting than the Sydney rock oyster. Wild populations, although sometimes harvested, are considered by some to be a noxious pest. Other information is provided for protein fingerprints (p. 373) and oil composition (p. 410).

Pearl oyster

Pteriidae

Previous names: wing shells, winged pearl oyster

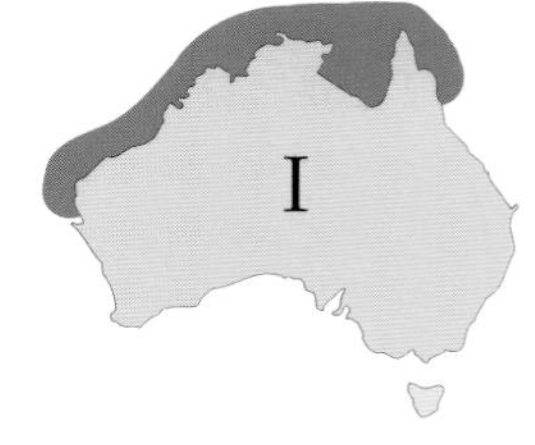

Pinctada maxima

Identifying features: ❶ dorsal profile (along hinge line) straight and ventral profile broadly rounded (shell sometimes almost circular and relative height increasing with size); ❷ triangular, wing-like projection usually bordering each end of the hinge line, sometimes very pronounced; ❸ valves of similar size and variable relative shape; ❹ outer surface usually brown (occasionally black), and sometimes with tinges or spots of darker brown, red or purple; ❺ inner surface with a white or silver lustre and sometimes with a gold band towards the margins; ❻ valve margins irregular.

Comparisons: Unlike other Australian commercial oysters, pearl oysters belong to the family Pteriidae, and have a distinctive shape. The goldlip pearl oyster (*Pinctada maxima*) has a light brown outer surface and a gold margin on the inner surface. Smaller species include the blacklip pearl oyster (*P. margaritifera*), which is darker both externally and internally, and the Shark Bay pearl oyster (*P. albina*), which has a much smoother outer shell.

Size: To 30 cm in greatest diameter (commonly to 20 cm). The smallest commercial species, the Shark Bay pearl oyster, reaches only 10 cm.

Habitat: Marine and estuarine; occupies a variety of habitats between low tide to depths of over 80 m. Some occur on muddy substrates and seagrass beds, others attach to rocks and coral, and many winged shells attach to seagrass and algae. Goldlip pearl oysters occur naturally on mud, sand or gravel and prefer areas with fast-flowing currents.

Fishery: The goldlip pearl oyster is the largest and most important commercially. Although mostly farmed for pearl and mother-of-pearl production, there is a small export market for its adductor muscle meat. Pearl production is centred in Broome (WA) but pearls are also produced elsewhere in Western Australia and in the Northern Territory and Queensland. Pearling is Australia's most valuable aquaculture industry and was worth over $185 million in 1997–98.

Remarks: Mainly tropical and subtropical but some occur in cooler southern Australian waters. The adductor muscle of the goldlip pearl oyster is sweet with a soft texture. Other information is provided for protein fingerprints (p. 373) and oil composition (p. 410).

Sydney rock oyster

Saccostrea glomerata

Previous names: commercial oyster, rock oyster, Sydney oyster, western rock oyster

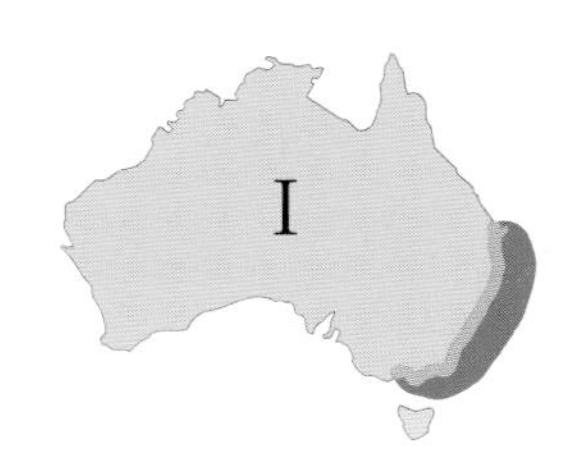

Identifying features: ① shell variable in shape, approximately triangular with a rounded ventral margin; ② valves of unequal size and shape; ③ outer surface dark purplish-blue or greyish-white; ④ inner surface white with blue or purple patches; ⑤ valves very irregular ventrally; ⑥ outer surface lacking spikey, transverse lobes along its ridges; ⑦ mantle edges pale in colour.

Comparisons: Has a smooth outer surface and pale mantle edges. These characteristics, among others, distinguish it from the Pacific oyster (*Crassostrea gigas*, p. 339).

Size: To 10 cm in length in the wild but some cultivated individuals have exceeded 25 cm. Usually marketed at 40–60 g whole weight.

Habitat: Marine and estuarine; lives intertidally to about 3 m water depth. Can tolerate a wide salinity range, and lives in sheltered bays and estuaries, particularly in mangrove areas. Prefers hard, rocky substrates but also found on soft substrates.

Fishery: Farming began nearly 130 years ago, now confined to southeastern Queensland and New South Wales. Larvae are almost entirely collected from natural spat fall and then transferred onto racks in suitable estuarine and riverine habitats. Farmers grade their oysters at regular intervals of 3–5 months over 2–3 years, until they reach market size. Growth rates can be increased by selective breeding but environmental factors are also very important. About 8.5 million dozen are produced for dinner tables each year, although production has declined in the last 20 years due to local water-quality problems. Recreational harvesting of wild stock is subject to regulations.

Remarks: Recent studies have shown that the New Zealand rock oyster, *S. glomerata*, and the Sydney rock oyster, previously known as *S. commercialis*, are the same species. Best consumed alive but can be prepared in numerous other ways. The flavour is excellent. Other information is provided for protein fingerprints (p. 373) and oil composition (p. 410).

Commercial scallop

Pecten fumatus

Previous names: king scallop, scallop, southern scallop, Tasmanian scallop, Tassie scallop

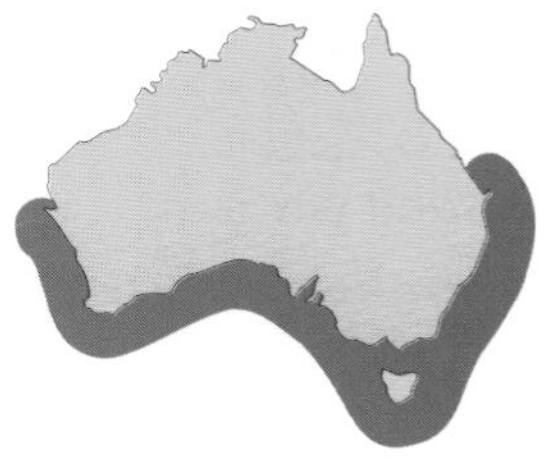

Identifying features: 1 valves almost the same size, ovate or oval in shape; 2 right (lower) valve gently convex, more inflated than the flat, almost concave left (upper) valve; 3 hinge line straight with equal-sized 'wings' on each side of dorsal midpoint; 4 outer valve surface sculptured with 12–18 broad, rounded or flat-topped radiating ribs; 5 colour off-white or brownish with pink or red areas, lower valve paler.

Comparisons: All of Australia's commercial scallops are bivalve molluscs belonging to the family Pectinidae. The commercial scallop can be distinguished from other scallops by having equal-sized 'wings', 12–18 radiating ribs on the outer surface, and by being off-white or brownish with pink or red areas. It is unique among harvested scallops in having a flat upper valve and deeply convex lower valve.

Size: Maximum length to over 14 cm (commonly 8–9 cm).

Habitat: Marine; benthic, found in or on soft sediments ranging from mud to coarse sand. Usually lies buried with only the flat valve visible. Congregates in discrete beds from depths of 1 m to at least 120 m. The larvae are planktonic.

Fishery: Harvested mainly by box-shaped, self-tipping 'mud dredges' operated from small or medium-sized fishing vessels and by diving. The major fisheries are in Tasmania and Victoria, while small fisheries operate in Jervis Bay (NSW) and Coffin Bay and Spencer Gulf (SA). The fishery in Victoria is valued at over $20 million in some years but scallop abundances vary widely from year to year. Commercial hatchery techniques have been developed in recent years and commercial scallops are now farmed in Tasmania. Also taken by recreational divers.

Remarks: Grows to a marketable size in about 18 months (or longer) and lives for up to 16 years. Flesh creamy or white with orange or red roe. Rich in flavour, popular and commands a high price. Other information is provided for protein fingerprints (p. 373) and oil composition (p. 410).

Saucer scallop

Amusium species

Previous names: Ballot's saucer scallop, mud scallop, northern saucer scallop, Queensland scallop

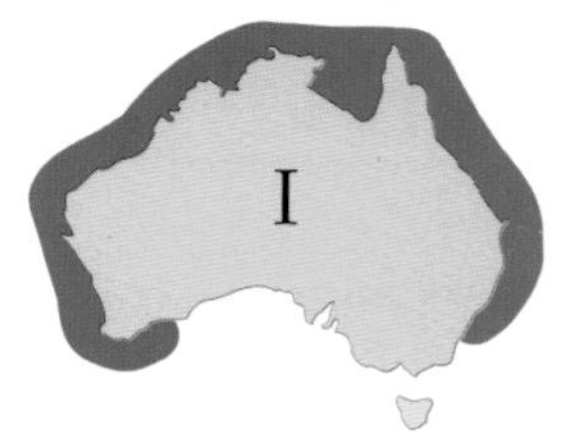

Amusium balloti

Identifying features: 1 valves the same size and almost round in shape; 2 valves flat to slightly convex, and thin; 3 hinge line straight with equal-sized 'wings' on each side of dorsal midpoint; 4 outer valve surface smooth and polished, inside surfaces with 22–54 radiating ribs; 5 upper (left) valve striped, banded or mottled in browns, pinks and reds, lower (right) valve generally pale or white.

Comparisons: Two species are taken commercially in Australia—the northern saucer scallop (*A. pleuronectes*) and Ballot's saucer scallop (*A. balloti*). The northern saucer scallop is medium brown or deep pink with radial rows of dark brown spots and about 22–34 internal ribs. Ballot's saucer scallop is darker, with dark brown concentric rings and random spotting, and 42–54 internal ribs. Saucer scallops are easily distinguished from other commercial species by having smooth and polished shells.

Size: Ballot's saucer scallop reaches 14 cm in maximum length (commonly 9–10 cm); the northern saucer scallop is smaller, to 8 cm in length.

Habitat: Marine; benthic, living on sand, rubble or mud in depths of 10–75 m. The northern saucer scallop is mostly found in waters less than 20 m deep.

Fishery: Ballot's saucer scallop is targeted by trawlers off Queensland and Western Australia. The Queensland fishery catches about 100 million scallops per year. Catch trends have been variable over the past eight years and Queensland stocks are believed to be heavily exploited. Northern saucer scallops are a bycatch of coastal trawl fisheries (mainly for prawns). Most of the catch is exported 'roe off' to south-east Asia and the USA. There is no recreational fishery.

Remarks: Ballot's saucer scallop may be a subspecies of the Japanese *A. japonicum*. It is fast-growing, reaching marketable size after 6–15 months. Highly regarded overseas. The flesh is generally paler in colour, sweeter and firmer than the commercial scallop (*Pecten fumatus*, p. 342). Other information is provided for protein fingerprints (p. 373).

Scallop

Pectinidae

Previous names: bay scallop, doughboy scallop, fan scallop, prickly scallop, queen scallop, saucer scallop, Tasmanian scallop

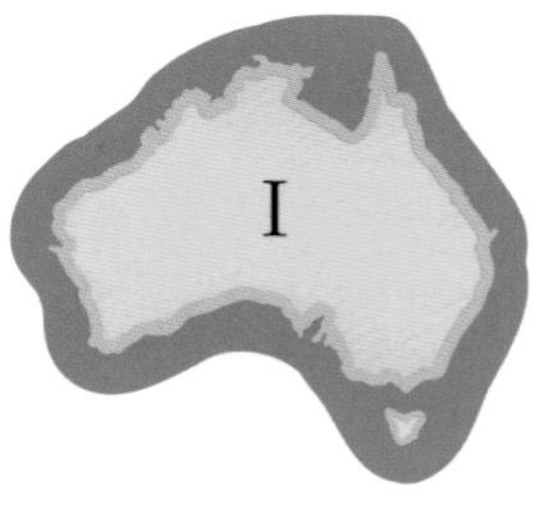

Annachlamys flabellata

Identifying features: 1 valves the same or slightly unequal size and ovate, oval or almost round in shape; 2 valves flat, convex, or occasionally concave; 3 hinge line straight with equal- or unequal-sized 'wings' on each side of dorsal midpoint; 4 outer valve surface with varying number of radiating ribs or smooth and polished; 5 valve colour variable, usually a shade of brown or pink with darker markings.

Comparisons: Pectinids other than the commercial (*Pecten fumatus*, p. 342) and saucer (*Amusium* species, p. 343) scallops are variably harvested around the Australian coastline. The fan scallop (*Annachlamys flabellata*), the doughboy scallop (*Mimachlamys asperrima*) and the queen scallop (*Equichlamys bifrons*) can be distinguished mainly by colouration (including eye colour) and the number of radial ribs. For example, the fan scallop has 17–20 radial ribs, the doughboy scallop 20–26, and the queen scallop 8–9.

Size: Harvested scallops usually reach at least 9 cm in maximum length.

Habitat: Marine and estuarine; benthic, living attached to firm substrates by byssal threads, or on sandy to silty bottoms near reefs, in depths of 2–120 m. Doughboy scallops are found in water temperatures of 9–25°C, where salinities are high and stable, and usually in currents.

Fishery: Doughboy and queen scallops are targeted by recreational fishers and also landed from commercial vessels (mainly trawlers) off southeastern Australia. Doughboy scallops, mostly caught in Bass Strait, have only a moderate market value as they have a low meat yield and require extra processing (sold as 'roe-off'). Fan scallops are taken regularly, but in small numbers, off Queensland. Scope for farming some species using suspended culture methods.

Remarks: Gregarious. Good swimmers, propelling themselves through the water by vigorously clapping their valves together and expelling jets of water. The edible qualities vary between species. The taste of doughboy scallops is sometimes tainted by commensal sponges on their shells if prepared carelessly. Other information is provided for protein fingerprints (p. 373) and oil composition (p. 410).

Bailer shell

Subfamily Zidoninae

Previous names: baler shell, false bailer shell, melon shell

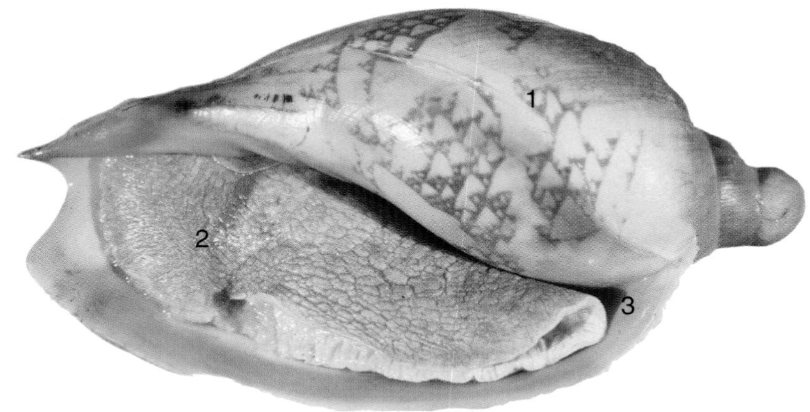

Livonia mamilla

Identifying features: ① shell large, ovate and smooth; ② no operculum on foot; ③ aperture oval-shaped but elongated longitudinally and pointed anteriorly.

Comparisons: Bailer shells are a type of sea snail. Like abalones, they are gastropod molluscs and the commercially harvested species all have their internal organs protected by a spirally-coiled shell. Unlike abalones, however, the shell is much taller (rather than flattened). Bailer shells belong to the family Volutidae. The main commercial species, the false bailer shell (*Livonia mamilla*), is large, ovate and smooth. The aperture is roughly oval-shaped but elongated longitudinally and pointed anteriorly. It lacks an operculum or 'door' which many gastropods use to protect themselves. The external shell is orange or brown and cream, with distinctive zigzag stripes. The melon shell (*Melo amphora*), common in tropical Australia, is similar but is armed with spines at the posterior end.

Size: The height of gastropod molluscs is measured from the anterior end to the apex of the spire. This measurement is often referred to as the shell's length. The false bailer shell reaches 30 cm in height (commonly about 25 cm) and the melon shell 50 cm (commonly about 30 cm).

Habitat: Marine; the false bailer shell prefers sand and mud bottoms to depths of 180 m; the melon shell is common on and near reefs to depths of 10 m.

Fishery: Now illegal to harvest in some areas. The false bailer shell is trawled and trapped in southeastern Australia and is regularly sold at the Sydney Fish Market. Other species are marketed sporadically.

Remarks: So named for their ability to hold water. They are often brightly coloured and more sought-after as ornaments or by shell collectors than for seafood. However, their meaty flesh is gaining popularity in Australia. Other information is provided for protein fingerprints (p. 374).

Periwinkle

Littorinidae, Neritidae, Trochidae & Turbinidae

Previous names: sea snail, turban, turban shell, turbo, warrener, wavy turbo

Turbo undulatus

Identifying features: ❶ single shell of variable shape and often rough or ridged; ❷ operculum present; ❸ aperture often circular or oval, rarely elongated longitudinally.

Comparisons: The four periwinkle families exhibit a variety of shell shapes from rounded neritids, to squat littorinids (the true periwinkles), to conical trochids. Turbo (*Turbo undulatus*) has a round aperture with a smooth operculum whereas trochus (*Trochus niloticus*, p. 347) has an oval aperture and horny operculum. Periwinkles are usually much smaller than commercial bailer shells (Zidoninae, p. 345) and always have an operculum (otherwise lacking).

Size: The green snail (*Turbo marmoratus*) is reported to reach over 20 cm in height. However, usually harvested at less than 5 cm; other periwinkles harvested under 7 cm.

Habitat: Marine; intertidal and shallow reefs, usually to depths of less than 20 m. Turbo species are benthic in pools on weed-covered reefs between tides.

Fishery: An abundant food and bait source right around Australia's coastline and have been harvested here for thousands of years. A dive-based fishery for turbo is located off north-eastern and southeastern Tasmania. Animals are sent live to mainland markets—primarily Melbourne—or sold to Tasmanian restaurants and processors. One conical species, trochus, is significant enough commercially to warrant a unique marketing name.

Remarks: Edible qualities of the group vary; they can be eaten fresh, in brine or pickled. Turbo species are among the best eating gastropods. Other information is provided for protein fingerprints (p. 374) and oil composition (p. 410).

Trochus

Trochus niloticus

Previous names: button shell, top shell

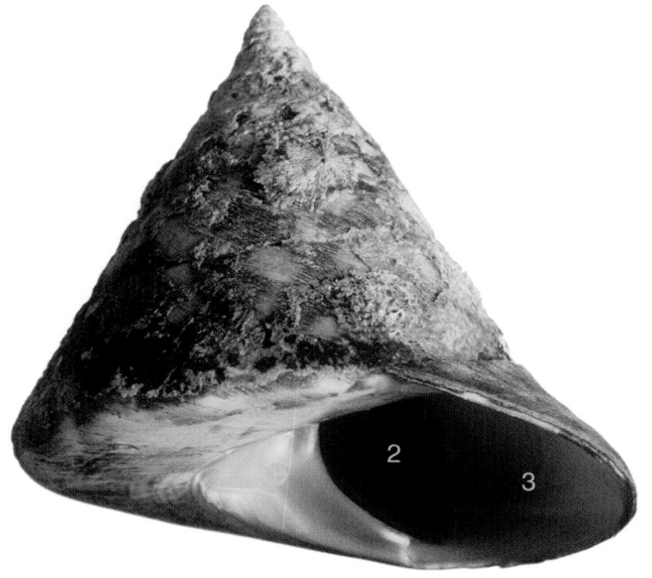

Identifying features: 1 shell conical, resembling a witch's hat; 2 operculum horny and thin; 3 aperture broadly oval; 4 inner surface pearly.

Comparisons: Distinguished from most other harvested gastropods by having a conical shell. Australia's other top shells (Trochidae), which includes genera other than *Trochus*, are smaller (attaining only about 4 cm shell diameter).

Size: In Australia, attains 16 cm in maximum shell diameter (commonly 8–12 cm).

Habitat: Marine; inhabits intertidal and shallow subtidal areas of coral reefs, mainly in exposed aspects.

Fishery: Harvested in tropical and subtropical Western Australia and Queensland. Traditionally taken for its shell, which is used to produce buttons and ornaments. Collected by walking on reef tops at low tide or by free diving. The harvesters use small dinghies and, in more remote areas, operate from a mother ship. Mariculture research is continuing in the Northern Territory and Western Australia.

Remarks: Widely distributed in Asia and has been introduced to some countries. Often maricultured, mainly to 'restock' denuded reefs, as its harvest provides substantial income to small village communities. The meat is of secondary importance to the shell but is sometimes consumed. After boiling the shell, the flesh is extracted to be eaten either immediately or frozen or dried. Other information is provided for protein fingerprints (p. 374).

Cuttlefish

Sepia species

Previous names: giant cuttlefish, golden cuttlefish, Pharaoh's cuttlefish, Smith's cuttlefish

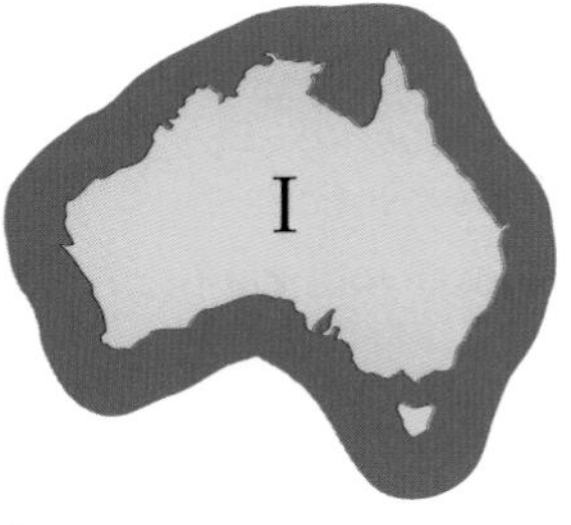

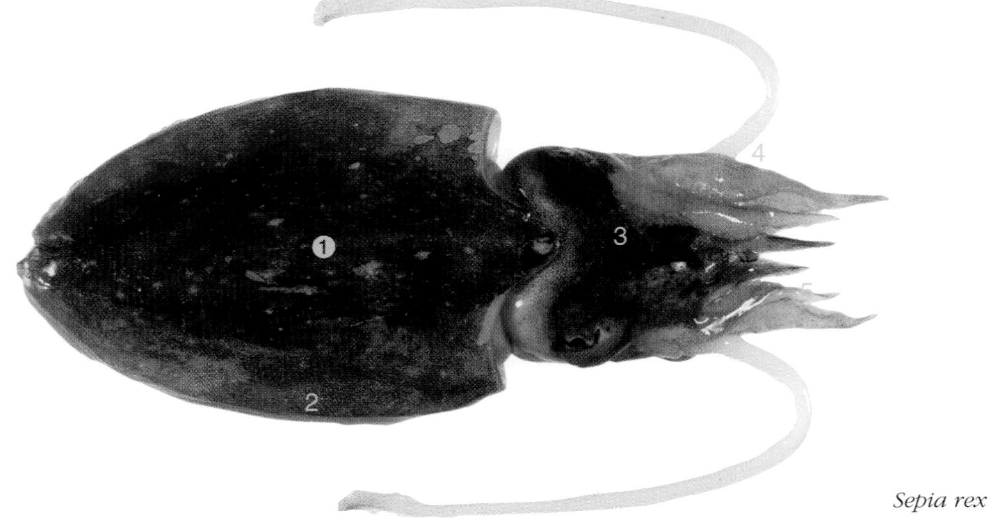

Sepia rex

Identifying features: ① calcified internal shell ('cuttlebone') along the back; ② body oval, with narrow fins around edges; ③ head relatively large; ④ 8 short arms and 2 long tentacles; ⑤ arm suckers with short teeth.

Comparisons: Cuttlefishes, like octopuses and squids, are cephalopod molluscs. There are about 10 *Sepia* species (family Sepiidae) in Australian inshore waters. They are readily distinguished from other cephalopods by the internal 'cuttlebone' and oval body. *Sepia* species are distinguished from each other mainly by the shape of the cuttlebone, body colouration and extent and shape of the mantle.

Size: To 52 cm mantle length and over 5 kg (usually marketed at a mantle length of 12 cm or less).

Habitat: Marine and estuarine. Giant cuttlefish (*S. apama*) live over reefs, seagrass beds and open grounds to 50 m depth; other species live in sheltered estuaries over seagrass and reefs, or offshore on the continental shelf to about 110 m depth.

Fishery: Pharaoh's cuttlefish (*S. pharaonis*), which was landed in large quantities by foreign pair-trawlers off northern Australia until the 1980s, is caught incidentally by prawn trawlers and sold on domestic markets. Giant cuttlefish and other smaller species are exploited in southern Australia, mainly by trawlers. Small quantities are landed using beach seines and fish traps, mostly from New South Wales and Queensland. Also taken in small quantities by recreational fishers using lures and baited hooks.

Remarks: Populations of giant cuttlefish in southern Spencer Gulf (SA) are considered to be susceptible to overfishing. Usually more flavoursome than squid. The cuttlebones are often fed to budgerigars as a source of calcium. Other information is provided for protein fingerprints (p. 374) and oil composition (p. 410).

Gould's squid

Nototodarus gouldi

Previous names: aero squid, aeroplane squid, arrow squid, seine boat squid, torpedo squid

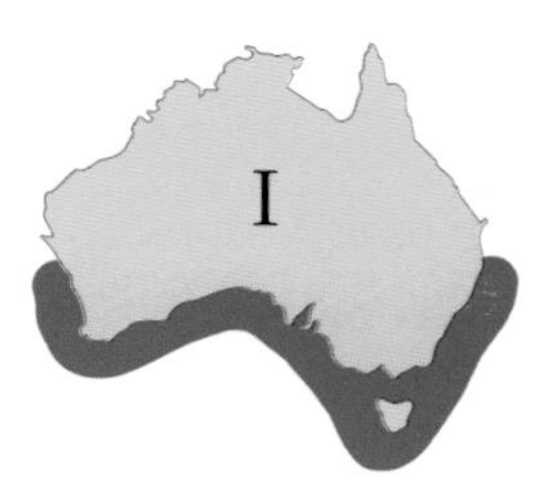

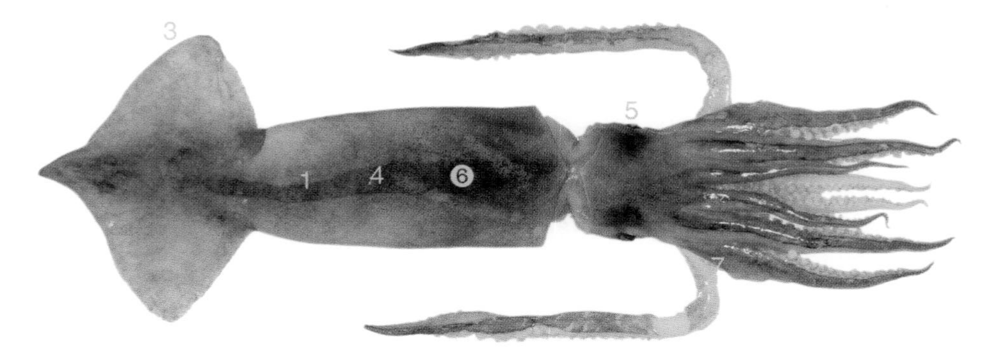

Identifying features: ① dorsal surface of mantle smooth, light brownish-pink with a blue or purple stripe; ② large arm suckers with 10–14, similar-sized, sharp, conical teeth; ③ fins short, rhomboidal; ④ torpedo-shaped body; ⑤ no transparent membrane over eyes; ⑥ a translucent, feather-like 'pen' running underneath the back; ⑦ 8 arms and 2 long tentacles.

Comparisons: Squids, like octopuses, are cephalopod molluscs. Commercial species can be distinguished from octopuses by the presence of 2 long tentacles, fins and an internal shell (the 'pen'). They are distinguished from calamaris (*Sepioteuthis* species, p. 350–351) by the length of the fins (usually less than half mantle length rather than as long as the mantle) and from cuttlefish (*Sepia* species, p. 348) by having a translucent, cartilage 'pen' rather than a calcified 'cuttlebone'. Gould's squid can be distinguished from the similar-looking Hawaiian arrow squid (*N. hawaiiensis*) by its smooth skin, and from other commercial squids by colour, fin shape and length, and number and relative size of sucker teeth.

Size: Females attain at least 40 cm mantle length and 1.6 kg; males are smaller. The average market size is about 23 cm mantle length and 0.3 kg.

Habitat: Marine; continental shelf and slope between the surface and 825 m, but most common in depths of 50–200 m. Schooling animals which aggregate near the sea bed during the day but at night disperse through the water column or to the surface to feed.

Fishery: Caught by jigging between October and April in Bass Strait and western Victoria. Lights are used to attract squids to the surface at night. Also taken as bycatch of otter trawlers and Danish seiners in the South East Fishery between Botany Bay (NSW) and western Victoria. Occasionally taken by beach seine and trolling as far north as southern Queensland. Targeted by recreational fishers using jigs.

Remarks: Fast growing and live for no more than 12 months. They are sold either frozen or chilled on domestic markets. The meat can be chewy if poorly prepared but is otherwise firm and dry with a mild flavour. Other information is provided for protein fingerprints (p. 374) and oil composition (p. 410).

Northern calamari

Sepioteuthis lessoniana

Previous names: calamary, squid, tiger squid

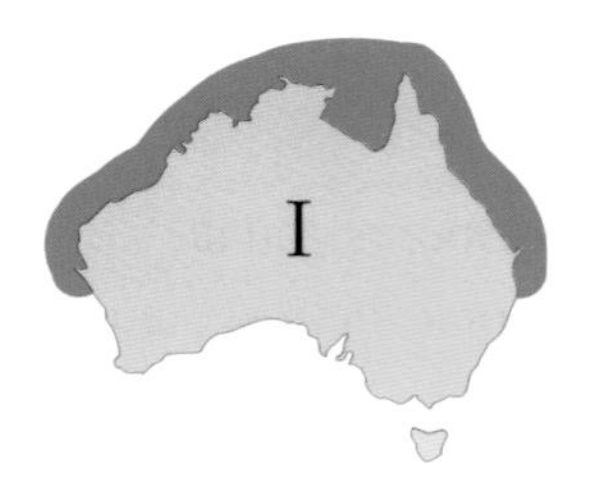

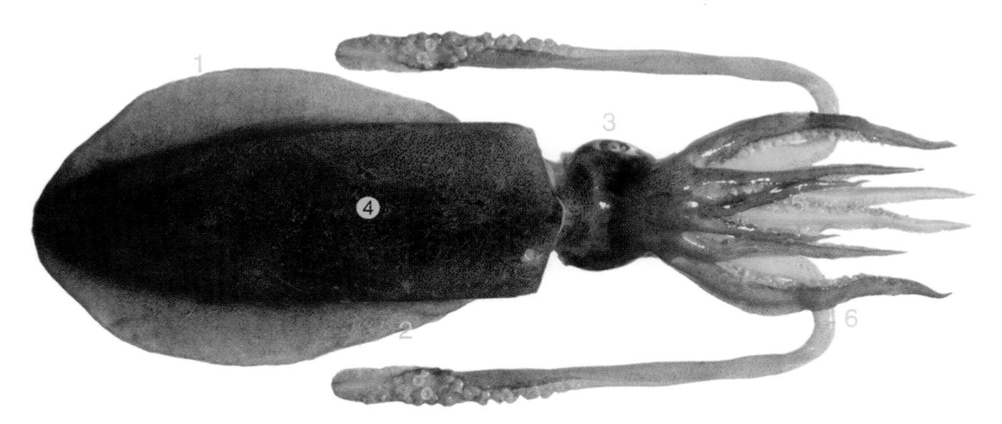

Identifying features: ① fins widest in their posterior third; ② fins long, rounded and extending nearly the whole length of the mantle; ③ eyes covered by a transparent membrane; ④ a translucent, feather-like 'pen' running underneath the back; ⑤ suckers with teeth; ⑥ 8 arms and 2 long tentacles.

Comparisons: Calamaris have much longer fins than other squids. The northern calamari has fins that extend nearly the whole length of the mantle and that are widest in their posterior third. The fins of the southern calamari (*S. australis*, p. 351) are similar, but widest at about half their length.

Size: To 42 cm in mantle length (commonly 20–30 cm) and nearly 2 kg.

Habitat: Marine; abundant in nearshore coastal waters but also occur on offshore reefs to depths of at least 100 m.

Fishery: Caught by jigging. Also a bycatch of the prawn trawl and inshore net fisheries off northern Australia. Commercial catches peak off southern Queensland during winter. Recreational anglers use baited jigs and lures.

Remarks: Many specimens are decorated with pale bands and locally called 'tiger squid'. Probably at least two species are currently lumped under the name *S. lessoniana* in northern Australian waters. Chilled or frozen, and sold on domestic markets for either human consumption or bait. The flesh is generally more tender than that of arrow squids. Other information is provided for protein fingerprints (p. 374).

Southern calamari

Sepioteuthis australis

Previous names: calamary, grass squid, squid

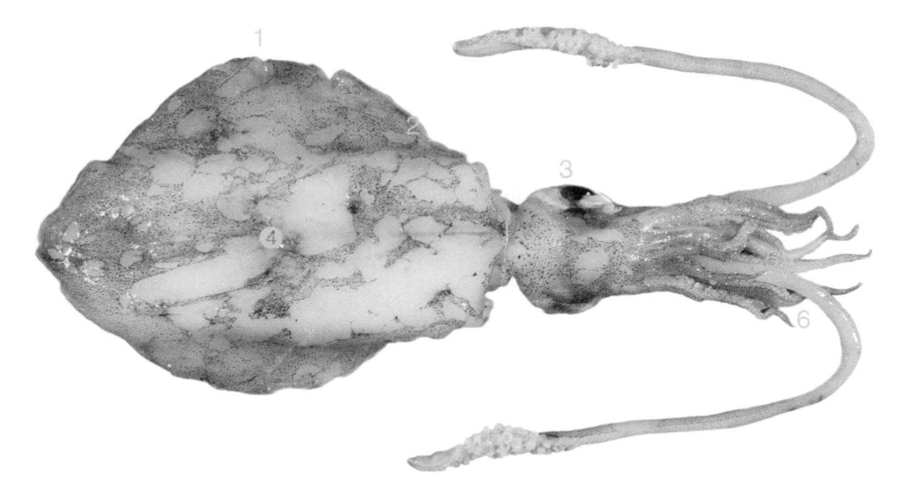

Identifying features: ① fins widest at about half their length; ② fins long and rounded and extending the whole length of the mantle; ③ eyes covered by a transparent membrane; ④ a translucent, feather-like 'pen' running underneath the back; ⑤ suckers with teeth; ⑥ 8 arms and 2 long tentacles.

Comparisons: May be confused with the northern calamari (*S. lessoniana*, p. 350) but has fins that are widest at about half their length (fins widest in their posterior third in northern calamari). Some deepwater squids have a similar body shape but these are rarely caught.

Size: Attains 55 cm in mantle length and 4 kg (commonly less than one-third that size).

Habitat: Marine; inshore animals that live in coastal waters in depths usually less than 100 m. Common in coastal bays and inlets.

Fishery: Caught mainly by jigging, between Shark Bay (WA) and south of Brisbane (Qld). Spawning aggregations are targeted. Also caught incidentally in tunnel nets, demersal trawls, inshore haul nets and beach seines. Targeted by recreational anglers using baited jigs and lures, and are taken from both boats and jetties.

Remarks: Spawn throughout the year in shallow water, laying their eggs on rocks or small animals attached to the bottom. After capture, they are sold on domestic markets either chilled or frozen, as flaps or rings. They are also used as bait but are becoming more highly regarded as human food. Calamari are highly regarded in southeastern states and are more tender than arrow squids. All squids can be tenderised using the juice of kiwi fruit or papaya. Other information is provided for protein fingerprints (p. 374) and oil composition (p. 410).

Squid

Order Teuthoidea

Previous names: broad squid, pencil squid, slender squid

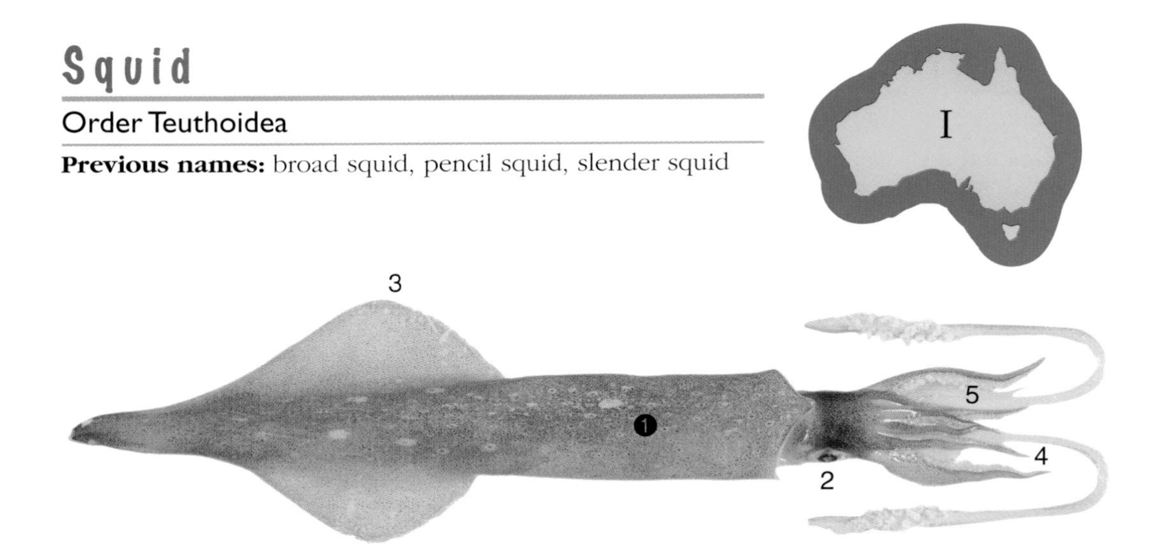

Photololigo sp.

Identifying features: ① a translucent, feather-like 'pen' running underneath the back; ② with or without a transparent membrane over the eyes; ③ fin length variable, often less than half mantle length; ④ 8 arms and 2 tentacles; ⑤ suckers with teeth.

Comparisons: Members of the order Teuthoidea have 8 arms, 2 tentacles and an internal cartilage 'pen' running underneath the back. The closely related cuttlefishes (*Sepia* species, p. 348) have an internal calcified 'cuttlebone'. Numerous characters—many technical and requiring expert knowledge—are used to distinguish squid species from one another. Some of the simpler ones are mantle shape, shape and extent of fins, number and arrangement of suckers, and number and shape of teeth-like sucker spines and hooks.

Size: Reach 60 cm in mantle length and can weigh up to 7 kg.

Habitat: Marine and estuarine. The red ocean squid (*Ommastrephes bartramii*) is oceanic and occurs between the surface and depths of 1500 m. However, most are pelagic over the continental shelf or the upper continental slope but travel freely through the water column.

Fishery: Mostly taken as a valuable bycatch of trawling but also targeted by commercial and amateur fishers using jigs and lures. Most Australian species are considered to be under exploited. The red ocean squid is the largest squid of commercial potential in Australian waters.

Remarks: Three species may be sold under separate marketing names. Squids vary enormously in edible qualities and, to reflect these differences, the general 'squid' category could be further subdivided. There are numerous undescribed species, such as the slender squid (*Photololigo* sp.) off northern Australia, which make excellent eating. However, others, particularly deepwater species taken as trawl bycatch, have large amounts of ammonia in the flesh and are not considered edible. Squid are becoming an increasingly popular seafood with a subtle flavour. The mantle, fins, arms and tentacles are all eaten. Squid ink can be used to flavour and colour food. Other information is provided for protein fingerprints (p. 374) and oil composition (p. 410).

Other invertebrates

G. K. Yearsley and P. R. Last

Jellyfish

Class Schyphozoa

Previous names: blue jellyfish, brown jelly blubber, brown jellyfish

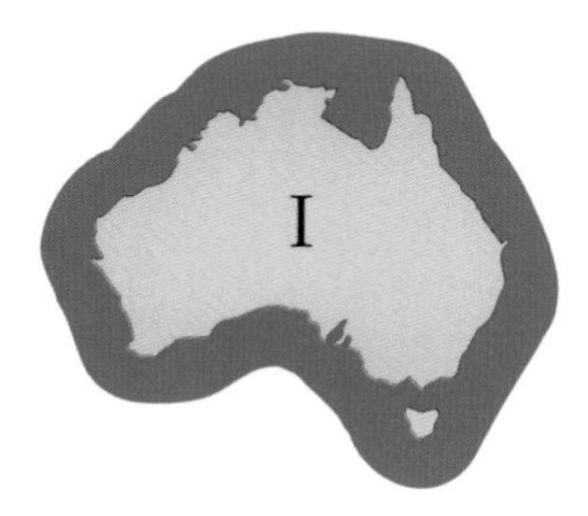

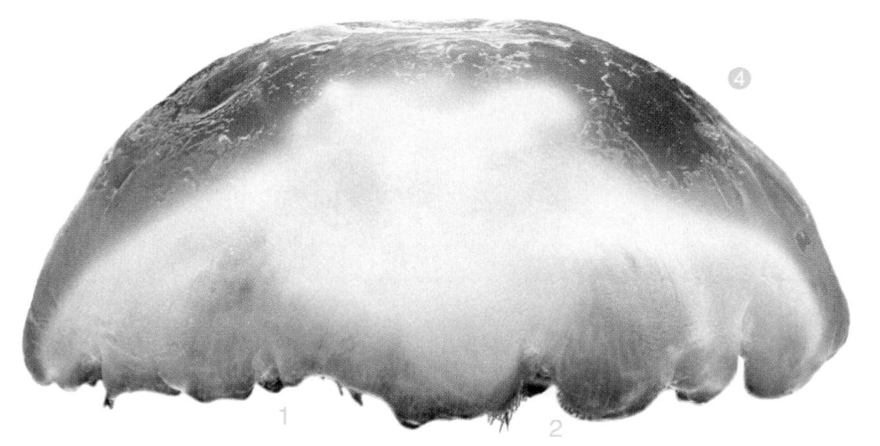

Aurelia aurita

Identifying features: ① tentacles sometimes present around bell margin; ② oral arms usually present under centre of bell; ③ distinct hemispherical-shaped bell containing gelatinous material; ④ radially symmetrical.

Comparisons: Whole jellyfish and associated processed products are unlike any other Australian seafood. A number of species are commercial targets. The most common one, jelly blubber or, more affectionately, 'the blubber' (*Catostylus mosaicus*), has 8 three-winged arms and an internal 'maltese' cross that can be seen through the top of the bell. It lacks tentacles around the bell margin. Other possible commercial species include brown jellyfish (*Phyllorhiza punctata*), which lacks tentacles and has a brown bell covered with small white spots, and moon jellyfish (*Aurelia aurita*), which has 4 transparent oral arms and 4 white, almost circular gonads usually visible through the body wall.

Size: Bell width of commercial species reaches 50 cm (commonly less than 30 cm).

Habitat: Marine and estuarine; coastal and offshore surface waters, mostly in depths to 30 m.

Fishery: Mostly developmental. Commercial catch netted in estuaries predominantly off eastern Australia although reasonable numbers were taken from the Swan River (WA) in 1989–90.

Remarks: Numbers vary enormously by season with most adults dying at summer's end. The majority of Australian product is exported to Asia where it is usually served as a condiment or salad ingredient. Colour variations may affect the price gained for the final product. Commonly called 'water mother' by the Chinese, the bell is more than 90 per cent water. It is usually salted and then dried for human consumption and in this form is not suitable for the protein fingerprint test used here. Despite a rubbery texture and only a subtle taste, its popularity in Australia is increasing.

Beche-de-mer

Holothuriidae & Stichopodidae

Previous names: black teatfish, cucumber fish, greenfish, leopard fish, lollyfish, prickly redfish, sandfish, sea cucumber, sea slug, surf redfish, tigerfish, trepang, white teatfish

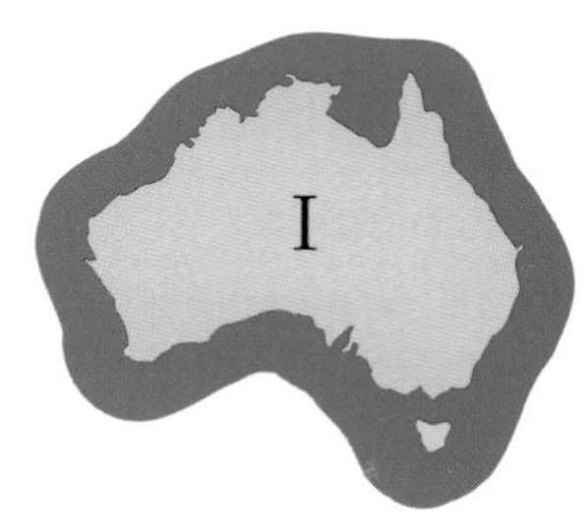

Holothuria scabra

Identifying features: 1 tapering, semi-cylindrical body; 2 circle of retractile tentacles around mouth; 3 body wall usually covered with small or minute spicules, giving surface a rough feel.

Comparisons: Closely related to other echinoderms such as sea urchins (Echinoidea, p. 356), but the almost cylindrical body shape is easily identifiable. Among the most commonly harvested species are sandfish (*Holothuria scabra*) identified by a stout, oval body, which is grey or greenish on the upper surface and covered with small, white-edged black spots, and prickly redfish (*Thelenota ananas*) which is reddish-orange and has numerous pointed 'teats' over the body.

Size: To about 70 cm in length (commonly 25–35 cm and up to 2 kg live weight).

Habitat: Marine and estuarine; benthic, from brackish inshore waters to depths of at least 2000 m. Most common on inshore reefs and in lagoons where they are found on sand or mud or in coral or rock crevices.

Fishery: Restricted to the tropics where animals are mostly collected by hand in shallow water but also as trawl bycatch. First collected in Australian waters probably in the 1600s and, at times, the fishery has earned considerable export income. In the late 1800s over 100 ships were fishing in tropical Australia. Today's fishery is small but increasing, with Australian product being sold locally and overseas both as food and for medicinal purposes.

Remarks: Of the 170 or so species in Australian waters, less than 10 are harvested. They are boiled before the thick body wall is dried and/or smoked to produce beche-de-mer, a delicacy among south-east Asians. Some species are able to expel their internal organs through their mouth or, more commonly, their cloaca. This rather drastic defence mechanism leaves some species without a gut for up to two months while their organs regenerate. Although high in protein, the boiled and dried flesh is not suitable for the protein fingerprint test used here.

Sea urchin

Class Echinoidea

Previous names: black urchin, purple sea urchin, spiny sea urchin, spiny urchin, white urchin

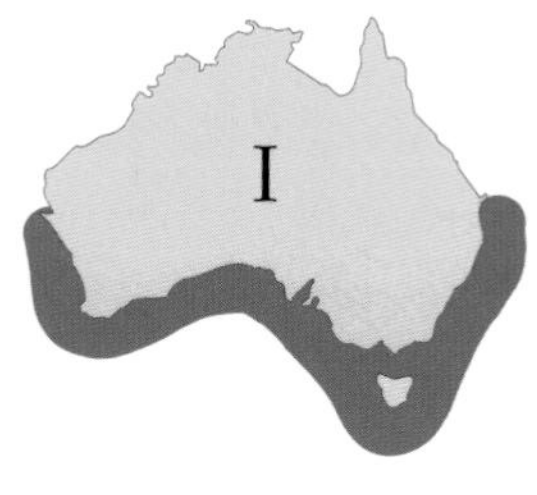

Heliocidaris erythrogramma

Identifying features: ① numerous long, sharp spines; ② spherical or subspherical, hard exoskeleton; ③ tube feet present.

Comparisons: Like their close relatives, such as sea stars, most are made of 5 symmetrical segments. Their unique body form is easily distinguishable from other Australian seafoods. The most important commercial species is the purple sea urchin (*Heliocidaris erythrogramma*) which, despite its common name, is variable in colour. It has sharp, solid spines. Black sea urchin (*H. tuberculata*) is orange red and has blunt spines that are oval in cross-section. The spines of two other harvested species, the longspine sea urchins (*Centrostephanus rodgersii* and *C. tenuispinus*) are hollow.

Size: Purple sea urchin to 9 cm diameter with 2.5 cm spines. Some smaller species have longer spines (up to 7.5 cm)

Habitat: Marine, bottom-dwelling. The purple sea urchin inhabits sheltered and moderately exposed coastal waters mostly close inshore but to depths of nearly 40 m. Other commercial species prefer more exposed areas.

Fishery: Based mainly in the south-east and small but expanding. Animals are collected by divers and then broken open to retrieve the roe (gonads). Also the target of a few recreational divers.

Remarks: Distribution of commercial species only shown on the map above. Product value is affected by roe colour, size, texture and taste; preferred colours are yellow and orange. The roe is not suitable for the protein fingerprint test used here. *Centrostephanus* species are not favoured by divers as their hollow spines break easily and can cause injury; they also produce less roe. Not commonly marketed in Australia. However, roe can fetch astronomical prices (sometimes as much as $400 per kg) in some parts of Asia.

Protein fingerprinting

R. D. Ward, R. K. Daley, J. Andrew and G. K. Yearsley

9

Introduction

Two fillets on a slab may look very similar, but how can you tell if they are from the same species, or from the species claimed? Colour, shape and muscle structure give important clues, but are often not enough to provide a firm identification. Genetic testing provides a way, because the genes of different species have changed over evolutionary time and are never identical.

There are many methods of genetic testing. These vary in their accuracy and in the time, labour and equipment required. The technique used here is inexpensive and simple. Results can be achieved in less than one hour and the equipment is portable. The procedure, known as protein electrophoresis, has been used previously for fish identification (e.g. Shaklee and Keenan 1986; Daley *et al.* 1997).

How protein fingerprinting works

Proteins are large molecules and are essential components of all living cells. They regulate chemical reactions in the cell, and provide components for membranes. Some carry essential substances through an organism (e.g. haemoglobin in our blood carries oxygen to cells) and others act as hormones (chemical messengers). There are thousands of different proteins and each one is 'built' from instructions carried in the DNA. Slight differences in these instructions in a gene can lead to small changes in either the size or electric charge of a particular protein molecule, even though the protein will still carry out its usual functions. Two similar proteins, which have different sizes or a different electric charge, move at different speeds when an electric current is applied to them. They can therefore be separated. This is the principle behind protein electrophoresis—difference in protein movement in an electric field indicates a difference in protein structure, which in turn indicates a genetic difference.

A sample of muscle tissue from a fish or invertebrate may contain thousands of proteins, but most will be present in only very small amounts. Because protein electrophoresis detects only the most abundant proteins, the protein patterns ('fingerprints') are normally quite simple. They usually differ between species, which makes identification of samples possible.

This method applies only to the identification of fresh or frozen material—any dried, canned or cooked product requires DNA tests. Protein fingerprints are therefore not available for the dried or smoked tissues from jellyfish or beche-de-mer. DNA tests may also be needed if problems arise with protein fingerprinting for species identification, or if the results need legal confirmation.

Checking an identification

If a fillet is claimed to be from species X, but the claim needs to be tested, that fillet's protein fingerprint can be determined and compared with the figure given for species X here. If there are obvious major differences, it's unlikely to be species X. However, slight differences can occur between electrophoretic tests run on different days, and a better procedure is to run a sample from a known specimen of species X beside the fillet to be tested. The protein fingerprints can be compared with one another and with the figure of species X given here. If the protein fingerprint of the fillet differs from that of species X, but is similar to that of another species, Y, the fillet sample should be run beside known samples of both species X and Y.

Reading protein fingerprint figures

The protein fingerprint of each species is compared with a 'standard' protein fingerprint—a mixture of chicken albumen and a crude protein extract from redfish (*Centroberyx affinis*). The standard gives seven well-defined bands: P5 is the fastest (the chicken albumen band), followed by P4, P3, P2 and P1—all located above the point of application of the protein mix at the origin (O); P0 is on or close to O; P-1 migrates below O.

Protein fingerprint descriptions in the text are approximate. For example, a band stated to be 'at P1.3', is about a third of the way between P1 and P2. These relative mobilities can vary a little, but not significantly. Faint staining bands not recorded in the diagrams may be seen; these sometimes occur in some individuals of a species but not others. They are disregarded here. The positions of bands migrating close to or below O can vary significantly, and are not generally used for diagnostic purposes.

Sometimes one or more bands vary among individuals of the same species. This usually reflects genetic variation (also called polymorphism) for that protein, and could result in misidentification. Where possible, at least eight individuals of each species were examined to determine the extent of such variation. Common patterns are included in the figures but others may occur. Such undescribed variants will be very similar to the described patterns (usually differing by only a single band) and will usually still be identifiable.

Some very closely related species have identical protein fingerprints. Supplementary allozyme tests (as in Daley *et al.* 1997) or DNA tests are then needed for accurate identification.

In the figures—diagrammatic representations of the gels—groups (angel sharks, dogfishes, *etc.*) are ordered alphabetically and species within groups by scientific name. Figures are labelled above by marketing name, with scientific names provided in the caption. The marketing name is given a suffix (1, 2, 3, *etc.*) where it covers more than one species. Common variants are further suffixed in the marketing name by the letters A, B or C. Common names, marked with a cross (†), are used for species that lack a marketing name.

Technical details

Protein separation used Titan III cellulose acetate gel plates, from Helena Laboratories. The equipment is modest: a small power pack, electrophoresis chamber, and an inexpensive microcentrifuge.

To test a sample, a small piece of white muscle was homogenised in a few drops of water in a 1.5 mL microcentrifuge tube. Fish tissue was taken from the 'shoulder'. Tissues tested in invertebrates were: crustaceans—abdominal muscle; abalones and sea snails—foot muscle; bivalves—adductor muscle; squids—mantle. The mixture was centrifuged at 10 000*g* at room temperature for three minutes, and the supernatant used for electrophoresis.

Up to twelve samples can be electrophoresed simultaneously on a single 76 x 76 mm plate using the Helena Super-Z12 system. Lanes 1 and 12 were left empty (outside samples can run unevenly), the standard protein mix (see above) was placed in lanes 2 and 11, and the test samples in lanes 3 to 10. A tris glycine buffer system was used (0.02 M tris, 0.192 M glycine; Hebert and Beaton 1993) for 25 minutes at 200 V at room temperature. The plates were then stained with a protein stain, Coomassie Blue (0.2% Coomassie Blue in a mixture of 6 parts water to 4 parts methanol to 1 part glacial acetic acid) for 5 to 15 minutes. Unbound stain was removed by washing in a destaining solution (the stain solution without Coomassie Blue). Gels were then digitally photographed and dried for future reference.

Sharks and rays

Angel sharks (Fig. 9.1)

Four species examined. Angel shark-1 (*Squatina australis*) is the only species with a major band close to P2, angel shark-4 (*S.* sp.) the only species with a major band at P1. The remaining two species, angel shark-2 (*S. tergocellata*) and angel shark-3 (*S.* sp.), cannot be separated.

Dogfishes (Figs 9.1 & 9.2)

Eleven species examined. Five species—Endeavour dogfish-1 (*Centrophorus moluccensis*), Endeavour dogfish-2 (*C. uyato*), leafscale gulper shark (*C. squamosus*), roughskin dogfish-1 (*Centroscymnus crepidater*) and roughskin dogfish-3 (*Deania calcea*)—lack bands in the O to P1 region and have major bands between P3 and P1; with the exception of *Centrophorus moluccensis* and *C. uyato*, which had identical patterns, these species could be clearly separated. A sixth species, roughskin dogfish-2 (*Centroscymnus plunketi*), can be identified by two major bands, one at P1.8 and one at P0.6. The five remaining dogfishes are all *Squalus* species—greeneye dogfish-1 (*S. mitsukurii*), greeneye dogfish-2 (*S.* sp.), greeneye dogfish-3 (*S.* sp.), spikey dogfish (*S. megalops*), and white-spotted dogfish (*S. acanthias*). None has a major band faster than P1.3, and all have a distinct band at P0.6. These five species are similar, but separable. There are unresolved taxonomic problems in this group, especially among greeneye dogfishes. More morphological and genetic work is required. *S.* sp. (greeneye dogfish-3) is similar to *S. megalops* but has a minor band at P1.6 (rather than P1.2). *S. acanthias* and most specimens of *S. mitsukurii* are similar, but other specimens provisionally identified as the latter show distinct patterns, especially in the O to P-1 region (not shown). Variation was noted for *S. megalops* (four Type A, four Type B), *S. acanthias* (one Type A, seven Type B) and *S.* sp., greeneye dogfish-3 (three Type A, one Type B).

Ghostsharks (Fig. 9.2)

Five species examined, all separable. There has been some debate as to whether ghostshark-3 (*Hydrolagus lemures*) and ghostshark-4 (*H. ogilbyi*) are morphologically sufficiently distinct to be considered different species. However, their protein fingerprints are clearly separable (especially in the P4 to P3 region) supporting their recognition as distinct species. The two undescribed *Chimaera* species are also distinct from each other and from the *Hydrolagus* species.

Hound sharks (Fig. 9.2)

Three species examined. They can be clearly discriminated—whiskery shark (*Furgaleus macki*) by the absence of major bands at P2.2 or P1, school shark (*Galeorhinus galeus*) by the absence of a major band at P2.2 and the presence of a major band at P1, and gummy shark-1 (*Mustelus antarcticus*) by the presence of a major band at P2.2. In addition, *G. galeus* has a distinct band at P3 which is weak or absent in *M. antarcticus*.

Rays & skates (Fig. 9.3)

Nine species examined. Two species lack distinct bands above P2; they can be distinguished by a major band at P1.1 in ray-3 (*Myliobatis australis*) while ray-2 (*Himantura toshi*) has no major bands near to or faster than P1. Ray-1 (*Gymnura australis*) can be identified by a band at P3, ray-4 (*Urolophus paucimaculatus*) by a band at P2.6, guitarfish-1 (*Rhynchobatus australiae*) by a band at P3.4, and guitarfish-2 (*Trygonorrhina* sp.) by a band at P4. The three skates examined all have major bands at P1 and P2.2; they cannot be clearly distinguished from each other.

Sharks & bony fishes

Sawsharks (Fig. 9.3)

The two species tested are very similar, but can be distinguished by the presence in sawshark-2 (*Pristiophorus nudipinnis*) of a band just below O (absent in sawshark-1, *P. cirratus*). The weak band just above O in both species is usually single but is sometimes double and sometimes too weak to see.

Sevengill sharks (Fig. 9.3)

The single species, broadnose shark (*Notorynchus cepedianus*), is characterised by a single major band, at P1.7.

Whaler sharks (Fig. 9.4)

Six species examined, all separable. Blue whaler shark (*Prionace glauca*) was variable, with three specimens observed of Type A, three of Type B and two of Type C. Only *P. glauca* Types A and B have a major band at or faster than P1. *P. glauca* Type C, bronze whaler shark-1 (*Carcharhinus obscurus*), blacktip shark-1 (*C. dussumieri*), blacktip shark-2 (*C. sorrah*), blacktip shark-3 (*C. tilstoni*) and blacktip shark-4 (*Rhizoprionodon acutus*) are distinguishable by bands between P1 and P-1, and often (and unusually) by bands close to O.

Australian salmons (Fig. 9.5)

Three species examined. Australian salmon-1 (*Arripis trutta*) and Australian salmon-2 (*A. truttaceus*) have the same patterns and the third species, Australian herring (*A. georgianus*), is similar. All have a pair of major bands at P1 and P0.7; a fainter band at P2 in *A. trutta* and *A. truttaceus* is weaker still or absent in *A. georgianus*. *A. georgianus* has a weak band at P0.4 which is weaker still or absent in the other two species.

Batfishes (Fig. 9.5)

Four of the five batfish species examined have similar but separable patterns. These four species—batfish-2 (*Platax batavianus*), butterfish-1 (*Scatophagus argus*), butterfish-2 (*S. multifasciatus*) and butterfish-3 (*S.* sp.)—all have major bands at P1.3 and P0.3, and a weak band at P2.4. *P. batavianus* is characterised by the combined presence of weak bands at P1, P1.8 and P2.4, *S. argus* by the absence of a band at P1, *S. multifasciatus* by the presence of a weak band at P1.1, and *S.* sp. by the absence of a band at P1.8. The fifth species, batfish-1 (*Drepane punctata*), has a major band at P1.7 rather than P1.3, and is further characterised by a group of three bands in the P3–P4 region.

Bigeyes (Fig. 9.5)

Two species, red bullseye-1 (*Priacanthus hamrur*) and red bullseye-2 (*P. tayenus*), were examined. They are clearly separable using bands in the P1–P2 region, but similar in that both show a weak band at P2.4 and a strong band at P0.2. One of eight *P. hamrur* showed the normal band at P1 together with an additional band at P1.2 (the position of a band in *P. tayenus*).

Bony fishes

Billfishes (Fig. 9.5)

Five species examined. Four marlin species—marlin-1 (*Makaira indica*), marlin-2 (*M. mazara*), marlin-3 (*Tetrapturus angustirostris*) and marlin-4 (*T. audax*)—cannot be unambiguously separated; they all have a major band which migrates just above O but just below O in swordfish (*Xiphias gladius*). DNA tests which distinguish all billfish species have been developed (Pepperell and Grewe 1998).

Boarfishes (Fig. 9.6)

All six species tested are distinguishable. Conway (*Oplegnathus woodwardi*) has no clear bands migrating faster than P2, bigspine boarfish (*Pentaceros decacanthus*) has a major band at P1.3 and no band at P1, boarfish (*Pentaceropsis recurvirostris*) has a band at P2.6 and no or a very faint band at P1, giant boarfish-1 (*Paristiopterus gallipavo*) has a single band at P3.1 and a band at P1, giant boarfish-2 (*Paristiopterus labiosus*) has a close coupled pair of bands at P3 (one of five specimens, not pictured, showed an additional band at P1.5), and the blackspot boarfish (*Zanclistius elevatus*) has a major band at P1.6 and no other bands between P1 and P2 (except one of eight specimens showed an additional band at P1.8).

Bonnetmouths (Fig. 9.6)

The two species examined share a major band at P1 and a minor band at P1.8; they can be separated by the presence of a band at P0.3 in rubyfish-1 (*Plagiogeneion macrolepis*) and a band at or just below O in redbait-1 (*Emmelichthys nitidus*).

Breams (Fig. 9.6)

Six species examined. Two, yellowfin bream (*Acanthopagrus australis*) and black bream (*A. butcheri*), are very similar but can usually be distinguished from other breams by three bands in the region P1.5–P2 (some specimens of *A. australis* may lack the fastest of these three bands). Pikey bream (*A. berda*) is characterised by a strong band at P2.4, frypan bream (*Argyrops spinifer*) by the absence of bands at P2 or faster, snapper (*Pagrus auratus*) by a major band at P1 and no bands between P1 and P2, and tarwhine (*Rhabdosargus sarba*) by only one band between P1 and P2 and no band at P2.4. Three of the eight *Acanthopagrus butcheri* (not pictured) showed an additional minor band between P1 and P1.5.

Cardinal fishes (Fig. 9.7)

The single species, *Epigonus telescopus*, is characterised by major invariant bands at P2.4 and P1.2. There is variation in the region P1.5–P2; four of six fish were Type A, two were Type B.

Carps (Fig. 9.7)

Two of the three species examined, goldfish (*Carassius auratus*) and European carp (*Cyprinus carpio*), cannot be separated. They have a major band at P2.1, whereas tench (*Tinca tinca*) has this band at P1.9.

Catfishes (Fig. 9.7)

Six species examined. All four *Arius* species have features in common (e.g. a band at P1.9) but all are separable. Catfish-1 (*A. bilineatus*) and silver cobbler (*A. midgleyi*) are very similar (with major bands at P1.5 and P1.9), but the latter has a major band at P3 which is weak or absent in the

Bony fishes

former. Catfish-2 (*A. mastersi*) and catfish-3 (*A. thalassinus*) are also similar to one another (both showing bands at P1 and P1.9), but the former has a major band at P1.5 and the latter a major band at P1.3. Cobbler (*Cnidoglanis macrocephalus*) is characterised by a pair of bands flanking P2, while freshwater catfish (*Tandanus tandanus*) has a pair of bands at P2 and P2.2, and another pair at P0.9 and P1.1.

Cods (Fig. 9.8)

Seven species examined. Cathodal bands (below O) are weak or absent. Southern rock cod-2 (*Pseudophycis bachus*) and southern rock cod-3 (*P. barbata*) have identical patterns; otherwise all species can be distinguished. *P. bachus, P. barbata* and *L. rhacina* have major bands at P2.1 and P0.2, with the former two species also having a weak band at P1.1. Ribaldo-1 (*Lepidion schmidti*) has a single major band, at P3. Ribaldo-2 (*Mora moro*) has major bands at P2.9, P1.9 and P0.6. Blue grenadier (*Macruronus novaezelandiae*) has major bands at P2.5 and P1.8. Southern hake (*Merluccius australis*) has a pair of major bands at P2.8 and P2.6, and another pair at P1.0 (this band sometimes weak) and P0.8.

Dories (Fig. 9.8)

Four species examined. All have a band at P1 and a weak band at P3 (sometimes this band may be too weak to be apparent), but the species can be distinguished. Only John dory (*Zeus faber*) and king dory (*Cyttus traversi*) have bands in the P2 region, *Z. faber* having a single weak band just below P2, while *C. traversi* has two strong bands at and just below P2. Mirror dory (*Zenopsis nebulosus*) has a strong band at about P0.6, silver dory (*C. australis*) has a weak band in that region. In addition, *Z. nebulosus* has a pair of bands at P2.8 and P2.4 which are not present in *C. australis*.

Drummers (Fig. 9.8)

The two species examined have a pair of bands at P1.8 and P1.6; these are equally intense in sweep-1 (*Scorpis lineolatus*) but in luderick (*Girella tricuspidata*) the P1.8 band is less intense. *S. lineolatus* has a weak band at P1 that is absent in *G. tricuspidata*.

Eels (Fig. 9.9)

Four species examined. The shortfin and longfin eels (*Anguilla* species) are not separable from each other; they have a major band at P1.2. The two conger eels (*Conger* species) are not separable from each other; their major band is at P1.7.

Emperors (Fig. 9.9)

Ten species examined. They all show some similarities (such as a weak band at P2 and a complex set of bands below the origin), but most can be differentiated from each other. Seven species, seabream-1 (*Gymnocranius grandoculis*), seabream-2 (*G. elongatus*), seabream-3 (*G.* sp.), grass emperor (*Lethrinus laticaudis*), redspot emperor (*L. lentjan*), spangled emperor-1 (*L. nebulosus*) and spangled emperor-2 (*L.* sp.) are particularly similar, sharing a major band at P1.2 and a band at P0.5 (strong in *L. laticaudis*, *L. lentjan*, and *L.* sp. and weak in *G. grandoculis*, *G. elongatus*, *G.* sp., and *L. nebulosus*). *L. laticaudis* and *L.* sp. share a distinct band at P2.5 which is weak or absent in all other emperors. One of eight *L. laticaudis* had a weak band at P0.5 plus an extra band at P0.7—Type B in the figure. *G. grandoculis* and *G.* sp. cannot be separated; they share a pair of major bands at P1.2 and P0.9 (the P0.9 band was absent in one of eight *G. grandoculis* examined). Emperor-1 (*L. atkinsoni*) and redthroat emperor (*L. miniatus*) are unique in having a major band at P1.6; *L. miniatus* also has a major band at P1.

Bony fishes

Flatfishes (Fig. 9.10)

Nine species examined; all are similar. They each have a major band at about P2 and most have a band at P2.4 and a band at P1. The position of the major band in the P2 region allows some separation: it is at P2.1 in flounder-1 (*Neoachiropsetta milfordi*), P2 in bay flounder-1 (*Ammotretis rostratus*) and greenback flounder (*Rhombosolea tapirina*), P1.9 in Australian halibut (*Psettodes erumei*), and P1.8 in flounder-2 (*Pseudorhombus arsius*), flounder-3 (*P. jenynsii*), flounder-4 (*P. spinosus*) and sole-2 (*Synaptura nigra*). Sole-1 (*Paraplagusia bilineata*) has a pair of bands of similar intensity flanking the P2 region. Within these groups, most species can be distinguished using minor bands; *A. rostratus* and *R. tapirina*, however, are very similar. There was variation for *N. milfordi* (four Type A individuals, three Type B) and *Psettodes erumei* (the weak band at P1.5 was sometimes absent).

Flatheads (Figs 9.10 & 9.11)

Fourteen species examined. They fall into seven groups depending on the mobilities of major bands in the P2 region: Group 1—bartail flathead (*Platycephalus indicus*), double band at P2 and P1.8; Group 2—tiger flathead-2 (*Neoplatycephalus richardsoni*), blue-spotted flathead (*P. caeruleopunctatus*), rock flathead (*P. laevigatus*), flathead-3 (*P. longispinis*) and southern flathead (*P. speculator*), double band at P1.8 and P1.6 (the P1.6 band is weak in *N. richardsoni*); Group 3—deepwater flathead (*N. conatus*), triple band at P2.1, P1.8 and P1.6 (the latter two being weaker than the P2.1 band); Group 4—flathead-1 (*Papilloculiceps nematophthalmus*), dusky flathead (*Platycephalus fuscus*) and flathead-4 (*P. marmoratus*), single band at P2; Group 5—northern sand flathead (*P. arenarius*) and flathead-2 (*P. endrachtensis*), single band at P1.8; Group 6—tiger flathead-1 (*N. aurimaculatus*), double band at P2.1 and P1.6; and Group 7—sand flathead (*P. bassensis*), single band at P1.6. Four of the species in Group 2 (*P. caeruleopunctatus*, *P. laevigatus*, *P. longispinis*, and *P. speculator*) cannot be distinguished from each other, neither can two of the species in Group 4 (*P. fuscus* and *P. marmoratus*). One of seven *N. richardsoni* showed a band at P2 which was absent in other specimens. Allozyme tests further separate the flatheads found in the South East Fishery (Daley *et al.* 1997).

Freshwater perches (Fig. 9.11)

Six species examined, all can be distinguished. Three of these—barramundi (*Lates calcarifer*), Murray cod-1 (*Maccullochella peelii*) and golden perch (*Macquaria ambigua*)—have the relatively unusual feature of an intense fast-migrating band, at P3. *Maccullochella peelii* has a very similar pattern to *Macquaria ambigua*, but has an intense rather than weak band at P0.3. *L. calcarifer* can be distinguished from these two species by a band at P0.7 rather than P0.3. Redfin (*Perca fluviatilis*) is readily distinguished by the absence of the P3 band and by a major band at P1.3. Estuary perch (*Macquaria colonorum*) is very similar to Australian bass (*M. novemaculeata*) but has a band at P2.4 (otherwise absent). They are both readily distinguished from other freshwater perches by major bands at P1.3 and P0.3.

Garfishes (Fig. 9.12)

Seven species examined; all are similar with bands at P1 and P2. Four species—garfish-1 (*Arrhamphus sclerolepis*), garfish-2 (*Hyporhamphus affinis*), garfish-3 (*H. quoyi*) and river garfish (*H. regularis*)—cannot be clearly separated; they all show, as well as the P1 and P2 bands, a weak band at P2.5. This band is weak or absent in the otherwise similar eastern sea garfish (*H. australis*). Southern garfish (*H. melanochir*) can be distinguished by a pair of major bands in the P1 region. Longtom-1 (*Tylosurus gavialoides*) is the only member of this group with a clear band at P3 (and no P2.5 band).

Bony fishes

Gemfishes (Fig. 9.12)

Six species examined. Longfin gemfish *(Rexea antefurcata)*, gemfish (*R. solandri*) and barracouta (*Thyrsites atun*) are similar, with a pair of strong bands at P1.3 and P1. Allozyme tests distinguish these species (Daley *et al.* 1997). Escolar-1 (*Lepidocybium flavobrunneum*) has a major band at P1.5 and escolar-2 (*Ruvettus pretiosus*) a major band at P1.6. Ribbonfish (*Lepidopus caudatus*) has a major band at P1.6 but a pair of bands tightly flanking P2 distinguishes it from *R. pretiosus*.

Goatfishes (Fig. 9.12)

Three species examined and clearly separable. Red mullet-1 (*Parupeneus heptacanthus*) has a pair of close-coupled major bands flanking P1, red mullet-2 (*P. indicus*) has a single major band at P1 and lacks a band at P2.4, red mullet-3 (*Upeneichthys vlamingii*) has a single major band at P1 and a clear band at P2.4. *U. vlamingii* is variable—six specimens of Type A and two of Type B were recorded.

Grunter breams (Fig. 9.13)

Three species examined. Grunter bream-1 (*Pomadasys kaakan*), with major bands at P1 and P1.5, is quite distinct from the two sweetlip breams, which are similar to each other—both lack bands in the P1 region and have a close-coupled band at P2. They can be distinguished by a band in the O–P1 region, which is at P0.3 in sweetlip bream-1 (*Diagramma labiosum*) and at P0.5 in sweetlip bream-2 (*Plectorhinchus flavomaculatus*).

Grunters (Fig. 9.13)

Three species examined and clearly separable. Striped perch-2 (*Pelates octolineatus*), a marine species, is distinguished from the two freshwater grunters by a major band at P1.3. Silver perch (*Bidyanus bidyanus*) has a distinct band between P2 and P3 (at P2.5) and a major band at P1.7, both of which are absent in striped perch-1 (*Hephaestus jenkinsi*).

Gurnards (Fig. 9.13)

Four species examined. These can be readily distinguished from each other, but all share a common band at P1.5. This band is strong in red gurnard (*Chelidonichthys kumu*), butterfly gurnard-1 (*Lepidotrigla mulhalli*) and butterfly gurnard-2 (*L. vanessa*), but weak in latchet (*Pterygotrigla polyommata*). *L. mulhalli* shows variation for the presence or absence of bands at P0.5 and P-1. Five Type A, one Type B and two Type C specimens were observed.

Herrings (Fig. 9.13)

Four of the five species examined are unusual in having most of their protein bands in the region of P1 to P-1; only bony bream-1 (*Nematalosa erebi*) has a major band migrating faster than P1. Sandy sprat (*Hyperlophus vittatus*) is the only herring examined with a major band at P1 and blue sprat (*Spratelloides robustus*) is the only species with a single major band at P0.6. Anchovy (*Engraulis australis*) can be distinguished from pilchard (*Sardinops neopilchardus*) by the presence of weak bands at P1 or faster (rather than all distinct bands below P1).

Icefishes (Fig. 9.14)

Patagonian toothfish (*Dissostichus eleginoides*) is characterised by strong bands at P1.9 and P0.8, together with weaker fast moving bands at P2.9 and P3.9.

Bony fishes

Jewfishes (Fig. 9.14)

Four species examined; all can be distinguished. Teraglin (*Atractoscion aequidens*) is the only jewfish with a single major band at P2; the other three species all have a pair of major bands at P2. Mulloway (*Argyrosomus hololepidotus*) is the only jewfish with a major band at P0.4. Jewfish-1 (*Johnius borneensis*) is similar to black jewfish (*Protonibea diacanthus*) but has a strong band at P3 (rather than weak or absent) and lacks a band at P4 (rather than often present). *A. hololepidotus* shows variation in the P1 to P1.5 region: three Type A, one Type B and two Type C specimens were recorded.

Leatherjackets (Fig. 9.14)

Five species examined: leatherjacket-1 (*Aluterus monoceros*), reef leatherjacket-1 (*Meuschenia freycineti*), ocean jacket (*Nelusetta ayraudi*), velvet leatherjacket (*Parika scaber*) and leatherjacket-2 (*Pseudomonacanthus peroni*). All can be readily distinguished from each other by bands in the P1 to P2.5 region. *Parika scaber* shows variation: four specimens were Type A, four were Type B.

Lings (Fig. 9.15)

Three species examined. Tusk (*Dannevigia tusca*) can be distinguished from the two lings (*Genypterus* species) by a pair of bands tightly flanking the P2 region, and by the absence of a major band below P1. The only separating feature of the two *Genypterus* species is a band at P3.8, present in pink ling (*G. blacodes*) but weak or absent in rock ling (*G. tigerinus*). These species are more reliably separated by an allozyme test (Daley *et al.* 1997).

Mackerels & tunas (Figs 9.15 & 9.16)

Twenty species examined, seven mackerels, twelve tunas and one butterfly mackerel.

The seven mackerel species form two major groups. The five members of Group 1—mackerel-1 (*Acanthocybium solandri*), Types A and B of Spanish mackerel (*Scomberomorus commerson*), spotted mackerel (*S. munroi*), school mackerel (*S. queenslandicus*) and grey mackerel (*S. semifasciatus*, Type A)—each have a band in the P1.8 region. Two of these, *S. munroi and S. queenslandicus*, cannot be reliably separated; other species are separable on the basis of the positions and presence or absence of bands in the O to P2 region, and their relative intensities. Two species showed variation—*S. semifasciatus* (seven specimens of Type A, one of Type B), and *S. commerson* (one specimen of Type A, five of Type B, two of Type C). Group 2 is characterised by the absence of a band in the P1.8 region, and consists of two species—shark mackerel (*Grammatorcynus bicarinatus*) and blue mackerel (*Scomber australasicus*)—plus Type B of *Scomberomorus semifasciatus* and Type C of *S. commerson*. Each of these can be distinguished from one other. *G. bicarinatus* is the only species with a major band at P0.8. *Scomber australasicus* has all major bands located close to O or between O and P-1. Type B of *Scomberomorus semifasciatus* has a major band at P1.5. Type C of *S. commerson* has three equidistant bands between O and P1.

The twelve tuna species fall into four groups. Group 1 consists of six *Thunnus* species: albacore (*T. alalunga*), yellowfin tuna (*T. albacares*), southern bluefin tuna (*T. maccoyii*), bigeye tuna (*T. obesus*), tuna-2 (*T. thynnus*), and longtail tuna (*T. tonggol*). These are very similar to one another, and are characterised by a pair of strong bands at or near O, and by a strong band at P-1. The band at P2.5, shown in the figure for some species, is sometimes weak or absent even in those species it is depicted in; it is not a reliable indicator of species identity. *T. maccoyii* and *T. tonggol* are both depicted as having three bands in the P1 to P1.5 region, but sometimes one or more of these bands

Bony fishes

are absent. The presence of a strong band at P1.4 indicates *T. alalunga*; the other species in this group cannot be reliably separated from one another. Group 2 consists of skipjack tuna (*Katsuwonus pelamis*) only. This lacks the pair of strong bands near O and also the strong band at P1.8 that identifies Group 3 species (see below); its major bands are at P1 and P-1. Four specimens of each of two Types, A and B, were observed. Group 3 consists of frigate mackerel (*Auxis thazard*) and mackerel tuna (*Euthynnus affinis*), and is characterised by strong bands at P1 and P1.8. These two species may be separated by differences in the O to P-1 region. Group 4 has three species, characterised by a pair of major bands in the O to P-1 region. Slender tuna (*Allothunnus fallai*) is the only species in this group with a major band in the P1 to P2 region (at P1.7, sometimes resolving into two close-coupled bands). Tuna-1 (*Cybiosarda elegans*) and bonito-1 (*Sarda australis*) can be separated from each other by bands in the O to P1 region.

The butterfly mackerel (*Gasterochisma melampus*) is distinguished from other mackerels and tunas by a pair of major bands at P1 and P1.2.

Milkfishes (Fig. 9.17)

The single species (*Chanos chanos*) is characterised by a major band at P0.7 and a pair of bands at P0.2 and P0.4.

Moonfishes (Fig. 9.17)

The two species have very similar protein patterns, although the weak bands migrating just above P1 appear more readily detectable in moonfish-2 (*Lampris immaculatus*) than moonfish-1 (*L. guttatus*). The two major bands depicted just above and at O sometimes fail to separate.

Morwongs (Fig. 9.17)

Six species examined; most can be distinguished from each other. Grey morwong (*Nemadactylus douglasii*) shows variation: four fish of Type A, two of Type B and two of Type C were observed. Morwong-1 (*N. macropterus*) and morwong-2 (*N.* sp.) share identical protein patterns, but can be distinguished from all other morwongs by having a band at P3. Red morwong (*Cheilodactylus fuscus*) and banded morwong (*C. spectabilis*) can be distinguished from other morwongs by a strong band at P1.8, and from each other by the presence of a band at P2.5 in *C. fuscus* (otherwise absent). *N. douglasii* and blue morwong (*N. valenciennesi*) are very similar, but any specimen with a band at P1.9 is *N. douglasii* Type A or B; *N. valenciennesi* cannot be readily distinguished from *N. douglasii* Type C.

Mullets (Fig. 9.18)

Five species examined. Patterns are rather similar across species, with bands in the P2, P1 and P0.5 regions, but all species are separable. Variation is present in mullet-1 (*Aldrichetta forsteri*): one of eight specimens showed three bands at P1.2, P1 and P0.8 (Type A), six showed only bands P1.2 and P1 (Type B), and one showed only bands P1.2 and P0.8 (Type C). Variation is also present in sea mullet (*Mugil cephalus*): all eight specimens from Perth and Sydney fish markets showed Type A (pronounced band at P2, band at P2.5 weak or absent), while two from a Brisbane fish market showed Type A and two showed Type B (band at P2 weak or absent, band at P2.5 strong).

Bony fishes

Ocean perches (Fig. 9.18)

Eight species examined. Their protein patterns fall into three groups. Group 1 consists of ocean perch-1 (*Helicolenus barathri*) and ocean perch-2 (*H. percoides*), which have identical patterns. They are distinguishable from other ocean perches by a major band at P1.5. Group 2 consists of coral perch-1 (*Neosebastes incisipinnis*), coral perch-2 (*N. scorpaenoides*) and coral perch-3 (*N. thetidis*). Their strongest-staining band is at P1.9. *N. incisipinnis* and *N. scorpaenoides* cannot be distinguished from each other, but show a weak band at P2.5 which is replaced in *N. thetidis* by a weak band at P3. Group 3 consists of coral perch-4 (*Scorpaena cardinalis*), coral perch-5 (*Trachyscorpia capensis*) and coral perch-6 (*T.* sp.), which show major bands at P2.5, P2 and P0.5. *T. capensis* and *T.* sp. are inseparable but are separated from *S. cardinalis* by lacking a weak band at P3.6 (otherwise present).

Oreos (Fig. 9.19)

Four species examined, all separable. Black oreo-1 (*Allocyttus niger*) and black oreo-2 (*A. verrucosus*) lack bands in the P2.5 region, and can be separated from each other by the presence of a major band at P0.5 in *A. niger* (otherwise absent). Smooth oreo (*Pseudocyttus maculatus*) has a pair of equally-staining tightly-coupled bands at P2.5, spikey oreo (*Neocyttus rhomboidalis*) has a single band in this region (sometimes a very weak slightly slower band can be seen). Spikey oreo shows variation in the P2 region. Most commonly a single major band at P2 is seen (Type A), but occasionally two or three bands are present (Type B).

Pearl perches (Fig. 9.19)

Four species examined. They are characterised by bands at P2 and P0.5, but all show distinguishing features. West Australian dhufish (*Glaucosoma hebraicum*) is the only pearl perch with a band at P1. Pearl perch-3 (*G. scapulare*) is distinguished by a pair of bands at P1.6 and P1.5. Pearl perch-1 (*G. buergeri*) is similar to pearl perch-2 (*G. magnificum*), but has a strong band below P-1 (otherwise absent). Some faint extra bands may be seen: *G. hebraicum* at P3, *G. scapulare* at P2.5 and *G. magnificum* at P3 and/or P2.5.

Pikes (Fig. 9.19)

Six species examined. All are quite different from each other and can be distinguished by the major bands alone.

Pomfrets (Fig. 9.20)

Three species examined; while they possess bands in common, they can all be distinguished. Ray's bream-1 (*Brama brama*) has a major band at P1 while the other two species have a major band at P0.3. The latter two species can be distinguished by Ray's bream-2 (*Taractichthys longipinnis*) having a minor band at P1.3 while Ray's bream-3 (*Xenobrama microlepis*) has a minor band at P1. One of three *T. longipinnis* examined showed an extra band at P0.4 (not figured).

Rabbitfishes (Fig. 9.20)

The single species tested is characterised by four equidistant bands between P1 and P1.8, the slowest two of which are the strongest. Usually the P1.6 band is stronger than the P1.8 band, but in two of eight specimens the reverse was true.

Bony fishes

Redfishes (Fig. 9.20)

Six species examined, all separable. Imperador (*Beryx decadactylus*) may be separated from alfonsino (*B. splendens*) by the presence of a major band at P1.6 (rather than P1.3) and they may both be distinguished from other redfishes by the presence of a minor band at P2.5 and a major band just below P2. Redfish (*Centroberyx affinis*) has a protein pattern identical to the control (except for the lack of a band at P5); yelloweye redfish (*C. australis*) is similar except for the replacement of the P1 band by one at P0.6. Bight redfish (*C. gerrardi*) has bands at both P1 and P0.6, and swallowtail (*C. lineatus*) is identified by a pair of bands at P1.1 and P0.9. *C. lineatus* showed variation for bands below O—three fish of Type A and three of Type B were observed.

Rockcods (Figs 9.21 & 9.22)

Nineteen species examined, classifiable into six groups. Group 1 contains coral trouts and a coral cod: coral trout-1 (*Plectropomus areolatus*), coral trout-2 (*P. leopardus*), coral trout-3 (*P. maculatus*) and coral cod-1 (*Cephalopholis cyanostigma*). These species cannot be readily separated. They lack a P2.5 band, and have a strong band at P2 (which may split into two sub-bands) and weak bands at P1 and P0.5. Group 2 contains yellow-spotted rockcod (*Epinephelus areolatus*), bar rockcod-1 (*E. ergastularius*) and white-spotted rockcod (*E. multinotatus*). These species have a double band at or just below the P2 region and the former two also have a strong band at P0.5. Three of seven *E. ergastularius* have the faster of the two bands in the P2 region stronger (Type B), four had the slower band stronger (Type A). Group 3 contains estuary rockcod (*E. coioides*) only. This is distinguished by a strong band at P1. Group 4 consists of eight species which are difficult to separate, and are characterised by a strong bands at P2.5, at or just below P2, and at P0.5. These species are coral cod-2 (*C. sonnerati*), barramundi cod (*Cromileptes altivelis*), blacktip rockcod (*E. fasciatus*), rockcod-1 (*E. merra*), rockcod-2 (*E. morrhua*), rockcod-3 (*E. ongus*), honeycomb rockcod (*E. quoyanus*), and Maori rockcod (*E. undulatostriatus*). *E. undulatostriatus* has a weak band at P-1 which is absent in the other six species. *Cephalopholis sonnerati* and *E. merra* have the P2 band at or only just below P2, whereas the other six species have it distinctly lower, at P1.8. Group 5 contains longfin perch (*Caprodon longimanus*) only. This is characterised by a strong band at P1.3, and by two close-coupled bands at P2.5 and P2.4. The three individuals examined showed different patterns in other gel regions: two had an extra strong band at P1.6, and one showed weak bands at P2 and P2.2 that were absent in the other two specimens (see Types A, B and C in the figure). Group 6 consists of two congeneric species, hapuku-1 (*Polyprion americanus*), which has a weak band at P1.9 and no band at P2.5, and hapuku-2 (*P. oxygeneios*), which has no band at P1.9 but a weak band at P2.5. These species have strong bands at P1.6 and P0.5.

Roughies (Fig. 9.23)

Three species examined, all separable. Darwin's roughy (*Gephyroberyx darwinii*) is characterised by the lack of a P1 band, orange roughy (*H. atlanticus*) by bands just above and just below P2, and the giant sawbelly (*Hoplostethus gigas*) by a band at P2.5. About half of all *H. atlanticus* also have a band at P2 (Type A) which the other half lack (Type B). The P1 band in *H. atlanticus* sometimes separates into a weak band at P1 and a strong band at P0.9.

Salmons (Fig. 9.23)

Five species examined. Trout-1 (*Oncorhynchus mykiss*), trout-2 (*Salmo trutta*) and Atlantic salmon (*Salmo salar*) are very similar; of these, *O. mykiss* has only a weak band at P1.8 compared with a strong band in the other two species, and *S. salar* appears to lack a P1 band that is evident in *O.*

Bony fishes

mykiss and *S. trutta*. Whitebait-1 (*Galaxias maculatus*) is the only species with a band between P2 and P3, at P2.4. Whitebait-2 (*Lovettia sealii*) is the only species with a band at P2. The band depicted for *G. maculatus* at P1.5 was present (Type A) in three but absent (Type B) from five specimens.

Seaperches (Figs 9.23 & 9.24)

Eighteen species examined. Most appear to be separable using the major bands alone; all are separable when the weaker bands are also used. Two species showed variation. King snapper (*Pristipomoides filamentosus*) had three Types: Type A—major bands at P1.1 and P0.9 (one of five fish); Type B—no P1.1 band, major band at P0.9 (three fish); and Type C—major band at P1.1, no P0.9 band (one fish). Goldband snapper-2 (*Pristipomoides typus*) also showed three Types (not shown), reflecting variable intensities at the P1 and P0.6 bands: Type A—P1 stronger than P0.6 (three of eight fish); Type B—P1 and P0.6 of similar intensity (two fish); and Type C—P1 weaker than P0.6 (three fish). Moses seaperch-1 (*Lutjanus russelli*) is very similar to Moses seaperch-2 (*L.* sp.) but has a major band at P2.2 (otherwise absent).

Silver biddies (Fig. 9.25)

Silver biddy-1 (*Gerres subfasciatus*) is characterised by a pair of bands at about P2 and P1.5 (the P1.5 band being the stronger), and another pair at about P1 and P0.5.

Stargazers (Fig. 9.25)

Stargazer-1 (*Kathetostoma canaster*) is characterised by a band at P3, a double band at about P2 (sometimes merging into a single somewhat diffuse band), a band at P1.7, and a band at P0.5.

Threadfin breams (Fig. 9.25)

Five species examined, falling into two groups along generic lines. Group 1 consists of threadfin bream-1 (*Nemipterus furcosus*), threadfin bream-2 (*N. hexodon*) and threadfin bream-3 (*N. peronii*). These species are very similar to one another, with a major band at P1.2. *N. furcosus* could be distinguished by a lack of clear bands faster than P1.5. *N. hexodon* specimens from the Gulf of Carpentaria (Type A) show a clear band at P0.6 which is less pronounced in specimens from the Great Barrier Reef (Type B). Group 2 consists of threadfin bream-4 (*Scolopsis monogramma*) and threadfin bream-5 (*S. taeniopterus*). These two species cannot be distinguished, and are characterised by a major band at P1.5.

Threadfin salmons (Fig. 9.25)

Two species examined. Blue threadfin (*Eleutheronema tetradactylum*) is easily separated from king threadfin (*Polydactylus sheridani*) by the presence of a major band at P1.5 and the absence of a major band at P2.

Tilefishes (Fig. 9.25)

The single species tested is characterised by a major band at P1.6 and minor bands at P2.4, P1.8, 1.0 and at or near O.

Bony fishes

Trevallas (Fig. 9.25)

Seven species examined; most are quite similar to one another. Blue-eye trevalla-1 (*Hyperoglyphe antarctica*) and blue-eye trevalla-2 (*Schedophilus labyrinthica*) are identical (but can be distinguished by an allozyme test, Daley *et al.* 1997). Silver warehou (*Seriolella punctata*) can be distinguished from these two species by its major band in the P1 region being a little slower; similarly, blue warehou (*S. brama*) can be distinguished from the two blue-eye trevallas by its major band in the P1 region being a little faster (and its minor band between P1 and P2 is closer to P2 than in these two species or *S. punctata*). White warehou (*S. caerulea*) is very similar to *H. antarctica* and *Schedophilus labyrinthica* but the band in the P0.5 region is faint or absent. Rudderfish-1 (*Centrolophus niger*) is the only species in this group lacking a band in the P1 region. Rudderfish-2 (*Tubbia tasmanica*) can be distinguished from other trevallas by a major band at P1.6 rather than very close to P1.

Trevallies (Figs 9.26 & 9.27)

Twenty four species examined. Most species can be clearly distinguished from each other. Jack mackerel-1 (*Trachurus declivis*) is very similar to jack mackerel-2 (*T. murphyi*) but the P1 band is more intense than the P0.5 band (rather than the two bands having similar intensity). Samson fish-1 (*Seriola dumerili*) and yellowtail kingfish (*S. lalandi*) are very similar to each other. Variation was observed for dart-1 (*Trachinotus botla*), with seven specimens of Type A and one of Type B recorded. Type B may be a different *Trachinotus* species. Variation was also observed for giant trevally (*Caranx ignobilis*); two of five specimens were Type A, three were Type B.

Trumpeters (Fig. 9.28)

Two species examined. They can be distinguished by a pair of major bands in striped trumpeter (*Latris lineata*) at P1.3 and P0.6, and by a major band at P1 in trumpeter-1 (*Latridopsis forsteri*). The P0.7 band in *L. forsteri* sometimes stains strongly, and sometimes weakly.

Whitings (Fig. 9.28)

Ten species examined, falling into four groups. Group 1 has seven species—school whiting-1 (*Sillago bassensis*), whiting-1 (*S. burrus*), sand whiting (*S. ciliata*), school whiting-2 (*S. flindersi*), trumpeter whiting (*S. maculata*), school whiting-3 (*S. robusta*) and whiting-3 (*S. sihama*)—which cannot be reliably separated. Allozyme tests distinguish five of these species (Daley *et al.* 1997). These species all have their major band at P1.3. Group 2 consists of whiting-2 (*S. ingenuua*). This species has the major band at P1.5. Group 3 consists of yellowfin whiting (*S. schomburgkii*). This has the major band at P1 and only a minor band at P1.3. All members of Groups 1–3 have a weak band at P2.4 and a weak band just below P2. A faint band or bands in the region of P4 to P5 may be seen. Group 4 contains King George whiting (*Sillaginodes punctata*). This has its major band at P1 and all faster bands are very weak or absent.

Wrasses (Fig. 9.29)

Sixteen species examined. Not all species can be readily distinguished. Tuskfish-1 (*Choerodon cauteroma*), baldchin groper (*C. rubescens*) and tuskfish-4 (*C. venustus*) are similar to one another, as are tuskfish-2 (*C. cephalotes*) and tuskfish-3 (*C. schoenleinii*). Wrasse-1 (*Achoerodus gouldii*) showed variation—one specimen was of Type A and two of Type B. Type B of *A. gouldii* cannot be distinguished from wrasse-2 (*A. viridis*).

Crustaceans

Bugs & rocklobsters (Fig. 9.30)

Six bug species examined, which fall into two groups. Group 1 consists of Balmain bug-2 (*Ibacus novemdentatus*) and Balmain bug-3 (*I. peronii*), and is characterised by a major band at P2.4. Group 2 consists of Balmain bug-1 (*I. ciliatus*), Moreton Bay bug-1 (*Thenus indicus*), Moreton Bay bug-2 (*T. orientalis*), and slipper lobster-1 (*Scyllarides squammosus*), and has a major band at about P2.1 (very slightly slower in *S. squammosus*). The weak band pictured at P1.7 for *T. orientalis* and *T. indicus* sometimes resolves into two close-coupled bands. Distinguishing species within groups is difficult.

Five lobster species examined. Champagne lobster-1 (*Linuparus trigonus*) has its major band at P2, whereas the other four species—southern rocklobster (*Jasus edwardsii*), eastern rocklobster (*J. verreauxi*), western rocklobster (*Panulirus cygnus*) and tropical rocklobster-1 (*Panulirus ornatus*)—all have their major band at P1.5 to P1.4. The latter group can be separated from one another on the basis of minor bands in the P2–P3 region.

Crabs (Fig. 9.31)

Nine species examined. They showed few clear distinguishing bands, and not all species could be clearly delineated. Most have a major band between P1 and P2, but two species have their major band at (giant crab, *Pseudocarcinus gigas*) or just below (spanner crab, *Ranina ranina*) P1. Coral crab (*Charybdis feriata*) and blue swimmer crab (*Portunus pelagicus*) can be distinguished from other crabs by a minor band at P0.5. *P. pelagicus* showed three Types: Type A (four specimens), Type B (three specimens and similar to *C. feriata*) and Type C (one specimen). Crab-3 (*P. sanguinolentus*) shows very similar variation to *P. pelagicus* (but lacks the P0.5 band)—Type A was seen in four individuals, Type B in one. Crabs with a major band at P1.4 are *P. pelagicus* or *P. sanguinolentus*. The remaining four species—crab-1 (*C. natator*), crab-2 (*Hypothalassia armata*), sand crab-1 (*Ovalipes australiensis*) and mud crab-1 (*Scylla serrata*)—all share a major band at about P1.8 and lack clear distinguishing features.

Prawns & scampi (Figs 9.32 & 9.33)

Sixteen prawn species examined. Major bands are mostly located in the P2–P3 region. Based on the position of these major bands, most species fall into one of two groups. Group 1 is characterised by a major band at P2.7, and consists of seven species: royal red prawn (*Haliporoides sibogae*), king prawn-1 (*Melicertus latisulcatus*), king prawn-2 (*M. plebejus*), bay prawn-1 (*Metapenaeus bennettae*), Endeavour prawn-1 (*M. endeavouri*), Endeavour prawn-2 (*M. ensis*), school prawn-1 (*M. macleayi*) and prawn-2 (*Metapenaeopsis rosea*). Many of these also have a weak band at about P2.3. These species are hard to differentiate using protein fingerprints. Group 2 consists of five species with a pair of protein bands at P2.2 and P2.3, the P2.2 band always being a major band. Sometimes these two bands fail to separate cleanly from one another. When these two bands do separate, banana prawn-1 (*Fenneropenaeus indicus*) and black tiger prawn (*Penaeus monodon*) show the P2.3 band as a weaker band, whereas the other species in the group show these two bands as equally intense. Some minor bands may allow some further separation of species. For example, banana prawn-2 (*F. merguiensis*) has a minor band at P1.5, whereas tiger prawn-1 (*P. esculentus*) and *P. monodon* have a minor band at P0.5. The fifth species in this group is tiger prawn-2 (*P. semisulcatus*). It is difficult to distinguish these species unambiguously. The three remaining species may be clearly identified: prawn-1 (*Aristaeopsis edwardsiana*) has a major band at P3, redspot king prawn (*Melicertus longistylus*) has a major band at P2.5, and freshwater prawn-1 (*Macrobrachium rosenbergii*) has a major band at P1.7 (as well as one at P2.5).

Crustaceans & molluscs

Four scampi species examined. These are all similar in that pronounced bands are only apparent in the P2.5–P4 region. Two groups, each of two species, can be distinguished. Group 1 consists of scampi-2 (*Metanephrops boschmai*) and scampi-3 (*M. velutinus*), with the slowest pronounced band at P2.7. Group 2 consists of scampi-1 (*M. australiensis*) and scampi-4 (*Nephropsis serrata*), with the slowest pronounced band at P2.5. Distinguishing within groups is difficult.

Yabbies (Fig. 9.33)

Three species examined. Redclaw (*Cherax quadricarinatus*) is easily distinguishable from marron (*C. tenuimanus*) and yabby-1 (*C. destructor*), having its major band at P1.4 rather than very close to P1. *C. tenuimanus* is very similar to *C. destructor* but has a slightly faster major band at P1 and a weak band at P2 (rather than weaker still or absent).

Abalones (Fig. 9.34)

Four species examined. Protein bands in all species are weak, diffuse, and hard to score; they are present in the P2–P5 region. The patterns presented are indicative only; no clear species-diagnostic features are apparent. We do not recommend our protein fingerprinting technique for identifying members of this group, but DNA-based forensic tests can distinguish most of them.

Clams (Fig. 9.34)

Five species examined. Protein bands in all species are weak, diffuse, and hard to score. The patterns presented are indicative only. Pipi-1 (*Donax deltoides*) appears to be the only species with a protein band in the O to P-1 region. We do not recommend our protein fingerprinting technique for identifying members of this group.

Mussels (Fig. 9.34)

The single species examined was characterised by having most protein bands in the O to P-1 region.

Octopuses (Fig. 9.34)

Three species examined. The most intense staining band is in the P4–P5 region; other bands are weak, diffuse, and hard to resolve. We do not recommend our protein fingerprinting technique for identifying members of this group.

Oysters (Fig. 9.34)

Four species examined; all can be readily separated. Pacific oyster (*Crassostrea gigas*) shows variation among individuals in the P2–P2.5 region: one of eight individuals showed Type A, four Type B, and three Type C.

Scallops (Fig. 9.35)

Four species examined; all can be separated. Two of the species have very similar variation in the P1.5 to P2 region: saucer scallop-2 (*Amusium pleuronectes*) showed two Type A individuals, five Type B and one Type C; commercial scallop (*Pecten fumatus*) showed six Type A and two Type B. These two species can be distinguished by a P3.4 band, present in *A. pleuronectes* but absent in *P. fumatus*.

Molluscs

Sea snails (Fig. 9.35)

Three species examined. All gave weak and rather indistinct bands. Bailer shell-1 (*Livonia mamilla*) has a faster band in the P2 to P1 region than periwinkle-1 (*Turbo undulatus*); no band is apparent in this region in trochus (*Trochus niloticus*).

Squids (Fig. 9.35)

Eight species examined, falling into four groups. Group 1 consists of squid-1 (*Nototodarus hawaiiensis*), with fast-migrating proteins. Group 2 consists of Gould's squid (*N. gouldi*) with a major band at P2.5. Group 3 consists of two *Sepia* species—cuttlefish-1 (*S. pharaonis*) and cuttlefish-2 (*S. rex*)—with the major band at P2.3. Group 4 consists of two *Sepioteuthis* species—southern calamari (*S. australis*) and northern calamari (*S. lessoniana*)—with the major band at P1.5. One of the two *S. lessoniana* examined had another major band at P1.8. Group 5 consists of squid-2 (*Photololigo* cf *chinensis*) and squid-3 (*P.* sp.); these have the major band at P1.1.

Sharks and rays

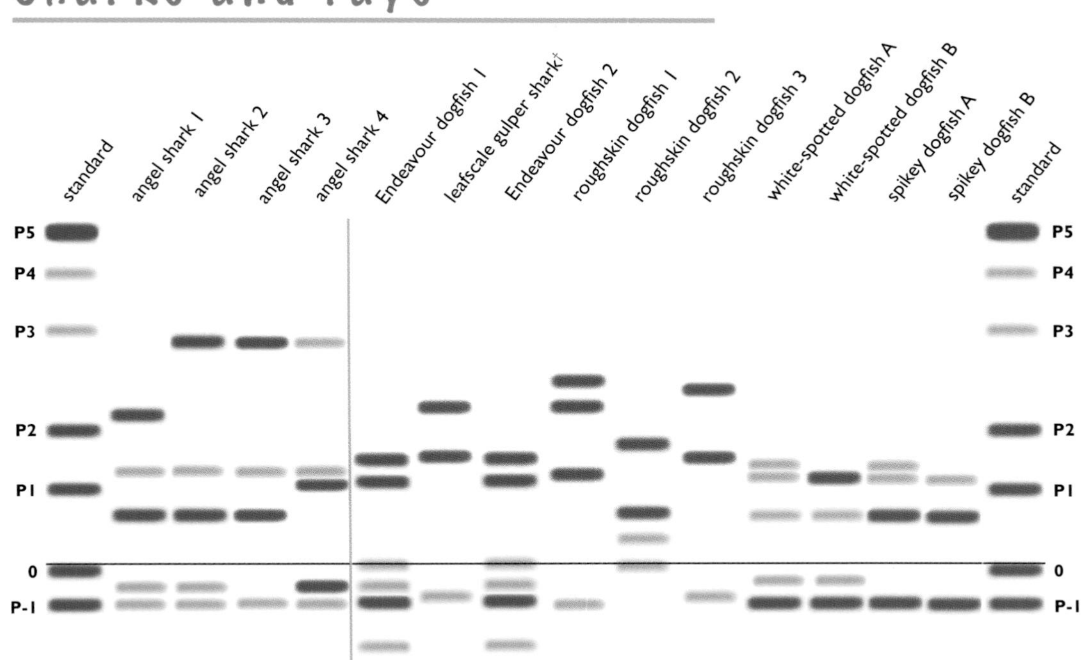

Figure 9.1—Protein fingerprints of angel sharks and dogfishes (part). Angel shark 1 (*Squatina australis*), angel shark 2 (*S. tergocellata*), angel shark 3 (*S.* sp., eastern angel shark), angel shark 4 (*S.* sp., western angel shark), Endeavour dogfish 1 (*Centrophorus moluccensis*), leafscale gulper shark† (*C. squamosus*), Endeavour dogfish 2 (*C. uyato*), roughskin dogfish 1 (*Centroscymnus crepidater*), roughskin dogfish 2 (*C. plunketi*), roughskin dogfish 3 (*Deania calcea*), white-spotted dogfish (*Squalus acanthias*), spikey dogfish (*S. megalops*).

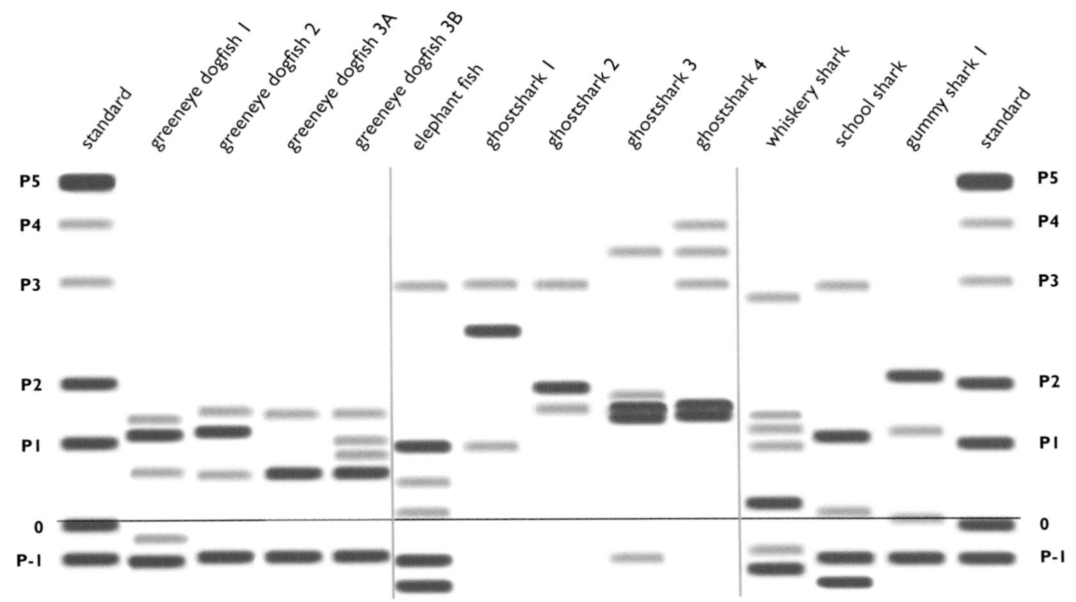

Figure 9.2—Protein fingerprints of dogfishes (cont.), ghostsharks and hound sharks. Greeneye dogfish 1 (*Squalus mitsukurii*), greeneye dogfish 2 (*S.* sp., western highfin dogfish), greeneye dogfish 3 (*S.* sp., eastern longnose dogfish), elephant fish (*Callorhinchus milii*), ghostshark 1 (*Chimaera* sp., giant chimaera), ghostshark 2 (*C.* sp., southern chimaera), ghostshark 3 (*Hydrolagus lemures*), ghostshark 4 (*H. ogilbyi*), whiskery shark (*Furgaleus macki*), school shark (*Galeorhinus galeus*), gummy shark 1 (*Mustelus antarcticus*).

Sharks and rays

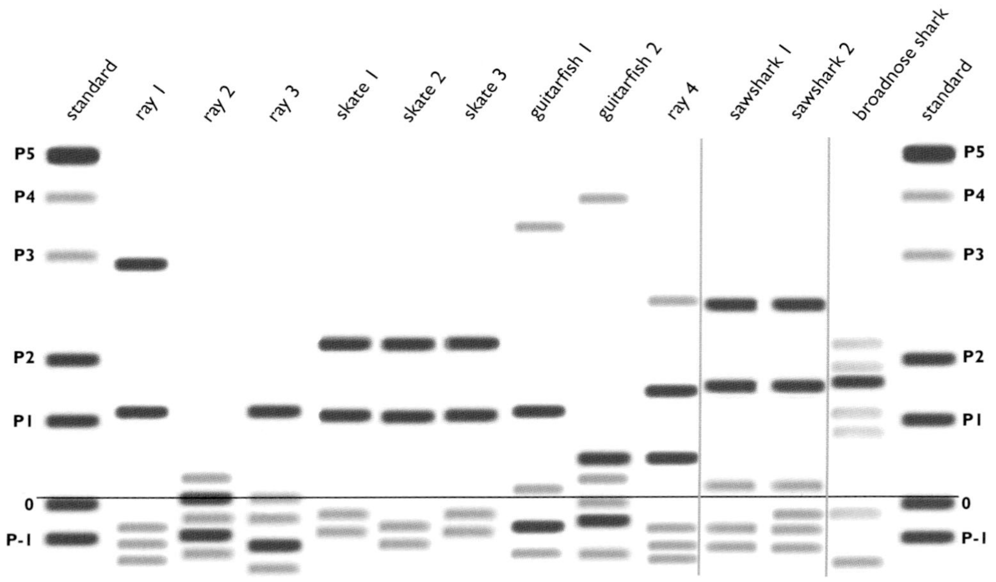

Figure 9.3—Protein fingerprints of rays, sawsharks and sevengill sharks. Ray 1 (*Gymnura australis*), ray 2 (*Himantura toshi*), ray 3 (*Myliobatis australis*), skate 1 (*Raja australis*), skate 2 (*R. whitleyi*), skate 3 (*R.* sp.), guitarfish 1 (*Rhynchobatus australiae*), guitarfish 2 (*Trygonorrhina* sp.), ray 4 (*Urolophus paucimaculatus*), sawshark 1 (*Pristiophorus cirratus*), sawshark 2 (*P. nudipinnis*), broadnose shark (*Notorynchus cepedianus*).

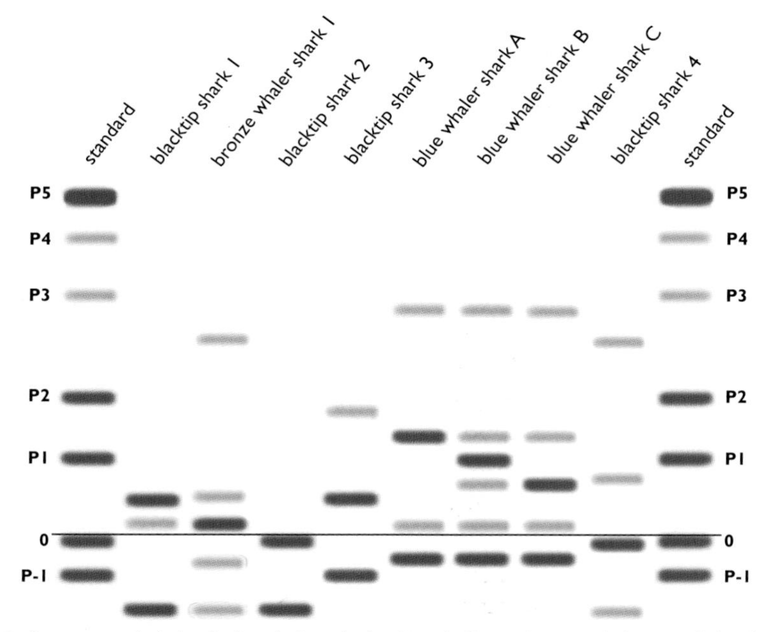

Figure 9.4—Protein fingerprints of whaler sharks. Blacktip shark 1 (*Carcharhinus dussumieri*), bronze whaler shark 1 (*C. obscurus*), blacktip shark 2 (*C. sorrah*), blacktip shark 3 (*C. tilstoni*), blue whaler shark (*Prionace glauca*), blacktip shark 4 (*Rhizoprionodon acutus*).

Bony fishes

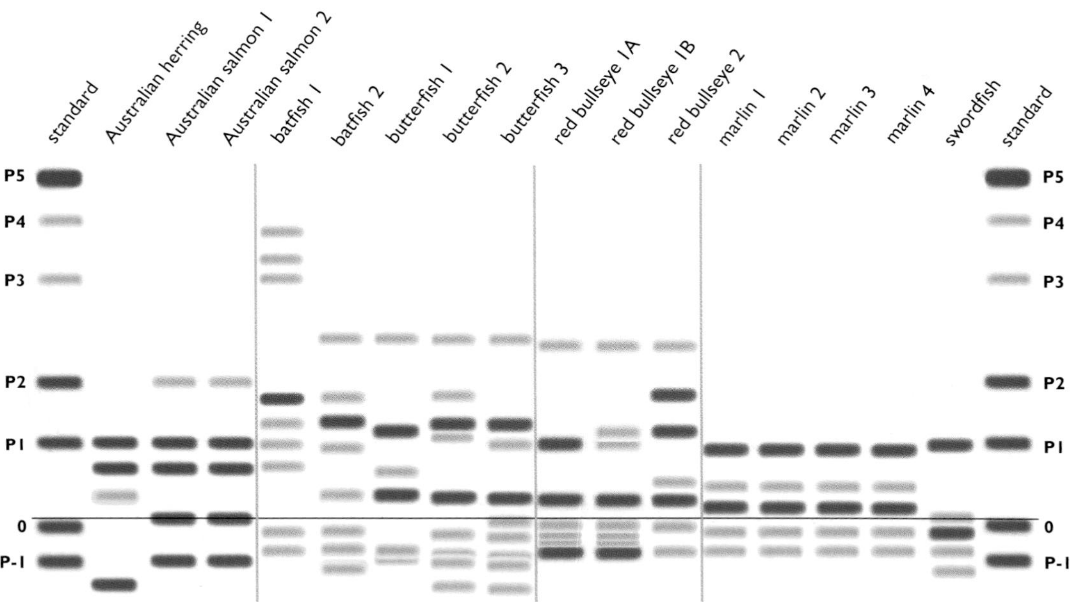

Figure 9.5—Protein fingerprints of Australian salmons, batfishes, bigeyes and billfishes. Australian herring (*Arripis georgianus*), Australian salmon 1 (*A. trutta*), Australian salmon 2 (*A. truttaceus*), batfish 1 (*Drepane punctata*), batfish 2 (*Platax batavianus*), butterfish 1 (*Scatophagus argus*), butterfish 2 (*S. multifasciatus*), butterfish 3 (*S.* sp.), red bullseye 1 (*Priacanthus hamrur*), red bullseye 2 (*P. tayenus*), marlin 1 (*Makaira indica*), marlin 2 (*M. mazara*), marlin 3 (*Tetrapturus angustirostris*), marlin 4 (*T. audax*), swordfish (*Xiphias gladius*).

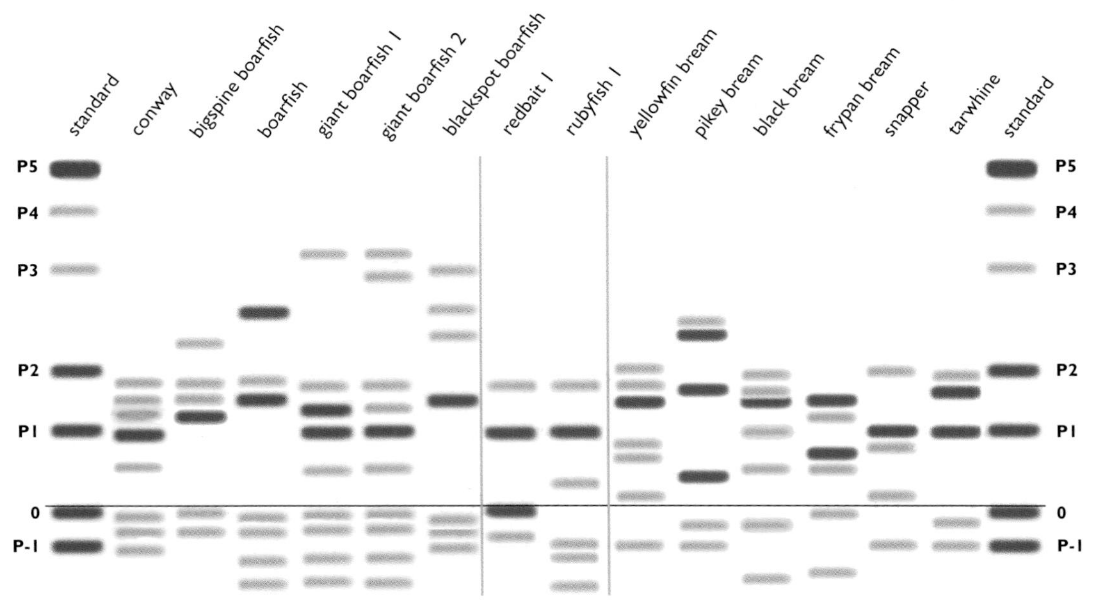

Figure 9.6—Protein fingerprints of boarfishes, bonnetmouths and breams. Conway (*Oplegnathus woodwardi*), bigspine boarfish (*Pentaceros decacanthus*), boarfish (*Pentaceropsis recurvirostris*), giant boarfish 1 (*Paristiopterus gallipavo*), giant boarfish 2 (*P. labiosus*), blackspot boarfish (*Zanclistius elevatus*), redbait 1 (*Emmelichthys nitidus*), rubyfish 1 (*Plagiogeneion macrolepis*), yellowfin bream (*Acanthopagrus australis*), pikey bream (*A. berda*), black bream (*A. butcheri*), frypan bream (*Argyrops spinifer*), snapper (*Pagrus auratus*), tarwhine (*Rhabdosargus sarba*).

Bony fishes

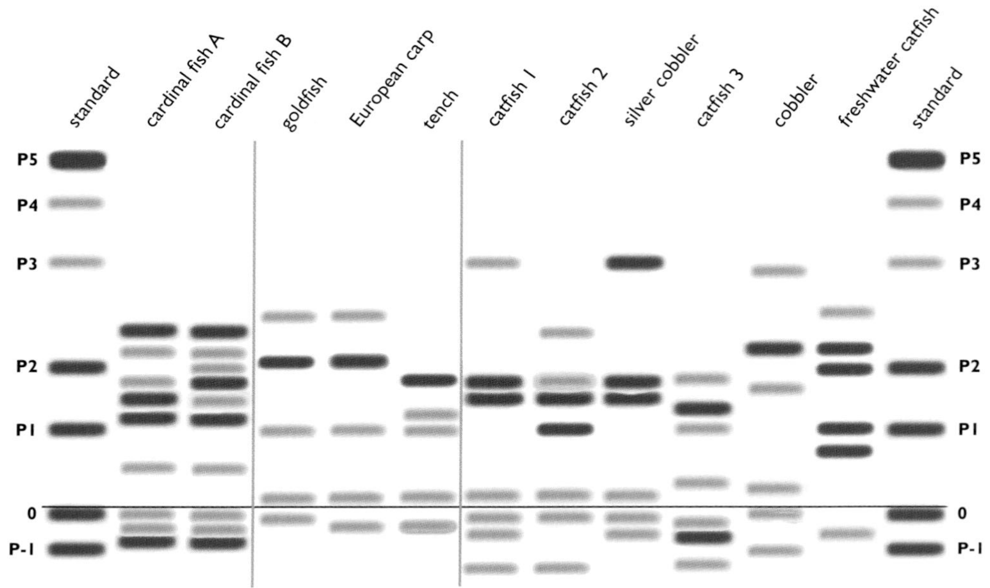

Figure 9.7—Protein fingerprints of cardinal fishes, carps and catfishes. Cardinal fish (*Epigonus telescopus*), goldfish (*Carassius auratus*), European carp (*Cyprinus carpio*), tench (*Tinca tinca*), catfish 1 (*Arius bilineatus*), catfish 2 (*A. mastersi*), silver cobbler (*A. midgleyi*), catfish 3 (*A. thalassinus*), cobbler (*Cnidoglanis macrocephalus*), freshwater catfish (*Tandanus tandanus*).

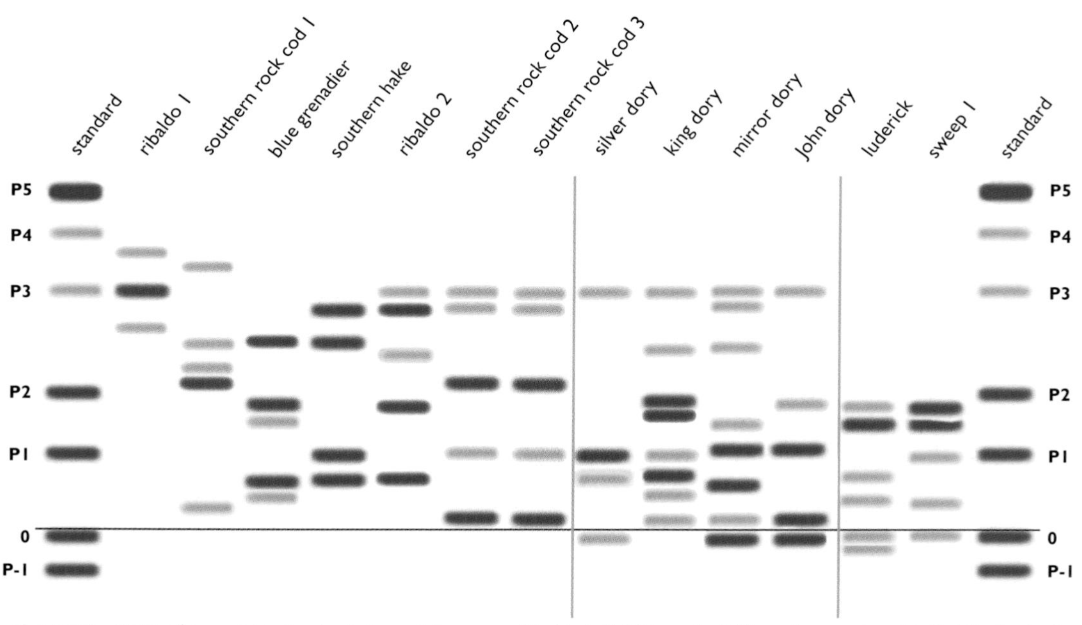

Figure 9.8—Protein fingerprints of cods, dories and drummers. Ribaldo 1 (*Lepidion schmidti*), southern rock cod 1 (*Lotella rhacina*), blue grenadier (*Macruronus novaezelandiae*), southern hake (*Merluccius australis*), ribaldo 2 (*Mora moro*), southern rock cod 2 (*Pseudophycis bachus*), southern rock cod 3 (*P. barbata*), silver dory (*Cyttus australis*), king dory (*C. traversi*), mirror dory (*Zenopsis nebulosus*), John dory (*Zeus faber*), luderick (*Girella tricuspidata*), sweep 1 (*Scorpis lineolatus*).

Bony fishes

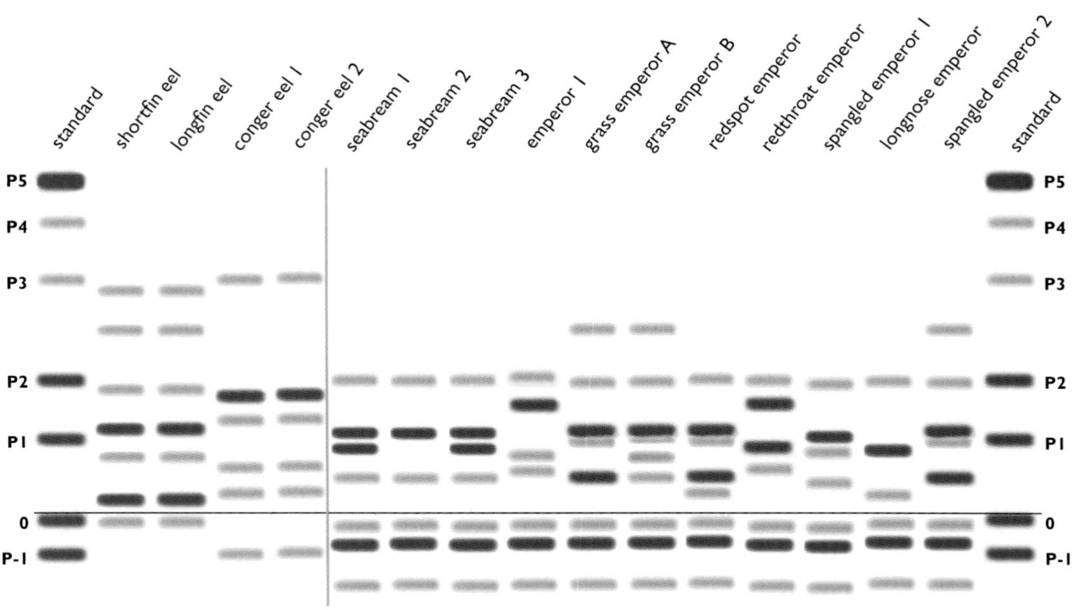

Figure 9.9—Protein fingerprints of eels and emperors. Shortfin eel (*Anguilla australis*), longfin eel (*A. reinhardtii*), conger eel 1 (*Conger verreauxi*), conger eel 2 (*C. wilsoni*), seabream 1 (*Gymnocranius grandoculis*), seabream 2 (*G. elongatus*), seabream 3 (*G.* sp.), emperor 1 (*Lethrinus atkinsoni*), grass emperor (*L. laticaudis*), redspot emperor (*L. lentjan*), redthroat emperor (*L. miniatus*), spangled emperor 1 (*L. nebulosus*), longnose emperor (*L. olivaceus*), spangled emperor 2 (*L.* sp.).

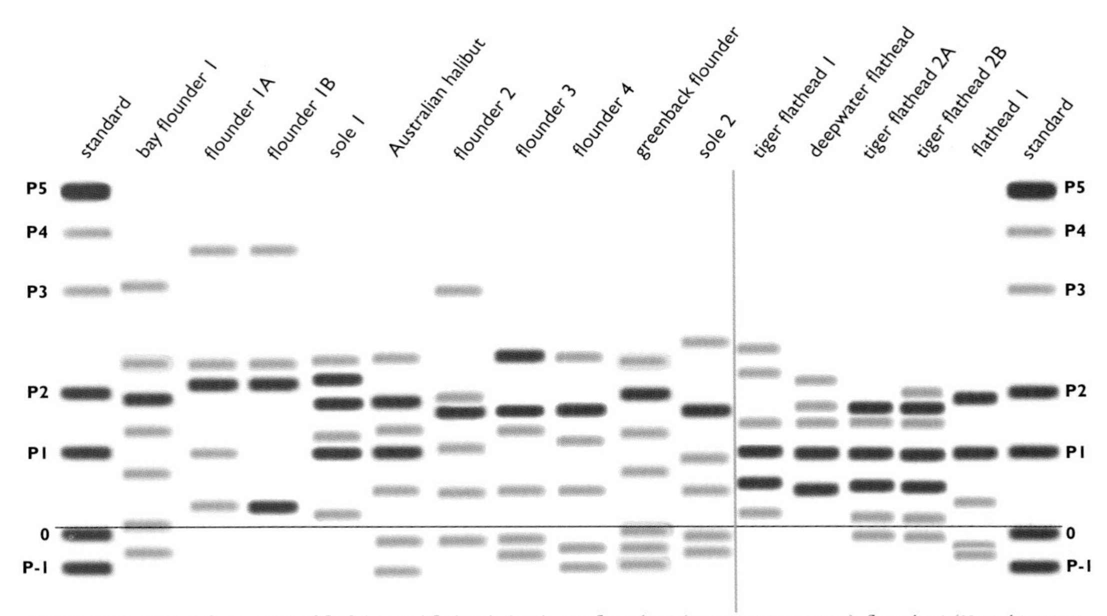

Figure 9.10—Protein fingerprints of flatfishes and flatheads (part). Bay flounder 1 (*Ammotretis rostratus*), flounder 1 (*Neoachiropsetta milfordi*), sole 1 (*Paraplagusia bilineata*), Australian halibut (*Psettodes erumei*), flounder 2 (*Pseudorhombus arsius*), flounder 3 (*P. jenynsii*), flounder 4 (*P. spinosus*), greenback flounder (*Rhombosolea tapirina*), sole 2 (*Synaptura nigra*), tiger flathead 1 (*Neoplatycephalus aurimaculatus*), deepwater flathead (*N. conatus*), tiger flathead 2 (*N. richardsoni*), flathead 1 (*Papilloculiceps nematophthalmus*).

Bony fishes

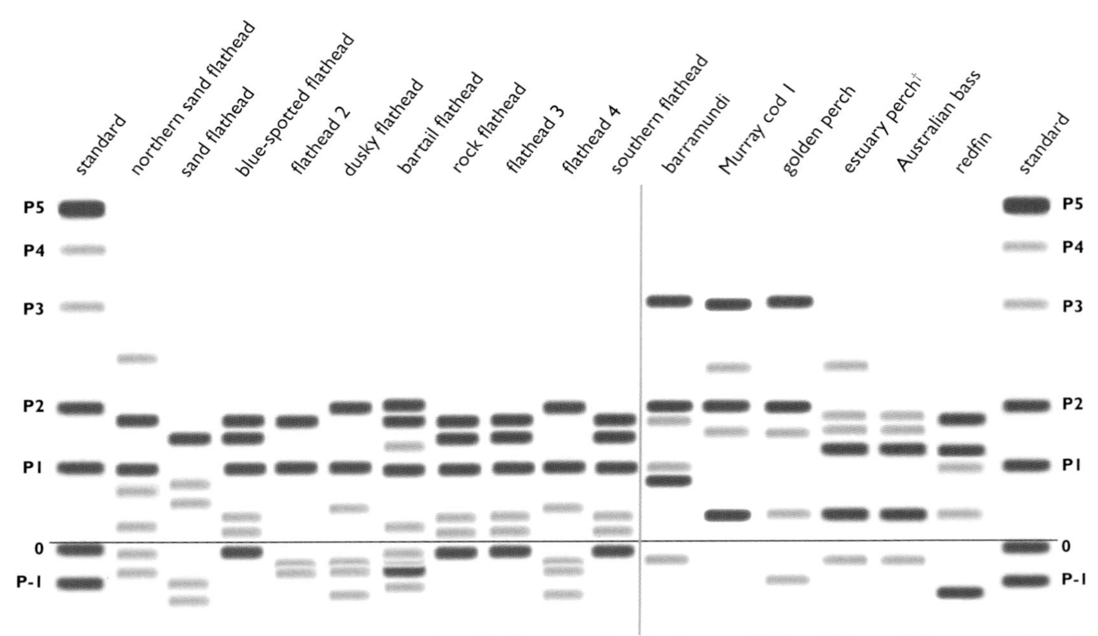

Figure 9.11—Protein fingerprints of flatheads (cont.) and freshwater perches. Northern sand flathead (*Platycephalus arenarius*), sand flathead (*P. bassensis*), blue-spotted flathead (*P. caeruleopunctatus*), flathead 2 (*P. endrachtensis*), dusky flathead (*P. fuscus*), bartail flathead (*P. indicus*), rock flathead (*P. laevigatus*), flathead 3 (*P. longispinis*), flathead 4 (*P. marmoratus*), southern flathead (*P. speculator*), barramundi (*Lates calcarifer*), Murray cod 1 (*Maccullochella peelii*), golden perch (*Macquaria ambigua*), estuary perch† (*M. colonorum*), Australian bass (*M. novemaculeata*), redfin (*Perca fluviatilis*).

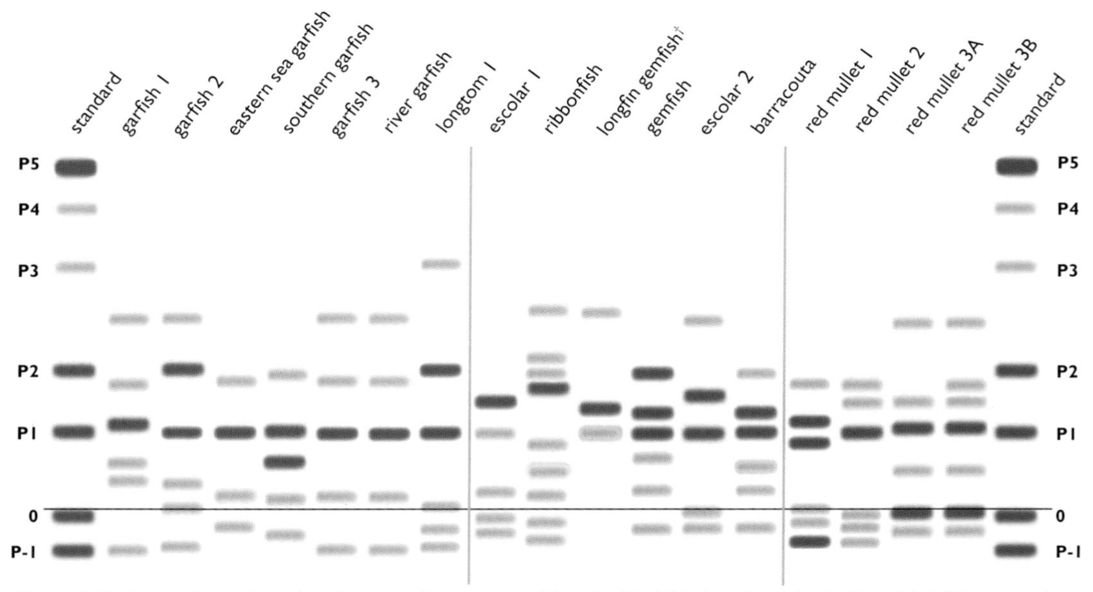

Figure 9.12—Protein fingerprints of garfishes, gemfishes and goatfishes. Garfish 1 (*Arrhamphus sclerolepis*), garfish 2 (*Hyporhamphus affinis*), eastern sea garfish (*H. australis*), southern garfish (*H. melanochir*), garfish 3 (*H. quoyi*), river garfish (*H. regularis*), longtom 1 (*Tylosurus gavialoides*), escolar 1 (*Lepidocybium flavobrunneum*), ribbonfish (*Lepidopus caudatus*), longfin gemfish† (*Rexea antefurcata*), gemfish (*R. solandri*), escolar 2 (*Ruvettus pretiosus*), barracouta (*Thyrsites atun*), red mullet 1 (*Parupeneus heptacanthus*), red mullet 2 (*P. indicus*), red mullet 3 (*Upeneichthys vlamingii*).

Bony fishes

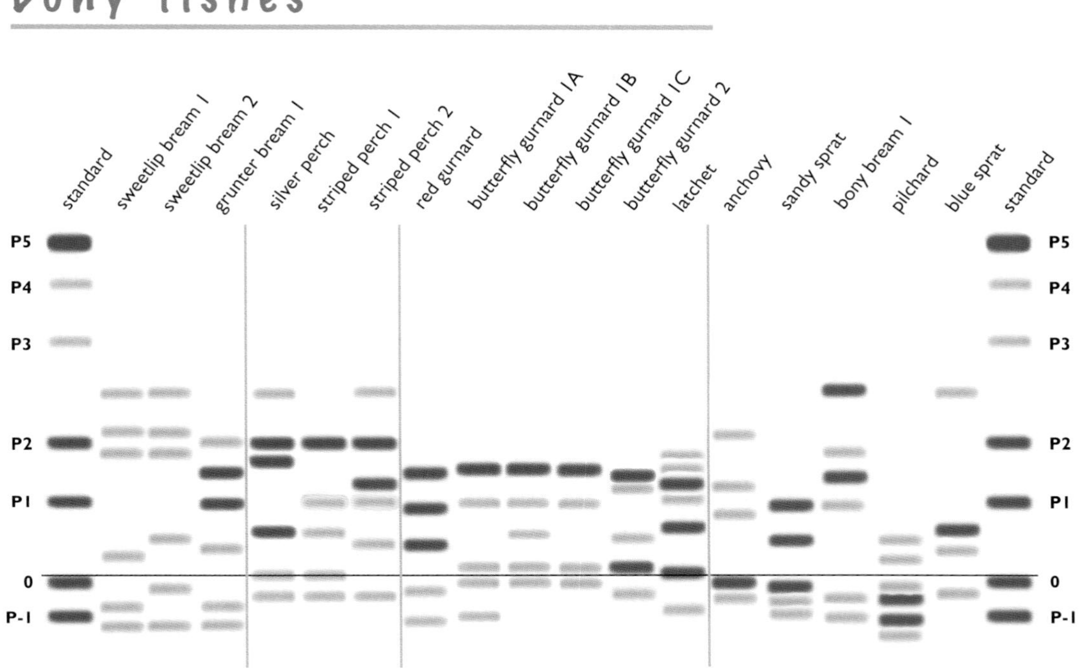

Figure 9.13—Protein fingerprints of grunter breams, grunters, gurnards and herrings. Sweetlip bream 1 (*Diagramma labiosum*), sweetlip bream 2 (*Plectorhinchus flavomaculatus*), grunter bream 1 (*Pomadasys kaakan*), silver perch (*Bidyanus bidyanus*), striped perch 1 (*Hephaestus jenkinsi*), striped perch 2 (*Pelates octolineatus*), red gurnard (*Chelidonichthys kumu*), butterfly gurnard 1 (*Lepidotrigla mulhalli*), butterfly gurnard 2 (*L. vanessa*), latchet (*Pterygotrigla polyommata*), anchovy (*Engraulis australis*), sandy sprat (*Hyperlophus vittatus*), bony bream 1 (*Nematalosa erebi*), pilchard (*Sardinops neopilchardus*), blue sprat (*Spratelloides robustus*).

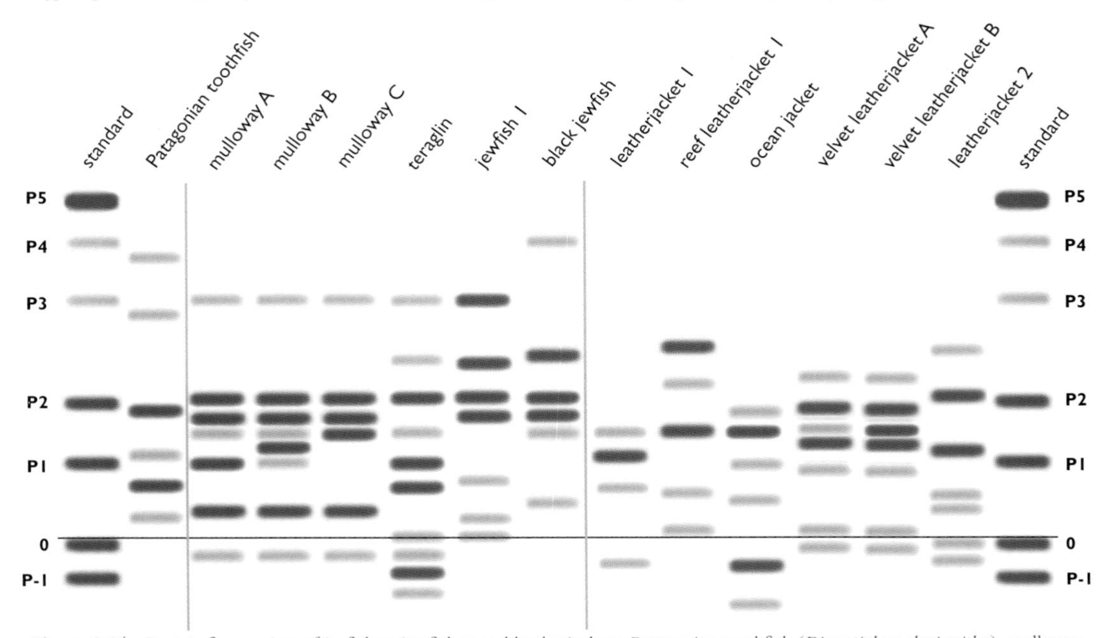

Figure 9.14—Protein fingerprints of icefishes, jewfishes and leatherjackets. Patagonian toothfish (*Dissostichus eleginoides*), mulloway (*Argyrosomus hololepidotus*), teraglin (*Atractoscion aequidens*), jewfish 1 (*Johnius borneensis*), black jewfish (*Protonibea diacanthus*), leatherjacket 1 (*Aluterus monoceros*), reef leatherjacket 1 (*Meuschenia freycineti*), ocean jacket (*Nelusetta ayraudi*), velvet leatherjacket (*Parika scaber*), leatherjacket 2 (*Pseudomonacanthus peroni*).

Bony fishes

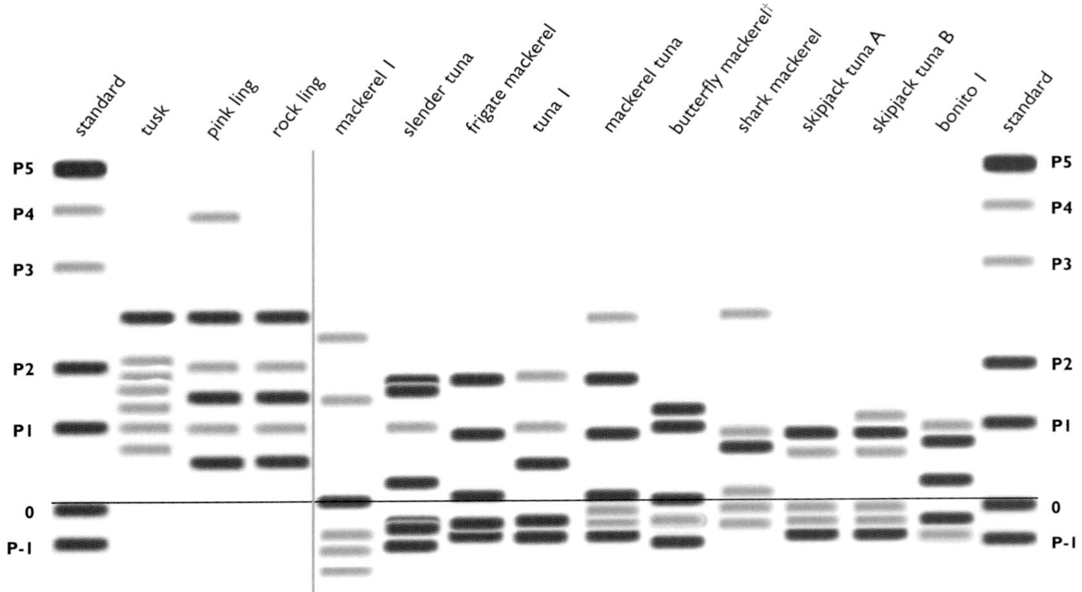

Figure 9.15—Protein fingerprints of lings and mackerels (part). Tusk (*Dannevigia tusca*), pink ling (*Genypterus blacodes*), rock ling (*G. tigerinus*), mackerel 1 (*Acanthocybium solandri*), slender tuna (*Allothunnus fallai*), frigate mackerel (*Auxis thazard*), tuna 1 (*Cybiosarda elegans*), mackerel tuna (*Euthynnus affinis*), butterfly mackerel† (*Gasterochisma melampus*), shark mackerel (*Grammatorcynus bicarinatus*), skipjack tuna (*Katsuwonus pelamis*), bonito 1 (*Sarda australis*).

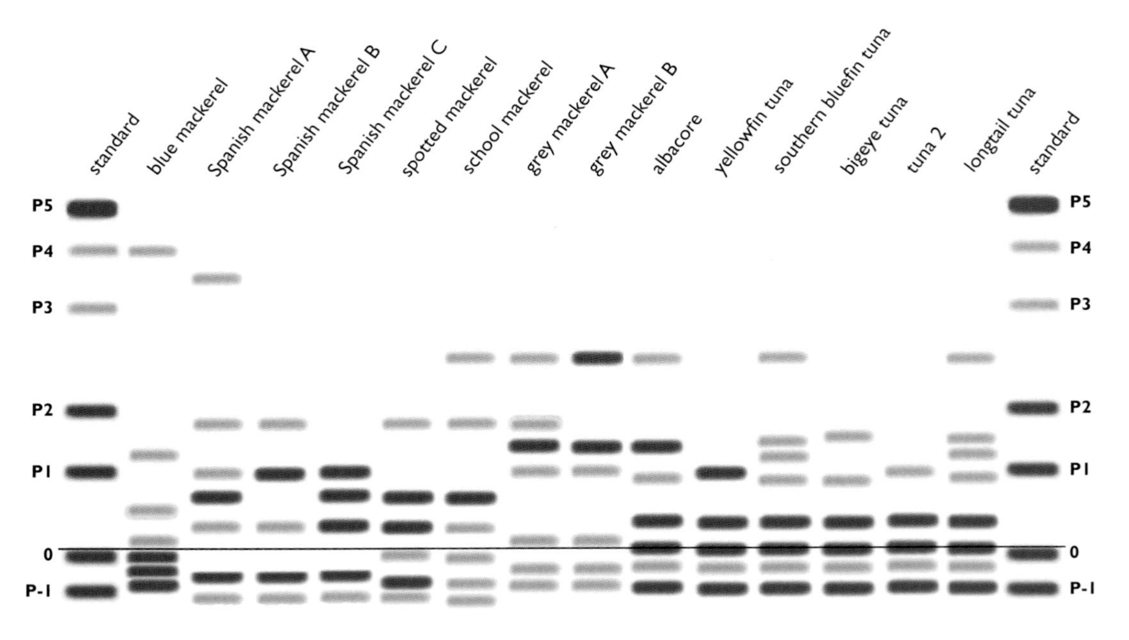

Figure 9.16—Protein fingerprints of mackerels (cont.). Blue mackerel (*Scomber australasicus*), Spanish mackerel (*Scomberomorus commerson*), spotted mackerel (*S. munroi*), school mackerel (*S. queenslandicus*), grey mackerel (*S. semifasciatus*), albacore (*Thunnus alalunga*), yellowfin tuna (*T. albacares*), southern bluefin tuna (*T. maccoyii*), bigeye tuna (*T. obesus*), tuna 2 (*T. thynnus*), longtail tuna (*T. tonggol*).

Bony fishes

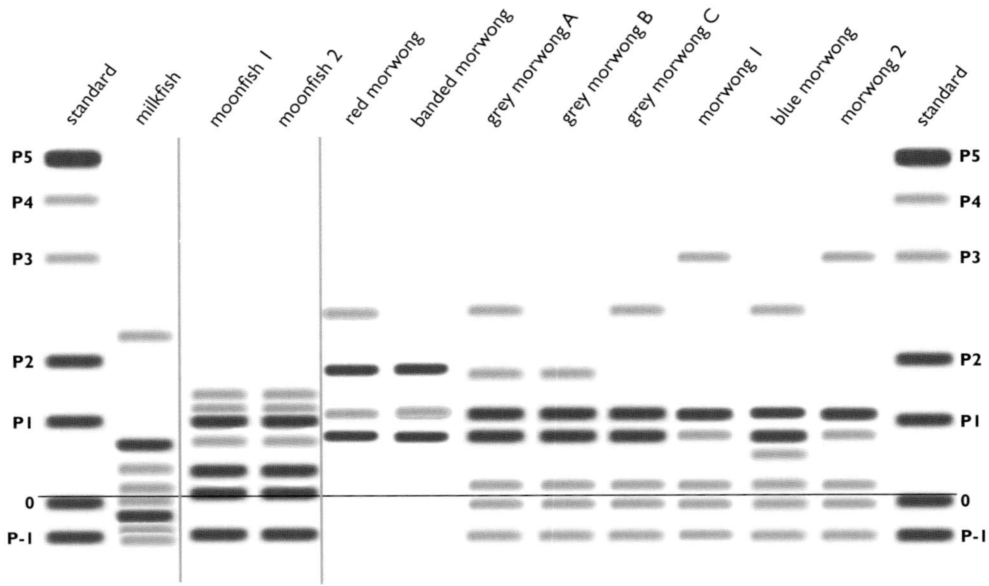

Figure 9.17—Protein fingerprints of milkfishes, moonfishes and morwongs. Milkfish (*Chanos chanos*), moonfish 1 (*Lampris guttatus*), moonfish 2 (*L. immaculatus*), red morwong (*Cheilodactylus fuscus*), banded morwong (*C. spectabilis*), grey morwong (*Nemadactylus douglasii*), morwong 1 (*N. macropterus*), blue morwong (*N. valenciennesi*), morwong 2 (*N.* sp.).

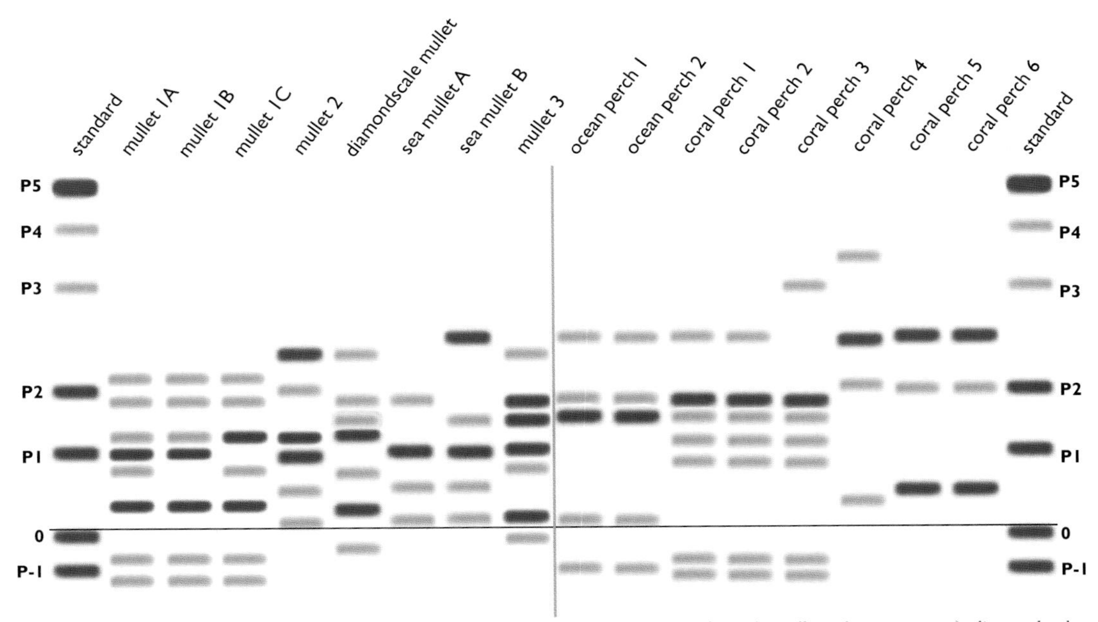

Figure 9.18—Protein fingerprints of mullets and ocean perches. Mullet 1 (*Aldrichetta forsteri*), mullet 2 (*Liza argentea*), diamondscale mullet (*L. vaigiensis*), sea mullet (*Mugil cephalus*), mullet 3 (*Valamugil seheli*), ocean perch 1 (*Helicolenus barathri*), ocean perch 2 (*H. percoides*), coral perch 1 (*Neosebastes incisipinnis*), coral perch 2 (*N. scorpaenoides*), coral perch 3 (*N. thetidis*), coral perch 4 (*Scorpaena cardinalis*), coral perch 5 (*Trachyscorpia capensis*), coral perch 6 (*T.* sp.).

Bony fishes

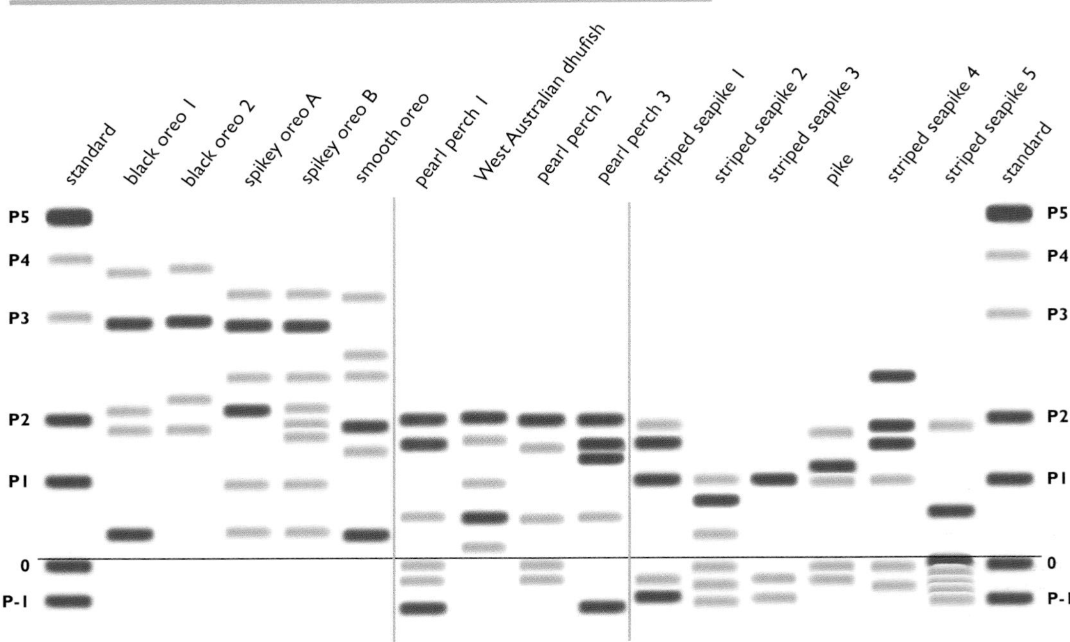

Figure 9.19—Protein fingerprints of oreos, pearl perches and pikes. Black oreo 1 (*Allocyttus niger*), black oreo 2 (*A. verrucosus*), spikey oreo (*Neocyttus rhomboidalis*), smooth oreo (*Pseudocyttus maculatus*), pearl perch 1 (*Glaucosoma buergeri*), West Australian dhufish (*G. hebraicum*), pearl perch 2 (*G. magnificum*), pearl perch 3 (*G. scapulare*), striped seapike 1 (*Sphyraena barracuda*), striped seapike 2 (*S. forsteri*), striped seapike 3 (*S. jello*), pike (*S. novaehollandiae*), striped seapike 4 (*S. obtusata*), striped seapike 5 (*S.* sp.).

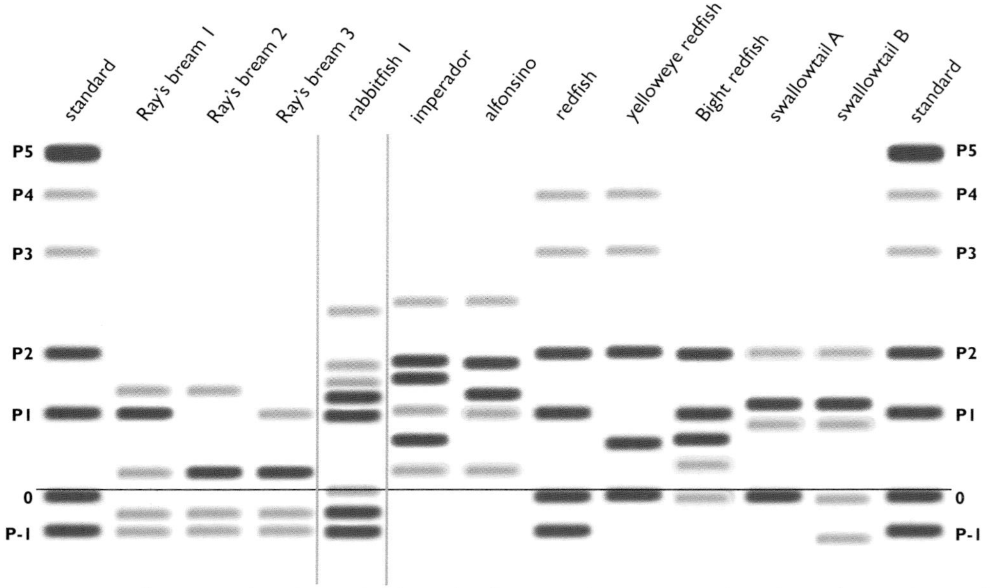

Figure 9.20—Protein fingerprints of pomfrets, rabbitfishes and redfishes. Ray's bream 1 (*Brama brama*), Ray's bream 2 (*Taractichthys longipinnis*), Ray's bream 3 (*Xenobrama microlepis*), rabbitfish 1 (*Siganus nebulosus*), imperador (*Beryx decadactylus*), alfonsino (*B. splendens*), redfish (*Centroberyx affinis*), yelloweye redfish (*C. australis*), Bight redfish (*C. gerrardi*), swallowtail (*C. lineatus*).

Bony fishes

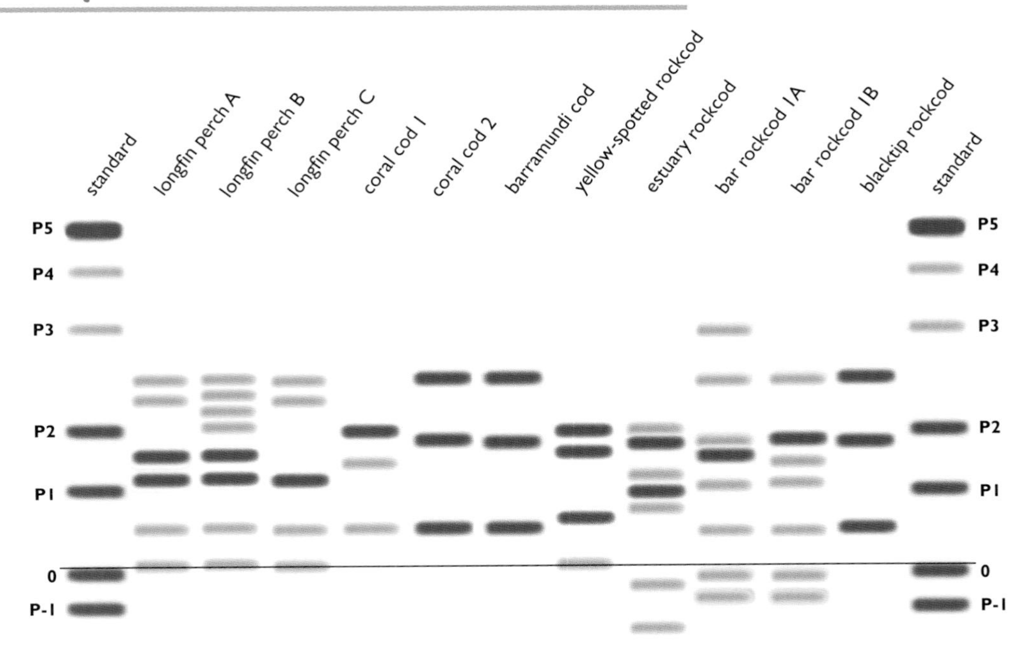

Figure 9.21—Protein fingerprints of rockcods (part). Longfin perch (*Caprodon longimanus*), coral cod 1 (*Cephalopholis cyanostigma*), coral cod 2 (*C. sonnerati*), barramundi cod (*Cromileptes altivelis*), yellow-spotted rockcod (*Epinephelus areolatus*), estuary rockcod (*E. coioides*), bar rockcod 1 (*E. ergastularius*), blacktip rockcod (*E. fasciatus*).

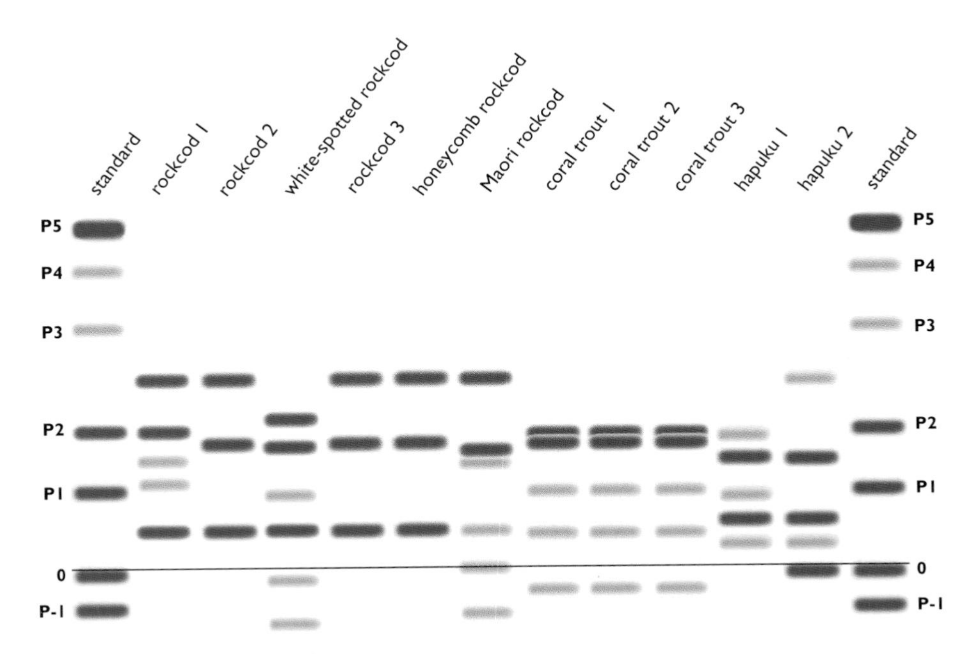

Figure 9.22—Protein fingerprints of rockcods (cont.). Rockcod 1 (*Epinephelus merra*), rockcod 2 (*E. morrhua*), white-spotted rockcod (*E. multinotatus*), rockcod 3 (*E. ongus*), honeycomb rockcod (*E. quoyanus*), Maori rockcod (*E. undulatostriatus*), coral trout 1 (*Plectropomus areolatus*), coral trout 2 (*P. leopardus*), coral trout 3 (*P. maculatus*), hapuku 1 (*Polyprion americanus*), hapuku 2 (*P. oxygeneios*).

Bony fishes

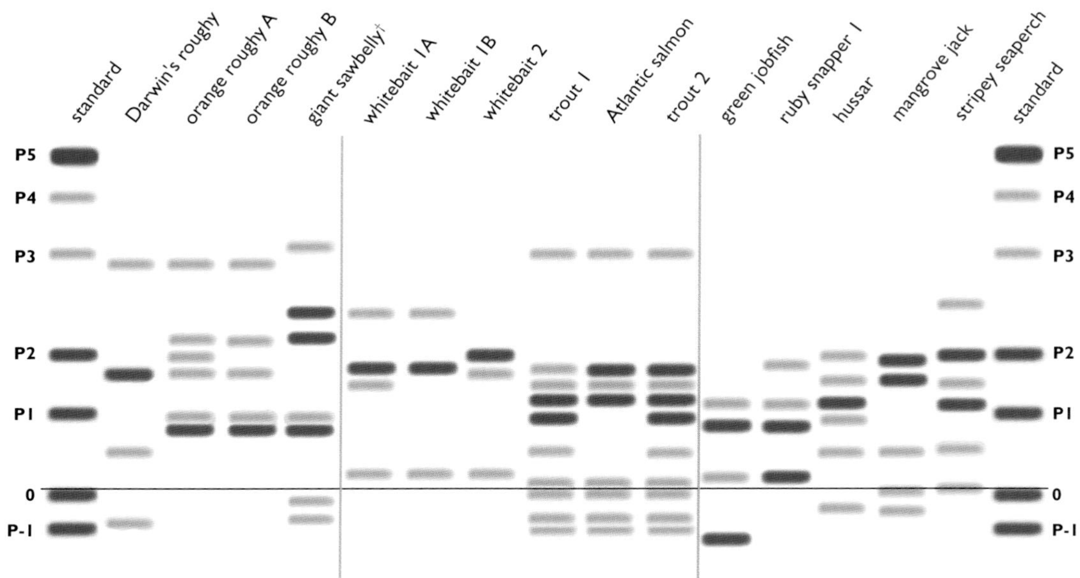

Figure 9.23—Protein fingerprints of roughies, salmons and seaperches (part). Darwin's roughy (*Gephyroberyx darwinii*), orange roughy (*Hoplostethus atlanticus*), giant sawbelly† (*H. gigas*), whitebait 1 (*Galaxias maculatus*), whitebait 2 (*Lovettia sealii*), trout 1 (*Oncorhynchus mykiss*), Atlantic salmon (*Salmo salar*), trout 2 (*S. trutta*), green jobfish (*Aprion virescens*), ruby snapper 1 (*Etelis coruscans*), hussar (*Lutjanus adetii*), mangrove jack (*L. argentimaculatus*), stripey seaperch (*L. carponotatus*).

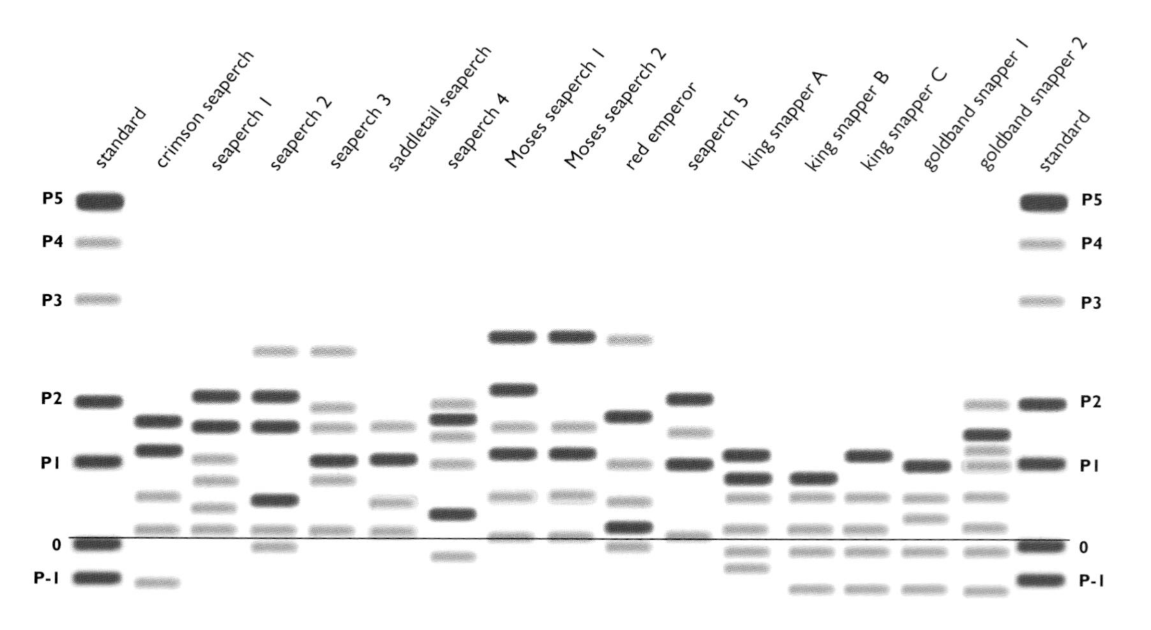

Figure 9.24—Protein fingerprints of seaperches (cont.). Crimson seaperch (*Lutjanus erythropterus*), seaperch 1 (*L. johnii*), seaperch 2 (*L. lemniscatus*), seaperch 3 (*L. lutjanus*), saddletail seaperch (*L. malabaricus*), seaperch 4 (*L. rivulatus*), Moses seaperch 1 (*L. russelli*), Moses seaperch 2 (*L.* sp.), red emperor (*L. sebae*), seaperch 5 (*L. vitta*), king snapper (*Pristipomoides filamentosus*), goldband snapper 1 (*P. multidens*), goldband snapper 2 (*P. typus*).

Bony fishes

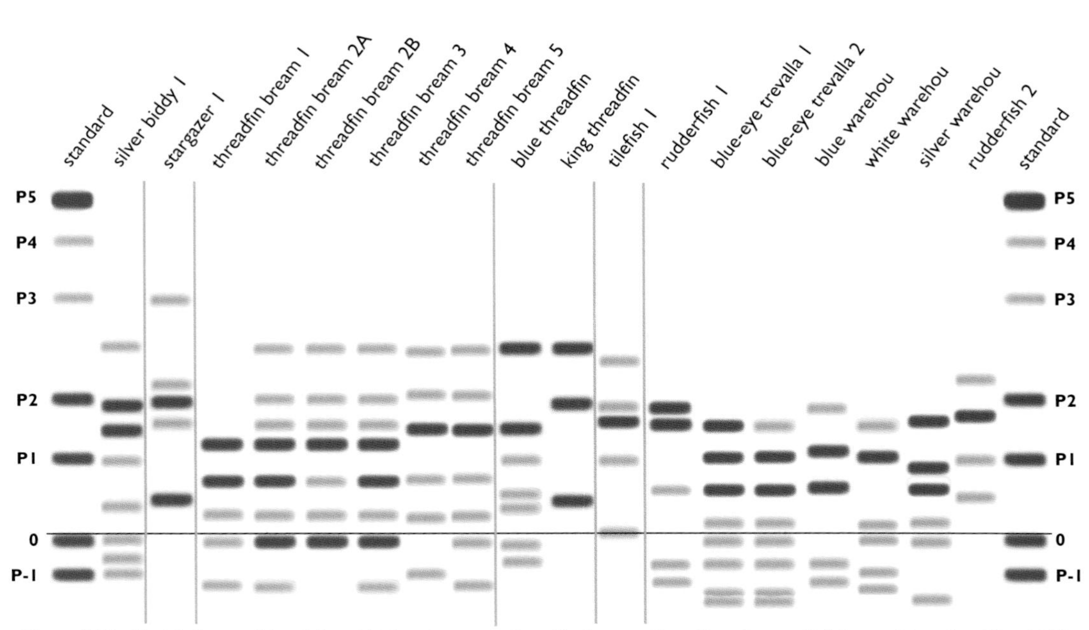

Figure 9.25—Protein fingerprints of silver biddies, stargazers, threadfin breams, threadfin salmons, tilefishes and trevallas. Silver biddy 1 (*Gerres subfasciatus*), stargazer 1 (*Kathetostoma canaster*), threadfin bream 1 (*Nemipterus furcosus*), 2 (*N. hexodon*), 3 (*N. peronii*), 4 (*Scolopsis monogramma*), 5 (*S. taeniopterus*), blue threadfin (*Eleutheronema tetradactylum*), king threadfin (*Polydactylus sheridani*), tilefish 1 (*Branchiostegus* sp.), rudderfish 1 (*Centrolophus niger*), blue-eye trevalla 1 (*Hyperoglyphe antarctica*), 2 (*Schedophilus labyrinthica*), blue warehou (*Seriolella brama*), white warehou (*S. caerulea*), silver warehou (*S. punctata*), rudderfish 2 (*Tubbia tasmanica*).

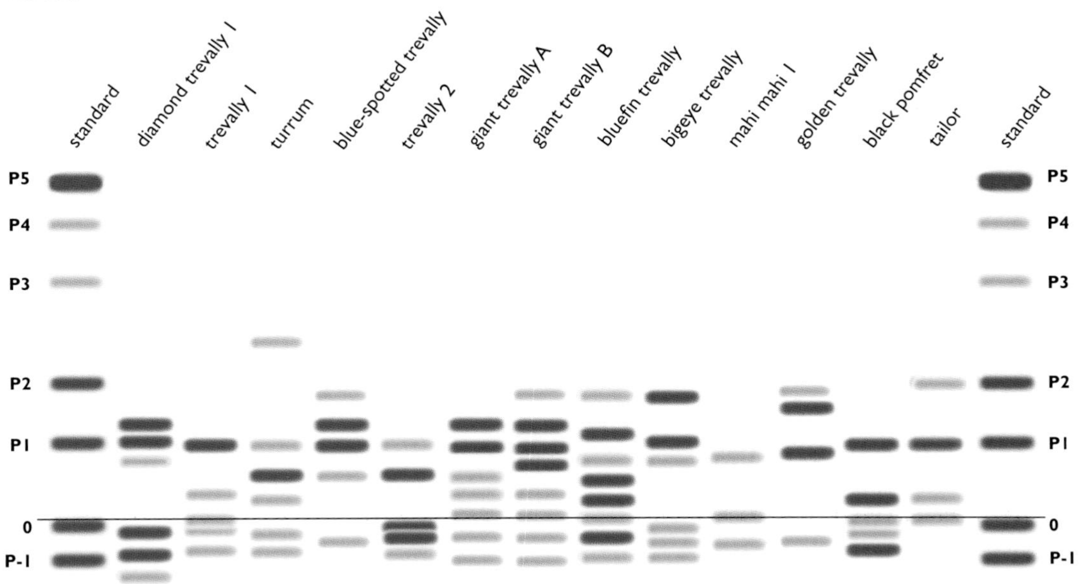

Figure 9.26—Protein fingerprints of trevallies (part). Diamond trevally 1 (*Alectis ciliaris*), trevally 1 (*Carangoides caeruleopinnatus*), turrum (*C. fulvoguttatus*), blue-spotted trevally (*Caranx bucculentus*), trevally 2 (*Carangoides gymnostethus*), giant trevally (*Caranx ignobilis*), bluefin trevally (*C. melampygus*), bigeye trevally (*C. sexfasciatus*), mahi mahi 1 (*Coryphaena hippurus*), golden trevally (*Gnathanodon speciosus*), black pomfret (*Parastromateus niger*), tailor (*Pomatomus saltatrix*).

Bony fishes

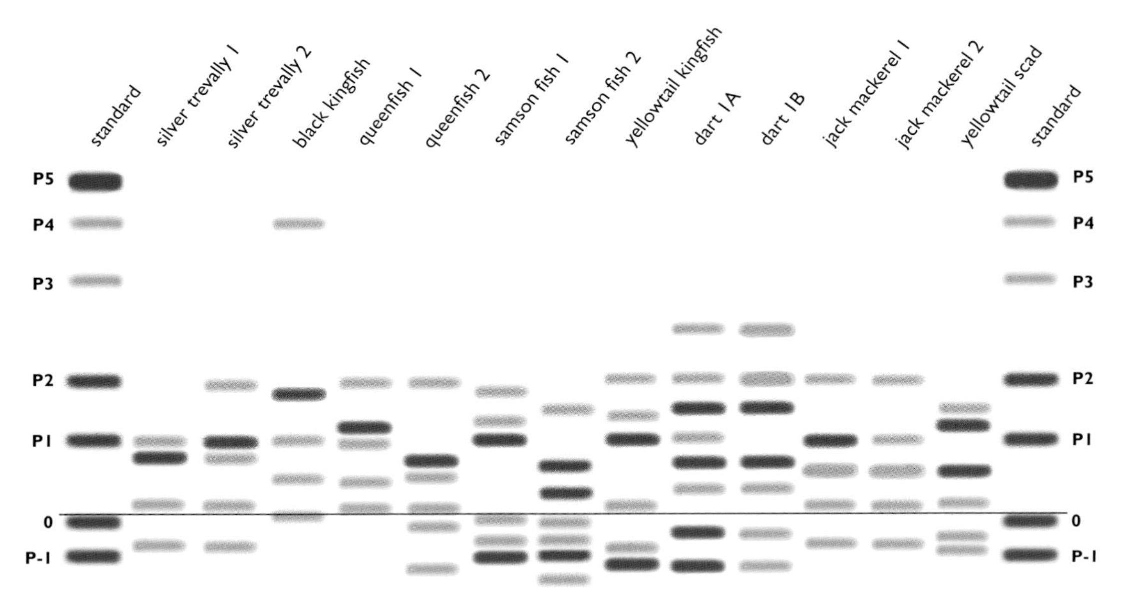

Figure 9.27—Protein fingerprints of trevallies (cont.). Silver trevally 1 (*Pseudocaranx dentex*), silver trevally 2 (*P. wrighti*), black kingfish (*Rachycentron canadum*), queenfish 1 (*Scomberoides commersonnianus*), queenfish 2 (*S. tol*), samson fish 1 (*Seriola dumerili*), samson fish 2 (*S. hippos*), yellowtail kingfish (*S. lalandi*), dart 1 (*Trachinotus botla*), jack mackerel 1 (*Trachurus declivis*), jack mackerel 2 (*T. murphyi*), yellowtail scad (*T. novaezelandiae*).

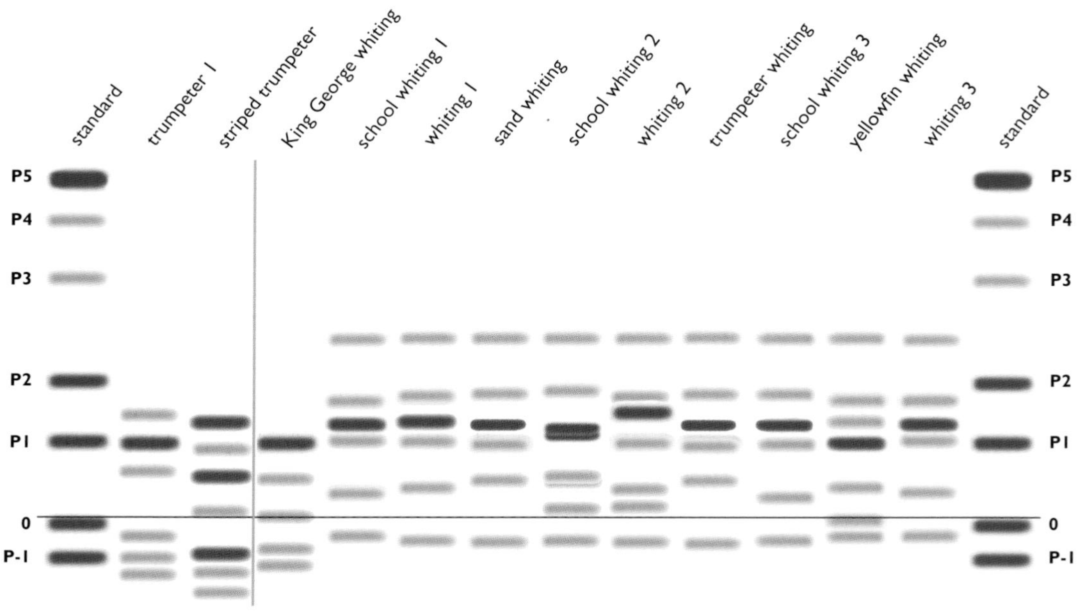

Figure 9.28—Protein fingerprints of trumpeters and whitings. Trumpeter 1 (*Latridopsis forsteri*), striped trumpeter (*Latris lineata*), King George whiting (*Sillaginodes punctata*), school whiting 1 (*Sillago bassensis*), whiting 1 (*S. burrus*), sand whiting (*S. ciliata*), school whiting 2 (*S. flindersi*), whiting 2 (*S. ingenuua*), trumpeter whiting (*S. maculata*), school whiting 3 (*S. robusta*), yellowfin whiting (*S. schomburgkii*), whiting 3 (*S. sihama*).

Bony fishes & crustaceans

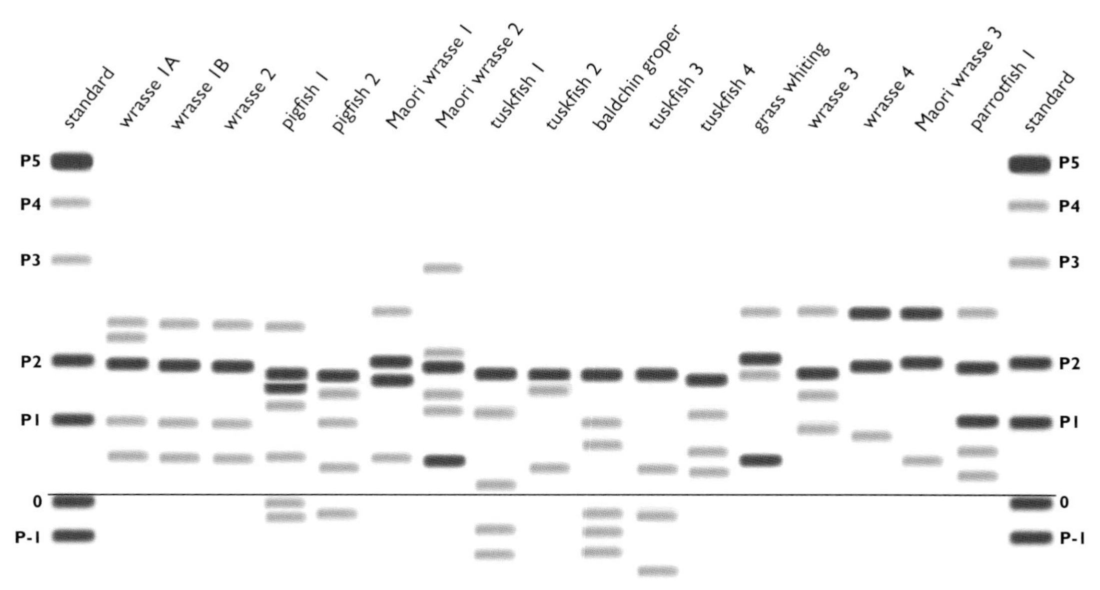

Figure 9.29—Protein fingerprints of wrasses. Wrasse 1 (*Achoerodus gouldii*), wrasse 2 (*A. viridis*), pigfish 1 (*Bodianus perditio*), pigfish 2 (*B. unimaculatus*), Maori wrasse 1 (*Cheilinus chlororus*), Maori wrasse 2 (*C. undulatus*), tuskfish 1 (*Choerodon cauteroma*), tuskfish 2 (*C. cephalotes*), baldchin groper (*C. rubescens*), tuskfish 3 (*C. schoenleinii*), tuskfish 4 (*C. venustus*), grass whiting (*Haletta semifasciata*), wrasse 3 (*Notolabrus gymnogenis*), wrasse 4 (*N. tetricus*), Maori wrasse 3 (*Oxycheilinus digrammus*), parrotfish 1 (*Scarus ghobban*).

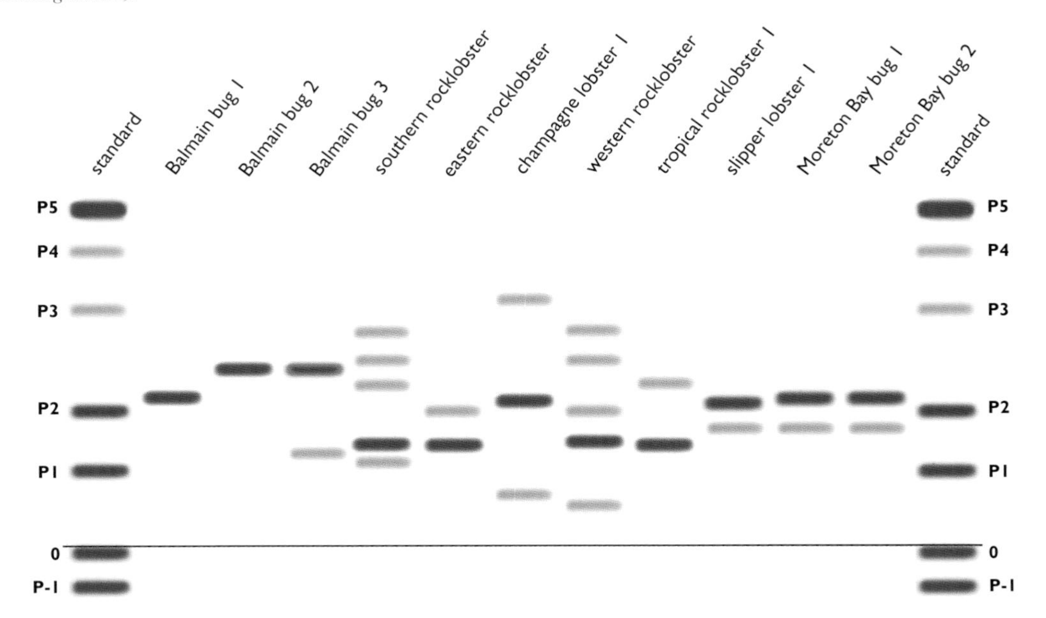

Figure 9.30—Protein fingerprints of bugs. Balmain bug 1 (*Ibacus ciliatus*), Balmain bug 2 (*I. novemdentatus*), Balmain bug 3 (*I. peronii*), southern rocklobster (*Jasus edwardsii*), eastern rocklobster (*J. verreauxi*), champagne lobster 1 (*Linuparus trigonus*), western rocklobster (*Panulirus cygnus*), tropical rocklobster 1 (*P. ornatus*), slipper lobster 1 (*Scyllarides squammosus*), Moreton Bay bug 1 (*Thenus indicus*), Moreton Bay bug 2 (*T. orientalis*).

Crustaceans

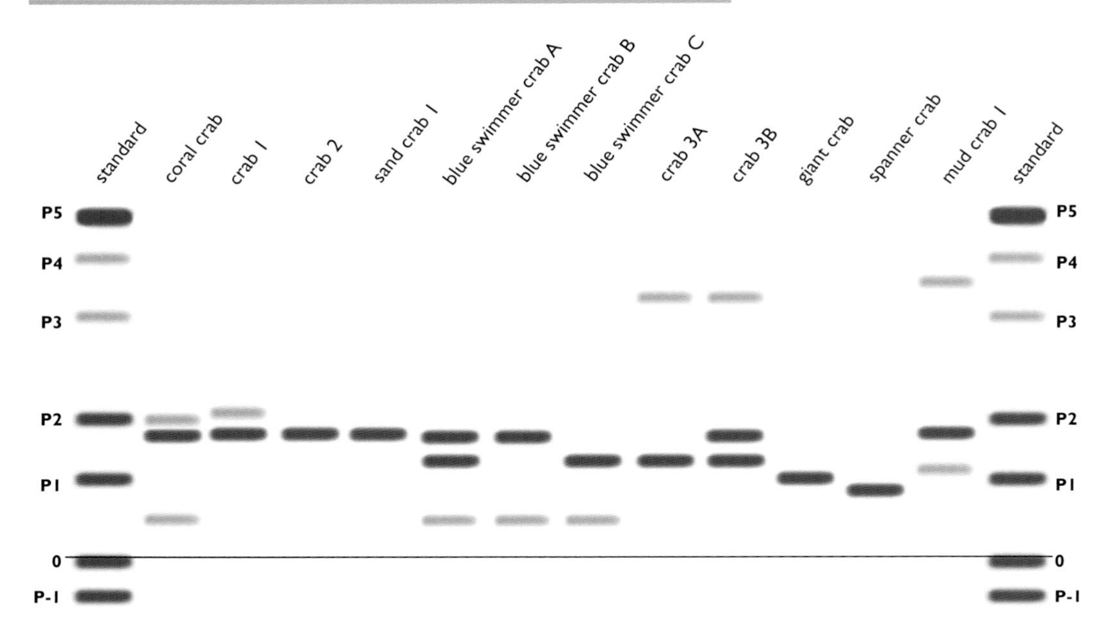

Figure 9.31—Protein fingerprints of crabs. Coral crab (*Charybdis feriata*), crab 1 (*C. natator*), crab 2 (*Hypothalassia armata*), sand crab 1 (*Ovalipes australiensis*), blue swimmer crab (*Portunus pelagicus*), crab 3 (*P. sanguinolentus*), giant crab (*Pseudocarcinus gigas*), spanner crab (*Ranina ranina*), mud crab 1 (*Scylla serrata*).

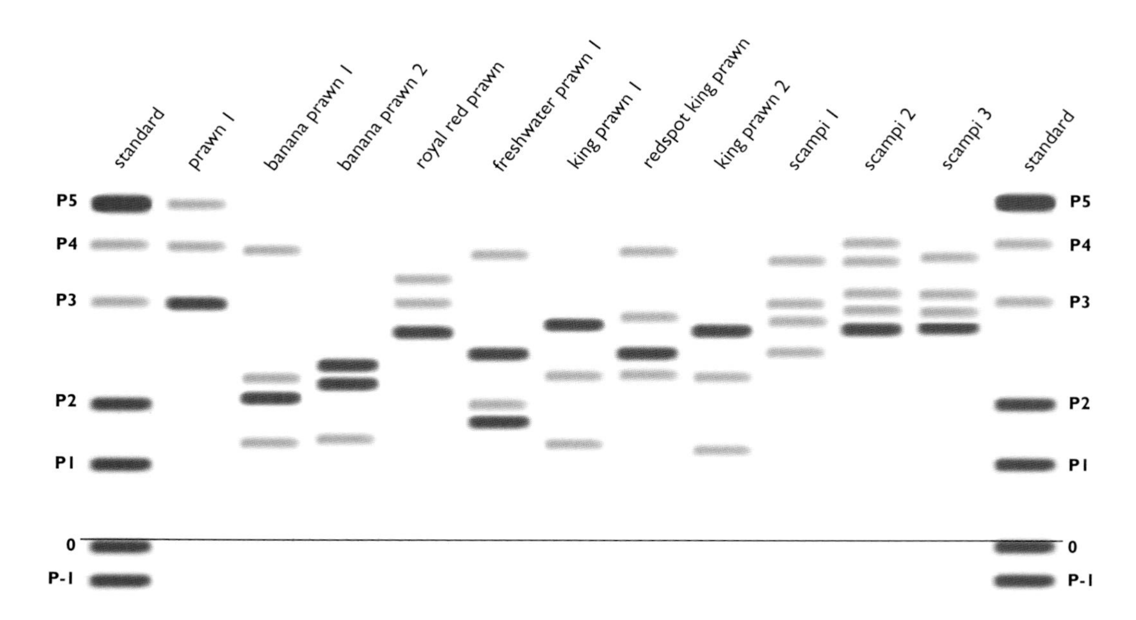

Figure 9.32—Protein fingerprints of prawns (part). Prawn 1 (*Aristaeopsis edwardsiana*), banana prawn 1 (*Fenneropenaeus indicus*), banana prawn 2 (*F. merguiensis*), royal red prawn (*Haliporoides sibogae*), freshwater prawn 1 (*Macrobrachium rosenbergii*), king prawn 1 (*Melicertus latisulcatus*), redspot king prawn (*M. longistylus*), king prawn 2 (*M. plebejus*), scampi 1 (*Metanephrops australiensis*), scampi 2 (*M. boschmai*), scampi 3 (*M. velutinus*).

Crustaceans & molluscs

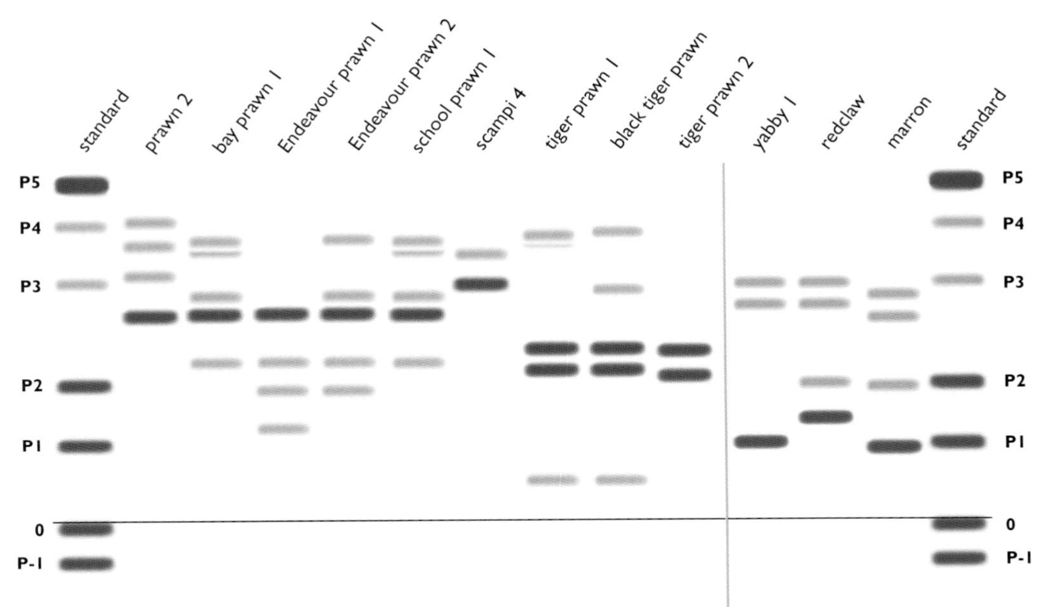

Figure 9.33—Protein fingerprints of prawns (cont.) and yabbies. Prawn 2 (*Metapenaeopsis rosea*), bay prawn 1 (*Metapenaeus bennettae*), Endeavour prawn 1 (*M. endeavouri*), Endeavour prawn 2 (*M. ensis*), school prawn 1 (*M. macleayi*), scampi 4 (*Nephropsis serrata*), tiger prawn 1 (*Penaeus esculentus*), black tiger prawn (*P. monodon*), tiger prawn 2 (*P. semisulcatus*), yabby 1 (*Cherax destructor*), redclaw (*C. quadricarinatus*), marron (*C. tenuimanus*).

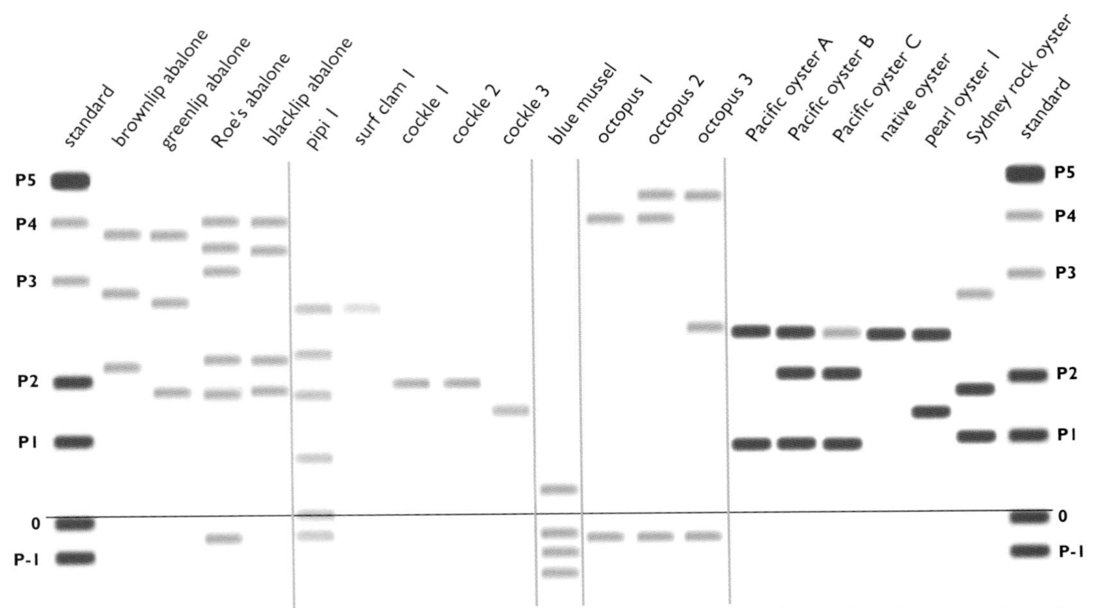

Figure 9.34—Protein fingerprints of abalones, clams, mussels, octopuses and oysters. Brownlip abalone (*Haliotis conicopora*), greenlip abalone (*H. laevigata*), Roe's abalone (*H. roei*), blacklip abalone (*H. rubra*), pipi 1 (*Donax deltoides*), surf clam 1 (*Dosinia caerulea*), cockle 1 (*Katelysia peronii*), cockle 2 (*K. rhytiphora*), cockle 3 (*K. scalarina*), blue mussel (*Mytilus edulis*), octopus 1 (*Octopus australis*), octopus 2 (*O. berrima*), octopus 3 (*O. pallidus*), Pacific oyster (*Crassostrea gigas*), native oyster (*Ostrea angasi*), pearl oyster 1 (*Pinctada maxima*), Sydney rock oyster (*Saccostrea glomerata*).

Molluscs

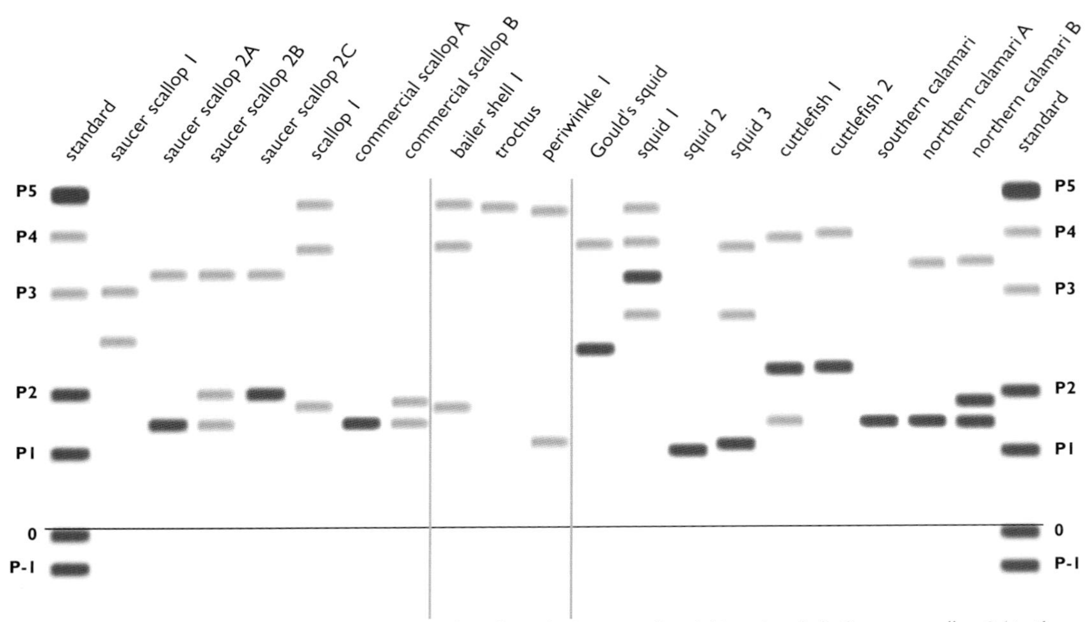

Figure 9.35—Protein fingerprints of scallops, sea snails and squids. Saucer scallop 1 (*Amusium balloti*), saucer scallop 2 (*A. pleuronectes*), scallop 1 (*Annachlamys flabellata*), commercial scallop (*Pecten fumatus*), bailer shell 1 (*Livonia mamilla*), trochus (*Trochus niloticus*), periwinkle 1 (*Turbo undulatus*), Gould's squid (*Nototodarus gouldi*), squid 1 (*N. hawaiiensis*), squid 2 (*Photololigo* cf *chinensis*), squid 3 (*P.* sp.), cuttlefish 1 (*Sepia pharaonis*), cuttlefish 2 (*S. rex*), southern calamari (*Sepioteuthis australis*), northern calamari (*S. lessoniana*).

Oil composition

P. D. Nichols, B. D. Mooney, P. Virtue and N. G. Elliott

10

Introduction

Australians eat less seafood than they do beef, lamb or chicken. However, consumption has increased in the 1990s, partly due to the effective promotion of the health benefits of eating both finfishes and shellfishes. Nutritionists and medical authorities are encouraging people to eat seafoods because the oils contain high levels of nutritionally important omega-3 polyunsaturated fatty acids (PUFA).

In seafood, oils are the second largest component after protein. The main oils are triglycerides and polar oils, with lower levels of cholesterol. Although excessive cholesterol is often considered a factor in coronary heart disease and other disorders, levels in seafood are generally lower than in other meats (Nichols *et al.* 1998a). Triglycerides serve as stores of energy, and both polar oils and cholesterol are structural components of cell membranes.

The main components of triglycerides and polar oils in seafood oils are 'saturated' and 'unsaturated' fatty acids—these include 'monounsaturated fatty acids' (MUFA, contain one double-bond) and 'polyunsaturated fatty acids' (PUFA, contain two or more double-bonds). Olive oil is a common monounsaturated plant oil, and evening primrose oil is a common polyunsaturated plant oil.

Seafood is an ideal source of the two long-chain omega-3 PUFA that are important to human health. EPA [eicosapentaenoic acid, 20:5(n-3)] and DHA [docosahexaenoic acid, 22:6(n-3)] are present in high amounts in seafood oil. The human body manufactures only small amounts of these and we must therefore rely on dietary sources such as seafood.

EPA and DHA are beneficial against various disorders (Kinsella 1987). Fish-derived EPA and DHA decrease the incidence of coronary heart disease (Suzukawa *et al.* 1995); for example, eating fish two or three times a week may prevent heart attacks (Siscovick *et al.* 1995). They are also beneficial in treating high blood pressure and inflammatory diseases such as rheumatoid arthritis, and in nutrition of infants, particularly for retina and brain development.

Seafood is also a good source of a long-chain omega-6 PUFA of great interest to nutritionists. AA [arachidonic acid, 20:4(n-6)] is important for growth and possibly for general good health and well-being. The benefits of plant- and fish-derived PUFA-rich oils in some forms of cancer therapy have been examined with positive effects noted, although further research is required.

The content and nutritional value of the oil from nearly 200 species of Australian fishes and shellfishes was recently examined (Nichols *et al.* 1998a,b). These two reports give information on the benefits of seafood oils, a list of selected recent nutritional studies, detailed profiles of oil composition and a comparative discussion of the oil profiles. The main findings were:

— Australian seafoods contain high levels of the long-chain omega-3 PUFA, typically ten times greater than in other food groups;

— Australian fishes generally have lower levels of oil than northern hemisphere fish species, but the relative level of PUFA is generally higher in fishes with less oil;

— Australian fishes generally have higher relative levels of the nutritionally important DHA than fishes from northern hemisphere waters and nutritional supplements currently available that are derived from those northern species.

Summary results are provided here on the amount of oil, the abundance of fatty acid families and the levels of the main omega-3 and omega-6 PUFA present in Australian seafoods.

Format of results and definitions

Oil parameters shown on the following pages are:

- oil (fat) content in per cent (%) wet weight of flesh (highlighted by blue column)—oil content definitions: low oil, <2%; moderate oil, 2–5%; high oil, >5%;
- content of arachidonic acid (AA), eicosapentaenoic acid (EPA) and docosahexaenoic acid (DHA) in milligrams per 100 gram wet weight of flesh (in each horizontal bar)—two species exceed the page scale and are given individual scale bars;
- average portion in families (expressed as a percentage of the total fatty acids) of saturated fatty acids (SAT), monounsaturated fatty acids (MUFA) and polyunsaturated fatty acids (PUFA) (adjacent to group headings).

PUFA are defined by the position of the first double-bond from the end of the molecule. The two main PUFA groups are the omega-3 (the first double-bond is three carbons from the terminal methyl end of the fatty acid) and omega-6 (the first double-bond is six carbons from the terminal methyl end of the fatty acid) series.

All data presented for fishes are for flesh without skin, and are the average of three specimens. Family groups are presented in alphabetical order within major categories. Marketing names with an asterix (*) cover more than one species. Remarks shown for families of seafood or for individual species highlight selected comparative features of the results on that page. Wider discussion is beyond the scope of this chapter and readers are referred to Nichols *et al.* (1998a) and references therein.

Methods

Sample collection

Commercially important finfish and shellfish species were selected following advice from industry and fisheries personnel in each state and territory. Animals were obtained from commercial fishers, aquaculture farms, seafood markets and research cruises. White muscle (flesh) tissue was sampled immediately after the animal's capture, after it was bought at the market or after it was frozen and sent to the laboratory. For most species 5–10 animals were available, from which three were randomly selected for analysis. All finfish muscle samples were taken from the right 'shoulder' region, an area normally included in a fillet. Skin and subcutaneous fat were excluded. Samples were taken from the abdomen (or 'tail') of crustaceans, the mantle of cephalopods, and either the whole body or abductor muscle of univalve and bivalve molluscs. All tissues were frozen until analysed.

Sample preparation and analysis

Briefly, oil (fat) was extracted with solvents, and the fatty acid fractions were prepared from a portion of the oil. The fatty acids were analysed by gas chromatography with additional mass spectrometric confirmation. Such instruments are used routinely for detecting and identifying drugs in medical research and in drug-testing. The analysis method used here is detailed in Nichols *et al.* (1998a).

Sharks and rays

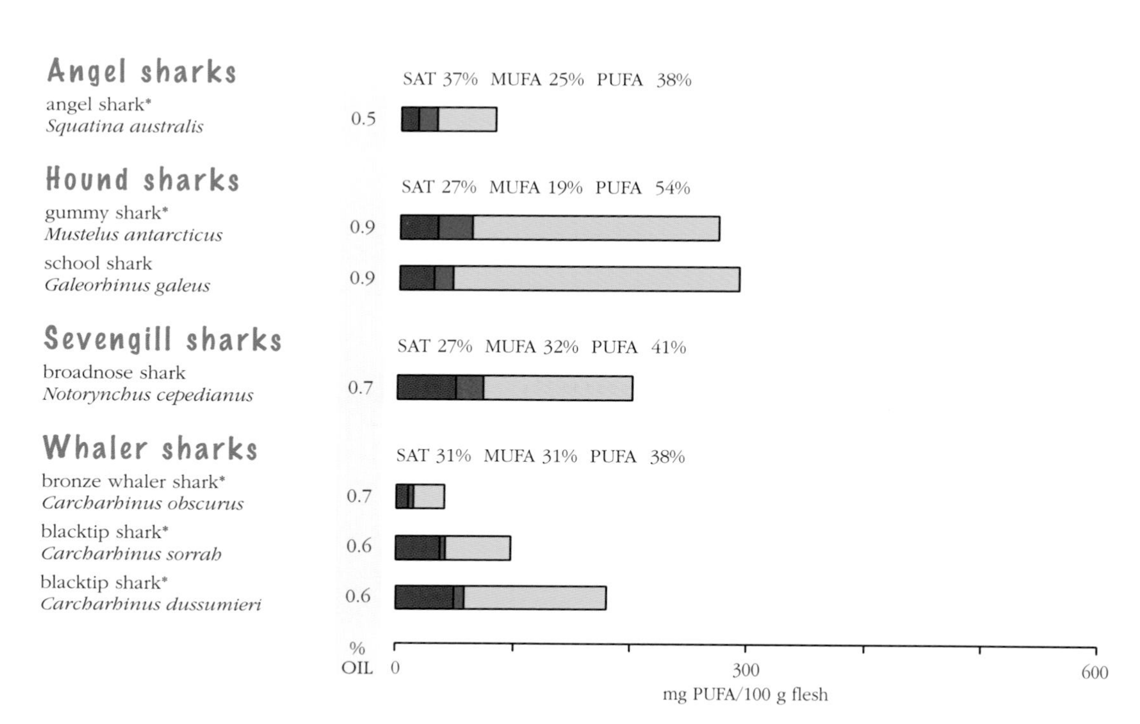

Remarks: Shark flesh generally contained low levels of oil. The two more common species, the school shark and the gummy shark, contained the highest oil content of the sharks analysed, the highest absolute and relative levels of total PUFA and the highest DHA to EPA ratio (15 and 7.6). The two tropical blacktip sharks also had high DHA to EPA ratios and, in addition, higher relative levels of the omega-6 fatty acid AA. The bronze whaler shark had several unusual non-methylene interrupted diunsaturated fatty acids (data not shown).

The oil results reported above are for fillets of sharks and rays. Interest also exists in shark liver oils which are used in human health supplements. The oil in these products is distinct from that in shark flesh. Shark liver oil is largely from deepwater species and the compounds of interest are the highly unsaturated hydrocarbon called squalene and a class of oils called diacylglyceryl ethers or 'alkoxyglycerols'.

Bony fishes

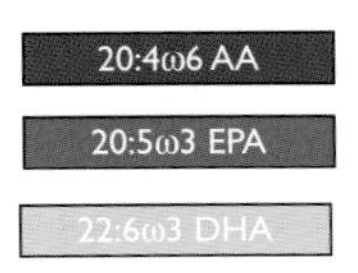

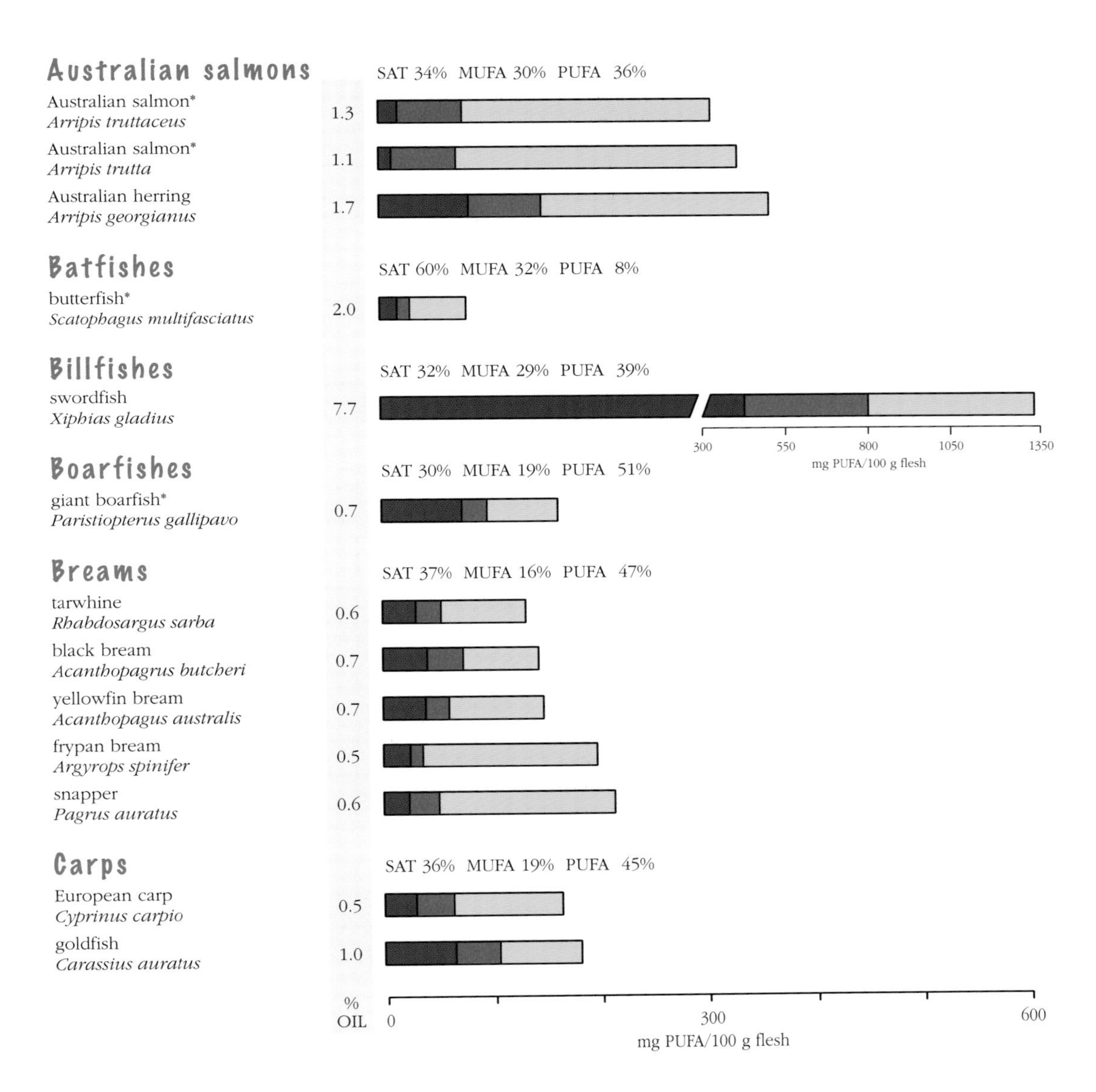

Remarks: The swordfish had the highest oil content of all fishes in this study; it also had the greatest variation between individuals, possibly reflecting its migratory behaviour. Other pelagic species (Australian salmons and the Australian herring), and the butterfish had low to moderate oil levels. The swordfish had the highest amount of PUFA of all fishes examined. The DHA to EPA ratio was higher in the Australian herring, Australian salmons, breams and the European carp. Butterfish had unusually low PUFA levels. The swordfish, Australian herring, giant boarfish and goldfish had the highest levels of the omega-6 fatty acid AA.

Bony fishes

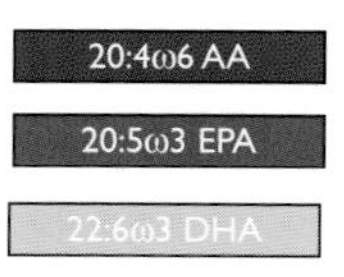

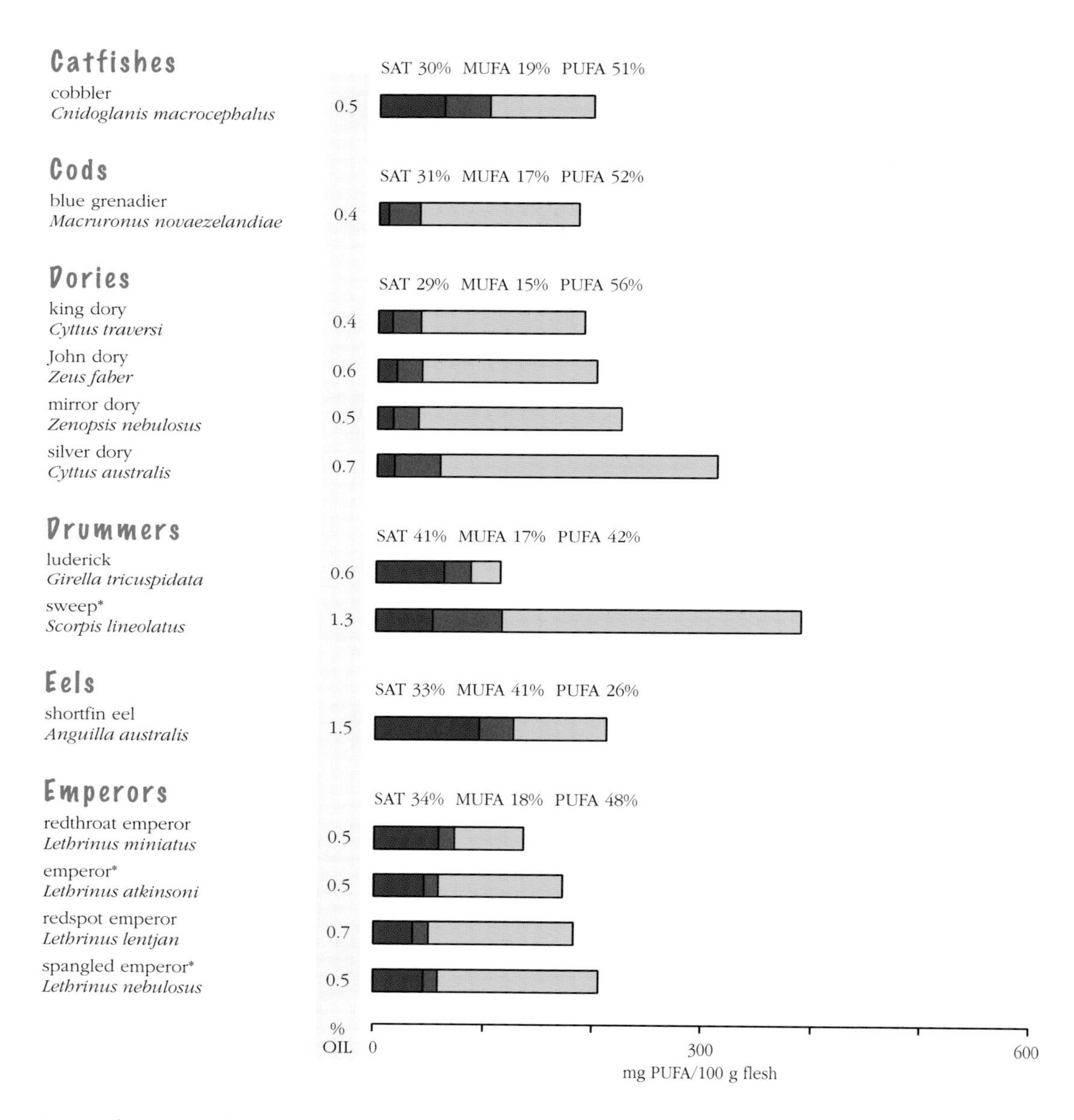

Remarks: The oil content in the fishes from these groups was generally low, with only the shortfin eel and silver sweep having >1%. The dories, blue grenadier and cobbler had relatively high PUFA levels (>50%), while the shortfin eel had lower PUFA (26%) and higher relative levels of MUFA (41%). The DHA to EPA ratio was highest in the dories, emperors and blue grenadier. The shortfin eel, luderick, cobbler and emperors had the highest relative levels of the omega-6 fatty acid AA.

Bony fishes

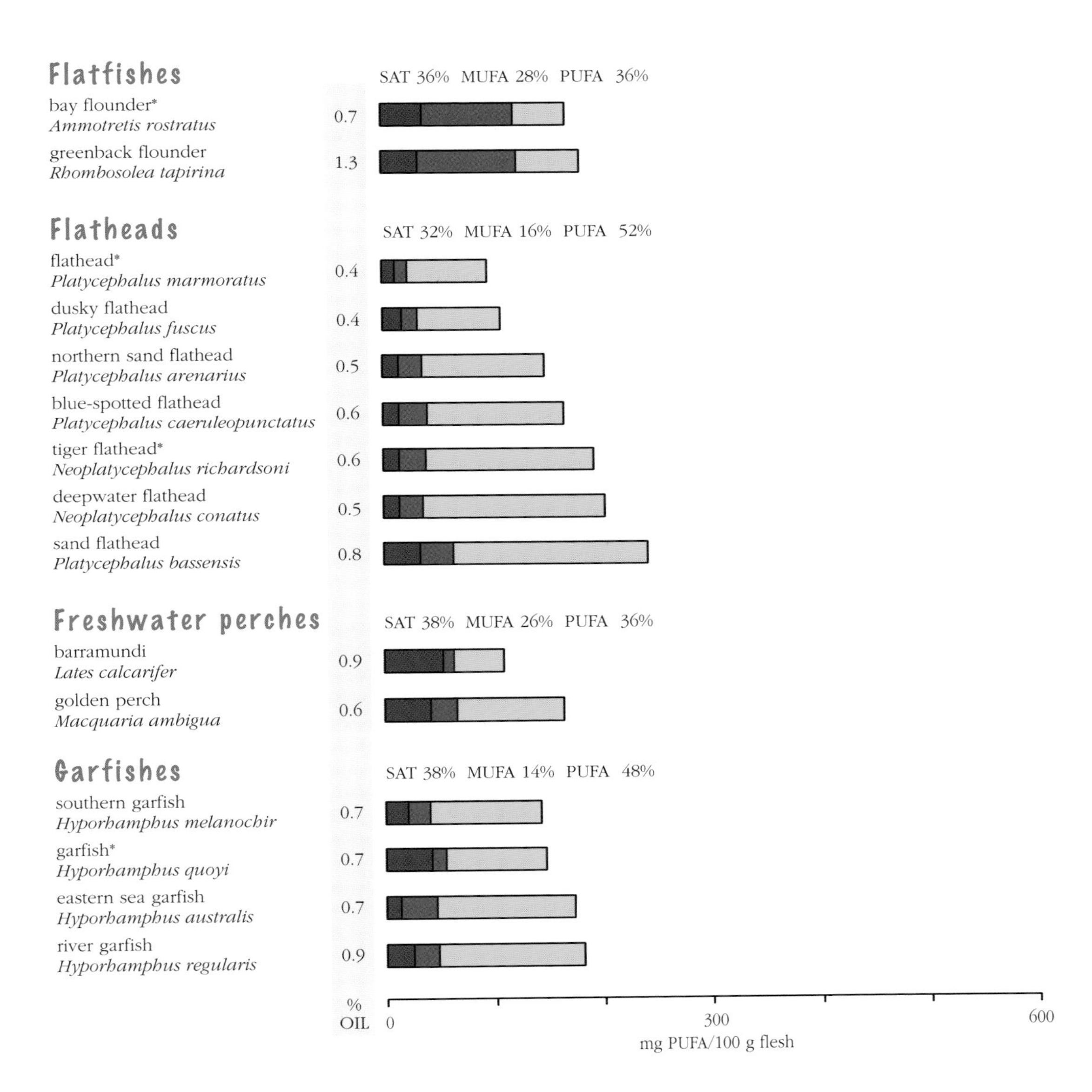

Remarks: Flounders, flatheads, freshwater perches and garfishes had low oil content; only the greenback flounder had >1%. PUFA levels in flatheads and garfishes were higher than in the other two families. The flounders and freshwater perches had higher relative levels of MUFA than the other species. The DHA to EPA ratio was highest in the flatheads and garfishes. The flounders had the highest relative levels of EPA, while the freshwater perches had the highest levels of the omega-6 fatty acid AA.

Bony fishes

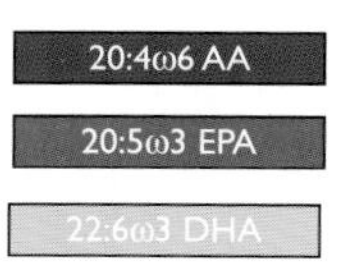

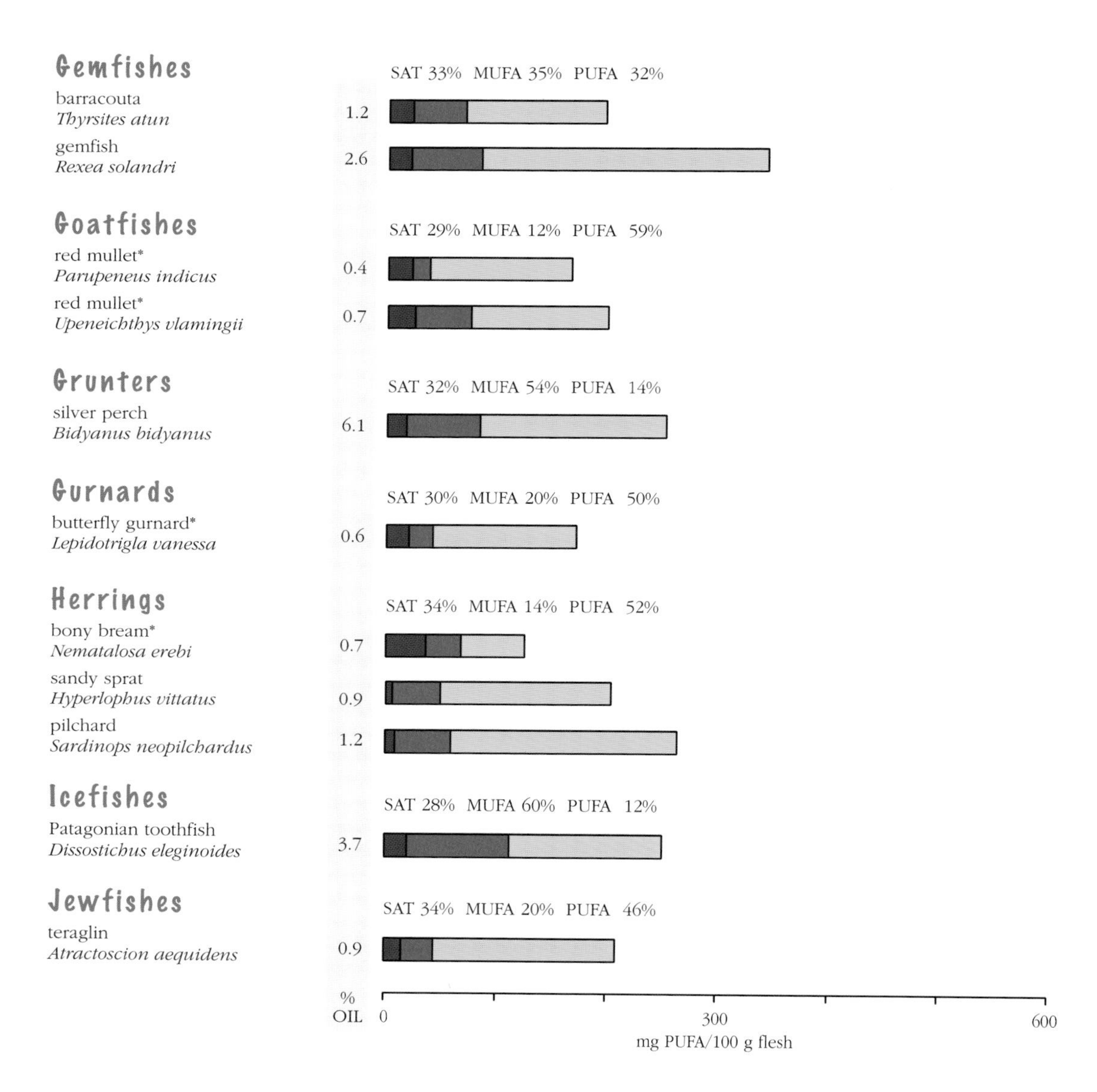

Remarks: Silver perch, harvested from an aquaculture farm, had high oil content (6.1%); the subAntarctic Patagonian toothfish and the gemfish had moderate oil content (3.7% and 2.6%). The barracouta and pilchard had >1% oil. The red mullet had very high PUFA levels (59%). The MUFA levels in these families ranged from 12% in red mullet to a very high 54% to 60% in silver perch and Patagonian toothfish. The DHA to EPA ratio was high in all groups, particularly the herrings, jewfishes and gurnards. The silver perch and Patagonian toothfish had the highest proportions of EPA. Except for bony bream, these families had low levels of AA.

Bony fishes

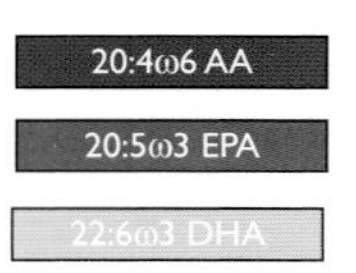

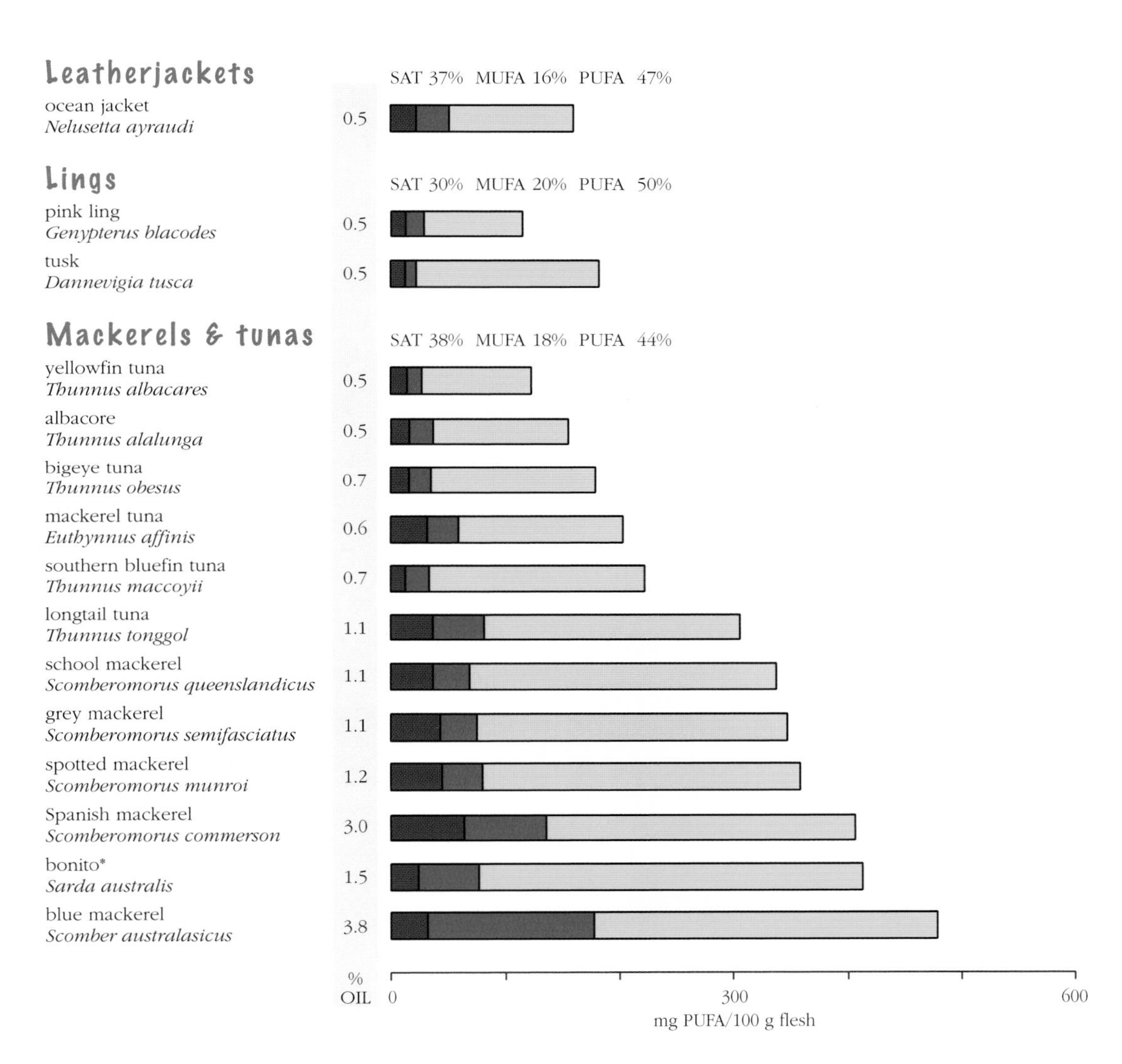

Remarks: Low oil content was noted in ocean jacket and lings (0.5%). Oil content varied across the mackerels and tunas, ranging from 0.5% in both yellowfin tuna and albacore to 3.0% and 3.8% (moderate oil content) in the Spanish mackerel and blue mackerel. PUFA levels were high in all species. The mackerels, in particular, are good sources of DHA due to an above-average oil content. DHA to EPA ratios were generally high, except for in blue mackerel which contained higher levels of EPA. Most species had low levels of AA.

Bony fishes

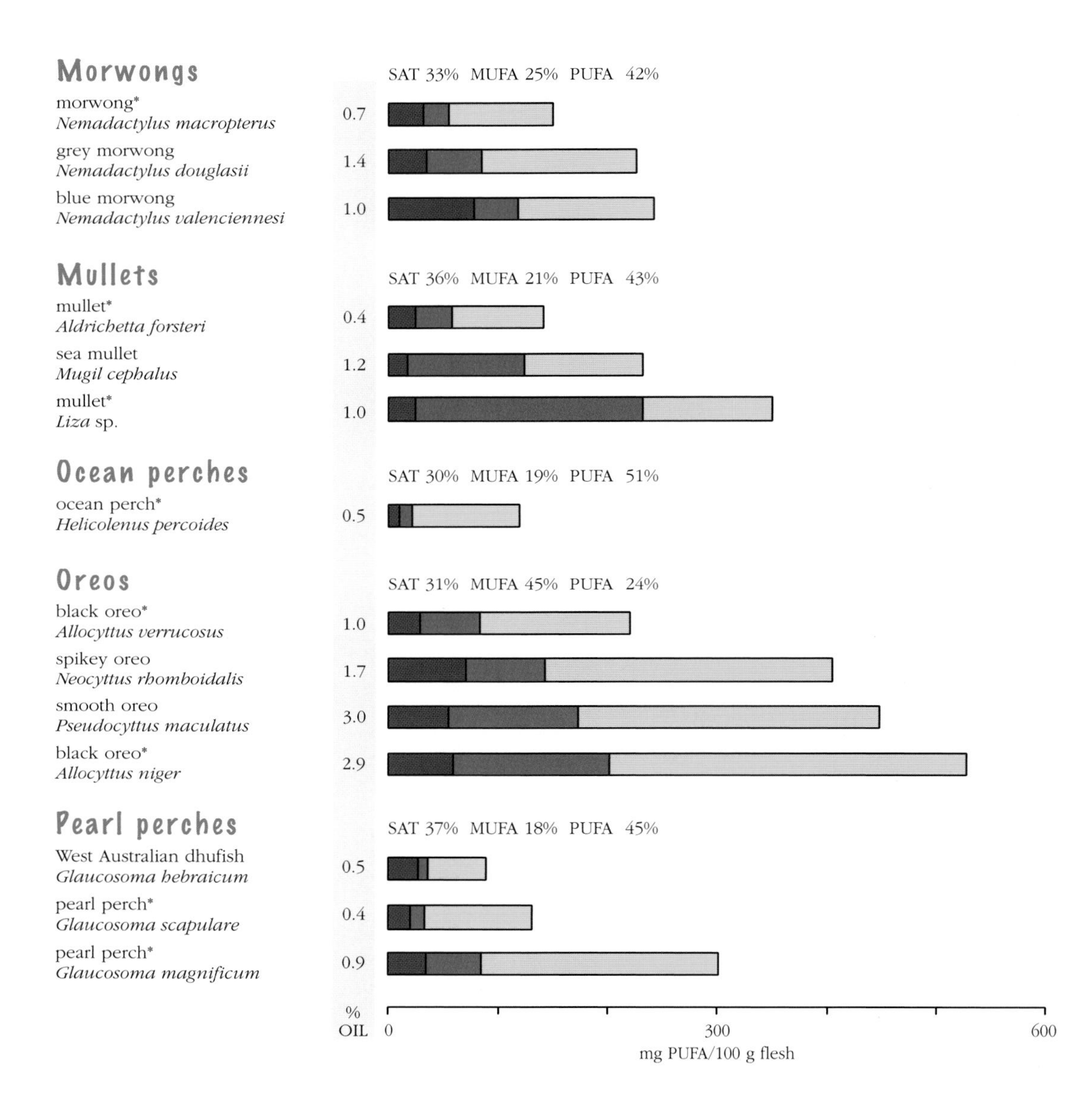

Remarks: The deepwater oreos generally had moderate oil content (up to 3.0%); most other species had low oil content. The deepwater oreos also had the lowest relative PUFA levels (24%) and the highest MUFA levels. Sea mullet and mullet (*Liza* sp.) had more EPA than DHA, and the oreos also had a higher proportion of EPA than other families. Blue morwong had the highest relative levels of the omega-6 fatty acid AA. Data for oreos from Bakes *et al.* (1995).

Bony fishes

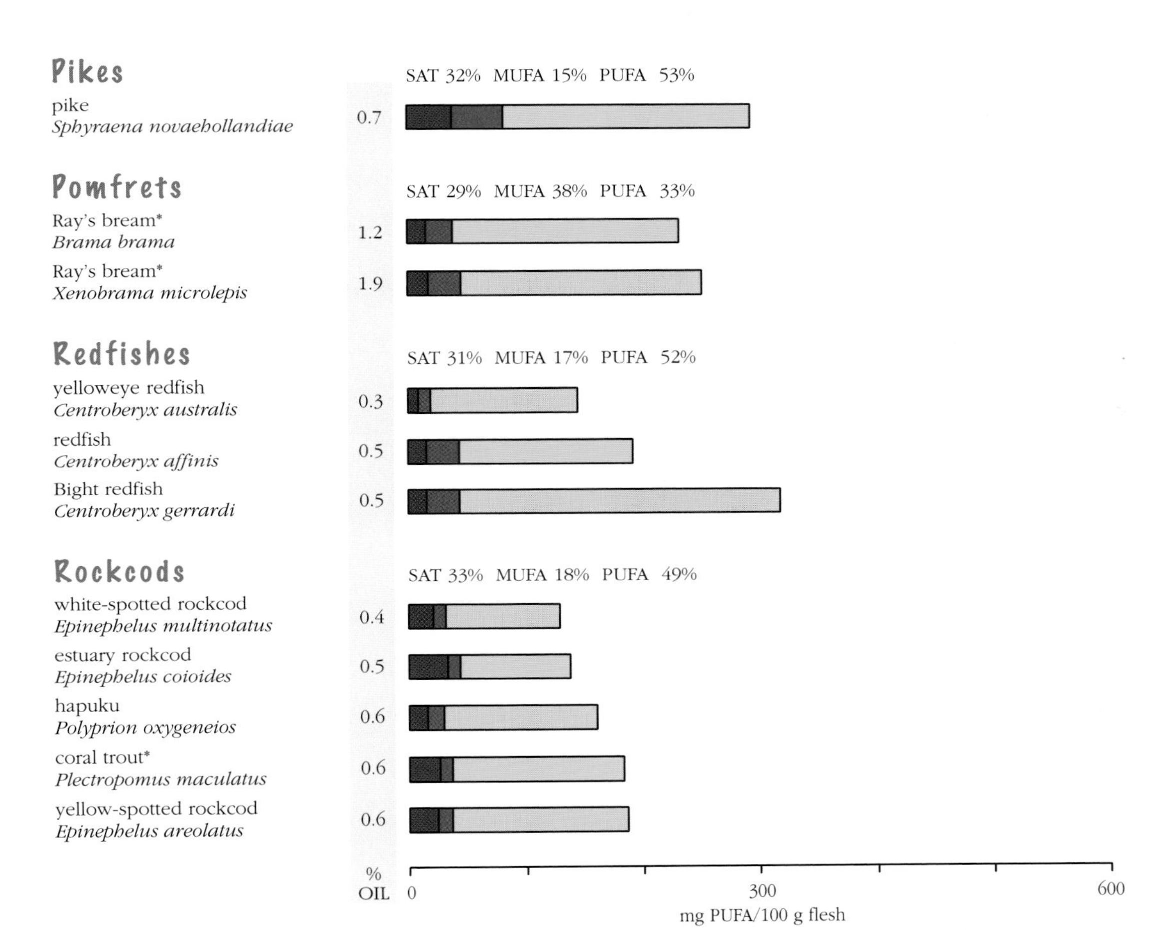

Remarks: Rockcods, redfishes and pike had low oil content; in Ray's breams it approached moderate levels. The pike, redfishes and rockcods had high PUFA levels; Ray's breams had high MUFA and lower PUFA levels. The DHA to EPA ratio was generally high (>5) across the four families. In the rockcod species, AA was at higher levels than EPA.

Bony fishes

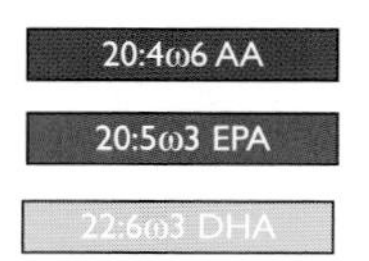

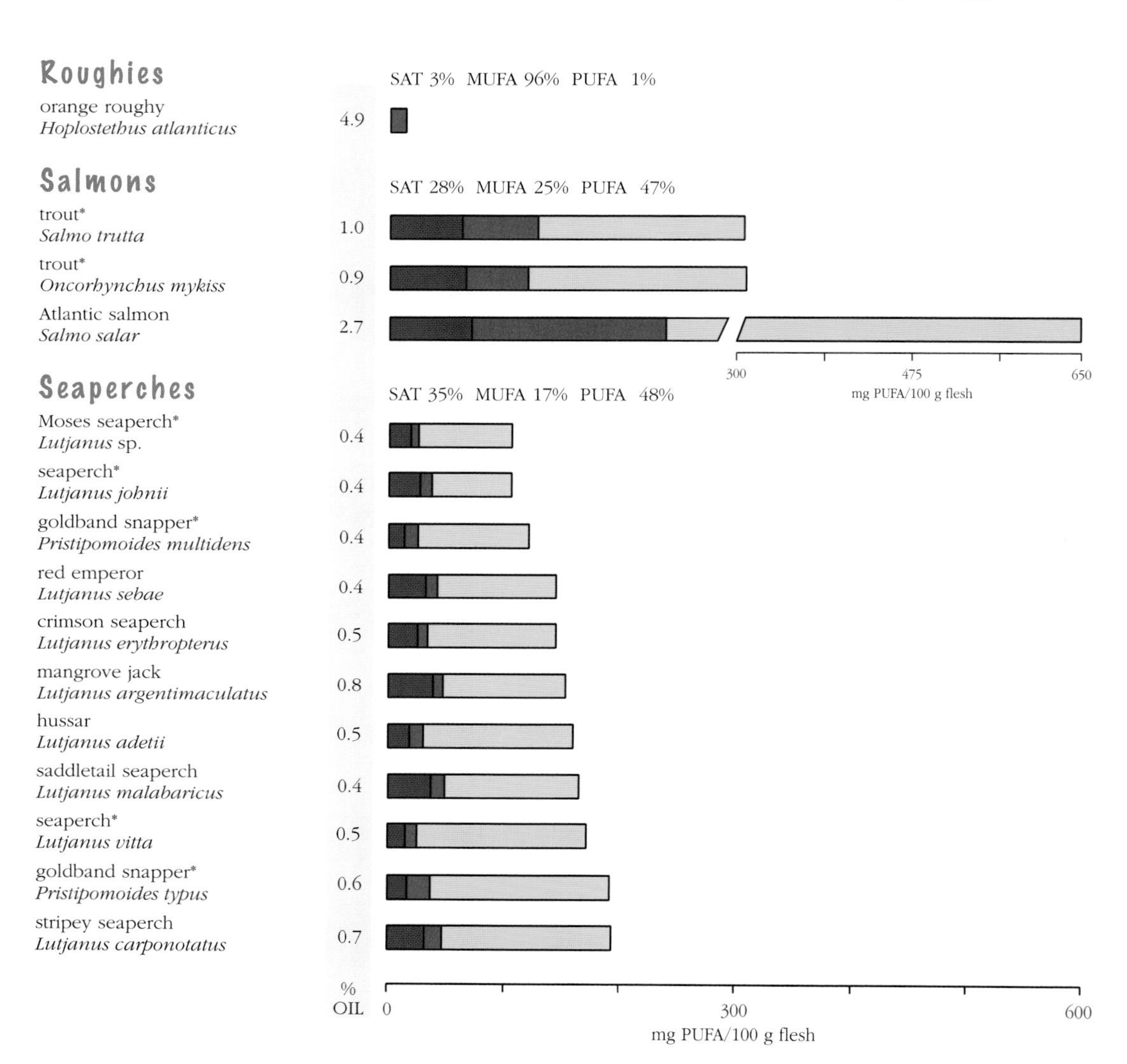

Remarks: Orange roughy had high oil content but an unusual oil profile, with little if any PUFA, and so very low omega-3 PUFA. Its oil is very distinct, as it has high proportions of wax esters and MUFA. The seaperches had low oil content. The wild trouts contained low oil content, although higher than in the seaperches, while the popular aquacultured Atlantic salmon had moderate oil content. PUFA levels were high in Atlantic salmon, trouts and seaperches; Atlantic salmon had the second highest amount of PUFA of all fishes examined. All seaperches had high DHA to EPA ratios (5–10). In Atlantic salmon, the ratio was close to that observed in mackerel oil used in commercial salmon feeds. The trouts had higher AA than the others in these groups. Variation in the level of AA in the seaperches may indicate differences in the diet of species in this family. Data for orange roughy from Bakes *et al.* (1995).

Bony fishes

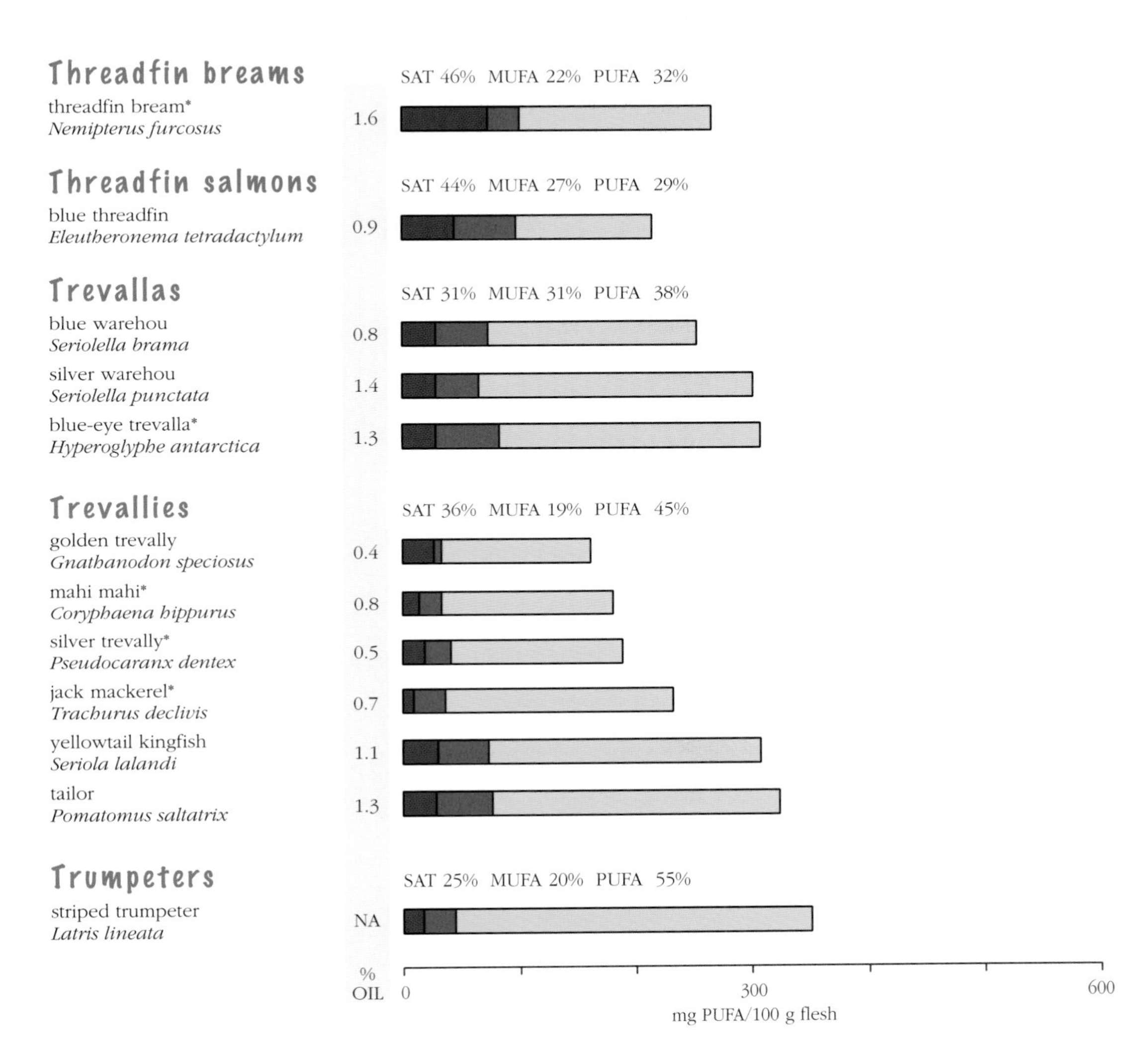

Remarks: Oil content in the trevallies and trevallas was low, but slightly higher in the trevallas. PUFA levels were highest in the trevallies and trumpeter, followed by the trevallas, threadfin bream and blue threadfin. MUFA was lower in trevallies, trumpeter and the threadfin bream than in blue threadfin and trevallas. All fishes had high DHA to EPA ratios, especially the striped trumpeter. Only threadfin bream had high levels of the omega-6 fatty acid AA. Data for striped trumpeter from Nichols *et al.* (1994).

Bony fishes

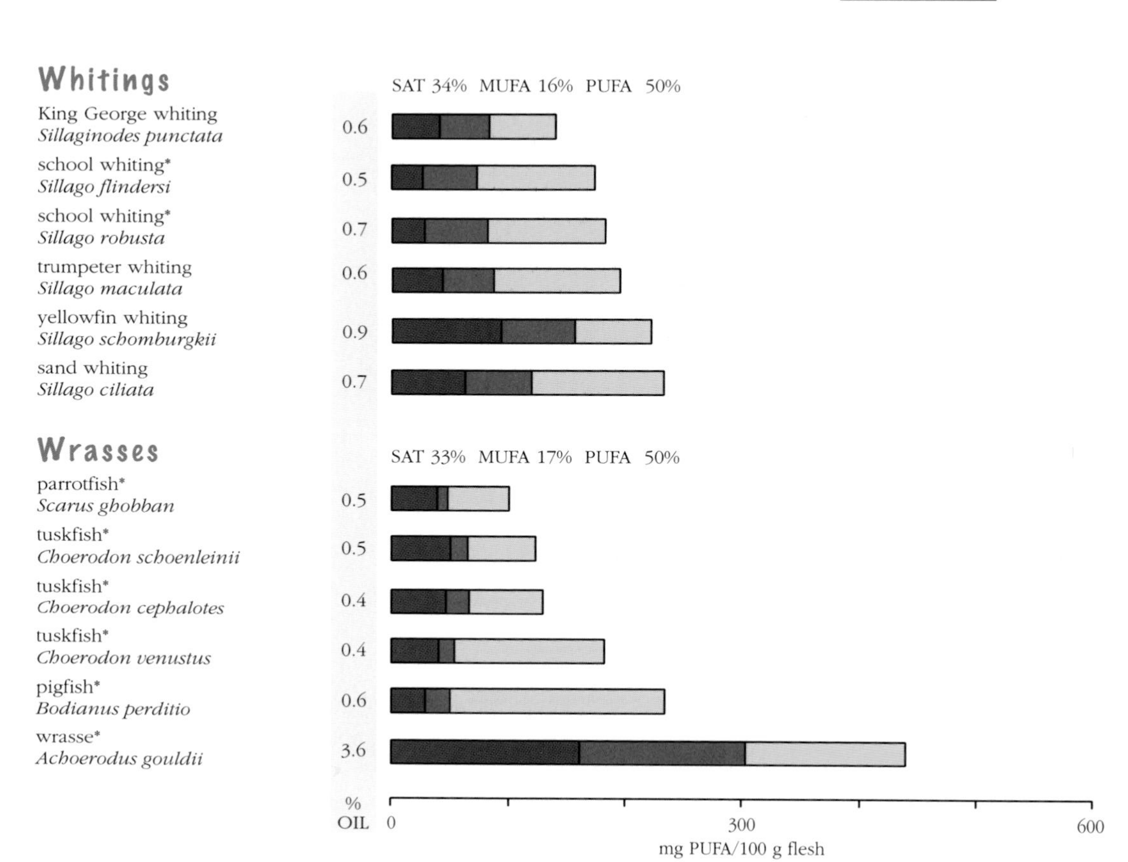

Remarks: Whitings showed low oil content. It was also generally low in the wrasse family, except in wrasse (3.6%) where it was in the moderate oil range. Both families had high PUFA (50%). DHA to EPA ratios were higher in the wrasses, except for wrasse which had them in equal proportions. The whitings generally had proportionally more EPA than the wrasses. The omega-6 AA levels were highest in the wrasse, and in yellowfin whiting and sand whiting.

Crustaceans

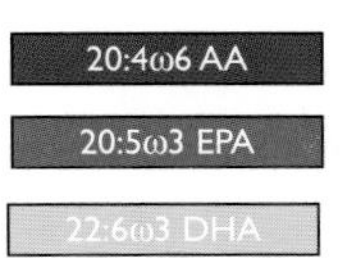

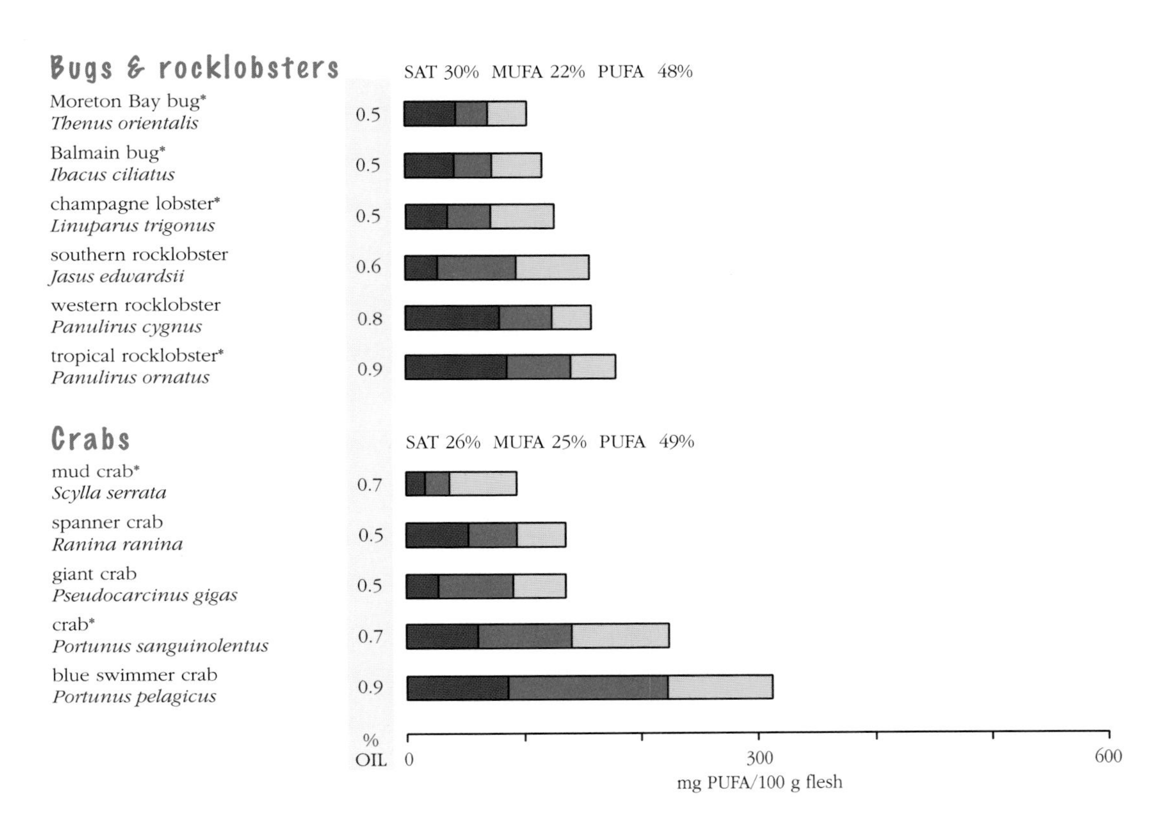

Remarks: Low oil content was observed in all the crustaceans. PUFA levels were high in both bugs and crabs. The DHA to EPA ratio was much lower than for fishes. EPA levels were higher than DHA in many crabs and lobsters. The omega-6 AA levels were also high in many species. AA was the main PUFA in Moreton Bay bug, Balmain bug, western and tropical rocklobsters, and spanner crab.

Crustaceans

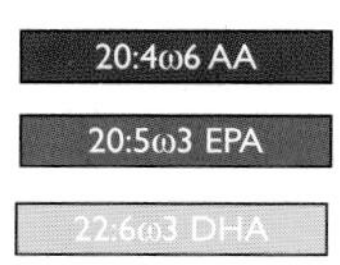

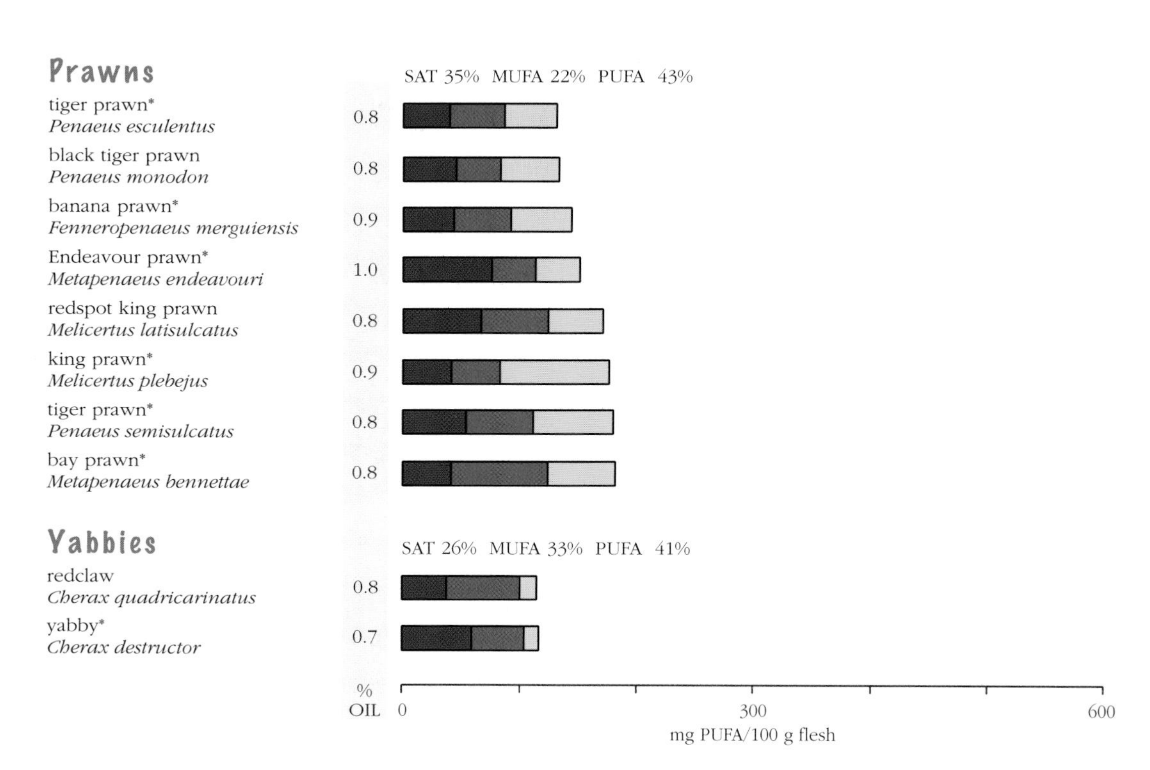

Remarks: The oil content of prawns and yabbies was low (generally <1%), but PUFA levels were moderate. The DHA to EPA ratio was much lower than for fishes; in many of the prawns and all the yabbies, EPA was at similar or greater levels than DHA. The omega-6 AA was the main PUFA in many species. Prawns contain the highest amount of cholesterol of all seafood tested (e.g. 120–160 mg/100 g; Nichols *et al.* 1998a,b).

Molluscs

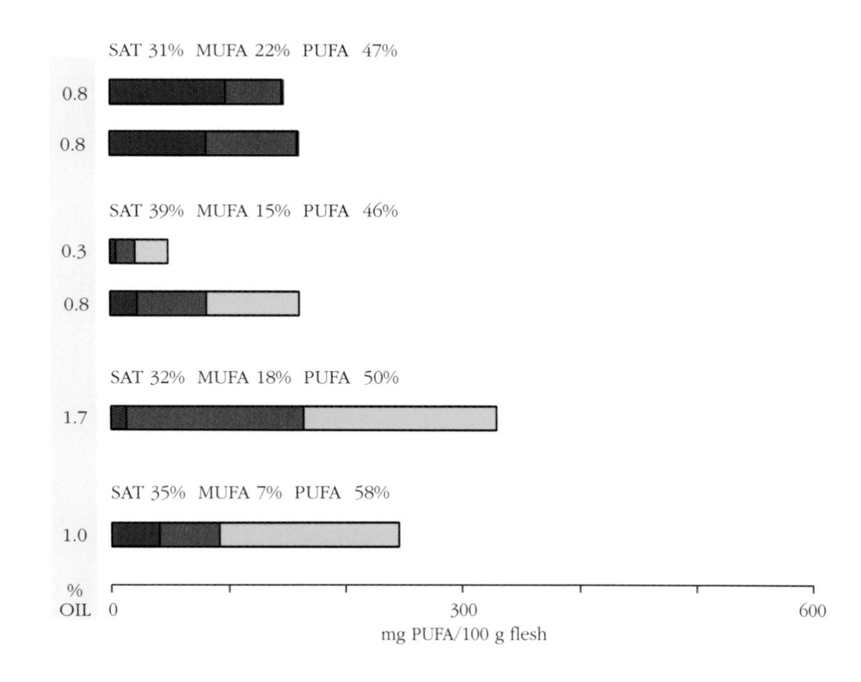

Remarks: Oil content was low in these four mollusc groups; only in the blue mussel did it approach moderate. PUFA levels were generally high, with the highest in octopus. The abalones had very low levels of DHA and, in contrast to all other seafood examined, their main PUFA was DPA [docosapentaenoic acid, 22:5(n-3)]. The omega-6 AA was also high in abalones. The blue mussel and pipi had similar levels of EPA and DHA, while the surf clam and octopus had more DHA than EPA. Data for abalones from Dunstan *et al.* (1996).

Molluscs

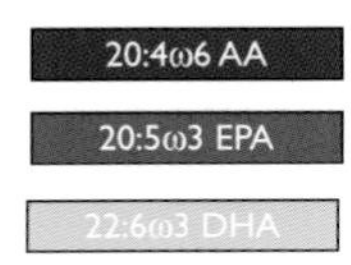

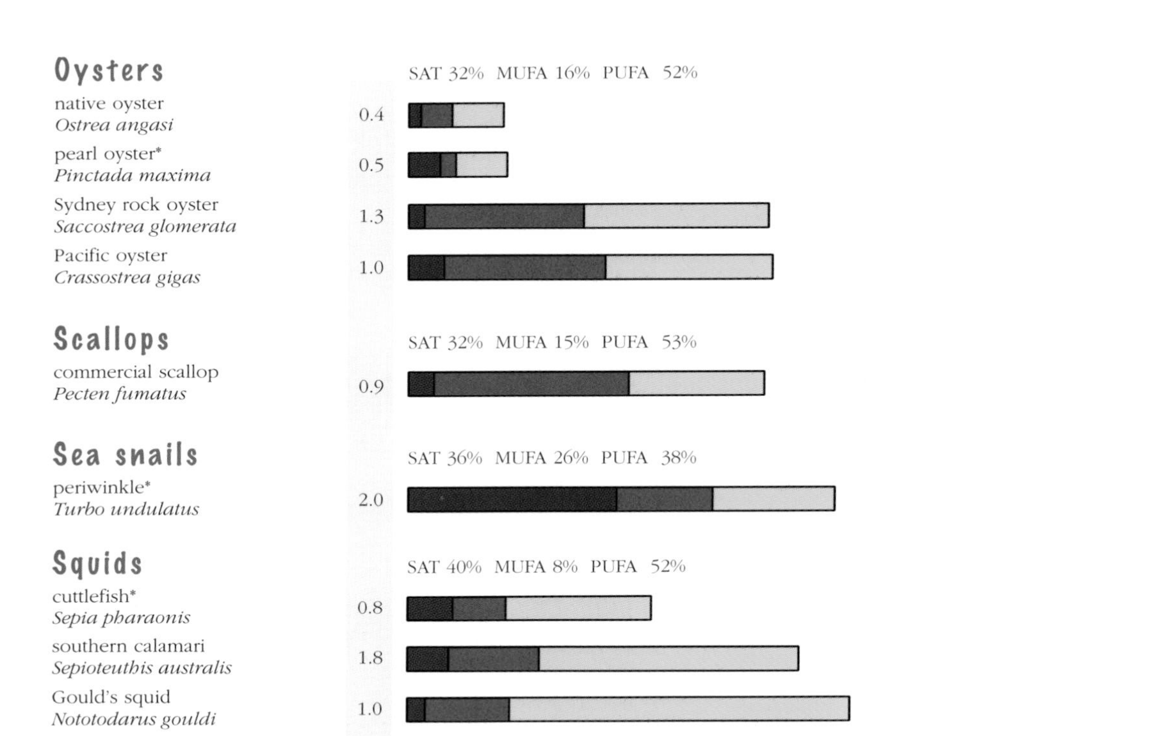

Remarks: Levels of oil were moderate in the periwinkle and the southern calamari. Although low, they were higher in the Sydney rock and Pacific oysters than in the other oysters. The PUFA levels were high in all species, except the periwinkle. The Sydney rock and Pacific oysters, periwinkle and commercial scallop had similar levels of EPA and DHA, while in the squid group and native and pearl oysters, DHA was higher than EPA. The omega-6 AA was the main PUFA in the periwinkle.

References

A: Seafood (Chapters 4–8)

ABARE (1998). *Australian Fisheries Statistics 1998.* Commonwealth of Australia, Canberra: 51 pp.

Beesley, P. L., Ross, G. J. B. and Wells, A. (eds) (1998). *Mollusca: the Southern Synthesis.* Fauna of Australia, Vol. 5. CSIRO Publishing, Melbourne: Part A, 563 pp; Part B, pp 565–1234.

Daley, R. K., Last, P. R., Yearsley, G. K. and Ward, R. D. (1997). *South East Fishery Quota Species: an Identification Guide.* CSIRO Division of Marine Research, Hobart: 91 pp.

Department of Primary Industries and Energy (1995). *Marketing Names for Fish and Seafood in Australia.* Commonwealth of Australia, Canberra: 170 pp.

Edgar, G. J. (1997). *Australian Marine Life: the Plants and Animals of Temperate Waters.* Reed Books, Kew: 544 pp.

Gomon, M. F., Glover, C. J. M. and Kuiter, R. H. (eds) (1994). *The Fishes of Australia's South Coast.* State Print, Adelaide: 992 pp.

Grey, D. L., Dall, W. and Baker, A. (1983). *A Guide to the Australian Penaeid Prawns.* Department of Primary Production of the Northern Territory, Darwin: 140 pp.

Helfman, G. S., Collette, B. B. and Facey, D. E. (1997). *The Diversity of Fishes.* Blackwell Science, Malden: 528 pp.

Hutchins, J. B. and Swainston, R. (1986). *Sea Fishes of Southern Australia.* Swainston Publishing, Daglish: 180 pp.

Jones, D. S. and Jones, G. J. (1994). *A Field Guide to Crustaceans of Australian Waters.* Reed, Chatswood: 216 pp.

Kailola, P. J., Williams, M. J., Stewart, P. C., Reichelt, R. E., McNee, A. and Grieve, C. (eds) (1993). *Australian Fisheries Resources.* Bureau of Resource Sciences, Department of Primary Industries and Energy, and the Fisheries Research and Development Corporation, Canberra: 422 pp.

Kane, J., Elliott, M., Warren, R. and Smith, K. (1994). *Australian Seafood Catering Manual.* Queensland Department of Primary Industries, Brisbane, and the Fisheries Research and Development Corporation, Canberra: Vol. 1, 133 pp; Vol. 2, 237 pp.

Kuiter, R. H. (1993). *The Complete Divers' and Fishermen's Guide to Coastal Fishes of South-eastern Australia.* Crawford House Press, Bathurst: 437 pp.

Lamprell, K. and Healy, J. M. (1998). *Bivalves of Australia, Vol 2.* Backhuys Publishers, Leiden: 288 pp.

Lamprell, K. and Whitehead, T. (1992). *Bivalves of Australia, Vol 1.* Crawford House Press, Bathurst: 182 pp.

Last, P. R., Scott, E. O. G. and Talbot, F. H. (1983). *Fishes of Tasmania.* Tasmanian Fisheries Development Authority, Hobart: 563 pp.

Last, P. R. and Stevens, J. D. (1994). *Sharks and Rays of Australia.* CSIRO Publications, Melbourne: 513 pp, 84 plates.

May, J. L. and Maxwell, J. G. H. (1986). *Field Guide to Trawl Fish from Temperate Waters of Australia.* CSIRO Division of Fisheries, Hobart: 492 pp.

Pérez Farfante, I. and Kensley, B. (1997). Penaeoid and Sergestoid Shrimps and Prawns of the World: Keys and Diagnoses for the Families and Genera. *Mémoires du Muséum National d'Histoire Naturelle, 175: 1–233.*

South Pacific Commission (1994). *Sea Cucumbers and Beche-de-Mer of the Tropical Pacific—a Handbook for Fishers.* Handbook No. 18, revised edition, South Pacific Commission, Noumea: 51 pp.

Tilzey, R. D. J. (ed.) (1994). *The South East Fishery: a Scientific Review with Reference to Quota Management.* Commonwealth of Australia, Canberra: 360 pp.

Wadley, V. A. and Dunning, M. C. (1998). *Cephalopods of Commercial Importance in Australian Fisheries.* CSIRO Marine Research, Hobart: 65 pp, 12 plates.

Wadley, V. A. and Evans, D. R. (1991). *Crustaceans from the Deepwater Trawl Fisheries of Western Australia.* CSIRO Division of Fisheries, Perth: 44 pp.

Yearsley, G. K., Last, P. R. and Morris, G. B. (1997). Codes for Australian Aquatic Biota (CAAB): an upgraded and expanded species coding system for Australian fisheries databases. *CSIRO Marine Laboratories Report 224: 125 pp.*

B: Protein fingerprinting (Chapter 9)

Daley, R. K., Last, P. R., Yearsley, G. K. and Ward, R. D. (1997). *South East Fishery Quota Species: an Identification Guide.* CSIRO Division of Marine Research, Hobart: 91 pp.

Hebert, P. D. N. and Beaton, M. J. (1993). *Methodologies for Allozyme Analysis Using Cellulose Acetate Electrophoresis.* Helena Laboratories, Beaumont, Texas: 32 pp.

Manchenko, G. P. (1994). *Handbook of Detection of Enzymes on Electrophoretic Gels.* CRC Press, Boca Raton: 341 pp.

Pepperell, J. and Grewe, P. M. (1998). *A Field Guide to the Indo–Pacific Billfishes.* CSIRO Marine Research, Hobart: 16 pp.

Richardson, B. J., Baverstock, P. R. and Adams, M. (1986). *Allozyme Electrophoresis: a Handbook for Animal Systematics and Population Studies.* Academic Press, North Ryde: 410 pp.

Shaklee, J. B. and Keenan, C. P. (1986). A practical laboratory guide to the techniques and methodology of electrophoresis and its application to fish fillet identification. *CSIRO Marine Laboratories Report 177: 59 pp.*

Whitmore, D. H. (1990). *Electrophoretic and Isoelectric Focusing Techniques in Fisheries Management.* CRC Press, Boca Raton: 350 pp.

C: Oil composition (Chapter 10)

Bakes, M. J., Elliott, N. G., Green, G. J. and Nichols, P. D. (1995). Within and between species variation in lipid composition of Australian and North Atlantic oreo and orange roughy: regional specific differences. *Comparative Biochemistry and Physiology, 111B: 633–642.*

Dunstan, G. A., Baillie, H. J., Barrett, S. M. and Volkman, J. K. (1996). Effect of diet on the lipid composition of wild and cultured abalone. *Aquaculture, 140: 115–127.*

Kinsella, J. E. (1987). *Seafoods and Fish Oils in Human Health and Disease.* Marcel Dekker, New York: 317 pp.

Nichols, D. S., Williams, R., Dunstan, G. A., Nichols, P. D. and Volkman, J. K. (1994). Fatty acid composition of Antarctic and temperate fish of commercial interest. *Comparative Biochemistry and Physiology, 107B: 357–363.*

Nichols, P. D., Mooney, B. D., Virtue, P. and Elliott, N. G. (1998a). *Nutritional value of Australian fish: oil fatty acid and cholesterol of edible species.* Fisheries Research and Development Corporation, Final Report Project 95/122. CSIRO Marine Research, Hobart: 113 pp.

Nichols, P. D., Virtue, P., Mooney, B. D., Elliott, N. G. and Yearsley, G. K. (1998b). *Seafood the Good Food: the Oil (Fat) Content and Composition of Australian Commercial Fishes, Shellfishes and Crustaceans.* CSIRO Marine Research, Hobart: 201 pp.

Siscovick, D.S., Raghunathan, T. E., King, I., Weinmann, S., Wicklund, K. G., Albright, J., Bovbjerg, V., Arbogast, P., Smith, H., Kushi, L., Cobb, L. A., Copass, M. K., Psaty, B. M., Lemaitre, R., Retzlaff, B., Childs, M. and Knopp, R. H (1995). Dietary intake and cell membrane levels of long-chain n-3 polyunsaturated fatty acids and the risk of primary cardiac arrest. *Journal of the American Medical Association, 274: 1363–1367.*

Suzukawa, M., Abbey, M., Howe, P. R. C. and Nestel, P. J. (1995). Effects of fish oil fatty acids on low density lipoprotein size, oxidisability and uptake by macrophages. *Journal of Lipid Research, 36: 473–484.*

Acknowledgements

While many people have helped in the preparation of this book, the editors would particularly like to thank several of our CSIRO Marine Research colleagues: Alastair Graham catalogued specimens, collected samples and gave considerable input to various other aspects of this project; Ross Daley constructed figures for the protein fingerprinting chapter and the glossary, etched and edited photographs, and helped with specimen collection and taxonomy; Nick Elliott carefully refereed the entire manuscript and assisted with specimen collection; Dave Evans (ex-CSIRO) went beyond the call of duty with specimen acquisition; Thor Carter contributed greatly by taking most of the photographs; Louise Bell gave considerable advice regarding design and layout; and Spikey Riddoch and Daniel Gledhill gave tremendous assistance and support, particularly during the final weeks of document compilation.

Fishery information was edited by the following fisheries experts who (sometimes with the assistance of their colleagues) kindly shared their knowledge: Kevin Rowling (NSW), Dave Smith (Vic.), Malcolm Dunning (Qld), Rod Lenanton (WA), Keith Jones (SA), Jeremy Lyle (Tas.) and Ric Fallu (NT). Their input greatly enhanced this finished product. Patricia Kailola gave generous assistance, particularly with researching the molluscs chapter.

The production of this handbook was dependent on the acquisition, accurate identification and cataloguing of thousands of samples. While specimen acquisition was not always easy, some industry members were exceptionally helpful and one deserves special mention. Bernie Taylor (A. Raptis and Sons, Pty. Ltd.) enthusiastically supplied literally hundreds of top-quality specimens and his contributions were essential to the completion of this handbook. We are also greatly indebted to the following industry and scientific personnel, and recreational fishers, for collecting, supplying, transporting, identifying or cataloguing samples:

> Gerry Allen, Peter Arnold, Tania Bardsley, Greg Barron, Mark Benson, Donna Bradley, Alan Bradshaw, Glen Bradshaw, Francene Brown, Clay Bryce, Adam Butcher, Roger Buttermore, Graham Caracciolo, Greg Carney, Kent Carpenter, Nick Catalano, Geoff Champion, Robert Chin, Viv Clements, John Collins, Phil Critchlow, Steve Crocker, Lara Damiani, Sian Damsche, Peter Davie, John Diggle, Chris Edwards, Bob Elliot, Pam Elliott, Brendan Etches, Brad Evans, Tim Farrell, Alan Faulkner, Wayne Fulton, Caleb Gardner, Jac Gibson, Neville Gill, Martin Gomon, Geoff Gooley, Sam Gordon, Ken Graham, John Hacking, Wayne Haggedorn, David Hales, Tracy Hay, Glen Haynes, David Heaslip, Barry Hutchins, Simon Johns, Jeff Johnson, Daniel Johnstone, Clive Jones, Liesl Jonker, Alan Jordan, Pheroze Jungalwalla, Brendan Kelleher, Bill Kyle, Kevin Lamprell, Chan Lee, Peter Lee, Matthew Locket, John Lynch, Denis Macgee, John Macich, Bruce Malcolm, Bernadette Manjaji, Loisette Marsh, Jamie Mason, Andrew Massey, Chris Massey, Brett Mawbey, Stan McDonald, Craig McDowall, Mark McGrouther, Robert McKenzie, Dave McKeown, Vicki Midgley, Kerry Miles, Glenn Murphy, David Needham, Stephen Newman, Terry Nichols, Mark Norman, Colin Ostle, Roy Palmer, John Paxton, Rod Pearn, Gretta Pecl, Kylie Pitt, Brett Poulson, Russell Reece, Tim Rieniets, Ian Riggs, John Roach, Arthur Rodas, Kevin Rowling, Andrew Sanger, Mandy Sansom, Richard Saul, Joel Scott, Glen Searle, Chris Shearer, Troy Sinclair, Steve Sly, Andrew Smith, Brad Smith, Errol Smith, Ken Smith, Paul Southgate, Peter Stephenson, Murray Stevenson, Neville Stewart, Tim Stranks, Denis Suckling, Wayne Sumpton, Calvin Terry, Lisa Terry, Albert Thorley, Andrew Tierney, Liz Turner, Goldie Tybell, Sven Uthicke, Terry Walker, Betty Warner, Lee Warner, Caren Willcox, Don Workman, and De-arne Yearsley.

Employees or representatives of the following companies or groups also assisted greatly:

> Angelakis Brothers, Ansett Air Freight, Australian Express, Australian Fisheries Management Authority, Christies Seafoods, Claudio's Quality Seafood, De-Costi Brothers Seafoods, Footscray wholesale fish market sellers, *FRV Southern Surveyor*, Marine Discovery Centre, Peter's Fish Market, Sydney Fish Market buyers and retailers, and numerous work experience students.

The following CSIRO Marine Research staff assisted with specimen collection and/or other aspects of the project:

> Denis Abbott, Franzis Althaus, Bruce Barker, Nic Bax, Paul Boult, Simon Braine, Nan Bray, Dave Brewer, Cathy Bulman, Peter Campell, Robert Campbell, Naomi Clear, Ian Cook, Darren Dennis, Geoff Dews, Graeme Dunstan, Kevin Eatt, Michael Gardner, Karen Gowlett-Holmes, Mark Green, Peter Grewe, John Gunn, Antonia Hodgman, Rob Kenyon, Mark Lewis, Clive Liron, Greg Lyden, Craig Mackinnon, Catriona Macleod, Janet Madsen, Tim Mangan, Vivienne Mawson, Don Michel, Paulette Midgley, Michael Moore, Meredith Newman, Anne Pirrone, Craig Proctor, Tony Rees, Tristan Richards, Keith Sainsbury, John Salini, Tim Skewes, John Stevens, Charles Sutherland, Brian Taylor, David Vance, Vicki Wadley, Ted Wassenberg, Angela Webb, Grant West, Wade Whitelaw, Alan Williams, and Owain Williams.

Photographs by Thor Carter, Gordon Yearsley, Alastair Graham, Ross Daley, Nick Elliott, Charles Sutherland, Peter Last and Graeme Johnson. The striped marlin photograph was kindly supplied by Julian Pepperell (Pepperell Research).

We also acknowledge colleagues (mostly museum taxonomists) involved in a recent Marine Bioregionalisation of Australia project. Their data provided the foundation for the distributional maps.

Comments on the text or other general assistance was provided by: Heather Coulston, Sheila and the late Jack Coulston, Nick and Ruth Freeman, Jeff and Karen Hogg, Bruce and Janet Hummerston, Jo Innes, Danny and Maria Milosavljevic, Michael Podagiel, Francis Ryan, Brian and Jennifer Yates, and Honor Yearsley.

Members of the Seafood Marketing Names Review Committee were exceptionally helpful; Roy Palmer, in particular, provided useful advice.

Funded by the Fisheries Research and Development Corporation and we thank all its staff, particularly Peter Dundas-Smith and Patrick Hone, for their continuous support, input and advice.

Appendix A: Marketing names contacts

Details of the role of the Seafood Marketing Names Review Committee, and the process by which a marketing name can be approved or changed, can be obtained from the organisations listed below. The Australian Seafood Extension and Advisory Service is the initial point of contact and is responsible for the processing of applications; the Executive Director of the Fisheries Research and Development Corporation chairs the Committee. In addition, CSIRO Marine Research provides technical advice.

Australian Seafood Extension and Advisory Service
19 Hercules St
Hamilton
Qld 4007

Tel: (07) 3406 8597
Fax: (07) 3406 8677
Email: auseas@dpi.qld.gov.au
Website: http://www.dpi.qld.gov.au/cft/auseas.html

Fisheries Research and Development Corporation
PO Box 22
Deakin West
ACT 2600

Tel: (02) 6285 0400
Fax: (02) 6285 4421
Email: frdc@frdc.com.au
Website: http://www.frdc.com.au

Appendix B: Table of species

This table lists systematically all the Australian seafood species mentioned in the text. Group number (No.) and feature-page number (p.) are provided for each marketing name. Group marketing names are in bold text, with component species indented.

No.	Marketing name	Scientific name	Authority	CAAB	Family or other	Group name	p.
		CARTILAGINOUS FISHES					
4.7	broadnose shark	*Notorynchus cepedianus*	(Péron, 1807)	37 005002	Hexanchidae	sevengill sharks	51
4.4	whiskery shark	*Furgaleus macki*	(Whitley, 1943)	37 017003	Triakidae	hound sharks	44
4.4	school shark	*Galeorhinus galeus*	(Linnaeus, 1758)	37 017008	Triakidae	hound sharks	43
4.4	**gummy shark**	***Mustelus* spp:**		37 017901	Triakidae	hound sharks	42
		Mustelus antarcticus	Günther, 1870	37 017001			
4.8	**blacktip shark**	***Carcharhinus, Loxodon* & *Rhizoprionodon* spp:**		37 018901	Carcharhinidae	whaler sharks	52
		Carcharhinus dussumieri	(Valenciennes, 1839)	37 018009			
		Carcharhinus sorrah	(Valenciennes, 1839)	37 018013			
		Carcharhinus tilstoni	(Whitley, 1950)	37 018014			
		Rhizoprionodon acutus	(Rüppell, 1837)	37 018006			
4.8	**bronze whaler shark**	***Carcharhinus brachyurus* & *C. obscurus***		37 018902	Carcharhinidae	whaler sharks	54
		Carcharhinus brachyurus	(Günther, 1870)	37 018001			
		Carcharhinus obscurus	(Lesueur, 1818)	37 018003			
4.8	blue whaler shark	*Prionace glauca*	(Linnaeus, 1758)	37 018004	Carcharhinidae	whaler sharks	53
4.2	**Endeavour dogfish**	***Centrophorus harrissoni, C. moluccensis* & *C. uyato***		37 020902	Squalidae	dogfishes	35
		Centrophorus harrissoni	McCulloch, 1915	37 020010			
		Centrophorus moluccensis	Bleeker, 1860	37 020001			
		Centrophorus uyato	(Rafinesque, 1810)	37 020011			
4.2	—	*Centrophorus squamosus*	(Bonnaterre, 1788)	37 020009	Squalidae	dogfishes	
4.2	**roughskin dogfish**	***Centroscymnus* & *Deania* spp:**		37 020904	Squalidae	dogfishes	37
		Centroscymnus crepidater	(Bocage & Capello, 1864)	37 020012			
		Centroscymnus plunketi	(Waite, 1910)	37 020013			
		Deania calcea	(Lowe, 1839)	37 020003			
4.2	white-spotted dogfish	*Squalus acanthias*	Linnaeus, 1758	37 020008	Squalidae	dogfishes	39

No.	Marketing name	Scientific name	Authority	CAAB	Family or other	Group name	p.
4.2	spikey dogfish	*Squalus megalops*	(Macleay, 1881)	37 020006	Squalidae	dogfishes	38
4.2	**greeneye dogfish**	***Squalus* spp:**		37 020901	Squalidae	dogfishes	36
		Squalus mitsukurii	Jordan & Snyder, 1903	37 020007			
		Squalus sp.		37 020018			
		Squalus sp.		37 020041			
4.6	**sawshark**	***Pristiophorus* spp:**		37 023900	Pristiophoridae	sawsharks	50
		Pristiophorus cirratus	(Latham, 1794)	37 023002			
		Pristiophorus nudipinnis	Günther, 1870	37 023001			
4.1	**angel shark**	***Squatina* spp:**		37 024900	Squatinidae	angel sharks	34
		Squatina australis	Regan, 1906	37 024001			
		Squatina sp.		37 024004			
		Squatina sp.		37 024005			
		Squatina tergocellata	McCulloch, 1914	37 024002			
4.5	**guitarfish**	**Rhinobatidae & Rhynchobatidae spp:**		37 990013	Rhinobatidae & Rhynchobatidae	rays	45–46
		Aptychotrema rostrata	(Shaw & Nodder, 1794)	37 027009	Rhinobatidae		
		Rhynchobatus australiae	Whitley, 1939	37 026001	Rhynchobatidae		
		Trygonorrhina sp.		37 027006	Rhinobatidae		
4.5	**skate**	**Rajidae spp:**		37 031000	Rajidae	rays	49
		Raja australis	Macleay, 1884	37 031002			
		Raja sp.		37 031005			
		Raja whitleyi	Iredale, 1938	37 031006			
4.5	**ray**	**Dasyatididae, Gymnuridae, Myliobatididae & Urolophidae spp:**		37 990001	Dasyatididae, Gymnuridae, *etc.*	rays	47–48
		Gymnura australis	(Ramsay & Ogilby, 1886)	37 037001	Gymnuridae		
		Himantura toshi	Whitley, 1939	37 035020	Dasyatididae		
		Myliobatis australis	Macleay, 1881	37 039001	Myliobatididae		
		Urolophus paucimaculatus	Dixon, 1969	37 038004	Urolophidae		
4.3	**ghostshark**	**Chimaeridae spp:**		37 042000	Chimaeridae	ghostsharks	41
		Chimaera sp.		37 042008			
		Chimaera sp.		37 042005			
		Hydrolagus lemures	(Whitley, 1939)	37 042003			
		Hydrolagus ogilbyi	(Waite, 1898)	37 042001			
4.3	elephant fish	*Callorhinchus milii*	(Bory de Saint-Vincent, 1823)	37 043001	Callorhinchidae	ghostsharks	40

No.	Marketing name	Scientific name	Authority	CAAB	Family or other	Group name	p.
		BONY FISHES					
5.14	shortfin eel	*Anguilla australis*	Richardson, 1841	37 056001	Anguillidae	eels	96
5.14	longfin eel	*Anguilla reinhardtii*	Steindachner, 1867	37 056002	Anguillidae	eels	95
5.14	**conger eel**	***Conger verreauxi* & *C. wilsoni***		37 067900	Congridae	eels	94
		Conger verreauxi	Kaup, 1856	37 067007			
		Conger wilsoni	(Bloch & Schneider, 1801)	37 067001			
5.25	sandy sprat	*Hyperlophus vittatus*	(Castelnau, 1875)	37 085005	Clupeidae	herrings	149
5.25	**bony bream**	***Nematalosa erebi* & *N. vlamingbi***		37 085902	Clupeidae	herrings	147
		Nematalosa erebi	(Günther, 1868)	37 085019			
		Nematalosa vlamingbi	(Munro, 1957)	37 085017			
5.25	pilchard	*Sardinops neopilchardus*	(Steindachner, 1879)	37 085002	Clupeidae	herrings	148
5.25	blue sprat	*Spratelloides robustus*	Ogilby, 1897	37 085003	Clupeidae	herrings	146
5.25	—	*Sprattus novaehollandiae*	(Valenciennes, 1847)	37 085004	Clupeidae	herrings	
5.25	anchovy	*Engraulis australis*	(White, 1790)	37 086001	Engraulididae	herrings	145
5.44	**trout**	***Oncorhynchus mykiss* & *Salmo trutta***		37 094900	Salmonidae	salmons	231
		Oncorhynchus mykiss	(Walbaum, 1792)	37 094003			
		Salmo trutta	Linnaeus, 1758	37 094004			
5.44	—	*Oncorhynchus tschawytscha*	(Walbaum, 1792)	37 094005	Salmonidae	salmons	
5.44	Atlantic salmon	*Salmo salar*	Linnaeus, 1758	37 094001	Salmonidae	salmons	230
5.44	**whitebait**	***Lovettia sealii* & *Galaxias* spp:**		37 990002	Aplochitonidae & Galaxiidae	salmons	232
		Galaxias maculatus	(Jenyns, 1842)	37 102006	Galaxiidae		
		Lovettia sealii	(Johnston, 1883)	37 103002	Aplochitonidae		
5.31	milkfish	*Chanos chanos*	(Forsskål, 1775)	37 142001	Chanidae	milkfishes	181
5.9	goldfish	*Carassius auratus*	(Linnaeus, 1758)	37 165001	Cyprinidae	carps	78
5.9	European carp	*Cyprinus carpio*	Linnaeus, 1758	37 165003	Cyprinidae	carps	77

No.	Marketing name	Scientific name	Authority	CAAB	Family or other	Group name	p.
5.9	tench	*Tinca tinca*	(Linnaeus, 1758)	37 165002	Cyprinidae	carps	79
5.10	silver cobbler	*Arius midgleyi*	Kailola & Pierce, 1988	37 188010	Ariidae	catfishes	83
5.10	**catfish**	***Arius* spp:**		37 188901	Ariidae	catfishes	80
		Arius bilineatus	(Valenciennes, 1840)	37 188002			
		Arius graeffei	Kner & Steindachner, 1867	37 188005			
		Arius leptaspis	(Bleeker, 1862)	37 188006			
		Arius mastersi	Ogilby, 1898	37 188007			
		Arius thalassinus	(Rüppell, 1837)	37 188001			
5.10	cobbler	*Cnidoglanis macrocephalus*	(Valenciennes, 1840)	37 192001	Plotosidae	catfishes	81
5.10	freshwater catfish	*Tandanus tandanus*	(Mitchell, 1838)	37 192006	Plotosidae	catfishes	82
5.11	**southern rock cod**	***Lotella & Pseudophycis* spp:**		37 224900	Moridae	cods	87
		Lotella rhacina	(Forster, 1801)	37 224005			
		Pseudophycis bachus	(Forster, 1801)	37 224006			
		Pseudophycis barbata	Günther, 1863	37 224003			
5.11	**ribaldo**	***Mora moro & Lepidion* spp:**		37 224901	Moridae	cods	85
		Lepidion schmidti	Svetovidov, 1936	37 224017			
		Mora moro	(Risso, 1810)	37 224002			
5.11	—	*Lyconus* sp.		37 227003	Macruronidae	cods	
5.11	blue grenadier	*Macruronus novaezelandiae*	(Hector, 1871)	37 227001	Macruronidae	cods	84
5.11	southern hake	*Merluccius australis*	(Hutton, 1872)	37 227002	Merlucciidae	cods	86
5.29	tusk	*Dannevigia tusca*	Whitley, 1941	37 228001	Ophidiidae	lings	162
5.29	pink ling	*Genypterus blacodes*	(Forster, 1801)	37 228002	Ophidiidae	lings	160
5.29	rock ling	*Genypterus tigerinus*	Klunzinger, 1872	37 228008	Ophidiidae	lings	161
5.19	**garfish**	**Hemiramphidae spp:**		37 234000	Hemiramphidae	garfishes	126–127
		Arrhamphus sclerolepis	Günther, 1866	37 234006			
		Hemiramphus far	(Forsskål, 1775)	37 234007			
		Hemiramphus robustus	Günther, 1866	37 234013			

No.	Marketing name	Scientific name	Authority	CAAB	Family or other	Group name	p.
5.19	**garfish**	**Hemiramphidae spp: (cont.)**		37 234000	Hemiramphidae	garfishes	126–127
		Hyporhamphus affinis	(Günther, 1866)	37 234016			
		Hyporhamphus quoyi	(Valenciennes, 1847)	37 234009			
5.19	eastern sea garfish	*Hyporhamphus australis*	(Steindachner, 1866)	37 234014	Hemiramphidae	garfishes	125
5.19	southern garfish	*Hyporhamphus melanochir*	(Valenciennes, 1847)	37 234001	Hemiramphidae	garfishes	130
5.19	river garfish	*Hyporhamphus regularis*	(Günther, 1866)	37 234012	Hemiramphidae	garfishes	129
5.19	**longtom**	**Belonidae spp:**		37 235000	Belonidae	garfishes	128
		Ablennes hians	(Valenciennes, 1846)	37 235001			
		Tylosurus crocodilus	(Péron & Lesueur, 1821)	37 235005			
		Tylosurus gavialoides	(Castelnau, 1873)	37 235002			
5.43	Darwin's roughy	*Gephyroberyx darwinii*	(Johnson, 1866)	37 255004	Trachichthyidae	roughies	228
5.43	orange roughy	*Hoplostethus atlanticus*	Collett, 1889	37 255009	Trachichthyidae	roughies	229
5.43	—	*Hoplostethus gigas*	McCulloch, 1914	37 255005	Trachichthyidae	roughies	
5.41	imperador	*Beryx decadactylus*	Cuvier, 1829	37 258001	Berycidae	redfishes	209
5.41	alfonsino	*Beryx splendens*	Lowe, 1834	37 258002	Berycidae	redfishes	207
5.41	redfish	*Centroberyx affinis*	(Günther, 1859)	37 258003	Berycidae	redfishes	210
5.41	yelloweye redfish	*Centroberyx australis*	Shimizu & Hutchins, 1987	37 258006	Berycidae	redfishes	212
5.41	Bight redfish	*Centroberyx gerrardi*	(Günther, 1887)	37 258004	Berycidae	redfishes	208
5.41	swallowtail	*Centroberyx lineatus*	(Cuvier, 1829)	37 258005	Berycidae	redfishes	211
5.41	—	*Centroberyx* sp.		37 258007	Berycidae	redfishes	
5.12	silver dory	*Cyttus australis*	(Richardson, 1843)	37 264002	Zeidae	dories	91
5.12	king dory	*Cyttus traversi*	Hutton, 1872	37 264001	Zeidae	dories	89
5.12	mirror dory	*Zenopsis nebulosus*	(Temminck & Schlegel, 1845)	37 264003	Zeidae	dories	90

No.	Marketing name	Scientific name	Authority	CAAB	Family or other	Group name	p.
5.12	John dory	*Zeus faber*	Linnaeus, 1758	37 264004	Zeidae	dories	88
5.36	**black oreo**	***Allocyttus niger* & *A. verrucosus***		37 266901	Oreosomatidae	oreos	195
		Allocyttus niger	James, Inada & Nakamura, 1988	37 266005			
		Allocyttus verrucosus	(Gilchrist, 1906)	37 266004			
5.36	spikey oreo	*Neocyttus rhomboidalis*	Gilchrist, 1906	37 266001	Oreosomatidae	oreos	197
5.36	smooth oreo	*Pseudocyttus maculatus*	Gilchrist, 1906	37 266003	Oreosomatidae	oreos	196
5.32	**moonfish**	***Lampris guttatus* & *L. immaculatus***		37 268900	Lampridae	moonfishes	182
		Lampris guttatus	(Brünnich, 1788)	37 268001			
		Lampris immaculatus	Gilchrist, 1904	37 268002			
5.35	**ocean perch**	***Helicolenus barathri* & *H. percoides***		37 287901	Scorpaenidae	ocean perches	194
		Helicolenus barathri	(Hector, 1875)	37 287093			
		Helicolenus percoides	(Richardson, 1842)	37 287001			
5.35	**coral perch**	**Scorpaenidae spp:**		37 287000	Scorpaenidae	ocean perches	192–193
		Neosebastes incisipinnis	Ogilby, 1910	37 287019			
		Neosebastes scorpaenoides	Guichenot, 1867	37 287005			
		Neosebastes thetidis	(Waite, 1899)	37 287006			
		Scorpaena cardinalis	Richardson, 1842	37 287066			
		Trachyscorpia capensis	(Gilchrist & von Bonde, 1924)	37 287046			
5.24	red gurnard	*Chelidonichthys kumu*	(Lesson, 1826)	37 288001	Triglidae	gurnards	144
5.24	**butterfly gurnard**	***Lepidotrigla* spp:**		37 288901	Triglidae	gurnards	142
		Lepidotrigla mulhalli	Macleay, 1884	37 288008			
		Lepidotrigla vanessa	(Richardson, 1839)	37 288003			
5.24	—	*Pterygotrigla andertoni*	Waite, 1910	37 288005	Triglidae	gurnards	
5.24	latchet	*Pterygotrigla polyommata*	(Richardson, 1839)	37 288006	Triglidae	gurnards	143
5.17	deepwater flathead	*Neoplatycephalus conatus*	(Waite & McCulloch, 1915)	37 296002	Platycephalidae	flatheads	112
5.17	**tiger flathead**	***Neoplatycephalus aurimaculatus* & *N. richardsoni***		37 296903	Platycephalidae	flatheads	119
		Neoplatycephalus aurimaculatus	(Knapp, 1987)	37 296035			
		Neoplatycephalus richardsoni	(Castelnau, 1872)	37 296001			

No.	Marketing name	Scientific name	Authority	CAAB	Family or other	Group name	p.
5.17	northern sand flathead	*Platycephalus arenarius*	Ramsay & Ogilby, 1886	37 296021	Platycephalidae	flatheads	115
5.17	sand flathead	*Platycephalus bassensis*	Cuvier, 1829	37 296003	Platycephalidae	flatheads	117
5.17	blue-spotted flathead	*Platycephalus caeruleopunctatus*	McCulloch, 1922	37 296007	Platycephalidae	flatheads	111
5.17	dusky flathead	*Platycephalus fuscus*	Cuvier, 1829	37 296004	Platycephalidae	flatheads	113
5.17	bartail flathead	*Platycephalus indicus*	(Linnaeus, 1758)	37 296033	Platycephalidae	flatheads	110
5.17	rock flathead	*Platycephalus laevigatus*	Cuvier, 1829	37 296006	Platycephalidae	flatheads	116
5.17	southern flathead	*Platycephalus speculator*	Klunzinger, 1872	37 296037	Platycephalidae	flatheads	118
5.17	**flathead**	**Platycephalidae spp:**		37 296000	Platycephalidae	flatheads	114
		Papilloculiceps nematophthalmus	(Günther, 1860)	37 296023			
		Platycephalus endrachtensis	Quoy & Gaimard, 1825	37 296020			
		Platycephalus longispinis	Macleay, 1884	37 296036			
		Platycephalus marmoratus	Stead, 1908	37 296038			
5.18	barramundi	*Lates calcarifer*	(Bloch, 1790)	37 310006	Centropomidae	freshwater perches	121
5.18	—	*Psammoperca waigiensis*	(Cuvier, 1828)	37 310001	Centropomidae	freshwater perches	
5.18	**Murray cod**	***Maccullochella* spp:**		37 311903	Percichthyidae	freshwater perches	123
		Maccullochella peelii	(Mitchell, 1838)	37 311076			
5.18	golden perch	*Macquaria ambigua*	(Richardson, 1845)	37 311075	Percichthyidae	freshwater perches	122
5.18	—	*Macquaria colonorum*	(Günther, 1863)	37 311033	Percichthyidae	freshwater perches	
5.18	Australian bass	*Macquaria novemaculeata*	(Steindachner, 1866)	37 311034	Percichthyidae	freshwater perches	120
5.42	**rockcod**	***Aethaloperca, Anyperodon & Epinephelus* spp:**		37 311901	Serranidae	rockcods	224–225
		Aethaloperca rogaa	(Forsskål, 1775)	37 311134			
		Epinephelus heniochus	Fowler, 1904	37 311019			
		Epinephelus lanceolatus	(Bloch, 1790)	37 311061			
		Epinephelus merra	Bloch, 1793	37 311063			
		Epinephelus morrhua	(Valenciennes, 1833)	37 311151			
		Epinephelus ongus	(Bloch, 1790)	37 311069			
		Epinephelus tauvina	(Forsskål, 1775)	37 311057			

No.	Marketing name	Scientific name	Authority	CAAB	Family or other	Group name	p.
5.42	longfin perch	*Caprodon longimanus*	(Günther, 1859)	37 311095	Serranidae	rockcods	222
5.42	**coral cod**	***Cephalopholis* spp:**		37 311904	Serranidae	rockcods	216
		Cephalopholis cyanostigma	(Valenciennes, 1828)	37 311136			
		Cephalopholis sonnerati	(Valenciennes, 1828)	37 311045			
5.42	barramundi cod	*Cromileptes altivelis*	(Valenciennes, 1828)	37 311044	Serranidae	rockcods	214
5.42	yellow-spotted rockcod	*Epinephelus areolatus*	(Forsskål, 1775)	37 311009	Serranidae	rockcods	227
5.42	estuary rockcod	*Epinephelus coioides*	(Hamilton, 1822)	37 311007	Serranidae	rockcods	219
5.42	**bar rockcod**	***Epinephelus ergastularius* & *E. septemfasciatus***		37 311910	Serranidae	rockcods	213
		Epinephelus ergastularius	Whitley, 1930	37 311147			
		Epinephelus septemfasciatus	(Thunberg, 1793)	37 311060			
5.42	blacktip rockcod	*Epinephelus fasciatus*	(Forsskål, 1775)	37 311014	Serranidae	rockcods	215
5.42	white-spotted rockcod	*Epinephelus multinotatus*	(Peters, 1877)	37 311010	Serranidae	rockcods	226
5.42	honeycomb rockcod	*Epinephelus quoyanus*	(Valenciennes, 1830)	37 311040	Serranidae	rockcods	221
5.42	Maori rockcod	*Epinephelus undulatostriatus*	(Peters, 1867)	37 311086	Serranidae	rockcods	223
5.42	**coral trout**	***Plectropomus* & *Variola* spp:**		37 311905	Serranidae	rockcods	217–218
		Plectropomus areolatus	(Rüppell, 1830)	37 311081			
		Plectropomus leopardus	(Lacépède, 1801)	37 311078			
		Plectropomus maculatus	(Bloch, 1790)	37 311012			
		Variola louti	(Forsskål, 1775)	37 311166			
5.42	**hapuku**	***Polyprion americanus* & *P. oxygeneios***		37 311902	Serranidae	rockcods	220
		Polyprion americanus	(Bloch & Schneider, 1801)	37 311170			
		Polyprion oxygeneios	(Forster, 1801)	37 311006			
5.37	West Australian dhufish	*Glaucosoma hebraicum*	Richardson, 1845	37 320004	Glaucosomatidae	pearl perches	200
5.37	**pearl perch**	***Glaucosoma* spp:**		37 320901	Glaucosomatidae	pearl perches	198–199
		Glaucosoma buergeri	Richardson, 1845	37 320001			
		Glaucosoma magnificum	(Ogilby, 1915)	37 320002			
		Glaucosoma scapulare	Macleay, 1881	37 320003			

No.	Marketing name	Scientific name	Authority	CAAB	Family or other	Group name	p.
5.23	silver perch	*Bidyanus bidyanus*	(Mitchell, 1838)	37 321008	Terapontidae	grunters	140
5.23	**striped perch**	**Terapontidae spp:**		37 321000	Terapontidae	grunters	141
		Hephaestus jenkinsi	(Whitley, 1945)	37 321015			
		Pelates octolineatus	(Jenyns, 1840)	37 321020			
5.3	**red bullseye**	***Priacanthus* spp:**		37 326901	Priacanthidae	bigeyes	60
		Priacanthus hamrur	(Forsskål, 1775)	37 326005			
		Priacanthus macracanthus	Cuvier, 1829	37 326001			
		Priacanthus tayenus	Richardson, 1846	37 326003			
5.8	cardinal fish	*Epigonus telescopus*	(Risso, 1810)	37 327035	Apogonidae	cardinal fishes	76
—		*Dinolestes lewini*	(Griffith, 1834)	37 327002	Dinolestidae	—	
5.18	redfin	*Perca fluviatilis*	Linnaeus, 1758	37 329001	Percidae	freshwater perches	124
5.53	**whiting**	**Sillaginidae spp:**		37 330000	Sillaginidae	whitings	284
		Sillago analis	Whitley, 1943	37 330003			
		Sillago burrus	Richardson, 1842	37 330004			
		Sillago ingenuua	McKay, 1985	37 330009			
		Sillago sihama	(Forsskål, 1775)	37 330006			
5.53	King George whiting	*Sillaginodes punctata*	(Cuvier, 1829)	37 330001	Sillaginidae	whitings	279
5.53	sand whiting	*Sillago ciliata*	Cuvier, 1829	37 330010	Sillaginidae	whitings	280
5.53	**school whiting**	***Sillago bassensis, S. flindersi & S. robusta***		37 330901	Sillaginidae	whitings	281–282
		Sillago bassensis	Cuvier, 1829	37 330002			
		Sillago flindersi	McKay, 1985	37 330014			
		Sillago robusta	Stead, 1908	37 330005			
5.53	trumpeter whiting	*Sillago maculata*	Quoy & Gaimard, 1824	37 330015	Sillaginidae	whitings	283
5.53	yellowfin whiting	*Sillago schomburgkii*	Peters, 1865	37 330012	Sillaginidae	whitings	285
—	**tilefish**	**Malacanthidae spp:**		37 331000	Malacanthidae	tilefishes	
		Branchiostegus sp.		37 331011			
5.51	tailor	*Pomatomus saltatrix*	(Linnaeus, 1766)	37 334002	Pomatomidae	trevallies	272

No.	Marketing name	Scientific name	Authority	CAAB	Family or other	Group name	p.
5.51	black kingfish	*Rachycentron canadum*	(Linnaeus, 1766)	37 335001	Rachycentridae	trevallies	258
5.51	—	*Remora remora*	(Linnaeus, 1758)	37 336002	Echeneidae	trevallies	
5.51	**diamond trevally**	***Alectis* spp:**		37 337909	Carangidae	trevallies	264
		Alectis ciliaris	(Bloch, 1787)	37 337018			
5.51	**trevally**	**Caranginae spp:**		37 337000	Carangidae	trevallies	273
		Atule mate	(Cuvier, 1833)	37 337024			
		Carangoides caeruleopinnatus	(Rüppell, 1830)	37 337021			
		Carangoides gymnostethus	(Cuvier, 1833)	37 337022			
		Caranx papuensis	Alleyne & Macleay, 1877	37 337064			
5.51	turrum	*Carangoides fulvoguttatus*	(Forsskål, 1775)	37 337037	Carangidae	trevallies	274
5.51	blue-spotted trevally	*Caranx bucculentus*	Alleyne & Macleay, 1877	37 337016	Carangidae	trevallies	262
5.51	giant trevally	*Caranx ignobilis*	(Forsskål, 1775)	37 337027	Carangidae	trevallies	265
5.51	black trevally	*Caranx lugubris*	Poey, 1860	37 337053	Carangidae	trevallies	260
5.51	bluefin trevally	*Caranx melampygus*	Cuvier, 1833	37 337050	Carangidae	trevallies	261
5.51	bigeye trevally	*Caranx sexfasciatus*	Quoy & Gaimard, 1825	37 337039	Carangidae	trevallies	257
5.51	golden trevally	*Gnathanodon speciosus*	(Forsskål, 1775)	37 337012	Carangidae	trevallies	266
5.51	black pomfret	*Parastromateus niger*	(Bloch, 1795)	37 337072	Carangidae	trevallies	259
5.51	**silver trevally**	***Pseudocaranx dentex* & *P. wrighti***		37 337910	Carangidae	trevallies	271
		Pseudocaranx dentex	(Bloch & Schneider, 1801)	37 337062			
		Pseudocaranx wrighti	(Whitley, 1931)	37 337063			
5.51	**queenfish**	***Scomberoides* spp:**		37 337905	Carangidae	trevallies	269
		Scomberoides commersonnianus	Lacépède, 1801	37 337032			
		Scomberoides tol	(Cuvier, 1832)	37 337044			
5.51	**samson fish**	***Seriola dumerili* & *S. hippos***		37 337911	Carangidae	trevallies	270
		Seriola dumerili	(Risso, 1810)	37 337025			
		Seriola hippos	Günther, 1876	37 337007			

No.	Marketing name	Scientific name	Authority	CAAB	Family or other	Group name	p.
5.51	yellowtail kingfish	*Seriola lalandi*	Valenciennes, 1833	37 337006	Carangidae	trevallies	275
5.51	**dart**	***Trachinotus* spp:**		37 337904	Carangidae	trevallies	263
		Trachinotus anak	Ogilby, 1909	37 337073			
		Trachinotus blochii	(Lacépède, 1801)	37 337075			
		Trachinotus botla	(Shaw, 1803)	37 337066			
5.51	**jack mackerel**	***Trachurus declivis* & *T. murphyi***		37 337912	Carangidae	trevallies	267
		Trachurus declivis	(Jenyns, 1841)	37 337002			
		Trachurus murphyi	Nichols, 1920	37 337077			
5.51	yellowtail scad	*Trachurus novaezelandiae*	Richardson, 1843	37 337003	Carangidae	trevallies	276
5.51	**mahi mahi**	***Coryphaena* spp:**		37 338901	Coryphaenidae	trevallies	268
		Coryphaena equiselis	Linnaeus, 1758	37 338002			
		Coryphaena hippurus	Linnaeus, 1758	37 338001			
5.39	**Ray's bream**	**Bramidae spp:**		37 342901	Bramidae	pomfrets	204–205
		Brama brama	(Bonnaterre, 1788)	37 342001			
		Taractichthys longipinnis	(Lowe, 1843)	37 342003			
		Xenobrama microlepis	Yatsu & Nakamura, 1989	37 342002			
5.1	Australian herring	*Arripis georgianus*	(Valenciennes, 1831)	37 344001	Arripidae	Australian salmons	56
5.1	**Australian salmon**	***Arripis trutta* & *A. truttaceus***		37 344900	Arripidae	Australian salmons	57
		Arripis trutta	(Bloch & Schneider, 1801)	37 344002			
		Arripis truttaceus	(Cuvier, 1829)	37 344004			
5.6	**redbait**	***Emmelichthys* spp:**		37 345901	Emmelichthyidae	bonnetmouths	68
		Emmelichthys nitidus	Richardson, 1845	37 345001			
5.6	**rubyfish**	***Plagiogeneion* spp:**		37 345900	Emmelichthyidae	bonnetmouths	69
		Plagiogeneion macrolepis	McCulloch, 1914	37 345002			
5.45	green jobfish	*Aprion virescens*	Valenciennes, 1830	37 346027	Lutjanidae	seaperches	235
5.45	**ruby snapper**	***Etelis* spp:**		37 346914	Lutjanidae	seaperches	241
		Etelis coruscans	Valenciennes, 1862	37 346038			
5.45	hussar	*Lutjanus adetii*	(Castelnau, 1873)	37 346033	Lutjanidae	seaperches	236

No.	Marketing name	Scientific name	Authority	CAAB	Family or other	Group name	p.
5.45	mangrove jack	*Lutjanus argentimaculatus*	(Forsskål, 1775)	37 346015	Lutjanidae	seaperches	238
5.45	stripey seaperch	*Lutjanus carponotatus*	(Richardson, 1842)	37 346011	Lutjanidae	seaperches	245
5.45	crimson seaperch	*Lutjanus erythropterus*	Bloch, 1790	37 346005	Lutjanidae	seaperches	233
5.45	saddletail seaperch	*Lutjanus malabaricus*	(Bloch & Schneider, 1801)	37 346007	Lutjanidae	seaperches	242
5.45	**Moses seaperch**	***Lutjanus russelli* & *L.* sp.**		37 346915	Lutjanidae	seaperches	239
		Lutjanus russelli	(Bleeker, 1849)	37 346065			
		Lutjanus sp.		37 346012			
5.45	red emperor	*Lutjanus sebae*	(Cuvier, 1829)	37 346004	Lutjanidae	seaperches	240
5.45	**seaperch**	***Lutjanus* spp:**		37 346905	Lutjanidae	seaperches	243–244
		Lutjanus bohar	(Forsskål, 1775)	37 346029			
		Lutjanus fulviflamma	(Forsskål, 1775)	37 346034			
		Lutjanus johnii	(Bloch, 1792)	37 346030			
		Lutjanus lemniscatus	(Valenciennes, 1828)	37 346010			
		Lutjanus lutjanus	Bloch, 1790	37 346008			
		Lutjanus quinquelineatus	(Bloch, 1790)	37 346006			
		Lutjanus rivulatus	(Cuvier, 1828)	37 346016			
		Lutjanus vitta	(Quoy & Gaimard, 1824)	37 346003			
5.45	king snapper	*Pristipomoides filamentosus*	(Valenciennes, 1830)	37 346032	Lutjanidae	seaperches	237
5.45	**goldband snapper**	***Pristipomoides multidens* & *P. typus***		37 346901	Lutjanidae	seaperches	234
		Pristipomoides multidens	(Day, 1870)	37 346002			
		Pristipomoides typus	Bleeker, 1852	37 346019			
5.48	**threadfin bream**	**Nemipteridae spp:**		37 347000	Nemipteridae	threadfin breams	248–249
		Nemipterus furcosus	(Valenciennes, 1830)	37 347005			
		Nemipterus hexodon	(Quoy & Gaimard, 1824)	37 347014			
		Nemipterus peronii	(Valenciennes, 1830)	37 347003			
		Scolopsis monogramma	(Kuhl & van Hasselt, 1830)	37 347006			
		Scolopsis taeniopterus	(Kuhl & van Hasselt, 1830)	37 347008			
5.46	**silver biddy**	**Gerreidae spp:**		37 349000	Gerreidae	silver biddies	246
		Gerres subfasciatus	Cuvier, 1830	37 349005			

No.	Marketing name	Scientific name	Authority	CAAB	Family or other	Group name	p.
5.22	**sweetlip bream**	***Haemulidae* except *Pomadasys* spp:**		37 350904	Haemulidae	grunter breams	138–139
		Diagramma labiosum	Macleay, 1883	37 350003			
		Plectorhinchus flavomaculatus	(Ehrenberg, 1830)	37 350007			
		Plectorhinchus polytaenia	(Bleeker, 1852)	37 350005			
5.22	**grunter bream**	***Pomadasys* spp:**		37 350902	Haemulidae	grunter breams	137
		Pomadasys kaakan	(Cuvier, 1830)	37 350011			
5.15	**seabream**	***Gymnocranius* & *Monotaxis* spp:**		37 351903	Lethrinidae	emperors	102
		Gymnocranius elongatus	Senta, 1973	37 351010			
		Gymnocranius grandoculis	(Valenciennes, 1830)	37 351005			
		Gymnocranius sp.		37 351023			
5.15	grass emperor	*Lethrinus laticaudis*	Alleyne & Macleay, 1877	37 351006	Lethrinidae	emperors	98
5.15	redspot emperor	*Lethrinus lentjan*	(Lacépède, 1802)	37 351007	Lethrinidae	emperors	100
5.15	redthroat emperor	*Lethrinus miniatus*	(Bloch & Schneider, 1801)	37 351009	Lethrinidae	emperors	101
5.15	**spangled emperor**	***Lethrinus nebulosus* & *L.* sp.**		37 351904	Lethrinidae	emperors	103
		Lethrinus nebulosus	(Forsskål, 1775)	37 351008			
		Lethrinus sp.		37 351001			
5.15	longnose emperor	*Lethrinus olivaceus*	Valenciennes, 1830	37 351004	Lethrinidae	emperors	99
5.15	**emperor**	***Lethrinus* spp:**		37 351902	Lethrinidae	emperors	97
		Lethrinus atkinsoni	Seale, 1910	37 351013			
		Lethrinus rubrioperculatus	Sato, 1978	37 351012			
5.7	yellowfin bream	*Acanthopagrus australis*	(Günther, 1859)	37 353004	Sparidae	breams	75
5.7	pikey bream	*Acanthopagrus berda*	(Forsskål, 1775)	37 353011	Sparidae	breams	72
5.7	black bream	*Acanthopagrus butcheri*	(Munro, 1949)	37 353003	Sparidae	breams	70
5.7	frypan bream	*Argyrops spinifer*	(Forsskål, 1775)	37 353006	Sparidae	breams	71
5.7	snapper	*Pagrus auratus*	(Bloch & Schneider, 1801)	37 353001	Sparidae	breams	73
5.7	tarwhine	*Rhabdosargus sarba*	(Forsskål, 1775)	37 353013	Sparidae	breams	74

No.	Marketing name	Scientific name	Authority	CAAB	Family or other	Group name	p.
5.27	mulloway	*Argyrosomus hololepidotus*	(Lacépède, 1802)	37 354001	Sciaenidae	jewfishes	153
5.27	teraglin	*Atractoscion aequidens*	(Cuvier, 1830)	37 354020	Sciaenidae	jewfishes	154
5.27	black jewfish	*Protonibea diacanthus*	(Lacépède, 1802)	37 354003	Sciaenidae	jewfishes	151
5.27	**jewfish**	**Sciaenidae spp:**		37 354000	Sciaenidae	jewfishes	152
		Johnius borneensis	(Bleeker, 1851)	37 354007			
		Otolithes ruber	(Bloch & Schneider, 1801)	37 354006			
5.21	**red mullet**	**Mullidae spp:**		37 355000	Mullidae	goatfishes	135–136
		Parupeneus heptacanthus	(Lacépède, 1801)	37 355004			
		Parupeneus indicus	(Shaw, 1803)	37 355005			
		Upeneichthys vlamingii	(Cuvier, 1829)	37 355029			
		Upeneus tragula	Richardson, 1846	37 355014			
5.13	luderick	*Girella tricuspidata*	(Quoy & Gaimard, 1824)	37 361007	Kyphosidae	drummers	92
5.13	—	*Girella zebra*	(Richardson, 1846)	37 361008	Kyphosidae	drummers	
5.13	—	*Atypichthys strigatus*	(Günther, 1860)	37 361010	Scorpididae	drummers	
5.13	**sweep**	**Scorpididae spp:**		37 361000	Scorpididae	drummers	93
		Scorpis aequipinnis	Richardson, 1848	37 361004			
		Scorpis lineolatus	Kner, 1865	37 361009			
5.2	**batfish**	**Ephippidae spp:**		37 362000	Ephippidae	batfishes	58
		Drepane punctata	(Linnaeus, 1758)	37 362005			
		Platax batavianus	Cuvier, 1831	37 362002			
5.2	**butterfish**	***Scatophagus* spp:**		37 363900	Scatophagidae	batfishes	59
		Scatophagus argus	(Linnaeus, 1766)	37 363002			
		Scatophagus multifasciatus	Richardson, 1846	37 363001			
		Scatophagus sp.		37 363003			
5.5	—	*Parazanclistius hutchinsi*	Hardy, 1983	37 367010	Pentacerotidae	boarfishes	
5.5	**giant boarfish**	***Paristiopterus gallipavo* & *P. labiosus***		37 367901	Pentacerotidae	boarfishes	67
		Paristiopterus gallipavo	Whitley, 1944	37 367001			
		Paristiopterus labiosus	(Günther, 1871)	37 367002			

No.	Marketing name	Scientific name	Authority	CAAB	Family or other	Group name	p.
5.5	boarfish	*Pentaceropsis recurvirostris*	(Richardson, 1845)	37 367003	Pentacerotidae	boarfishes	65
5.5	bigspine boarfish	*Pentaceros decacanthus*	Günther, 1859	37 367004	Pentacerotidae	boarfishes	63
5.5	—	*Pseudopentaceros richardsoni*	(Smith, 1844)	37 367009	Pentacerotidae	boarfishes	
5.5	blackspot boarfish	*Zanclistius elevatus*	(Ramsay & Ogilby, 1888)	37 367005	Pentacerotidae	boarfishes	64
5.5	conway	*Oplegnathus woodwardi*	(Waite, 1900)	37 369002	Oplegnathidae	boarfishes	66
5.33	red morwong	*Cheilodactylus fuscus*	Castelnau, 1879	37 377009	Cheilodactylidae	morwongs	187
5.33	banded morwong	*Cheilodactylus spectabilis*	(Hutton, 1872)	37 377006	Cheilodactylidae	morwongs	183
5.33	grey morwong	*Nemadactylus douglasii*	(Hector, 1875)	37 377002	Cheilodactylidae	morwongs	185
5.33	**morwong**	***Nemadactylus macropterus* & *N.* sp.**		37 377901	Cheilodactylidae	morwongs	186
		Nemadactylus macropterus	(Bloch & Schneider, 1801)	37 377003			
		Nemadactylus sp.		37 377014			
5.33	blue morwong	*Nemadactylus valenciennesi*	(Whitley, 1937)	37 377004	Cheilodactylidae	morwongs	184
5.52	**trumpeter**	***Latridopsis* spp:**		37 378900	Latrididae	trumpeters	278
		Latridopsis ciliaris	(Bloch & Schneider, 1801)	37 378003			
		Latridopsis forsteri	(Castelnau, 1872)	37 378002			
5.52	striped trumpeter	*Latris lineata*	(Schneider, 1801)	37 378001	Latrididae	trumpeters	277
5.34	diamondscale mullet	*Liza vaigiensis*	(Quoy & Gaimard, 1824)	37 381008	Mugilidae	mullets	188
5.34	sea mullet	*Mugil cephalus*	Linnaeus, 1758	37 381002	Mugilidae	mullets	191
5.34	**mullet**	**Mugilidae spp:**		37 381000	Mugilidae	mullets	189–190
		Aldrichetta forsteri	(Valenciennes, 1836)	37 381001			
		Liza argentea	(Quoy & Gaimard, 1825)	37 381004			
		Liza sp.		—			
		Mugil broussonnetii	Valenciennes, 1836	37 381018			
		Valamugil sebeli	(Forsskål, 1775)	37 381017			
5.38	pike	*Sphyraena novaehollandiae*	Günther, 1860	37 382002	Sphyraenidae	pikes	201

No.	Marketing name	Scientific name	Authority	CAAB	Family or other	Group name	p.
5.38	**striped seapike**	***Sphyraena* spp:**		37 382901	Sphyraenidae	pikes	202–203
		Sphyraena barracuda	(Walbaum, 1792)	37 382008			
		Sphyraena forsteri	Cuvier, 1829	37 382005			
		Sphyraena jello	Cuvier, 1829	37 382004			
		Sphyraena obtusata	Cuvier, 1829	37 382001			
		Sphyraena sp.		—			
5.49	blue threadfin	*Eleutheronema tetradactylum*	(Shaw, 1804)	37 383004	Polynemidae	threadfin salmons	250
5.49	king threadfin	*Polydactylus sheridani*	(Macleay, 1884)	37 383005	Polynemidae	threadfin salmons	251
5.54	**pigfish**	***Bodianus* spp:**		37 384904	Labridae	wrasses	291
		Bodianus perditio	(Quoy & Gaimard, 1834)	37 384007			
		Bodianus unimaculatus	Günther, 1862	37 384061			
5.54	**Maori wrasse**	***Cheilinus* & *Oxycheilinus* spp:**		37 384903	Labridae	wrasses	288–289
		Cheilinus chlorourus	(Bloch, 1791)	37 384064			
		Cheilinus undulatus	Rüppell, 1835	37 384038			
		Oxycheilinus digrammus	(Lacépède, 1801)	37 384065			
5.54	baldchin groper	*Choerodon rubescens*	(Günther, 1862)	37 384039	Labridae	wrasses	286
5.54	**tuskfish**	***Choerodon* spp:**		37 384902	Labridae	wrasses	292
		Choerodon cauteroma	Gomon & Allen, 1987	37 384005			
		Choerodon cephalotes	(Castelnau, 1875)	37 384004			
		Choerodon schoenleinii	(Valenciennes, 1839)	37 384010			
		Choerodon venustus	(de Vis, 1884)	37 384042			
5.54	**wrasse**	**Labridae spp:**		37 384000	Labridae	wrasses	293–294
		Achoerodus gouldii	(Richardson, 1843	37 384002			
		Achoerodus viridis	(Steindachner, 1866)	37 384043			
		Anampses lennardi	Scott, 1959	37 384016			
		Notolabrus gymnogenis	(Günther, 1862)	37 384041			
		Notolabrus tetricus	(Richardson, 1840)	37 384003			
5.54	grass whiting	*Haletta semifasciata*	(Valenciennes, 1840)	37 385009	Odacidae	wrasses	287
5.54	**parrotfish**	**Scaridae spp:**		37 386000	Scaridae	wrasses	290
		Scarus ghobban	Forsskål, 1775	37 386001			

No.	Marketing name	Scientific name	Authority	CAAB	Family or other	Group name	p.
5.47	**stargazer**	**Uranoscopidae spp:**		37 400000	Uranoscopidae	stargazers	247
		Kathetostoma canaster	Gomon & Last, 1987	37 400018			
5.26	Patagonian toothfish	*Dissostichus eleginoides*	Smitt, 1898	37 404792	Nototheniidae	icefishes	150
5.40	**rabbitfish**	***Siganus* spp:**		37 438902	Siganidae	rabbitfishes	206
		Siganus nebulosus	(Quoy & Gaimard, 1824)	37 438001			
5.20	—	*Rexea antefurcata*	Parin, 1989	37 439009	Gempylidae	gemfishes	
5.20	gemfish	*Rexea solandri*	(Cuvier, 1832)	37 439002	Gempylidae	gemfishes	133
5.20	**escolar**	***Lepidocybium flavobrunneum* & *Ruvettus pretiosus***		37 439901	Gempylidae	gemfishes	132
		Lepidocybium flavobrunneum	(Smith, 1849)	37 439008			
		Ruvettus pretiosus	Cocco, 1829	37 439003			
5.20	barracouta	*Thyrsites atun*	(Euphrasen, 1791)	37 439001	Gempylidae	gemfishes	131
5.20	ribbonfish	*Lepidopus caudatus*	(Euphrasen, 1788)	37 440002	Trichiuridae	gemfishes	134
5.20	—	*Trichiurus lepturus*	Linnaeus, 1758	37 440004	Trichiuridae	gemfishes	
5.30	slender tuna	*Allothunnus fallai*	Serventy, 1948	37 441021	Scombridae	mackerels	175
5.30	frigate mackerel	*Auxis thazard*	(Lacépède, 1800)	37 441009	Scombridae	mackerels	167
5.30	mackerel tuna	*Euthynnus affinis*	(Cantor, 1849)	37 441010	Scombridae	mackerels	171
5.30	—	*Gasterochisma melampus*	Richardson, 1845	37 441019	Scombridae	mackerels	
5.30	shark mackerel	*Grammatorcynus bicarinatus*	(Quoy & Gaimard, 1825)	37 441025	Scombridae	mackerels	173
5.30	skipjack tuna	*Katsuwonus pelamis*	(Linnaeus, 1758)	37 441003	Scombridae	mackerels	174
5.30	**bonito**	***Sarda* spp:**		37 441910	Scombridae	mackerels	166
		Sarda australis	(Macleay, 1880)	37 441020			
		Sarda orientalis	(Temminck & Schlegel, 1844)	37 441006			
5.30	blue mackerel	*Scomber australasicus*	Cuvier, 1832	37 441001	Scombridae	mackerels	165

No.	Marketing name	Scientific name	Authority	CAAB	Family or other	Group name	p.
5.30	**mackerel**	**Scomberomorini & Scombrini spp:**		37 441911	Scombridae	mackerels	170
		Acanthocybium solandri	(Cuvier, 1831)	37 441024			
		Grammatorcynus bilineatus	(Rüppell, 1836)	37 441028			
		Rastrelliger kanagurta	(Cuvier, 1816)	37 441012			
5.30	Spanish mackerel	*Scomberomorus commerson*	(Lacépède, 1800)	37 441007	Scombridae	mackerels	177
5.30	spotted mackerel	*Scomberomorus munroi*	Collette & Russo, 1980	37 441015	Scombridae	mackerels	178
5.30	school mackerel	*Scomberomorus queenslandicus*	Munro, 1943	37 441014	Scombridae	mackerels	172
5.30	grey mackerel	*Scomberomorus semifasciatus*	(Macleay, 1884)	37 441018	Scombridae	mackerels	168
5.30	**tuna**	**Sardini & Thunnini spp:**		37 441912	Scombridae	mackerels	179
		Cybiosarda elegans	(Whitley, 1935)	37 441008			
		Thunnus thynnus	(Linnaeus, 1758)	37 441026			
5.30	albacore	*Thunnus alalunga*	(Bonnaterre, 1788)	37 441005	Scombridae	mackerels	163
5.30	yellowfin tuna	*Thunnus albacares*	(Bonnaterre, 1788)	37 441002	Scombridae	mackerels	180
5.30	southern bluefin tuna	*Thunnus maccoyii*	(Castelnau, 1872)	37 441004	Scombridae	mackerels	176
5.30	bigeye tuna	*Thunnus obesus*	(Lowe, 1839)	37 441011	Scombridae	mackerels	164
5.30	longtail tuna	*Thunnus tonggol*	(Bleeker, 1851)	37 441013	Scombridae	mackerels	169
5.4	swordfish	*Xiphias gladius*	Linnaeus, 1758	37 442001	Xiphiidae	billfishes	62
5.4	**marlin**	**Istiophoridae spp:**		37 444000	Istiophoridae	billfishes	61
		Makaira indica	(Cuvier, 1832)	37 444006			
		Makaira mazara	(Jordan & Snyder, 1901)	37 444003			
		Tetrapturus angustirostris	Tanaka, 1915	37 444007			
		Tetrapturus audax	(Phillipi, 1887)	37 444002			
5.50	**blue-eye trevalla**	***Hyperoglyphe antarctica & Schedophilus labyrinthica***		47 445902	Centrolophidae	trevallas	252
		Hyperoglyphe antarctica	(Carmichael, 1818)	37 445001			
		Schedophilus labyrinthica	McAllister & Randall, 1975	37 445014			
5.50	blue warehou	*Seriolella brama*	(Günther, 1860)	37 445005	Centrolophidae	trevallas	253

No.	Marketing name	Scientific name	Authority	CAAB	Family or other	Group name	p.
5.50	white warehou	*Seriolella caerulea*	Guichenot, 1848	37 445011	Centrolophidae	trevallas	256
5.50	silver warehou	*Seriolella punctata*	(Forster, 1801)	37 445006	Centrolophidae	trevallas	255
5.50	**rudderfish**	***Centrolophus, Schedophilus & Tubbia* spp:**		37 445903	Centrolophidae	trevallas	254
		Centrolophus niger	(Gmelin, 1789)	37 445004			
		Tubbia tasmanica	Whitley, 1943	37 445002			
5.16	Australian halibut	*Psettodes erumei*	(Bloch & Schneider, 1801)	37 457001	Psettodidae	flatfishes	104
5.16	**flounder**	**Bothidae & Pleuronectidae spp:**		37 990014	Bothidae & Pleuronectidae	flatfishes	106–107
		Neoachiropsetta milfordi	(Penrith, 1965)	37 460052	Bothidae		
		Pseudorhombus argus	Weber, 1913	37 460038	Bothidae		
		Pseudorhombus arsius	(Hamilton, 1822)	37 460009	Bothidae		
		Pseudorhombus jenynsii	(Bleeker, 1855)	37 460002	Bothidae		
		Pseudorhombus spinosus	McCulloch, 1914	37 460011	Bothidae		
5.16	**bay flounder**	***Ammotretis* spp:**		37 461900	Pleuronectidae	flatfishes	105
		Ammotretis lituratus	(Richardson, 1843)	37 461004			
		Ammotretis rostratus	Günther, 1862	37 461001			
5.16	greenback flounder	*Rhombosolea tapirina*	Günther, 1862	37 461003	Pleuronectidae	flatfishes	108
5.16	**sole**	**Cynoglossidae & Soleidae spp:**		37 990015	Cynoglossidae & Soleidae	flatfishes	109
		Paraplagusia bilineata	(Bloch, 1787)	37 463001	Cynoglossidae		
		Synaptura nigra	Macleay, 1881	37 462017	Soleidae		
5.28	**reef leatherjacket**	***Meuschenia* spp:**		37 465902	Monacanthidae	leatherjackets	158
		Meuschenia freycineti	(Quoy & Gaimard, 1824)	37 465036			
5.28	**leatherjacket**	**Monacanthidae spp:**		37 465000	Monacanthidae	leatherjackets	155–156
		Aluterus monoceros	(Linnaeus, 1758)	37 465022			
		Eubalichthys mosaicus	(Ramsay & Ogilby, 1886)	37 465003			
		Pseudomonacanthus peroni	(Hollard, 1854)	37 465020			
		Thamnaconus degeni	(Regan, 1903)	37 465037			
5.28	ocean jacket	*Nelusetta ayraudi*	(Quoy & Gaimard, 1824)	37 465006	Monacanthidae	leatherjackets	157
5.28	velvet leatherjacket	*Parika scaber*	(Schneider, 1801)	37 465005	Monacanthidae	leatherjackets	159

No.	Marketing name	Scientific name	Authority	CAAB	Family or other	Group name	p.
		CRUSTACEANS					
6.3	**prawn**	**Caridea & Penaeoidea spp:**		00 701926	Caridea & Penaeoidea	prawns	318
		Aristaeopsis edwardsiana	(Johnson, 1867)	00 701003	Aristaeidae		
		Metapenaeopsis rosea	Racek & Dall, 1965	00 701313	Penaeidae		
6.3	**banana prawn**	***Fenneropenaeus indicus* & *F. merguiensis***		00 701925	Penaeidae	prawns	312
		Fenneropenaeus indicus	(Milne Edwards, 1837)	00 701306			
		Fenneropenaeus merguiensis	(de Man, 1888)	00 701901			
6.3	royal red prawn	*Haliporoides sibogae*	(de Man, 1907)	00 701004	Solenceridae	prawns	320
6.3	**freshwater prawn**	***Macrobrachium* spp:**		00 701073	Palaemonidae	prawns	316
		Macrobrachium rosenbergii	(de Man, 1879)	00 701113			
6.3	**tiger prawn**	***Marsupenaeus japonicus*, *Penaeus esculentus* & *P. semisulcatus***		00 701900	Penaeidae	prawns	323
		Marsupenaeus japonicus	(Bate, 1888)	00 701305			
		Penaeus esculentus	Haswell, 1879	00 701308			
		Penaeus semisulcatus	de Haan, 1844	00 701307			
6.3	**king prawn**	***Melicertus latisulcatus* & *M. plebejus***		00 701919	Penaeidae	prawns	317
		Melicertus latisulcatus	(Kishinouye, 1896)	00 701904			
		Melicertus plebejus	(Hess, 1865)	00 701304			
6.3	redspot king prawn	*Melicertus longistylus*	(Kubo, 1943)	00 701303	Penaeidae	prawns	319
6.3	**bay prawn**	***Metapenaeus bennettae* & *M. insolitus***		00 701340	Penaeidae	prawns	313
		Metapenaeus bennettae	Racek & Dall, 1965	00 701323			
		Metapenaeus insolitus	Racek & Dall, 1965	00 701324			
6.3	**Endeavour prawn**	***Metapenaeus endeavouri* & *M. ensis***		00 701903	Penaeidae	prawns	315
		Metapenaeus endeavouri	(Schmitt, 1926)	00 701320			
		Metapenaeus ensis	(de Haan, 1844)	00 701322			
6.3	**school prawn**	***Metapenaeus* spp:**		00 701341	Penaeidae	prawns	322
		Metapenaeus dalli	Racek, 1957	00 701325			
		Metapenaeus macleayi	(Haswell, 1879)	00 701321			
6.3	**scampi**	**Nephropidae spp:**		00 703905	Nephropidae	prawns	321
		Metanephrops australiensis	(Bruce, 1965)	00 703001			
		Metanephrops boschmai	(Holthuis, 1964)	00 703006			

No.	Marketing name	Scientific name	Authority	CAAB	Family or other	Group name	p.
6.3	**scampi**	**Nephropidae spp: (cont.)**		00 703905	Nephropidae	prawns	321
		Metanephrops velutinus	Chan & Yu, 1991	00 703005			
		Nephropsis serrata	Macpherson, 1993	00 703018			
6.3	black tiger prawn	*Penaeus monodon*	Fabricius, 1798	00 701908	Penaeidae	prawns	314
6.1	**Balmain bug**	***Ibacus* spp:**		00 703036	Scyllaridae	bugs	296
		Ibacus ciliatus	(von Siebold, 1824)	—			
		Ibacus novemdentatus	Gibbes, 1850	00 703039			
		Ibacus peronii	Leach, 1815	00 703028			
		Ibacus pubescens	Holthuis, 1960	00 703031			
6.1	southern rocklobster	*Jasus edwardsii*	(Hutton, 1875)	00 703014	Palinuridae	bugs	301
6.1	eastern rocklobster	*Jasus verreauxi*	(Milne Edwards, 1834)	00 703013	Palinuridae	bugs	298
6.1	**champagne lobster**	***Linuparus* spp:**		00 703041	Palinuridae	bugs	297
		Linuparus sordidus	Bruce, 1965	00 703009			
		Linuparus trigonus	(von Siebold, 1824)	00 703040			
6.1	western rocklobster	*Panulirus cygnus*	George, 1962	00 703999	Palinuridae	bugs	303
6.1	**tropical rocklobster**	***Panulirus* spp except *P. cygnus*:**		00 703904	Palinuridae	bugs	302
		Panulirus ornatus	(Fabricius, 1798)	00 703034			
6.1	**slipper lobster**	***Scyllarides* spp:**		00 703907	Scyllaridae	bugs	300
		Scyllarides haanii	(de Haan, 1841)	00 703040			
		Scyllarides squammosus	(Milne Edwards, 1837)	00 703030			
6.1	**Moreton Bay bug**	***Thenus* spp:**		00 700005	Scyllaridae	bugs	299
		Thenus indicus	Leach, 1815	00 700004			
		Thenus orientalis	(Lund, 1793)	00 700002			
6.4	redclaw	*Cherax quadricarinatus*	(von Martens, 1868)	00 704003	Parastacidae	yabbies	325
6.4	**yabby**	***Cherax* spp except *C. quadricarinatus* & *C. tenuimanus*:**		00 704005	Parastacidae	yabbies	326
		Cherax albidus	Clark, 1951	00 704006			
		Cherax destructor	Clark, 1936	00 704002			
		Cherax glaber	Riek, 1967	00 704007			
		Cherax plebejus	(Hess, 1865)	00 704008			
		Cherax preissii	(Erichson, 1846)	00 704009			

No.	Marketing name	Scientific name	Authority	CAAB	Family or other	Group name	p.
6.4	marron	*Cherax tenuimanus*	(Smith, 1912)	00 704004	Parastacidae	yabbies	324
6.2	**crab**	**Brachyura spp:**		00 702904	Brachyura	crabs	306–307
		Charybdis natator	(Herbst, 1789)	00 710048	Portunidae		
		Hypothalassia armata	(de Haan, 1833)	00 702009	Eriphiidae		
		Nectocarcinus tuberculosus	Milne Edwards, 1860	00 710049	Portunidae		
		Portunus sanguinolentus	(Herbst, 1783)	00 710042	Portunidae		
6.2	coral crab	*Charybdis feriata*	(Linnaeus, 1758)	00 710041	Portunidae	crabs	305
6.2	**sand crab**	***Ovalipes* spp:**		00 702901	Portunidae	crabs	310
		Ovalipes australiensis	Stephenson & Rees, 1968	00 702905			
6.2	blue swimmer crab	*Portunus pelagicus*	(Linnaeus, 1766)	00 702003	Portunidae	crabs	304
6.2	giant crab	*Pseudocarcinus gigas*	(Lamarck, 1818)	00 701001	Eriphiidae	crabs	308
6.2	spanner crab	*Ranina ranina*	(Linnaeus, 1758)	00 702002	Raninidae	crabs	311
6.2	**mud crab**	***Scylla* spp:**		00 702013	Portunidae	crabs	309
		Scylla olivacea	(Herbst, 1796)	00 702006			
		Scylla serrata	(Forsskål, 1775)	00 702001			
			MOLLUSCS				
7.1	—	*Haliotis asinina*	Linnaeus, 1758	00 662006	Haliotidae	abalones	
7.1	brownlip abalone	*Haliotis conicopora*	Péron, 1816	00 662004	Haliotidae	abalones	329
7.1	greenlip abalone	*Haliotis laevigata*	Donovan, 1808	00 662002	Haliotidae	abalones	330
7.1	Roe's abalone	*Haliotis roei*	Gray, 1827	00 662003	Haliotidae	abalones	331
7.1	blacklip abalone	*Haliotis rubra*	Leach, 1814	00 662001	Haliotidae	abalones	328
7.7	**periwinkle**	**Littorinidae, Neritidae, Trochidae & Turbinidae spp:**		00 663900	Turbinidae	sea snails	346
		Turbo marmoratus	Linnaeus, 1758	00 663005			
		Turbo undulatus	Lightfoot, 1786	00 663004			
7.7	trochus	*Trochus niloticus*	(Linnaeus, 1758)	00 665001	Trochidae	sea snails	347

No.	Marketing name	Scientific name	Authority	CAAB	Family or other	Group name	p.
7.7	**bailer shell**	**Zidoninae spp:**		00 661900	Volutidae	sea snails	345
		Livonia mamilla	(Gray, 1844)	00 661001			
		Melo amphora	(Lightfoot, 1786)	00 661002			
7.2	**cockle**	***Anadara* & *Katelysia* spp:**		00 657901	Arcidae & Veneridae	clams	332
		Anadara trapezius	(Deshayes, 1840)	00 657002	Arcidae		
		Katelysia peronii	Lamarck, 1818	00 657003	Veneridae		
		Katelysia rhytiphora	Lamy, 1937	00 657004	Veneridae		
		Katelysia scalarina	(Lamarck, 1818)	00 657004	Veneridae		
7.2	**pipi**	***Donax* spp:**		00 654001	Donacidae	clams	333
		Donax deltoides	Lamarck, 1818	00 654002			
7.2	**surf clam**	***Dosinia* spp:**		00 657902	Veneridae	clams	334
		Dosinia caerulea	Reeve, 1850	00 657001			
7.3	blue mussel	*Mytilus edulis*	Linnaeus, 1758	00 652001	Mytilidae	mussels	335
7.6	**saucer scallop**	***Amusium* spp:**		00 651006	Pectinidae	scallops	343
		Amusium balloti	(Bernardi, 1861)	00 651002			
		Amusium pleuronectes	(Linnaeus, 1758)	00 651003			
7.6	commercial scallop	*Pecten fumatus*	Reeve, 1852	00 651001	Pectinidae	scallops	342
7.6	**scallop**	**Pectinidae spp:**		00 651000	Pectinidae	scallops	344
		Annachlamys flabellata	(Lamarck, 1819)	00 651007			
		Equichlamys bifrons	(Lamarck, 1819)	00 900203			
		Mimachlamys asperrima	(Lamarck, 1819)	00 900202			
7.5	Pacific oyster	*Crassostrea gigas*	(Thunberg, 1793)	00 653002	Ostreidae	oysters	339
7.5	native oyster	*Ostrea angasi*	Sowerby, 1871	00 653004	Ostreidae	oysters	337
7.5	**oyster**	**Ostreidae & Pteriidae spp:**		00 653003	Ostreidae & Pteriidae	oysters	338
		Crassostrea gigas	(Thunberg, 1793)	00 653002	Ostreidae		
		Saccostrea cucullata	(Born, 1778)	00 653006	Ostreidae		
		Saccostrea echinata	(Quoy & Gaimard, 1835)	00 653007	Ostreidae		
7.5	**pearl oyster**	**Pteriidae spp:**		00 653005	Pteriidae	oysters	340
		Pinctada albina	(Lamarck, 1819)	00 658002			
		Pinctada margaritifera	(Linnaeus, 1758)	00 658003			
		Pinctada maxima	(Jameson, 1901)	00 658001			

No.	Marketing name	Scientific name	Authority	CAAB	Family or other	Group name	p.
7.5	Sydney rock oyster	*Saccostrea glomerata*	(Gould, 1850)	00 653001	Ostreidae	oysters	341
7.4	**octopus**	***Octopus* spp:**		00 601001	Octopodidae	octopuses	336
		Octopus australis	Hoyle, 1885	00 645006			
		Octopus berrima	Stranks & Norman, 1992	00 645001			
		Octopus maorum	Hutton, 1880	00 645002			
		Octopus pallidus	Hoyle, 1885	00 645003			
7.8	Gould's squid	*Nototodarus gouldi*	(McCoy, 1888)	00 600001	Ommastrephidae	squids	349
7.8	**cuttlefish**	***Sepia* spp:**		00 610008	Sepiidae	squids	348
		Sepia apama	Gray, 1849	00 610003			
		Sepia pharaonis	Ehrenberg, 1831	00 610001			
		Sepia rex	(Iredale, 1926)	00 610007			
7.8	southern calamari	*Sepioteuthis australis*	Quoy & Gaimard, 1832	00 600003	Loliginidae	squids	351
7.8	northern calamari	*Sepioteuthis lessoniana*	Lesson, 1830	00 620011	Loliginidae	squids	350
7.8	**squid**	**Teuthoidea spp:**		00 600000	Loliginidae	squids	352
		Nototodarus hawaiiensis	Berry, 1912	00 636005	Ommastrephidae		
		Ommastrephes bartramii	(Lesueur, 1821)	00 636010	Ommastrephidae		
		Photololigo sp.		00 620002	Loliginidae		
		Photololigo cf chinensis	(Gray, 1849)	00 600004	Loliginidae		
			OTHER INVERTEBRATES				
8.1	**jellyfish**	**Schyphozoa spp:**		00 670001	Schyphozoa	jellyfishes	354
		Aurelia aurita	(Linnaeus, 1758)	00 670002			
		Catostylus mosaicus	(Quoy & Gaimard, 1824)	00 670003			
		Phyllorhiza punctata	von Lendenfeld, 1884	00 670004			
8.2	**beche-de-mer**	**Holothuriidae & Stichopodidae spp:**		00 708100	Holothuriidae & Stichopodidae	sea cucumbers	355
		Holothuria scabra	Jaeger, 1833	00 708103	Holothuriidae		
		Thelenota ananas	(Jaeger, 1833)	00 708108	Stichopodidae		
8.3	**sea urchin**	**Echinoidea spp:**		00 708001	Echinoidea	sea urchins	356
		Centrostephanus tenuispinus	Clark, 1914	00 711004	Diadematidae		
		Centrostephanus rodgersii	(Agassiz, 1863)	00 711001	Diadematidae		
		Heliocidaris erythrogramma	(Valenciennes, 1846)	00 711002	Echinometridae		
		Heliocidaris tuberculata	(Lamarck, 1816)	00 711003	Echinometridae		

Scientific names index

N

O

P

Marketing and common names index

G

H

I

J

R

S

Proudly printed in Australia by

COURTNEY

COLOUR GRAPHICS
PTY LTD

ACN 005 755 224

Tel: 03 9735 5811